Ecological Risk Assessment for Contaminated Sites

Ecological Risk Assessment for Contaminated Sites

Glenn W. Suter II
Rebecca A. Efroymson
Bradley E. Sample
Daniel S. Jones

LEWIS PUBLISHERS
Boca Raton London New York Washington, D.C.

Library of Congress Cataloging-in-Publication Data

Ecological risk assessment for contaminated sites/ Glenn W. Suter II … [et al.].
p. cm.
Includes bibliographical references and index.
ISBN 1-56670-525-8 (alk. paper)
1. Ecological risk assessment. 2. Hazardous waste sites—Environmental aspects.
I. Suter, Glenn W.
QH541.15.R57 E257 2000
628.5′028′7—dc21 99-088049
CIP

Direct all inquiries to CRC Press LLC, 2000 N.W. Corporate Blvd., Boca Raton, Florida 33431.

Lewis Publishers is an imprint of CRC Press LLC

International Standard Book Number 1-56670-525-8
Library of Congress Card Number 99-088049
Printed in the United States of America 1 2 3 4 5 6 7 8 9 0
Printed on acid-free pape

Preface

This book is written for practitioners of ecological risk assessment. The existing guidance from the U.S. Environmental Protection Agency (EPA) for ecological risk assessment of contaminated sites was written for site managers and remedial project managers, not the scientists and engineers who actually perform the assessments (EPA, 1989a; Sprenger and Charters, 1997). Hence, the EPA guidance emphasizes procedure and provision of a basic understanding of ecological risk assessment concepts. This book presents only a minimal discussion of procedure and presents it in a general form that can be related to a variety of national and local procedures. Instead, the emphasis is on techniques for ecological risk assessment that are applicable to contaminated sites. Because the intended audience is practitioners, this book does not present basic environmental science. We assume that we do not need to explain concepts such as biological population or toxicity test. Therefore, the techniques are presented in a manner that should be understood by a reader with a good grounding in the environmental sciences. The techniques presented are state of the art, but no more. Advanced ecological and toxicological methods such as individual-based population modeling are not included. Rather, this book is meant to provide a stake in the ground to mark the technical progress achieved by good practitioners of ecological risk assessment as of 2000. We believe that this book is timely. Although programs to remediate contaminated sites have been ongoing for some years in the United States and elsewhere, only recently have ecological risks become important considerations in remedial decisions.

In our experience, new practitioners of ecological risk assessment often have a basic grounding in mathematics, statistics, and the natural sciences, as well as specialized knowledge of one or more of the environmental sciences (ecology, ecotoxicology, or environmental chemistry). However, they often do not have a good grounding in the general principles and concepts of ecological risk assessment. For that, we refer them to the EPA guidelines (EPA, 1998) and to a general ecological risk assessment text (Suter, 1993a).

This book builds on prior publications by the authors. In particular, the U.S. Department of Energy (DOE) Oak Ridge Operations Office has supported the development of a series of guidance documents for the performance of ecological risk assessments at the DOE Oak Ridge, TN, Portsmouth, OH, and Paducah, KY sites. We have scavenged those documents while writing this book. However, readers who are familiar with those documents will find that this book greatly expands and updates the material in those documents and treats topics for which there are no Oak Ridge guidance documents. A particular concern has been that this book be applicable to all sorts of contaminated sites and not simply to the relatively large and undisturbed DOE sites.

The authors have biases that must be revealed. First, we prefer measurement to models or assumptions and site-specific measurements to surrogates. This amounts to a preference for realism. While few will argue with this bias in principle, recommendations to take measurements are often dismissed because of their cost. However, the costs of measurement are small relative to the costs of remediation. Because of the tendency to use conservative assumptions and models in the absence of measurements, tens of thousands of dollars in measurement can save millions of dollars in remediation and prevent damage to the environment from the physical disturbance associated with unnecessary remediation. In addition, results of measurement are more credible to a skeptical public than are results of models. Finally, standard models or assumptions may be advocated, particularly by regulators, because they are more consistent. However, we believe that it is more important that decisions be correct for a site than that they be consistent across sites.

We also have a strong bias for clarity. While few would argue against being clear and explicit, there are strong forces acting in favor of obscurity. For example, few decision makers want to go before the public and say, "As a result of the contamination, five fish species are missing from the stream, which is 15% of the expected total, but no remedial action will be taken because that effect is not sufficient to justify a $3 million remedial action." Most decision makers would be happier to make the ambiguous statement, "There is no significant reduction in ecosystem health." Given common definitions of ecosystem health, the latter statement is consistent with the first, but is much less clear. One virtue of the ecological risk assessment paradigm is the separation of risk assessment from risk management. Assessors do not decide what is important, but rather provide a clear statement of what is estimated and a clear explanation of how it was estimated. This division of labor allows and even encourages the assessors to be clear about their technical methods and results, leaving the decision makers and stakeholders to provide the value-laden interpretations concerning significance. However, assessors may be tempted to be obscure when they make technical assumptions that are difficult to defend (e.g., contaminant levels in foliage-feeding insects are the same as those in earthworms) or conclusions that are difficult to justify (e.g., although ambient waters were never found to be toxic, the low species richness of the fish community is judged to be due to copper and selenium). In such situations, clear exposition may save the assessors from later difficulties when they are challenged in court or public hearings. In addition, the discipline of writing clearly compels clear thinking, a benefit in itself.

The relative emphasis placed on the various topics in this volume may seem peculiar to some. Some topics and issues that are very important in practice at waste sites are treated briefly, and some that are given little attention at most sites are treated extensively. In general, the topics that are relatively slighted are those that are treated in detail in numerous sources and are not peculiar to ecological risk assessment. For example, sampling and analysis of chemically contaminated media are treated in detail in EPA reports, commercial publications, and standard methods from standards organizations. Media sampling and analysis are important issues, but they are not peculiar to ecological risk assessment. Therefore, we have simply explained their role in the remedial process and pointed the reader to other sources. On the other hand, we have devoted a relatively large space to the issue of setting boundaries on the site to be assessed and then dividing that area into units and subunits for the assessment. This is an important issue for which there is little existing guidance, and which is more important for ecological risk assessment than for human health risk assessment. We believe that these apparent imbalances in subject matter actually increase the utility of the book to ecological risk assessors by focusing on the areas where ecological risk assessors need the most help.

The recommendations in this book are consistent with guidance available from the EPA Office of Emergency and Remedial Response, EPA Region IV, and the Tennessee Department of Environment and Conservation, at the time the book was written. However, other states and EPA regions have their own guidance, and regulatory personnel may change their policies without notice, so adherence to the guidance in this book does not guarantee an acceptable ecological risk assessment. Finally, the reader should bear in mind that all generic guidance should be superseded by good scientific judgment when provisions of the guidance are not applicable to a given situation. The first rule of good risk assessment, which supersedes everything else you will read in this book, is "Don't do anything stupid."

About the Authors

Glenn W. Suter II is currently science advisor in the U.S. Environmental Protection Agency, National Center for Environmental Assessment–Cincinnati, and was formerly a senior research staff member in the Environmental Sciences Division, Oak Ridge National Laboratory. He has a Ph.D. in ecology from the University of California, Davis, and 24 years of professional experience including 19 years of experience in ecological risk assessment. He is the editor and principal author of the major text in the field of ecological risk assessment, and has edited another book and authored more than 80 open literature publications.

Dr. Suter has served on an International Institute of Applied Systems Analysis Task Force on Risk and Policy Analysis, the Board of Directors of the Society for Environmental Toxicology and Chemistry, an Expert Panel for the Council on Environmental Quality, the U.S. EPA Risk Assessment Forum, and the editorial boards of *Environmental Toxicology and Chemistry, Environmental Health Perspectives,* and *Human and Ecological Risk Assessment*, among other positions. His research experience includes development and application of methods for ecological risk assessment, soil microcosm tests, aquatic toxicity tests, and environmental monitoring.

Rebecca A. Efroymson is a research staff member in the Environmental Sciences Division, Oak Ridge National Laboratory. She has a Ph.D. in environmental toxicology from Cornell University. She has led and provided technical support for ecological risk assessments of contaminated burial grounds, streams, ponds, and watersheds for U.S. Department of Energy facilities in Oak Ridge, TN. Her research experience includes development of frameworks, toxicity benchmarks, and models for ecological risk assessment, with emphases on contaminated soils, air pollutants, plants, microorganisms, and soil invertebrates.

Dr. Efroymson has led an ecological risk assessment for land application of sewage sludge in forests and arid ecosystems, and is developing an ecological risk assessment framework for military aircraft overflights (e.g., impacts of noise) and contributing to a broader risk assessment framework for military training and testing activities. Prior to working in Oak Ridge, she was an American Association for the Advancement of Science Diplomacy Fellow at the U.S. Agency for International Development, where she was involved in comparative risk assessment and pollution prevention programs. Her graduate research was on rates and mechanisms of biodegradation of hydrocarbons.

Bradley E. Sample, an environmental scientist with CH2M HILL, was formerly a research staff member of the Environmental Science Division of Oak Ridge National Laboratory. He received his Ph.D. in wildlife toxicology from West Virginia University and has authored more than 70 publications and presentations on wildlife toxicology and ecological risk assessment topics. His research interests are extensive and focus on development and application of methods for ecological risk assessment emphasizing organism- and population-level risks to avian and mammalian wildlife. Areas of research in which he has worked include wildlife exposure assessment, modeling of bioaccumulation and food-web transfer of contaminants, extrapolation and estimation of interspecies toxicity, development of toxicity reference values for wildlife based on contaminant body burdens, estimation of population-level effects based on individual-level effects, indirect effects of contaminants, assessment of toxicity of petroleum constituents to wildlife and other ecological endpoints, and landscape ecotoxicology.

Dr. Sample has performed screening and baseline assessments of ecological risks to terrestrial wildlife species at Department of Energy facilities and in support of the activities of other federal agencies. Prior to joining Oak Ridge National Laboratory, he conducted research on the effects of large-scale aerial applications of insecticides to forests for control of the gypsy moth.

Daniel S. Jones is a research staff member in the Environmental Sciences Division at Oak Ridge National Laboratory. He earned an M.Sc. in environmental sciences from the University of Massachusetts, Amherst. As the Ecological Risk Assessment Team Leader for two U.S. Department of Energy facilities, he led a team of ecologists and coordinated remedial investigation activities with project managers, state and federal regulators, and other technical experts. He has conducted assessments of hazardous waste site impacts on aquatic and sediment-associated receptors in systems ranging from headwater streams to a 30-mile stretch of the Clinch River in Tennessee.

Mr. Jones also led efforts to develop standardized, quantitative methods for prioritizing Oak Ridge National Laboratory environmental restoration projects based on ecological risks. He is currently developing ecological risk assessment frameworks for ocean range weapons testing and for the evaluation of on-site wastewater systems. Prior to developing and applying ecological risk assessment methodologies, he coordinated a multimedia sampling campaign investigating the nature and extent of contamination in the watershed downstream of Oak Ridge National Laboratory. His graduate research was on the impacts of chlorinated wastewater discharges on benthic invertebrates in the upper hyporheic zone of receiving streams.

Acknowledgments

We gratefully acknowledge the support of our life partners: Linda Suter, Bill Hargrove, Stephanie Ashburn, and Ellen Tierney. Thanks to them, none of us fits the stereotype of the lonely and misunderstood author.

We would like to acknowledge the Environmental Sciences Division of Oak Ridge National Laboratory for supporting our efforts to advance the science of ecological risk assessment while performing assessments of contaminated Department of Energy sites. Although two of us have left Oak Ridge, we have all benefited from the lab's culture of excellence and innovation. Current and past colleagues at Oak Ridge who have contributed to the development of ecological risk assessment for contaminated sites include Tom Ashwood, Larry Barnthouse, Lisa Baron, Steve Bartell, Gordon Blaylock, Bob Cook, Owen Hoffman, Ruth Hull, Barbara Jackson, Yetta Jager, David Macintosh, Dennis Opresko, Allen Tsao, and Elizabeth Will. Jim Loar, Frances Sharples, and Steve Hildebrand provided the sort of management support that this effort required; they cleared obstacles, provided encouragement, and otherwise stayed out of the way.

Portions of the following chapters were previously published in U.S. Government reports: Chapter 5 in Suter (1995a), Chapter 6 in Suter (1996c), Chapter 7 in Suter (1997), and Chapter 8 in LMES (1997).

Acronymns

As practiced in the United States, environmental assessment and regulation are laden with acronyms. The following list is offered not only as a guide to acronyms used in this book, but also as a guide to acronyms that the reader will encounter in reports, regulations, and meetings.

AChe	Acetyl Cholinesterase
AEC	Ambient Exposure Concentration
ALAD	Aminolevulinic Acid Dehydrogenase
ARAR	Applicable or Relevant and Appropriate Requirement
ASTM	American Society for Testing and Materials
ATP	Adenosine Triphosphate
BERA	Baseline Ecological Risk Assessment
BTAG	Biological Technical Assistance Group
CEAM	Center for Exposure Assessment Modeling, Athens, GA
CEC	Cation Exchange Capacity
CERCLA	Comprehensive Environmental Response, Compensation, and Liability Act of 1980
CFR	Code of Federal Regulations
CLP	Contract Laboratory Program
COC	Chemicals of Concern
COEC	Chemicals of Ecological Concern
COPC	Chemicals of Potential Concern
COPEC	Chemicals of Potential Ecological Concern
CV	Chronic Value
DQO	Data Quality Objective
DTPA	Diethly Triamine Pentaacetic Acid
EC50	Median Effective Concentration (formerly EC_{50})
EPA	(U.S.) Environmental Protection Agency
EqP	Equilibrium Partitioning
ERA	Ecological Risk Assessment
ERAGS	Ecological Risk Assessment Guidance for Superfund
ER-L	Effects Range–Low
ER-M	Effects Range–Median
EROD	7-Ethoxyresorufin-*o*-deethylase
FFA	Federal Facility Agreement
FS	Feasibility Study
FWS	(U.S.) Fish and Wildlife Service
GIS	Geographic Information System
HC_p	Hazardous Concentration for the *p*th percentile of species
HI	Hazard Index
HQ	Hazard Quotient

IFERAGS	Interim Final Ecological Risk Assessment Guidance for Superfund
INM	In Need of Management
LC50	Median Lethal Concentration (formerly LC_{50})
LOAEL	Lowest Observed Adverse Effects Level
LOEC	Lowest Observed Effects Concentration
LOEL	Lowest Observed Effects Level
MDL	Method Detection Limit
NAPL	Nonaqueous-Phase Liquid
NAWQC	National Ambient Water Quality Criteria
NCP	National Oil and Hazardous Substances Pollution Contingency Plan
NESHAP	National Environmental Standards for Hazardous Air Pollutants
NOAA	National Oceanic and Atmospheric Administration
NOAEL	No Observed Adverse Effects Level
NOEC	No Observed Effects Concentration
NOEL	No Observed Effects Level
NPL	National Priority List
NRDA	Natural Resource Damage Assessment
OC	Organic Carbon
OM	Organic Matter
ORNL	Oak Ridge National Laboratory
ORR	Oak Ridge Reservation
PAH	Polycyclic Aromatic Hydrocarbon
PCB	Polychlorinated Biphenyl
PQLs	Practical Quantification Limits
PRG	Preliminary Remedial Goal
QA/QC	Quality Assurance and Quality Control
QSAR	Quantitative Structure–Activity Relationship
RBCA	Risk-Based Corrective Action
RG	Remedial Goal
RGO	Remedial Goal Option
RI	Remedial Investigation
RI/FS	Remedial Investigation/Feasibility Study
ROD	Record of Decision
RPM	Remedial Project Manager
SCV	Secondary Chronic Values
SERA	Screening Ecological Risk Assessment
SSD	Species Sensitivity Distribution
T&E	Threatened and Endangered
TCDD	2,3,7,8-Tetrachlorodibenzo-*p*-dioxin
TEC	Toxicologically Effective Concentration
TIE	Toxicity Identification Evaluation
TPH	Total Petroleum Hydrocarbons
TU	Toxic Unit
UCL	Upper Confidence Limit
VOCs	Volatile Organic Compounds
ΣTU	Sum of Toxic Units

Table of Contents

Chapter 4

1 Introduction: Definitions and Concepts

When it comes to conceptual issues, scientists are no more immune to confusion than lay people.

— D. C. Dennett (1991)

1.1 SCOPE AND DEFINITIONS

This book presents a conceptual approach and specific methods for assessing the ecological risks posed by contaminated sites. Contaminated sites are those areas that have been contaminated by waste disposal or by spillage or improper use of chemicals. This definition includes lands that have been directly contaminated by surface or subsurface disposal or spillage, surface waters and sediments that have been directly contaminated by release of aqueous wastes or dumping of solid wastes, and lands and waters that have been secondarily contaminated by aqueous or atmospheric transport. In the United States these sites have been the subject of great controversy. Under the Comprehensive Environmental Response, Compensation, and Liability Act of 1980 (CERCLA or Superfund), they have been the subject of debate, assessment, litigation, and remediation by scientists, engineers, managers, lawyers, and regulators.

Ecological risk assessment (ERA) is a process for collecting, organizing, and analyzing information to estimate the likelihood of undesired effects on nonhuman organisms, populations, or ecosystems. It was developed in the early to mid-1980s to provide a basis for environmental decision making equivalent to human health risk assessment (Barnthouse et al., 1982; Barnthouse and Suter, 1986). It was derived from practices in human health risk assessment, environmental hazard assessment, and environmental impact assessment (Munn, 1975; Cairns et al., 1979; National Research Council, 1983). However, the concept of estimating risks as a means of managing financial hazards through insurance, options, and other instruments dates at least to the late 17th century (Bernstein, 1996). All varieties of risk assessment are based on the recognition that decisions must be made under conditions of uncertainty and that the desirability of alternative outcomes depends on their likelihood as well as their utility. The application of ecological risk assessment to contaminated sites was stimulated by the EPA requirement that this analytical approach be applied at Superfund sites (EPA, 1989a). The practice of ecological risk assessment took off after 1992 when two texts and the EPA framework were published (Bartell et al., 1992; EPA, 1992a; Suter, 1993a). A text specifically addressing ecological assessment of hazardous waste sites was published the following year (Maughan, 1993).

The primary purpose for conducting ecological risk assessments of contaminated sites is to provide information needed to make decisions concerning site remediation.

That information consists of estimates of the likelihood that the exposures of ecological receptors to contaminants are sufficient or will become sufficient to cause specified effects. This information is not the only possible basis for remedial decisions. The alternatives include cleaning up to background concentrations of chemicals, declaring that the site is a sacrifice area (i.e., if risks exist, they do not matter), remediating in a manner dictated by public pressure, remediating as far as available funds allow, or cleaning up to standards that are not based on risk. However, it is now generally recognized that human health and ecological risks provide a major rationale for remedial decisions that, at minimum, should be considered along with other considerations such as public pressure and financial limitations.

There are at least two purposes for assessing ecological risks at contaminated sites other than guiding the remedial decision. The first is disclosure. The public wants to know what effects are being caused by contamination and what improvements may be expected from remediation. The other is damage assessment. In the United States those who contaminate a site must compensate the trustee for any injury to natural resources that is not made good by the remediation (Chapter 10).

There are at least three means of expressing ecological risks that have been used to support remedial decisions.

Determine if risk-based standards are exceeded — The only available ecological risk-based standards for the United States are the National Ambient Water Quality Criteria (EPA, 1985a). They are risk based in the sense that they are intended to correspond to contaminant concentrations that would reasonably be expected to cause minimal effects on aquatic communities in most waters most of the time. They have the disadvantage that they are available for relatively few chemicals, do not apply to media other than water, do not correspond to any particular effect, and have little predictive power at particular sites. Standards for water and other media have been produced in Canada, Europe, and elsewhere. Because the use of standards is quick, easy, and easily enforced, it has been popular. However, the limited scope of available criteria makes them inadequate for assessment of most receptors and contaminants.

Estimate the likelihood of exceeding some risk level that is a threshold for regulatory significance — For example, the parties to a remedial decision may agree that remediation is called for if the abundance of fish in a stream contaminated by leachate is reduced by at least 20%. This type of expression of assessment results has been called for by the EPA Office of Solid Waste and Emergency Response (OSWER) (Quality Assurance Management Staff, 1994). The ease of generating this assessment result depends on the magnitude of site effects relative to the threshold. For example, if the fish abundance in the contaminated stream is less than 0.1/m of stream reach and all reference streams have at least 5/m, the level of effect clearly exceeds the 20% threshold.

Estimate the risks — That is, estimate the probability distribution of effects or the levels of effects with certain probabilities of occurrence. For example, one might estimate that the most likely (50th percentile) effect is a 10% reduction in fish abundance, but the worst case estimate (90th percentile) is that fish abundance is reduced by as much as 60% relative to an uncontaminated state. This is clearly the most difficult form of output to generate with any confidence. However, if remedial risks are to be balanced against contaminant risks (Chapter 9) or if damages are to be collected for the effects of contamination (Chapter 10), the levels of effects must be estimated.

The primary goal of all this analysis is to support a remedial action. Remedial alternatives include removal of the wastes and contaminated media for disposal or treatment, treatment of the contaminants *in situ* through chemical treatment or enhanced biological degradation, isolation or redirection (e.g., grout curtains), immobilization *in situ* (e.g., grout injection), land-use controls to regulate human or ecological exposures, covering of the contaminated area with a cap, or no action (remediation through natural degradation and attenuation). To support the remedial decision, the risk assessment must provide information about what media or wastes may require remediation and what criteria (e.g., concentrations achieved or area capped) would result in reductions of risk to prescribed levels.

1.2 REGULATORY CONTEXT

In the United States, remediation of contaminated sites is performed primarily under the provisions of CERCLA. The regulatory basis for implementing that law is the EPA *National Oil and Hazardous Substances Pollution Contingency Plan* (NCP) (EPA, 1990a). These regulations call for protection of the nonhuman environment as well as human health. They establish a process called the Remedial Investigation/Feasibility Study (RI/FS) for determining the need for remediation (RI) and for choosing a remedial alternative (FS). Both components require assessment of risks. The implementation of these regulations is performed by the EPA regional offices, which have considerable autonomy to develop and apply their own interpretations of the regulations. Further, some contaminated sites are regulated under other laws including the Resource Conservation and Recovery Act (RCRA) and the Oil Pollution Act of 1990 (OPA). In addition, many of the states have their own laws and regulations for contaminated sites. As a result, the regulatory context of ecological risk assessments for contaminated sites is highly complex and variable and will not be treated in any detail here. The best general source of information concerning the federal regulations is the EPA web site (http://www.epa.gov), and a recent review of U.S. federal and state and some non-U.S. guidance is provided by Sorenson and Margolin (1998).

Guidance for ecological risk assessment in the RI/FS process was provided in "Risk Assessment Guidance for Superfund" (EPA, 1989a). However, the guidance is very general and is intended for site managers, not assessment scientists. More recently, an unofficial guidance document, "Interim Final Ecological Risk Assessment Guidance for Superfund" (IFERAGS), has been developed by the EPA Environmental Response Team to replace the ecological portion of the 1989 guidance (Sprenger and Charters, 1997). Like the earlier guidance, it is intended for site managers, but it provides more detailed procedures plus appendixes containing examples of assessment methods. It is being used by some EPA regions, but other regions have their own guidance (Callahan and Steele, 1998). In practice, the procedural details are seldom consistently followed. Some states have developed or are developing guidance for ecological risk assessment for contaminated sites. Particularly noteworthy are those developed by the states of Massachusetts and California (Cal EPA, 1996; Environmental Risk Characterization Work Group, 1996). The EPA has provided a site with links to ecorisk guidance and tools for Superfund (www.epa.gov/superfund/programs/risk/ecolgc.htm).

1.3 ECORISK FRAMEWORK

All of the current guidance for remediation of contaminated sites incorporates ERA as the means for determining what effects on the environment may need to be remediated. One of the defining features of ecological risk assessment is the standard framework which has evolved from the National Research Council framework for human health risk assessment (National Research Council, 1983) to the current version portrayed in Figure 1.1. It consists of a problem formulation phase, an analysis phase, and a risk characterization phase. The specific components of the framework vary, but the version in the EPA guidelines consists of the following components (EPA, 1998).

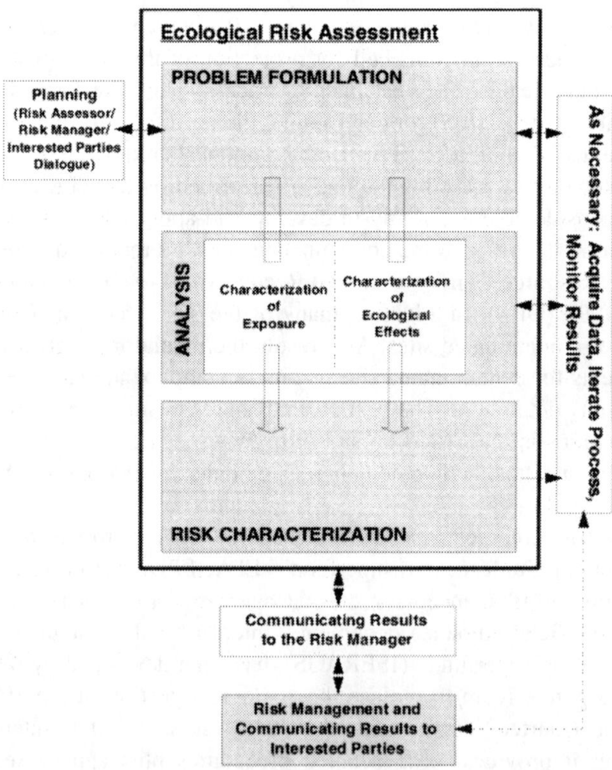

FIGURE 1.1 The EPA framework for ecological risk assessment (EPA, 1998).

Problem formulation is the planning and problem definition phase. It includes:

- Integrate available information — Assemble and summarize information concerning sources, contaminants, effects, and the receiving environment.
- Assessment endpoints — Define in operational terms the environmental values that are to be protected.
- Conceptual model — Develop a description of the hypothesized relationships between the wastes and the endpoint receptors.
- Analysis plan — Develop a plan for obtaining the needed data and performing the assessment.

Analysis is the phase in which a technical evaluation of the data concerning exposure and effects is performed.

The *characterization of exposure* component consists of:

- Measures of exposure — Results of measurements indicating the nature, distribution, and amount of the waste and its components at points of potential contact with receptors
- Exposure analysis — A process of estimating the spatial and temporal distribution of exposure to the contaminants
- Exposure profile — A summary of the results of the exposure analysis

The *characterization of effects* component consists of:

- Measures of effect — Results of measurements or observations indicating the responses of assessment endpoints to variation in exposure
- Ecological response analysis — A quantitative analysis of the effects data
- Stressor–response profile — The component of the ecological response analysis that specifically deals with defining a relationship between the magnitude and duration of exposure and relevant effects

Risk characterization is the phase in which the results of the analysis phase are integrated to estimate and describe risks. It consists of:

- Risk estimation — The process of using the results of the analysis of exposure to paramaterize and implement the exposure–response model and estimate risks and of analyzing the associated uncertainty
- Risk description — The process of describing and interpreting the results of the risk estimation for communication to the risk manager

Risk management is, in this context, the process of making a decision concerning the need for remediation. It appears at two points in the framework.

- At the beginning of the assessment, the risk manager provides policy input to the problem formulation.
- At the end of the assessment, the risk manager learns the results of the risk analysis and makes a decision.

1.4 THE REMEDIAL PROCESS

Ecological risk assessment is a component of a remedial process that includes a variety of planning, sampling, analysis, remediation, and engineering activities, all intended to convert the site to an acceptable state. Ecological risk assessors must be involved in all stages of the process. A generalized version of that process is diagrammed in Figure 1.2, and is briefly discussed in the following paragraphs. This diagram does not include identification of candidate sites or the document review and revision processes, which typically occur after the production of each plan or assessment document. Some schemes, such as the IFERAGS, include numerous reviews (Sprenger and Charters, 1997). Although it is valuable in theory to assure that the assessors are not going off on a wrong course, it is the authors' experience that risk managers and the staffs of regulatory agencies are overcommitted and often perform few or superficial reviews of intermediate products.

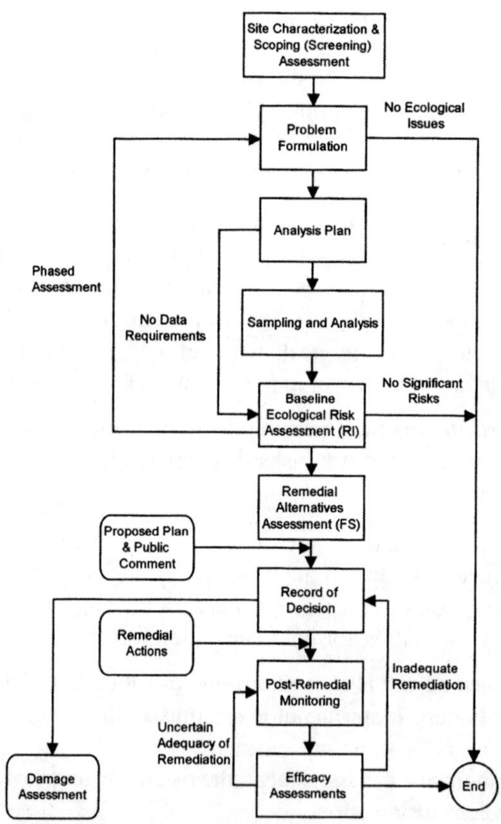

FIGURE 1.2 A flowchart of the general remedial process for contaminated sites.

Note that this general remedial process is not appropriate for emergency responses. If a site is contaminated by a spill of a hazardous material, spending a period of months planning, seeking regulator review comments, gathering ecological data, etc. while the contaminant spreads from the spill site would be inappropriate. Rather, existing emergency response plans should be implemented to contain and remove the material quickly. If there is residual contamination following the emergency response, this process may be used to assess and remediate it.

The scheme presented here for performing the remedial process does not exactly correspond to any of the procedures defined by the EPA Office of Solid Waste and Emergency Response, by EPA regions, or by any of the states. However, it captures the major features of all of them. That is, any scheme for assessing risks at contaminated sites is likely to include all of the activities discussed below, but they may not be identified as distinct steps, and they are likely to be given different names. The steps in the process may vary among sites, particularly depending on the amount of existing data and information that are available concerning a site. At a particularly well-characterized site, one might skip the scoping and screening assessments. As indicated in IFERAGS, any logical and technically valid process is acceptable as long as it meets the needs of the risk manager and is clearly documented (Sprenger and Charters, 1997).

1.4.1 Preliminary Site Characterization and Scoping Assessment

Before an ecological risk assessment can be performed, it is necessary to have a basic understanding of the nature and extent of contamination, potential routes of exposure, and ecological resources present at the site. Site visits and a preliminary site survey (reconnaissance) must be conducted, and available information pertaining to the state and history of the site must be accumulated. A reconnaissance involves visiting the site and describing its physical characteristics (topography, geology, hydrology, anthropogenic structures, and other features that would affect transport and fate); potential sources of contaminants such as drums, lagoons, and pipes; and the types, extent, and condition of plant and animal communities present. A useful set of data sheets for site description is presented as Appendix A of Environmental Response Team (1997), which is appended to Sprenger and Charters (1997). In addition to the field reconnaissance, local experts should be consulted, local residents and employees should be interviewed, and information should be collected from existing sources such as aerial photographs, maps such as wetland inventories and endangered species range maps, and prior assessments of the site or nearby sites. Previous remedial actions taken at the site that have affected the environment, such as capping of landfills, should also be described. If data concerning the composition and concentration of the contaminants in site media are unavailable from existing information, reconnaissance sampling and analysis may be performed. In such cases, the sampling should focus on a broad-spectrum analysis of samples of the sources or of the most contaminated areas. Alternatively, portable devices such as photoionization detectors (PIDs) may be used to provide rapid, broad-spectrum analytical data. Data collected during the preliminary site characterization should be used to perform a scoping assessment in preparation for the assessment planning process.

That is, the available information should be used to determine whether there is a potential hazard. If there is sufficient information concerning the concentrations of contaminants on the site, a screening assessment may be performed, instead of the scoping assessment, to identify the hazard and narrow the scope of the assessment. (These types of assessments are described below.)

1.4.2 The Problem Formulation Process

The problem formulation defines the purpose, scope, and goals of the assessment, as well as the methods by which it will be performed. Its products include the assessment endpoints, the conceptual model, and the analysis plan. The problem formulation should be conducted collaboratively. The ultimate arbiter of the scope, methods, and content of the assessment is the decision maker who is responsible for the remedial decision. The other necessary participants in the problem formulation are the technical experts who will prepare the analysis plan and carry out the assessment. Potential additional participants include representatives of the responsible parties, resource agencies, and local citizens. Typically, a screening assessment is performed to determine the nature and scope of the assessment problem (Section 1.5). The problem formulation process is discussed in more detail in Chapter 2.

1.4.3 The Analysis Plan

The analysis plan (RI work plan) explains how the decisions made during the problem formulation process will be implemented. A screening ecological risk assessment is typically included in the analysis plan to provide a technical justification for the scope of ecological sampling, analysis, and assessment activities. The analysis plan is a plan for obtaining the data needed for the baseline assessment, as identified in the problem formulation process, and a plan for using those data to assess ecological risks. The ecological risk assessor must ensure that the proposed activities will in fact fill the needs of the assessment by consulting with the authors of the sampling and analysis sections and by authoring a plan for performing the baseline ecological risk assessment using that data set.

If the site is poorly characterized by existing data, the assessment may be carried out in phases. That is, a Phase 1 sampling and analysis program may be defined and carried out, which provides the basis for planning a more-focused second phase. This sequence requires that the results of the Phase 1 studies be used to conduct a screening assessment, which is used to identify any data gaps that must be filled in Phase 2. In theory, any number of iterations of planning, sampling and analysis, and risk assessment could be performed. However, few schedules or budgets permit more than two phases, and few sites are so complex as to require them. Analysis plans are described in more detail in Chapter 2.

1.4.4 Baseline Ecological Risk Assessment

In contrast to the screening assessment, which defines the scope of the baseline assessment, the baseline assessment uses new and existing data to evaluate the risk of leaving the site unremediated. The purposes of the baseline assessment are to determine (1) if significant ecological effects are occurring at the site, (2) the causes of these effects, (3) the source of the causal agents, and (4) the potential future risks from leaving the system unremediated. The baseline assessment provides the ecological basis for determining the need for remediation. Under CERCLA, the baseline risk assessment is part of the remedial investigation.

Because the baseline assessment focuses on a smaller number of chemicals and receptors than the screening assessment, it can provide a higher level of characterization of toxicity to the populations and communities at the site. In the baseline ecological risk assessment, a weight-of-evidence approach should be employed to determine if and to what degree ecological effects are occurring or may occur.

Most baseline ecological risk assessments focus primarily on current risks. However, future baseline risks should be characterized when:

- Contaminant exposures are expected to increase in the future (e.g., a contaminated ground water plume will intersect a stream)
- Biological succession is expected to increase risks (e.g., a forest will replace a lawn)
- Natural processes will result in significant recovery in the near term without remedial actions (i.e., the expense and ecological damage associated with remedial actions may not be justified)

1.4.5 Remedial Alternatives Assessment

If significant risks to humans or ecological receptors are found by the baseline risk assessments, it is necessary to assess the remedial alternatives. Under CERCLA, this is referred to as the feasibility study. For the no-action alternative and for those alternatives that simply limit human access to and use of the site, the risks are those identified in the baseline ecological risk assessment. Other remedial alternatives actively reduce site contamination but also cause physical disturbances, which have their own ecological risks. These remedial risks include destruction of the biotic communities on the site and on uncontaminated sites for borrow pits, landfills, roads, laydown areas, parking lots, etc. (Chapter 8). The remedial ecological risk assessment must consider these direct effects, secondary effects such as erosion and habitat fragmentation, and the expected rate and degree of recovery of the disturbed areas given the site management and expected land uses. The importance of assessing remedial risks is demonstrated by the record of decision (ROD) for Lower East Fork Poplar Creek in Oak Ridge, in which contaminant levels that are estimated to constitute a significant risk to shrews and wrens, but not humans, were left in place because the potential remedial actions would result in destruction of riparian forests and wetlands. That is, the remedial risks exceeded the baseline risks.

The results of this assessment are used to develop a proposed plan for remediation. That plan is presented to the public for review and comment. The results of this process are used to revise the plan, which is then incorporated into the ROD. Guidance for this process of developing and presenting the decision documents is contained in EPA (1999).

1.4.6 Efficacy Assessment

Ecological risk assessments are performed after completion of remedial actions to meet legal requirements or for purposes of environmental stewardship. In the United States, if the remedial actions leave contaminants in place rather than removing or destroying them, the responsible party is required under CERCLA to monitor the remediated site and, every 5 years, to assess the efficacy of the remedial actions in terms of the protection of human health and the environment, until unimpaired use of the site is possible. The nature of this monitoring and assessment activity is specified in the ROD. However, the monitoring required by the EPA is usually minimal and focused on human health risks. Therefore, the site owner must, to meet planning and site-stewardship needs, determine whether additional monitoring is needed to ensure that risks to environmental resources have been sufficiently reduced.

Both the legal requirement and the environmental stewardship goals require the collection of data to characterize the post-remedial condition and to estimate levels of effects. All of the techniques that were applied to earlier stages of the assessment could be employed at this stage. However, efficacy assessments are typically based on a minimal monitoring program that is focused on one or a few critical issues. The data analyses are also typically minimal and are focused on clearly defined criteria for acceptable risk. For example, despite significant risks to the benthic invertebrate community, the postremedial monitoring of Watts Bar Reservoir, below Oak Ridge, was limited to fish fillets, and the analysis was limited to determining whether fishing advisories are required. The alternative outcomes from an efficacy assessment are (1) risks are minimal, and the assessment and remediation process can end; (2) risks are significant, and more remediation is needed; or (3) risks are not minimal but do not require additional remediation, so monitoring continues. Efficacy assessments are discussed further in Chapter 10.

1.4.7 Damage Assessment

The Natural Resource Damage Assessment (NRDA) provisions of CERCLA require that the residual injuries, following remediation, be assessed so that natural resource trustees can be compensated for lost natural resource services. The purpose is to use the funds to restore the environment to provide the lost natural resource services. To be efficient and to ensure that remedial actions are taken which avoid excessive payments of natural resource damages, NRDA should be integrated with the rest of the CERCLA process. Therefore, results of the baseline and remedial alternatives risk assessments are important input to the NRDA, and selection of assessment endpoints which are also "natural resource services" could save assessment costs and damage payments. While this integration is logical and would be cost-effective, it is complicated by the fact that, except for informing the natural resource trustees, the EPA does not participate in NRDA activities. Therefore, NRDA is not explicitly addressed in the EPA Data Quality Objectives (DQO) process or in RI/FS-related documents that are reviewed by the EPA. Damage assessments are discussed further in Chapter 10.

1.5 TIERS: SCOPING, SCREENING, AND DEFINITIVE ASSESSMENTS

In general, ecological risk assessments are performed in phases or tiers, including scoping assessments, screening assessments, and definitive baseline assessments. The successive tiers each require more time and effort. Scoping assessments determine whether an ecological risk assessment is needed; screening assessments determine what contaminants, media, and receptors need to be assessed; and definitive assessments determine the nature and magnitude of risks. Each tier of risk assessment should include all three phases: problem formulation, analysis, and risk characterization. Even the most hurriedly performed scoping assessment should define the nature of the problem, analyze available information on exposure and effects, and explain how they are potentially related. The purpose of tiers of assessment is to perform the assessment as efficiently as possible by performing only the number of assessment activities needed to reach an informed decision. At any point the assessment process might end because the risks are clearly unacceptable or are clearly negligible.

Scoping assessments determine whether a formal ecological risk assessment is needed by ascertaining whether there is a potential for current or future exposure of ecological receptors. There are several questions to be answered. (1) Is there a source of contaminants? The apparent waste may be innocuous, there may actually be no waste, or the waste may have been removed by prior actions. At one Oak Ridge unit, the waste was merely construction rubble. (2) Are there currently or might there be in the future ecological receptors on the contaminated site? In some cases, there is no complete pathway from the contaminants to an ecological receptor, because the contamination is limited to an industrial facility or an inactive industrial site that will be returned to industrial use or converted to commercial use (i.e., a brownfield). In those cases the site has little ecological value and is not expected to have significant ecological value under future land uses. There is no complete pathway on the site because there are no significant receptors. (3) Could movement of the contaminants result in significant exposure of ecological receptors off the site? The principal concern is with contamination of surface waters and wetlands through runoff or lateral groundwater movement. Groundwater contamination is not normally a basis for performing an ecological risk assessment unless the groundwater intersects the surface or the near-surface root zone.

Screening assessments go beyond scoping assessments by asking whether the identified pathways from the contaminants to ecological receptors could result in toxicologically significant exposures. Screening allows assessors to focus resources by first applying rapid and conservative assessment techniques to screen out sites, portions of sites, media, or chemicals that clearly pose minimal risks. More time and effort can then be devoted to definitive assessments that provide risk estimates by applying more realistic and site-specific analyses to those hazards that have not been screened out. Chemicals that are retained by the screening process are termed chemicals of potential ecological concern (COPECs). Screening assessments should be conservative and as broad as is reasonably possible so that no potential hazards are overlooked.

The screening phase may itself be performed in tiers. For example, a screening assessment may be performed using a relatively small set of preliminary data for the purpose of determining what sampling and analysis needs to be performed. The new data should then be used in a screening assessment to determine which hazards should be modeled and analyzed in detail in the definitive assessment.

In general, data gaps identified in screening assessments should be treated as a basis for including a hazard in the definitive assessment. For example, if a class of chemicals has not been measured in a medium and potentially occurs in the source, it should be included as a COPEC. Similarly, receptors that may occur on the site but are unconfirmed and routes of exposure that are credible but have not been investigated should be retained. Treating unknowns in this manner results in a more credible assessment.

If screening assessments have identified credible hazards to ecological receptors, a definitive assessment must estimate risks and identify preliminary remedial goals. Definitive assessments should replace conservative assumptions with best estimates of exposures and effects and associated uncertainties. Because previous tiers of assessment will have reduced the scope of the definitive assessment by identifying contaminants, media, and receptors that constitute credible hazards, it should be possible to devote sufficient time and resources to their assessment. In general, additional testing and analysis reduce uncertainties and increase realism, thereby reducing the need to overremediate to ensure protection. Definitive assessments are not normally performed in tiers unless the assessment process reveals potentially significant pathways and receptors that were missed by the screening assessments.

The phased assessments performed at contaminated sites require a different logical approach from the tiered assessment schemes developed to assess risks from new chemicals. The latter schemes are based on the performance of brief and inexpensive tests and simple exposure models in the first tier and performance of more expensive and realistic tests in higher tiers if assessments based on lower tiers are inconclusive (Cairns et al., 1979; Urban and Cook, 1986). However, the acute lethality tests used in the early tiers of those assessment schemes cannot be used to determine acceptability of risks at a contaminated site. For example, if a soil from a contaminated site is acutely lethal to plants, one can conclude that there are potentially significant risks. However, if the soil is not acutely lethal, one cannot conclude that there are no potentially significant risks. The soil may in fact be lethal in longer exposures or may reduce growth or seed production. In hazard assessment schemes, the insensitivity of traditional early tier tests is compensated for by applying safety factors to the test endpoints (e.g., median lethal concentration, LC50s) and by using conservative assumptions in the transport and fate models. However, when testing contaminated media, those techniques are not applicable. Therefore, one must begin with a sensitive test in order to avoid falsely concluding that there are no significant risks.

Tiered assessment schemes for ecological risk assessment are more complex than those for human health risk assessment because of the multiple lines of evidence that are available to ecological assessors. For example, the American Society for Testing and Materials (ASTM, 1994) Risk Based Corrective Action (RBCA) scheme prescribes that the assessment is performed by comparing modeled or measured exposure levels for individual chemicals in air, soil, and groundwater to defined levels that constitute thresholds for significant human risk. In RBCA, three tiers of exposure values are defined by use of (1) generic values, (2) easily derived site-specific values, and (3) values derived by complex and extensive site-specific analyses. Ecological risk assessors perform similar analyses but also employ toxicity tests of contaminated media from the site and surveys of biota at the site. Surveys range from analyses of body burdens and biomarkers to surveys of species composition and abundance. Any of these lines of evidence may be employed in any tier of the assessment, depending on the availability of data, the availability of time and resources to obtain data, or the appropriateness to the source and receiving environment. This possibility makes development of standard tiers, like those in RBCA, a difficult proposition.

1.6 RELATIONSHIP TO HUMAN HEALTH RISK ASSESSMENT

Because EPA policy now places equal emphasis on assessing the potential impacts of hazardous waste sites on human health and the environment, human health assessments and ecological assessments are performed concurrently in the United States. Some information and data are likely to be relevant for assessing both human and environmental threats. Common data needs identified during problem formulation should be met in such a way as to avoid duplication of effort. It is imperative that human health and ecological risk assessors coordinate their activities and communicate throughout the process so that all relevant data are accessible to all parties concerned.

Common data needs for human and ecological risk assessments are determined by the individual characteristics of the site and by the scale of the assessments. In general, the following data are likely to be useful for both human health and ecological risk assessments. Differences between ecological and human health effects data needs are noted in parentheses.

- Chemical concentrations in media including:
  - Soils/sediments (concentrations in the pore water are useful in ecological risk assessments)
  - Surface water (concentrations of dissolved forms of chemicals are useful for ecological risk assessments)
  - Groundwater (not usually needed for ecological risk assessment)
  - Air (not usually needed for health or ecological risk assessments of waste sites)
  - Biota, including those consumed by humans, such as fish, geese, deer (whole body concentrations will be needed for ERAs, not just "edible parts"), plus food for ecological endpoint species such as mice for hawks and foxes and earthworms for thrushes and shrews
- Chemical inventories (including chemicals known or thought to have been deposited on the site)
- Operational history and current practices at the site
- Factors affecting fate and transport of chemicals; for example:

  - Physical parameters — geohydrologic setting, soil properties, topography
  - Bioaccumulation models, particularly for exposure pathways involving indirect exposures to humans via the food chain
- Background concentrations of chemicals

All of these factors should be considered as possible common data needs during problem formulation.

In some cases, ecological assessments may require samples to be analyzed in a specific way. For example, the list above notes that sediment pore water concentrations are required for ecological risk assessments. In addition, minimum required detection limits for some chemicals are lower for ecological concerns than for human health concerns. Chemicals that pose a greater ecological than human health risk need to be analyzed with more precision than would otherwise be required. Certain water quality parameters, such as pH, hardness, and oxygen levels, are also more important for ecological than human health risk assessments.

The fact that ecological and human health risk assessments have common data needs should be considered during the prioritization of sites for risk assessments. For example, a site of low ecological priority would not normally require an immediate environmental evaluation. If, however, the site is of immediate concern for human health reasons, ecological risks should also be assessed immediately. This is because the analysis of remedial action alternatives must simultaneously consider the acceptability of remedial alternatives in terms of both ecological and human health impacts.

When the assessment of risks is completed and remedial action goals are developed, ecological risks must be compared to human health risks to determine which will drive the selection of remedial options. At this point, apparent discrepancies in results of the two types of risk assessment will require explanation. Some of these discrepancies will be due to differences in the assessment approaches or assumptions, and others may be due to unexpectedly greater sensitivities of nonhuman receptors.

1.6.1 Differences in Human Health and Ecological Risk Assessment

Ecological risk assessments are more complex than human health risk assessments and are fundamentally different in their inferential approaches. The greater complexity is mainly due to the large number of species and the diversity of routes of exposure that must be considered in ecological risk assessments. However, the differences in inferential approach and part of the greater complexity are due to the fact that ecological risk assessments for waste sites may be based on epidemiological approaches while human health risk assessments for the waste sites are nearly always based on modeling. This discrepancy raises the question, why not just model ecological risks as well? The reasons are as follows.

- Epidemiological approaches, when they are feasible, are fundamentally more reliable than modeling, because they address real observed responses of real receptors. Human health risk assessments are based on epidemiology when possible, but epidemiology is (fortunately) not feasible for most sites because there are no observable effects in human populations.

- Ecological epidemiology is feasible in practice, because nonhuman organisms reside on most sites and are, in some cases, experiencing observable exposures and effects.
- Ecological epidemiology is feasible in principle, because the levels of effects that are deemed to be significant by most regulatory agencies are observable in many populations and communities.
- Because of the assumptions that must be made to model risks, the uncertainties in model-generated risk estimates are large. These uncertainties can be accepted in practice by human health assessors because the effects are not observable. However, it is common for modeled ecological risks to be manifestly incorrect because the predicted effects are not occurring or effects are observed where they are not predicted. Therefore, it is incumbent to use an epidemiological approach to avoid mistakes and embarrassment.
- Because of the great value placed on human life, remedial actions may be taken on the basis of highly uncertain estimates of hypothetical risks. However, because of the lesser value placed on nonhuman organisms and natural ecosystems, decision makers are somewhat reluctant to spend millions of dollars to remediate highly uncertain ecological risks. Therefore, if ecological risk assessments are to be useful, they must be compelling.
- Biological surveys and ecological toxicity tests are highly cost-effective, because they are inexpensive relative to chemical analyses and provide more direct evidence concerning ecological risks.

Even in those cases when ecological epidemiology is not feasible, the process of determining that to be the case is instructive and aids the interpretation of modeled risks. For example, if contaminants on a site would cause reproductive failure in robins feeding on that site, counting robins would not reveal that effect, because the number of breeding pairs is limited by habitat availability relative to territory size, and the loss of production on the site would be easily replaced by birds produced elsewhere. This example also suggests that the hypothesized effect (i.e., reduced reproduction on the site) would not be significant at the population level for a species that is not limited by production. Therefore, if epidemiological techniques are inappropriate, it may be a sign that the endpoint itself is inappropriate. The epidemiological approach is also not directly applicable to the estimation of potential future risks, because future conditions cannot be observed. For example, a contaminated groundwater plume may be predicted to reach surface water in the future. However, even in those cases, the set of assessment tools available to the ecological assessor is greater than those available to the human health risk assessor. For example, samples of the contaminated groundwater can be subject to toxicity testing both full-strength and in a dilution series using the surface water with which it would be mixed in the future as diluent. Also, if the future risk simply involves expansion of the contaminated area, then studies of the current risks can be used in the prediction of future risks.

1.6.2 Why Ecological Entities May Be More Sensitive than Humans

It is commonly assumed by the public and by some individuals involved in site remediation that protection of human health will also result in protection of nonhuman organisms. For this reason, when ecological risks, but not human risks, are estimated to be significant, the apparent discrepancy should be explained. Despite the highly protective endpoints in human health risk assessments and the use of large safety factors that exaggerate risks, the risks to nonhuman organisms are often so much greater that, for a variety of reasons, ecological risks are estimated to be greater (Suter, 1993a). When greater ecological risks are found, they must be explained. Types of explanations include the following:

- Ecological receptors experience modes of exposure, such as respiration of water, consumption of sediment, or drinking from waste sumps, that do not occur in humans.
- Ecological receptors experience quantitatively greater exposure such as a diet of 100% local fish.
- Ecological receptors include particular taxa with inherently greater sensitivity than humans.
- Secondary effects such as reduced production of herbivore populations due to loss of plant production are not important for humans.

1.7 SCALE IN HUMAN HEALTH AND ECOLOGICAL RISK

The issue of scale is treated differently in human health and ecological risk assessment. Because human health risks are estimated for hypothetical individuals, they can be calculated for the points in space at which samples are collected. For example, risks from contaminants in the water of White Oak Creek on the Oak Ridge Reservation are calculated at an integration point, the weir of the dam, where an individual is assumed to collect his or her 2 liters of drinking water every day for 30 years. However, the endpoints for ecological risk assessments (except for those involving threatened or endangered species) are largely defined at the population or community level. Therefore, it is not reasonable to estimate ecological risks at a specific point in space, except as a screening technique. For example, for the RI, risks to the fish community were assessed in the reaches of White Oak Creek, not at a point at the end of the creek. Similarly for assessments of contaminated soils, the human health assessment may assume that a human lives for 30 years on a small site, but the ecological risk assessment must acknowledge that vertebrate animal populations have large ranges, of which the contaminated area may constitute a small fraction.

This difference means that the data for the two types of assessments should be averaged differently and the results will not be point-to-point comparable. However, such comparability is not required by regulations or guidance and is not necessary for risk assessments to be useful. Rather, both human health and ecological risk assessments must produce defensible estimates of risks to their respective endpoints.

1.8 PARTICIPANTS IN ECOLOGICAL RISK ASSESSMENTS

The generic framework for ecological risk assessment identifies two participants in the process: risk assessor and risk manager. In the simplest case, a site owner may wish, in the absence of regulatory requirements, to assess and potentially remediate a site for purposes of site stewardship. In that case the owner is the risk manager, and the technical staff or consultants are the risk assessors. However, the risk assessor/risk manager dichotomy is often more complicated at contaminated sites. Typical cases may involve representatives of:

- The party(s) that contaminated the site
- The current owner of the site or the owner's site manager
- One or more regulatory agencies
- Natural resource agencies
- Local citizens groups
- Environmental advocacy groups

The roles of these various entities depend on the regulatory context. For example, on U.S. government lands remediated under CERCLA, a federal facility agreement may be negotiated in which the EPA regional office, the appropriate state agency, and the federal land manager act by consensus. More commonly, the responsible parties perform the assessment, and the remedial decision is made by a representative of a regulatory agency, possibly after consultation or negotiation with others. In some cases, a regulatory agency both performs the assessment and makes the remedial decision. The organizations that participate in the remedial decision making are referred to as the parties to the decision. Other organizations such as citizens' groups and natural resource agencies that have a stake in the outcome but are not formally parties to the decisions are termed *stakeholders*. The line between parties and stakeholders may become blurred when the parties involve the stakeholders in the assessment and decision-making process. Stakeholder involvement is becoming increasingly common as regulators realize that the public does not necessarily accept that the government is acting in its interest. In such cases, the public or members of environmental advocacy groups may wish to be involved in making remedial decisions. Although the singular term *risk manager* will be used through the rest of this book, readers should remember that there may be collective responsibility for remedial decisions.

In addition to the representatives of agencies and groups, each organization may have its own technical experts (staff or contractors). The technical experts for one group, usually either the responsible party or a regulatory agency, will perform the assessment. Scientists acting for the other parties and stakeholders review plans, assessments, and decision documents and may perform independent analyses of data or even collect their own data. The technical experts performing the assessment should comprise a multidisciplinary team including experts in analysis of human and ecological risks and scientists and engineers needed to collect and analyze samples, perform toxicity tests, and manage and assure the quality of the data. In the remainder of the book, all of these people will be referred to as *assessors*.

The last group that may become involved in the remedial process is lawyers. For any number of reasons which will not be discussed here, a remedial action may be adjudicated. Therefore, assessors should bear in mind that they may have to defend their data, analyses, and inferences while under attack from hostile attorneys supported by their own technical experts.

2 Problem Formulation

Upon this gifted age, in its dark hour,
Rains from the sky a meteoric shower
Of facts...they lie unquestioned, uncombined.
Wisdom enough to leach us of our ill
Is daily spun; but there exists no loom
To weave it into fabric...

— Edna St. Vincent Millay, "Sonnet 137"

Problem formulation is a process of defining the nature of the problem to be solved and specifying the risk assessment needed to solve the problem. In the poet's metaphor, the problem formulation attempts to build and string a fact loom. The rest of the assessment process is fact weaving. The principal results of the problem formulation are the assessment endpoints, a conceptual model of the induction of ecological risks on the site, and an analysis plan. If the problem formulation is done in a haphazard manner, the resulting assessment is unlikely to be useful to the risk manager. The process should be taken as seriously as the performance of toxicity tests or the creation of a hydrologic model and should be done with at least as much care.

2.1 RISK MANAGERS AND RISK ASSESSORS

The primary purpose of performing ecological risk assessments for contaminated sites is to provide information needed for a decision concerning remediation. Therefore, the participation of the individuals who will make the decisions, the risk managers, is imperative. Many of the decisions made in the problem formulation involve values rather than facts and therefore are policy judgments rather than scientific decisions. There are several questions: What should be protected? What is the appropriate spatial and temporal scale? What future scenarios are relevant? What expressions of risk are useful for the decision? However, the form and extent of participation by risk managers are highly variable. There are at least three ways in which their participation can occur.

First, the risk manager may provide input prior to the problem formulation. This option is suggested by the EPA framework for ecological risk assessment, which shows the risk manager outside the problem formulation box and suggests that the risk manager's contribution is policy goals (EPA, 1998). The risk manager's input may be statements about goals for the particular site (e.g., ultimate uses) or may simply be generic policies for site remediation. If policies are ambiguous, risk assessors should look for precedents that would indicate what sorts of ecological issues and evidence have been sufficiently compelling to lead to remediation in the past, and which have not.

The second possibility is that the risk manager's input may come in the form of a review of the analysis plan (Section 2.7). This option is popular with regulatory agencies. However, when it is the only form of substantive input, it is undesirable for two reasons. First, the risk manager may not know or be willing to state what is wanted, but will say that what is offered is wrong (the infamous "bring me a rock" approach). This form of communication can lead to frustrating and wasteful iterations of writing, review, rewriting, and rereview. Second, the reviews are often performed by the risk manager's technical experts rather than the risk manager. For example, CERCLA documents are often reviewed by contractors for the EPA regional offices rather than by the EPA Remedial Project Manager. This substitution can lead to risk management input that bears little relation to the actual decision-making process.

The final possibility is that the risk managers collaborate with the risk assessors in the problem formulation. That is, the risk manager, in collaboration with the assessors, decides how the problem should be formulated. The EPA has developed a procedure for this activity called the Data Quality Objectives (DQO) process, which is the primary operational innovation of their Superfund Accelerated Cleanup Model (SACM) (Blacker and Goodman, 1994a,b; Quality Assurance Management Staff, 1994). This process is outlined in Box 2.1. One or more meetings are held, each of which may take more than a day. If multiple risk managers are involved or if stakeholders are included in the process, a professional facilitator can be essential to success.

For large, complex sites, it may be efficient to address some generic issues for the entire site and then address more specific issues at each unit. For example, an ecological DQO meeting for the Oak Ridge site established generic conceptual models and a list of generic assessment endpoints, including the levels of effects (Suter et al., 1994). Then endpoints for individual units were selected from this list as appropriate.

The DQO process has the tremendous advantage of ensuring that assessment resources are focused on providing exactly the information that is needed to make a defined, risk-based decision. However, the DQO process was designed for human health risk assessment, and has been difficult to apply to ecological assessments. Part of the problem is simply the complexity of ecological risks relative to human health risks, discussed above. It is difficult to define a "bright line" risk level like a 10^{-4} human cancer risk for the various ecological endpoints. A probability of exceeding a bright line significance level is not even the best expression of the results of an ecological risk assessment. In most cases, it is better to express results as an estimate of the effects level and associated uncertainty (Suter, 1996a; EPA, 1998). In addition, ecological risks are assessed by weighing multiple lines of evidence, so the uncertainty concerning a decision about the level of ecological risk is often not quantifiable. It is directly applicable if only one line of evidence is used, as in many wildlife risk assessments, and if, as in human health risk assessments, one is willing to assume that the decision error is exclusively a result of variance in sampling and analysis as is required by the DQO process. Also, in the authors' experience, risk managers have been reluctant to identify a quantitative decision rule for ecological risks. This is in part because there is little policy or precedent for decisions based

BOX 2.1
The Steps in the Data Quality Objectives Process

1. State the Problem — Clearly specify the problem to be resolved through the remediation process. For example, the sediment of a stream has been contaminated with mercury and is believed to be causing toxic effects in consumers of fish. The ecological assessment endpoint entity is the local population of belted kingfishers.
2. Identify the Decision — Identify the decision that must be made to solve the problem. For example, should the sediment be dredged from some portion of the stream?
3. Identify Inputs to the Decision — Identify the information that is needed in order to make the decision and the measurements and analyses that must be performed to provide that information. For example, the diet and range of kingfishers, the relationship between concentrations of mercury in food and reproductive decrement in kingfishers, the distributions of mercury concentrations in sediment, etc.
4. Define the Study Boundaries — Specify the conditions to be assessed, including the spatial area, the time period, and the site-use scenarios to which the decision will apply and for which the inputs must be generated. For example, the kingfisher population of concern is that occurring in the entire stream from its headwaters to its confluence with the river.
5. Develop Decision Rules — Define conditions under which an action will be taken to remove, degrade, or isolate the contaminants. This is usually in the form of an "if ... then ..." statement. For example, if the average production of the population is estimated to be reduced by at least 20%, the stream will be remediated sufficiently to restore production.
6. Specify Acceptable Limits of Decision Error — Define the error rates that are acceptable to the decision maker, based on the relative desirability of outcomes. For example, the acceptable rate for falsely concluding that production is not reduced by as much as 20% is 10% and for falsely concluding that it is reduced by at least 20% is 25%.
7. Optimize the Design — Based on the expected variance in the measurements and the exposure and effects models, design the most resource-efficient program that will provide an acceptable error rate for each decision rule. For example, on the basis of Monte Carlo analysis of the kingfisher exposure model, the species composition of the kingfisher's diet should be determined by 10 h of observation during each of four seasons for each bird inhabiting the stream or a maximum of 6 birds, the mercury composition of the fish species comprising at least 80% of the diet should be determined twice a year in 10 individuals with a limit of detection of 0.1 µg/kg, etc. (*Steps cited from*: Quality Assurance Management Staff, 1994.)

on quantitative ecological risks (Troyer and Brody, 1994). Finally, the remedial decision is not dichotomous. There may be a number of remedial alternatives with different costs, different public acceptability, and different levels of physical damage to the ecosystems. Therefore, the remedial decision typically does not depend simply on whether a certain risk level is exceeded, but also on the magnitude of exceedence, how many endpoints are in exceedence, the strength of evidence for exceedence, etc.

These issues, however, do not completely negate the utility of using an adaptation of the DQO process for ecological risk assessment. Steps 1 through 4 of the process (Box 2.1) correspond to conventional problem formulation. Therefore, even if only those steps are completed, the risk managers and assessors should be able to develop assessment endpoints, a conceptual model, and measures of exposure and effects in a manner that leads to a more useful assessment because of the collaboration and the emphasis on the future remedial decision. Further, even if the risk manager will not specify decision rules, for the sake of planning, he or she should be willing to specify what effects should be detected with what precision using what techniques. Discussions of the use of the DQO process in ecological risk assessment can be found in Barnthouse (1996) and Bilyard et al. (1997).

In practice, more than one of these forms of risk management input may be applied to a site. Ideally, risk assessors would prepare for the problem formulation by reviewing policy and precedents, they would then meet with the risk manager to perform the problem formulation through the DQO process or some equivalent process, and finally the risk manager would review the analysis plan to ensure that it reflects the manager's intent.

The assessors' role in a DQO process is fourfold. First, they must organize existing information and present it in a useful manner. Second, they must be prepared to answer questions about the potential risks, including the relative susceptibilities of the receptors and the likelihood of various future exposure scenarios. Third, they must be prepared to answer questions about the options for performing the assessment, including the costs and time required to provide different types and qualities of information and the uncertainties that will be associated with different assessment methods. Finally, they must translate the results of the interactions with the risk manager into an operational plan for performing the assessment.

Risk assessors must be aware that not all representatives of agencies with risk management responsibilities are risk managers. For example, in the United States the EPA risk managers for CERCLA are the Remedial Project Managers (RPMs). However, the EPA input to the ecological risk problem formulation may come from staff of the EPA national Office of Emergency and Remedial Response (OERR); from a group of federal employees in each EPA region termed the Biological Technical Assistance Group (BTAG); or from an EPA regional staff member who heads this group, the BTAG coordinator (Office of Emergency and Remedial Response, 1991). While these technical experts may apply more scientific expertise to the problem and have knowledge of agency policies that is useful to the problem formulation, they are no substitutes for the actual risk manager, the RPM. Only the RPM knows what information he or she needs to make the decision and what form will be most useful.

2.2 PHYSICAL SCOPE

Defining the physical scope of the assessment presents two problems: including the entire area that is potentially affected and then dividing that area into manageable and relevant units. These problems are particularly severe for large, complex sites like the Oak Ridge Reservation, but they are relevant to all sites.

2.2.1 Spatial Extent

The spatial extent of the site may be established based on one or more of the following criteria:

The areas in which wastes deposited — The site must at minimum include all areas within which the wastes were spilled or deposited, such as the total area of a landfill or waste burial ground.

The areas believed to be contaminated — The site must also include areas that are believed to be contaminated, including those areas where contaminants are detected by inspection or by sampling and analysis.

The area owned or controlled by the responsible party — Often, when the area contaminated is not well specified, the entire area controlled by the responsible party is designated to be the waste site. For example, the entire Oak Ridge Reservation was declared a Superfund site although most of it is uncontaminated.

The extent of transport processes — The site should include all areas to which transport processes may have carried significant amounts of the contaminants or to which they may be transported in the future. Hydrological processes are the major concern at most sites, including flow patterns, exchange between groundwaters and surface waters, confluence of contaminated streams with waters that have significant dilution volumes, and barriers to transport. For example, the Oak Ridge Reservation contaminants entered streams which drain into the Clinch River. The river was deemed not to have sufficient dilution volume to assure negligible risks, so it was added to the site. However, the reservoir created by the first dam downstream retained most contaminants because they were largely particle associated. Therefore, the Oak Ridge site was deemed to extend downstream to the Watts Bar dam.

Buffer zones — When the extent of transport or the distance from which endpoint organisms travel to the site is unknown, it may be appropriate to extend the site bounds to include a prescribed area beyond the directly contaminated site. For example, California requires characterization of an area extending 1 mile beyond the designated site (Polisini et al., 1998).

Much of the information needed to define the site bounds can be obtained from records of waste disposal, from the site inspection, and by inference. Site inspections should look for visible evidence of contamination, olfactory evidence of contamination, and evidence of transport processes. For example, the contamination of a stream at the Portsmouth, OH, Gaseous Diffusion Plant was identified by hydrocarbon smells associated with seeps. However, sampling and analysis are usually required to establish the actual extent of contamination. The extent of contamination may best be determined using field analytical techniques. Bounds may need to be extended as more information is gathered over the course of the assessment.

2.2.2 Spatial Units

If a site is relatively small, it may be assessed and remediated as a single unit, but large sites generally must be subdivided for practical reasons. Large, complex sites cannot be investigated and remediated all at once because of funding and staff limitations. Given those limitations, early efforts should be directed to units that are likely to pose the greatest risk. In addition, some areas such as burial grounds and spill sites are sources of contamination, whereas others such as streams and wetlands are receptors that integrate all contaminant sources within their watersheds. Logically, these integrators should not be remediated until after source remediation is complete. Otherwise, they could become recontaminated. The decision about how to divide a site into units must be based on two considerations: the location of contaminants and the dynamics of the site. The manner in which the definition of units is performed depends on the available knowledge about the site.

For most sites the information that is available prior to new sampling is that certain wastes were deposited in certain locations in some manner. The locations may include waste ponds or sumps, burial pits or trenches, landfills, soil contaminated by direct deposition (e.g., spills or land farms), or simple dumps. A distinct area where wastes have been deposited can be termed a source unit. There may be numerous source units on a site. In many cases they are identified in advance of the initiation of assessment activities, but in others it may be necessary to search records, interview former employees or local residents, and survey the site for signs of waste disposal.

Having identified the source units within the site, one must delimit areas to be assessed within the rest of the site. Movement of contaminants out of the source units secondarily contaminates other areas. The most obvious such areas are the streams and associated riparian areas that receive drainage from the source units. These areas are obvious units for assessment of risks to aquatic biota. In addition, riparian areas may contain wetlands or other distinct terrestrial communities that may be contaminated and would constitute logical units for assessment. Examples include the East Fork Poplar Creek in Oak Ridge, where flooding contaminated the floodplain with mercury, and the Clark Fork River in Montana, where wetlands created by sediment deposition in a reservoir were contaminated with mine tailings (Pascoe and DalSoglio, 1994). Each watershed that is contaminated or may become contaminated if the site is not remediated should be identified as a unit to be assessed and potentially remediated. The lateral extent of these units may be defined by the extent of the 100-year flood plain, the extent of contaminated riparian soils, or by the extent of distinct riparian vegetation or soils.

Another type of spatial unit is groundwater aquifers. Aquifers are typically secondarily contaminated by leachate or by losing reaches of contaminated streams, but may be directly contaminated by waste injection. Aquifers may vertically overlap, and their spatial extent may bear little relation to watersheds or other surface features. Aquifers, like watersheds, may be contaminated by multiple sources, and different strata may be contaminated by different sources. At simple sites with a single source unit that is relatively new, defining the immediately underlying aquifer as an assessment unit may be straightforward. However, at complex sites, considerable effort may be expended on investigating geohydrology. Each distinct aquifer that is contaminated or may become contaminated if the site is not remediated, and which may cause ecological exposures, should be identified as a unit to be assessed and potentially remediated.

In addition to the hydrological dynamics that define the watersheds and aquifers, the dynamics of organisms may create assessment units. Animal populations may extend across areas that encompass multiple source or watershed units, and individuals of the more mobile species may in a day feed on one unit, drink from another, and rest on a third. The size of these units depends on the mobility of the organisms and the extent and quality of habitat. On the Oak Ridge Reservation, the entire 17,000 ha reservation has been treated as an assessment unit for highly mobile organisms such as deer and turkey. For less mobile organisms such as small mammals, watersheds may be used as assessment units. In some cases, waste sites may constitute distinct habitats which can serve as assessment units. For example, a grassy and rarely mowed waste burial ground surrounded by forest or industry may support distinct populations of small mammals.

As a result of these considerations, four classes of units may be recognized: sources, watersheds, aquifers, and wildlife units. Each of these units may be the subject of a separate assessment or they may be aggregated in various ways depending on budgets, schedules, and other management considerations. The nature of these classes of units and the relationships among them are discussed in the following text. In general, each assessment for each unit must address the ecological values that are distinct to that unit. However, the assessment for each unit must also characterize its ongoing contributions to risks on other units. These risks are due to fluxes of contaminants out of the unit (e.g., leachate or emergent mayflies), uses of a unit by animals that are not distinct to that unit (e.g., deer grazing on a source unit), or physical disturbances that extend off the unit (e.g., deposition of silt or construction of facilities for the remedial action off the site).

2.2.2.1 Source Units

Source units are sites where wastes were directly deposited. Because the source units are typically highly modified systems, they often have low ecological value; some of them are entirely industrialized. Many waste burial grounds are vegetated, but the vegetation is frequently maintained as a mowed lawn to reduce erosion while minimizing use of the sites by native plants and animals that might disturb, mobilize, accumulate, and transport the wastes.

The intensity of effort devoted to ecological risk assessment for a source unit should depend on its current character and its assumed future use. A paved unit would have negligible ecological value and would normally require minimal or no assessment. A waste pond or sump may be treated as a waste source to be removed or destroyed or as a receptor ecosystem to be remediated. Waste ponds and sumps may support a tolerant aquatic community, but toxicological risks to that community need not be assessed, because destruction or removal of the liquid wastes would destroy the community. However, organisms that drink from the pond or consume aquatic organisms would be the appropriate endpoint species, because they might benefit from removal of a source of toxic exposure. Source units maintained as large lawns may support a distinct plant community (the lawn) and the associated soil heterotrophic community and herbivorous and predatory arthropods characteristic of such plant communities. In such a situation, the ERA for the unit would address the toxicity of the soil to plants and soil heterotrophs. At sites with multiple source units, risks to wider-ranging organisms that occasionally use the unit could not be evaluated in the risk assessment for the unit because neither their exposure nor their response could be associated with a single unit. However, the sources of exposure of these animals must be characterized as input to assessments of wildlife units (below). The appropriate assessment endpoints (Section 2.5) for source units should be discussed during the DQO process.

Some ecological expertise must be applied to evaluating these managed communities. For example, the low-level waste burial grounds at Oak Ridge National Laboratory (ORNL) are frequently mowed, so they do not support small mammals except around the edges where adjacent natural vegetation supplies cover (Talmage and Walton, 1990). In contrast, waste sites associated with other facilities in Oak Ridge are seldom mowed and are surrounded by forest or industrial facilities, so it is likely that they support distinct small mammal populations.

The appropriate assumptions concerning future states of the source units are a matter to be decided by the risk manager. Typically in the United States, regulatory agencies have employed worst-case assumptions. For human health risk assessments this often implies a homesteader scenario with a resident family that drinks from its own well, raises its own food, etc. For ecological risk assessments, the corresponding assumption is that natural succession of vegetation is allowed to occur unimpeded until the native flora and fauna are reestablished. Such assumptions reflect a desire to return the site to its full potential for unimpeded use. Even when there is no realistic expectation that these scenarios could occur, they provide a benchmark against which to compare the remedial alternatives. However, the trend in the United States is toward more realistic scenarios. In particular, urban industrial sites known as brownfields are being assessed and remediated on the assumption that they will be returned to industrial use. In such cases, the ecological risk assessment may be limited to relevant off-site risks, such as risks to aquatic communities from runoff and leachate. Alternatively, the biotic communities associated with the lawns and shrubs used to landscape industrial sites may be considered to have ecological value.

2.2.2.2 Watershed Units

Watershed units are streams and their associated floodplains. These units receive contaminants from all of the source units in their watersheds; incorporate them into sediments, floodplain soils, and biota; and pass a portion of them along to the next unit downstream.

The watershed units generally have much greater ecological value than the source units. They support stream communities, and, except in reaches that are channelized, riparian communities that are diverse and provide ecosystem services such as hydrologic regulation. Although the inventories of contaminants are greater in most source units, the communities of watershed units are likely to be more susceptible to contaminants than the communities of source units, because the contaminants are in the surface soils and waters, and because the biological diversity is greater. Future land-use scenarios may change exposures in some portions of watershed units. For example, White Oak Creek on the grounds of ORNL is channelized and riprapped. If it were assumed that ORNL will be removed, and no new industrial or residential development is allowed to replace it, the stream would eventually develop a natural channel and riparian community, leading to a more diverse and abundant aquatic community.

2.2.2.3 Groundwater Units

Groundwater units are the major spatial units of human health risk assessments because of the leaching of wastes into deep aquifers that potentially provide drinking water. In contrast, ecological assessors usually consider these units only when groundwater is sufficiently near the surface to affect vegetation or when they intersect the surface to contribute to streams or to form wetlands. However, aquifers constitute ecosystems that contain microbes and multicellular organisms that occur incidentally in aquifers (stygoxenes), that occur in aquifers as well as other habitats (stygophiles), or that are restricted to and highly adapted to life in aquifers (stygobites). Aquifer ecosystems are not normally subject to ecological risk assessment, because they have not been protected by regulators. However, they are receiving increasing attention and may be assessed and protected in the future (Committee on Pesticides and Groundwater, 1996). A related problem that is more likely to lead to regulatory action is the exposure of cave organisms and ecosystems to groundwater contaminants. Finally, groundwater may be used for irrigation. This practice may result in accumulation of toxic concentrations of contaminants in soil and drain waters.

2.2.2.4 Wildlife Units

Most of the area of large sites such as the Oak Ridge Reservation or the Rocky Mountain Arsenal lies outside the source units or the contaminated streams and floodplains of watershed units. However, wildlife populations extend beyond these units, and individual animals visit and use multiple units. The process of defining a terrestrial integrator unit depends heavily on the endpoint, the nature of the environment surrounding the source units, the distribution of contaminants, and factors such as property boundaries. In Oak Ridge, the entire DOE reservation was declared a terrestrial integrator unit based on concerns for wide-ranging wildlife. In addition to being a property boundary, the reservation constitutes the limits of a relatively undisturbed area of forest and supports distinct populations of large wildlife species such as deer and wild turkey. Although the reservation will not be remediated as a unit, assessments have been performed of the risks to populations of wide-ranging species on the reservation (Sample et al., 1996a). These reservation-wide assessments have provided a context for actions on individual source and watershed units and eliminated the need to assess risks to those wildlife species at every source unit.

2.2.3 Spatial Subunits

The division of the site into units is intended to identify potentially contaminated areas that constitute logical units for assessment. However, for a variety of reasons the units often need to be subdivided and treated separately during the risk assessment. Subdivision is required by the following considerations.

1. Units are not uniformly contaminated, so it is not reasonable to average contaminant concentrations across the entire unit. Rather, considerations of sampling design require that areas termed the *sampling units* be identified within which samples may be considered to have come from a single statistical population.
2. Ecological risk assessments require that measurements of chemical concentrations, physical properties, and biological properties be related to each other. However, for various reasons, measurements are not all made at identical locations. Therefore, spatial units had to be established that are sufficiently uniform for different types of measurements to be associated to investigate causal relationships.
3. Receptor populations and communities do not exist at single points, but, because of limited mobility or habitat differences, most of them do not occupy an entire unit. Therefore, it is necessary to identify subunits within which it may be assumed that the receptors are exposed.
4. The sources of contamination and the structures and processes controlling contaminant fate often do not result in a simple gradient of contamination. Rather, because of discontinuities, it is often reasonable to use discrete subunits.
5. Because of the large size and variable contamination of many units, it is unreasonable to assume that any engineered remedial action would be uniformly applied. Subunits with relatively uniform risks would be logical areas for remedial actions. An example is the subdivision of the depositional areas of the Milltown Reservoir, MT into 12 subunits based on their physiography and metal concentrations (Pascoe and DalSoglio, 1994).

In general, watershed units should not be assessed as single undifferentiated units, because they are large and vary significantly in their structure and degree of contamination. Rather, they must be divided into reaches. The Clinch River and Poplar Creek assessment provides an example (Cook et al., 1999). The reaches can be defined as distinct and reasonably uniform units for assessment and remediation by applying the following criteria:

- Sources of contamination should be used as bounds on reaches. Examples include contaminated tributaries, outfalls, and sets of seeps associated with drainage from a source unit.

- Tributaries that provide sufficient input to change the hydrology or basic water quality (e.g., pH or hardness) of a stream significantly should serve as bounds of reaches.
- Physical structures that divide a stream, particularly if they limit the movement of animals or trap contaminated sediments, should be used as bounds of reaches. Examples include dams, weirs, and some culverts.
- Changes in land use should be used to delimit reaches. Clearly, ecological risks are different where floodplains have commercial or agricultural land uses than where they are forested.
- Reaches should not be so finely divided that they do not constitute ecological units. Reaches that are too short will contain fish or small mammmals that cannot be clearly associated with the reach because they move in and out.

Some source units are too large and diverse to be assessed and remediated as a unit. In those cases, the unit should be divided into subunits. Although these divisions are likely to be based primarily on the types of wastes present and the manner of their disposal, such divisions may also take ecological differences in the site into consideration. For example, boundaries between distinctly different vegetation types (e.g., lawn and forest) may serve as bounds of subunits.

Wildlife units are seldom so large relative to population ranges that they require division into subunits. For example, the Oak Ridge Reservation 17,000 ha is large for a Superfund site but is not so large that it supports multiple distinct populations of birds or of those amphibians, reptiles, or mammals that are sufficiently wide-ranging to require assessment at the scale of an integrator unit rather than a source unit. However, it is important to recognize that the endpoint species will use only those parts of the unit that meet its habitat needs. In general, these habitat distinctions pertain and are best applied on a species-by-species basis within the unit. However, if there are a few very distinct habitat types that are applicable to nearly all endpoint species, habitat-based subunits may be appropriate.

2.3 SITE DESCRIPTION

The site description should be limited to those features of the site that are important to the estimation of risks from the contaminants. Extensive descriptive material that adds so much bulk to many environmental impact assessments should be avoided. For example, if the contaminants of potential concern have low phytotoxicity or if the relative sensitivity of site species is unknown, there is no need for a plant species list, much less abundance data for plant species. That information would not be used to estimate the risks to plants. In that case, it would be necessary only to indicate what vegetation types are present and whether there are endangered plant species or other species of special concern. While it is important that the assessors perform a site survey, this is not the only useful source of information for the site description. Other sources include natural resource agencies, people living or working near the site, and prior documents describing the site. Information that should be included in the site description is listed in Box 2.2.

BOX 2.2

Information Normally Included in the Site Description for Ecological Risk Assessments of Contaminated Sites

Location and boundary — Latitude and longitude, political units, and boundary features (e.g., bounded on the south by the Clinch River) should be described and mapped.

Topography and drainage — Gradients of elevation and patterns of surface and subsurface drainage determine the hydrologic transport of the contaminants. Floodplains and other depositional areas are particularly important.

Important climatic and hydrological features — For example, if flows are highly variable due to infrequent storms or spring snowmelt or if groundwater seasonally rises to contaminated strata, those features should be noted and characterized.

Current and past site land use — Land use suggests what sorts of contaminants may be present, what sorts of physical effects (e.g., soil compression) may have occurred, and what sorts of ecological receptors may be present.

Surrounding land use — Land use in the vicinity of the site determines to a large extent the potential ecological risks. A site in a city surrounded by industry will not have the range of ecological receptors of a site surrounded by forest.

Nearby areas of high environmental value — Parks, refuges, critical habitat for endangered species, and other areas with high natural value that may be affected by the site should be identified, described, and the physical relation to the site characterized.

Habitat types — On terrestrial portions of the site, habitat types correspond to vegetation types. Aquatic habitats should be described in appropriate terms such as ephemeral stream, low-gradient stream with braided channel, or farm pond. In general, a habitat map should be presented along with a brief description of each type and the proportion of the site that it occupies. The map should include anthropogenic as well as natural habitats (e.g., waste lagoons).

Wetlands — Wetlands are given special attention because of the legal protections afforded wetlands in the United States. Wetlands on the site or receiving drainage from the site should be identified.

Species of special concern — These include threatened and endangered species; recreationally or commercially important species; and culturally important species.

Dominant species — Species of plants or animals that are abundant on the site and may be indicative of the site's condition or value should be noted.

Observed ecological effects — Areas with apparent loss of species or species assemblages (e.g., stream reaches with few or no fish) or apparent injuries (e.g., sparse and chlorotic plants or fish with deformities or lesions) should be identified.

Spatial distribution of features — A map should be created, indicating the spatial distribution of the features discussed above.

2.4 SOURCE DESCRIPTION

For many assessments of contaminated sites, the source will have been adequately characterized by site records or other activities prior to the assessment. However, in some cases the contaminants will not have been characterized. In such cases, it may be appropriate to obtain and analyze samples of the material at the source. In other cases, the source may be unknown, and characterization of the source may serve not only the analysis of risks but also the determination of responsibility. In these cases, the assessors should seek out potential sources and characterize them. If indicator chemicals or fingerprinting techniques are to be used to associate ambient contamination with the source, then analyses of the sources and the contaminated media must be coordinated. The description should include the physical state of the source (e.g., liquid in leaking drums on the land surface, or tailings sluiced behind an earth dam), the composition of the source, and its history, including prior remedial actions (e.g., deposited 15 years ago and covered with clean soil 10 years ago).

In some cases the source may be obscure. For example, contaminants may be detected in water or may have caused specific effects (e.g., fish kills), but the source may be unknown (e.g., leaking buried drums of waste). In such cases, environmental information such as drainage patterns, locations of kills, and groundwater flow directions collected for the site description may help to track down the source.

2.5 ASSESSMENT ENDPOINT SELECTION

Assessment endpoints are the explicit expressions of the environmental values to be protected (Suter, 1989; EPA, 1998). They are the ecological equivalent of the lifetime cancer risk to a reasonable maximally exposed individual in human health risk assessments. Therefore, the endpoint must be an important property of the system that can be estimated, not a policy goal such as fishable waters or some vague desire, such as healthy ecosystems. The selection of the assessment endpoints depends on knowledge of the receiving environment and the contaminants, provided by the assessment scientists, as well as the values that will drive the decision, provided by the risk manager. At a minimum, an assessment endpoint includes an entity, such as a vascular plant community, and a property of that entity, such as net production. These concepts are explained below. If the results of the risk assessment are to be expressed as the likelihood that a threshold for significant risk is exceeded, as in the DQO process, the threshold level of effects must be specified (e.g., a 15% reduction in production relative to reference communities). To design a sampling program for the direct estimation of the endpoint, as in the DQO process, a desired degree of statistical confidence must also be specified (e.g., a maximum Type II error of 20%, or 95% confidence bounds within a factor of 5 of the mean). The area for which the risk is estimated should also be defined. For example, the change in plant production may be averaged over the entire site, or species richness of the fish community may be estimated for specified reaches. All assessment endpoints should at least specify the entity and property. This must be done on a site-specific basis since the EPA is only beginning to consider standard entities and properties (Barton et al., 1997).

2.5.1 Selection of Endpoint Entities

Criteria for selection of endpoint entities and properties are listed in Box 2.3, and common problems with assessment endpoints are listed in Box 2.4. Classes of potential assessment endpoint entities are discussed below.

BOX 2.3

Criteria for Selection of Assessment Endpoints for Ecological Risk Assessments

1. Policy goals and societal values — Because the risks to the assessment endpoint are the basis for decision making, the choice of endpoint should reflect the policy goals and societal values that the risk manager is expected to protect.
2. Ecological relevance — Entities and properties that are significant determinants of the properties of the system of which they are a part are more worthy of consideration than those that could be added or removed without significant system-level consequences. Examples include a keystone predator species or the process of primary production.
3. Susceptibility — Entities that are potentially highly exposed and responsive to the exposure should be preferred, and those that are not exposed or do not respond to the contaminant should be avoided.
4. Appropriate scale — Ecological assessment endpoints should have a scale appropriate to the site being assessed. This criterion is related to susceptibility in that populations with large ranges relative to the site have low exposures. In addition, the contamination or responses of organisms that are wide-ranging relative to the scale of an unit may be due to sources or other causes not associated with the unit.
5. Operationally definable — Without an unambiguous operational definition of the assessment endpoints, it would not be possible to determine what must be measured and modeled in the assessment, and the results of the assessment would be too vague to be balanced against costs of regulatory action or against countervailing risks.
6. Practical considerations — Some potential assessment endpoints are impractical because good techniques are not available for use by the risk assessor. For example, there are few toxicity data available to assess effects of contaminants on lizards, no standard toxicity tests for any reptile are available, and lizards may be difficult to survey quantitatively. Therefore, lizards may have a lower priority than other, better-known taxa. Practicality should be considered only after the other criteria are evaluated. If, for example, lizards are included because of evidence of particular sensitivity or policy goals and societal values (e.g., presence of an endangered lizard species), then some means should be found to deal with the practical difficulties. (*Sources*: Suter, 1989; EPA, 1992a.)

BOX 2.4
Common Problems with Assessment Endpoints

- Endpoint is a goal rather than a property (e.g., maintain and restore endemic populations).
- Endpoint is vague (e.g., estuarine integrity rather than eelgrass abundance and distribution).
- Endpoint is a measure of an effect that is not a valued property (e.g., midge emergence when the concern is production of fish which depends in part on midge production).
- Endpoint is not directly or indirectly exposed to the contaminant (e.g., fish community when there is no surface water contamination).
- Endpoint is irrelevant to the site (e.g., a species for which the site does not offer habitat).
- Endpoint does not have an appropriate scale for the site (e.g., golden eagles on a 1000-m^2 site).
- Value of an entity is not sufficiently considered (e.g., rejection of benthic invertebrates at a site where crayfish are harvested).
- Property does not include the value of the endpoint entity (e.g., number of species when the community is valued for game fish production).
- Property is not sufficiently sensitive to protect the value of the endpoint entity (e.g., survival when the entity is valued for its production).

(*Source*: Modified from EPA, 1998.)

Ecosystems — Ecosystems are assessment endpoint entities if they are valued as ecosystems (e.g., wetlands) or if the properties to be protected are ecosystem properties. A component of an ecosystem that is valued for its functional properties rather than its community or population properties may also be considered an ecosystem-level endpoint entity. The soil "ecosystems," which degrade natural and anthropogenic organic materials, recycle nutrients, and support plant growth, are the major example.

Community — Fishes, benthic macroinvertebrates, and upland plants typically have community-level assessment endpoints. That is, the intent is to protect fishes, macroinvertebrates, or plants as a group rather than individual populations (Stephan et al., 1985; Van Leeuwen, 1990; Solomon, 1996). (Some readers will correctly object that these are assemblages, not communities, but this usage is well established in ecology.) In cases in which components of the community such as benthic-feeding fish or trees are believed to differ from the rest in their susceptibility, the functional or other group should be distinguished in the conceptual model and may be considered a separate assessment endpoint. Each community or subcommunity should be described both in biological terms (e.g., all benthic macroinvertebrates) and in operational terms (e.g., all invertebrates collected by a Surber sampler and retained by a 1-mm-mesh screen).

Population — Wildlife are conventionally assessed as population-level assessment endpoints. The populations used are usually chosen to represent a particular trophic group and a taxonomic class (i.e., birds and mammals). The conceptual model should identify these receptors both in terms of the species and assumed range of the population (e.g., short-tailed shrews in Waste Area 2) and the group that they represent (e.g., ground invertebrate feeding mammals). Some trophic/taxonomic groups may have more than one representative species (e.g., kingfishers and osprey for piscivorous birds). Others such as reptiles may have none because of the paucity of toxicological information concerning those species. The narrative for these receptors should indicate why the representative species was chosen and exactly what group of species it represents. The issue of selecting representative species is discussed more fully in Box 2.5. In some cases, populations are chosen for their importance per se rather than as representatives. For example, an important species such as a game fish may be selected as a population-level endpoint entity, even if it is a component of a community-level endpoint.

Organism — The only organisms that are legally protected as individuals are threatened and endangered species. Hence, individuals of these species are automatic candidates for endpoint entities if they are potentially present on the site. Although wildlife species that are not threatened or endangered are managed as populations, such populations are commonly protected as individuals by regulators. For example, in the EPA interim guidance for ecological risk assessment for Superfund, two out of three examples had protection of the fecundity of individual birds as the assessment endpoint (Sprenger and Charters, 1997). Similarly, in Oak Ridge, regulators called for poisoning a fish community and draining a pond to protect individual kingfishers. Such use of individuals of common species as assessment endpoint entities is not encouraged by the authors, but may be required by risk managers.

The definition of ecosystems, communities, and populations requires setting a spatial boundary on the entity. For ecosystems, the boundaries should be based on features that delimit the processes for which the ecosystem is valued or that demarcate recognizable ecosystem types. An example of the former is watershed boundaries, and an example of the latter is the extent of wetlands. Where possible, bounds on communities should be based on the extent of a distinct community type, or on changes in species composition. In terrestrial systems, communities are conventionally defined by the form of the dominant plants (e.g., meadow or hardwood forest), but in aquatic systems one may need to use the extent of specified species. Populations are defined in terms of actual or potential interbreeding, a process that is not readily observed. For mobile species, population boundaries may be inferred from the occurrence of features that are likely to limit movement and therefore interbreeding. These include natural features such as streams and vegetation transitions and anthropogenic features such as highways, dams, or industrial areas. Such boundaries are also likely to limit the spread of gametes and propagules of plants and other relatively immobile organisms. Ideally, the features used to define the boundaries of units of a large site will also serve to define bounds on endpoint populations or communities.

Note that none of these boundaries is absolute. For example, emergent adults of aquatic insects may fly from one stream to another to breed. However, most breeding will occur between individuals from the same stream and even from the

BOX 2.5
Representative Species

It is common practice when selecting endpoints for wildlife to designate a representative species (Hampton et al., 1998). That is, one may choose the meadow vole as a representative herbivore or the red fox as a representative carnivore. This practice can lead to confusion unless it is clear what category of organisms is represented and in what sense the species is representative. For example, the meadow vole may represent all herbivores, all small mammals, all herbivorous small mammals, or all microtine rodents. A representative species may be representative in the sense that it is judged likely to be sensitive, because its activities are confined to the site (e.g., the vole rather than deer as representative herbivore), its behavior is likely to result in high levels of exposure (e.g., birds feeding on soil invertebrates rather than on herbivorous invertebrates or seeds), it is known to be inherently sensitive to the contaminant of concern (e.g., mink and PCBs), or it is predicted to be sensitive by application of allometric models. A species may also be representative in an ecological sense if it is the most abundant representative of the category of organisms on the site. Finally, a representative species may be chosen because it is particularly amenable to sampling and analysis or to demographic surveys.

The groups that representative species represent are commonly defined in terms of higher taxa or broad trophic groups. However, if the characteristics that control exposure and toxicological or ecological sensitivity can be defined, endpoint groups may be defined by cluster analysis of those traits. This approach was applied to birds at Los Alamos, NM using only diet and foraging strategy to generate "exposure guilds" (Myers, 1999). This approach is more objective than the typical subjective grouping of species, and the hierarchy of clusters provides a basis for increasing the level of detail in the analysis as the assessment progresses.

In general, it is not a good idea to select highly valued species as representative species because it tends to confuse the roles of endpoint species and representative of a community or taxon. For example, if bald eagles occur on a site, they are likely to be an endpoint species protected at the organism level. If piscivorous wildlife as a trophic group are also an endpoint, bald eagles might be thought to also serve to represent that group. However, because bald eagles cannot be sampled except under exceptional circumstances and they are not likely to be highly exposed due to their wide foraging area, it would be advisable to choose a species that is more exposed, more abundant on the site, or less protected as a representative (e.g., kingfishers or herons). By using a different species to represent the trophic group, one could perform a better assessment of the trophic group and could clearly distinguish the two endpoints in the risk communication process.

When using a representative species, it is essential to determine how the risks to the represented category of organisms will be estimated. The method may range from assuming that all members of the category are equivalent, to using mechanistic extrapolation models to estimate risks to all members of the category once risk to the representative species is demonstrated to be significant.

same stream reach. Therefore, one should not hesitate to define a stream as having a distinct aquatic insect community or a distinct population of a mayfly species.

It is common practice to define the boundaries of endpoint entities in terms of site boundaries or unit boundaries. However, this practice should be discouraged, unless the site has features that make its boundaries reasonably correspond to ecosystem process, habitat, or dispersal boundaries. Otherwise, biological realities may conflict with assumptions. For example, regular movement of individuals into and out of a "population" will render survey results meaningless. If there are concerns that natural boundaries of populations and communities will dilute out the toxic effects, then perhaps less extensive populations or communities should be considered. Alternatively, one may lower the level of organization at which an endpoint is defined. For example, rather than defining an endpoint population of red-tailed hawks on a 1 ha site, one may use individual hawks occurring on the site as the endpoint entity, if their significance can be justified.

Entities that should be considered when selecting assessment endpoints because of policy goals or societal value include the following:

- Endangered, threatened, or rare species
- Species with special legal protection
- Rare community or ecosystem types
- Protected ecosystem types (e.g., wetlands)
- Species with recreational or commercial value
- Species with particular aesthetic or cultural value

Entities that should be considered when selecting assessment endpoints because of their ecological relevance include the following:

- Taxa that are major contributors to energy or nutrient dynamics
- Taxa that provide important habitat structure
- Assemblages or taxa that regulate physical or biogeochemical processes
- Consumers that regulate the relative abundance of their prey species (i.e., keystone species)

When selecting entities based on their susceptibility, the following points should be considered:

- The sensitivity of a species is most likely to be predicted by the sensitivity of the most closely related tested species (Suter et al., 1983; Fletcher et al., 1990; Suter, 1993a).
- When the contaminant is a pesticide, species belonging to the same taxon as the target species are likely to be sensitive.
- No species or taxa are consistently most sensitive, but daphnids are on average more sensitive than other aquatic species (Host et al., 1991).
- Living systems cannot be more sensitive than their most sensitive component, and, because of compensatory mechanisms, they are often less sensitive (O'Neill et al., 1986; Suter, 1995b).
- When relative inherent sensitivity is unknown, differences in exposure determine relative susceptibility.

2.5.2 Selection of Assessment Endpoint Properties

Generically appropriate properties of the entities selected by the criteria above can be identified based on the level of organization of the entity and the criteria that led to its selection.

Organism Level — In general, protection of individual organisms is appropriate only for threatened and endangered species. For those species, individual survivorship and reproductive success are appropriate endpoint properties. Individual organisms of common species may be selected as endpoint entities by risk managers (see above). The same properties, survival and reproduction, may be used with them as well.

Population Level — In general, the appropriate endpoint properties for populations of endpoint species are abundance and production.

Community Level — In general, the appropriate endpoint properties for endpoint communities are species richness and abundance. The measure of abundance varies among communities. For example, the abundance of the fish community is determined as numbers of all component species, whereas herbaceous plant community abundance may be expressed as biomass per unit area. Various diversity and "integrity" indexes have been used, but they are generally less sensitive to toxic effects than species richness (Dickson et al., 1992; Adams and Ryon, 1994; Hartwell et al., 1995) and are less understandable by decision managers and stakeholders.

Ecosystem Level — Some ecosystems such as wetlands are valued for their properties as ecosystems rather than for their composition as communities. Properties of wetlands that are specifically protected in the United States are provision of habitat for wetland-dependent species, regulation of hydrology, and retention or cycling of nutrients. Some components of ecosystems are clearly ecologically relevant for their role in ecosystem processes but not for their population or community properties. The soil heterotrophic community is a prominent example.

Properties of specific classes of receptors that might be endpoint properties are discussed below.

Soil ecosystem properties — Given the importance of soil as a biogeochemical system supporting all terrestrial life, it seems obvious that assessment endpoints for contaminated soils should include appropriate soil properties. An example of a soil property that would usually be desirable for a hazardous waste site would be a high rate of biodegradation of organic contaminants. However, other appropriate properties are not always self-evident. Reduced nitrification is sometimes proposed as an endpoint property, but if the rate of nitrate production is too high, the nitrate may leach to groundwater, posing a risk to human health. Similarly, a reduction in the rate of litter decomposition is not always undesirable (Efroymson and Suter, 1999). Many of the properties of soil ecosystems, such as reduced nutrient availability and changes in the relative abundance of microbial taxa, which change in soils following contamination with organic contaminants such as petroleum, are results of biodegradation, a desirable process. In other words, many of the changes occur because the contaminant acts as an organic substrate as well as a toxicant. As a result, many of the soil processes and properties that have been proposed as test endpoints would not be appropriate for use at sites contaminated with organic chemicals (Health

Council of the Netherlands, 1991). For example, soil respiration increases as organic chemicals degrade, and net nitrogen mineralization is reduced due to immobilization. These effects can mask any toxic effects on mineralization of native organic carbon and nitrogen. In addition, to most decision makers and stakeholders, the soil is a black box which is acceptable if it supports plants and animals. Therefore, soil properties are less likely to be drivers for decision making than are other potential assessment endpoints.

Plant properties — Plant production is one of the clearest and most generally accepted assessment endpoints for contaminated soils. The biological and societal importance of plant production is clear. Also, plants have a scale of exposure that is appropriate to contaminated sites in that plants do not wander out of the contaminated area, and many contaminated sites are large enough to encompass a population of herbaceous plants. Although plants do not appear to be particularly sensitive to soil contaminants on average, their sensitivity is not well predicted by other receptors, and they are highly sensitive to some chemicals. Although various other properties might be used for the assessment endpoint (e.g., mortality or species richness), the common use of tests of plant growth suggests that production should be the endpoint property.

Properties of soil fauna — Soil invertebrates are ecologically important in terms of soil structure and nutrient cycling and as food for wildlife. They are potentially sensitive to soil contaminants due to their intimate contact with and consumption of the contaminated soil. In addition, because of their low mobility, they have an appropriate scale of exposure for any contaminated terrestrial site. Their societal significance is less clear. A review of bases for regulatory decisions by the EPA found that aquatic and benthic invertebrates, fish, birds, mammals, reptiles, amphibians, and plants were considered, but soil invertebrates and microorganisms were not (Troyer and Brody, 1994). Therefore, if risk managers are willing to make remedial decisions on the basis of effects on soil invertebrates, they are appropriate assessment endpoint organisms. The appropriate property is less clear. The common use in the United States of earthworm survival, growth, and reproduction as test endpoints suggests that the assessment endpoint should be population abundance or production of earthworms, or of all invertebrates as represented by earthworms. Risk assessors in the Netherlands have used protection of 95% of species of soil invertebrates as an endpoint (van Straalen and Denneman, 1989), as well as survival, production, and abundance of earthworms and collembola (Health Council of the Netherlands, 1991).

Properties of terrestrial vertebrates — Mammals and birds are common endpoint entities for contaminated terrestrial sites. However, vertebrates in general are less ecologically important than plants, invertebrates, and microbes. In addition, they typically have an inappropriate scale for contaminated sites. That is, all bird populations and many other vertebrate populations have much larger ranges than typical contaminated sites. Even individual vertebrates often have ranges that are larger than contaminated areas. As a result, the susceptibility of vertebrates is often low if risks are realistically assessed because the exposure is diluted over the entire range of organisms and the effects are diluted over the range of the population. Shrews and moles are potentially important exceptions, because they have relatively small ranges and they have high dietary and direct exposures. Terrestrial salamanders and burrowing anurans and reptiles are also potentially highly susceptible, but their responses to chemical exposures are poorly known, and no standard toxicity tests exist for them. Common endpoint properties for terrestrial vertebrates include survival of individuals and abundance or production of populations.

Properties of aquatic vertebrates — Fish are the most common endpoint entity for assessments of aquatic contaminants. Precedent suggests that the community properties of species richness and absolute and relative abundance are the most important properties in terms of regulatory policy (Plafkin et al., 1989). However, where fish harvesting occurs, the abundance and production of game or commercial species have clear societal significance. Another property that is societally significant where fish are harvested is the presence of gross pathologies or deformities. Common properties used in regulatory assessments of fish communities are indexes of heterogeneous variables such as the Index of Biotic Integrity (IBI) (Karr et al., 1986). Because of their many undesirable properties, which are too numerous and complex to describe here, these indexes should not be used unless mandated by the risk manager (Suter, 1993b). Often some portion of the fish community is resident, and another portion is migratory or seasonally present. In such cases, it is important to define community endpoint properties in terms of resident species. Note that although contaminant concentrations are often measured in fish, they do not constitute an endpoint property for fish. Rather, they are a measure of internal exposure for fish and of dietary exposure for fish eaters.

Properties of aquatic invertebrates — Although not societally valued like fish, aquatic macroinvertebrates are common assessment endpoints when aquatic systems are contaminated. Precedent suggests that the community properties of species richness and absolute and relative abundance are the most important properties in terms of regulatory policy (Plafkin et al., 1989). Where aquatic invertebrates such as crayfish and mussels are harvested, their abundance and production have clear societal significance. Finally, as with fish, indexes of heterogeneous variables are used as endpoint properties of macroinvertebrate communities. Like fish community indexes, they are not good assessment endpoints and should be avoided.

Properties of aquatic plants — Unlike terrestrial plants, aquatic plant production is not a generally accepted assessment endpoint. However, the biological and societal importance of plant production is clear. Also, with the exception of phytoplankton, aquatic plants have a scale of exposure that is appropriate to contaminated sites, in that plants do not wander out of the contaminated area, and many contaminated sites are large enough to encompass a population of aquatic plants. The neglect of aquatic plants as assessment endpoints is apparently due to their general insensitivity to most chemicals relative to aquatic animals, and the lack of interest of many risk managers in "pond weeds and green scum." However, as with terrestrial plants, the sensitivity of aquatic plants is not well predicted by other receptors, and they are highly sensitive to some chemicals (e.g., herbicides). Although various other properties might be used for the assessment endpoint (e.g., mortality or species richness), the common use of tests of algal or macrophyte growth suggests that production should be the endpoint property.

These general properties should be selected, modified, or supplemented for site-specific assessments, as appropriate, based on properties of the contaminants, the modes of exposure, and the receptors. However, care should be taken to avoid excessive specificity. For example, DDT and some other chemicals cause thinning of avian eggshells, which reduces reproductive success. In that case, the measure of effects is the concentration of the chemical that causes sufficient thinning to reduce reproductive success, and the assessment endpoint is individual reproduction or population production. This distinction is made because shell thickness is not ecologically or societally important per se, but it is important as a measure of a particular mode of action by which individual reproduction or population production may be reduced.

2.5.3 Selection of Levels of Effect on Properties of Endpoint Entities

The levels of effects on endpoint properties that should be detected and may constitute grounds for remedial action have not been specified on a national basis for ERAs as they have been for human health risk assessment (Troyer and Brody, 1994). Although levels of effects are seldom specified in ecological risk assessments, they are valuable for the following reasons. First, if the DQO process or conventional sampling statistics are used to plan sampling and analysis, the level of effect to be detected must be specified. Second, a level of effect provides a basis of comparison of (1) lines of evidence in the risk characterization and (2) different risks such as risks at different sites or risks due to contaminants vs. those due to remedial actions. Third, specification of a level of effect provides a basis for informed risk management and informed input by stakeholder groups.

For the Oak Ridge Reservation, a level of effect that is considered potentially significant has been inferred on the basis of analysis of EPA and Tennessee regulatory practice (Suter et al., 1994). The clearest ecological criteria for regulation in the United States are those developed for the regulation of aqueous effluent under the National Pollution Discharge Elimination System (NPDES). NPDES permitting may be based on any of three types of evidence — water quality criteria, effluent toxicity tests, and biological surveys — and the use of each of these implies that a 20% reduction in ecological parameters is acceptable.

1. The Chronic National Ambient Water Quality Criteria (NAWQC) for Protection of Aquatic Life are based on thresholds for statistically significant effects on individual responses of fish and aquatic invertebrates. Those thresholds correspond to approximately 25% reductions in the parameters of fish chronic tests (Suter et al., 1987). Because of the compounding of individual responses across life stages, the chronic NAWQC concentrations are estimated to correspond to much more than 20% effects on a continuously exposed fish population (Barnthouse et al., 1990). Hence, while the EPA did not intend to design the NAWQC to correspond to a 20% effect or any other particular level of effect, the consequence of the procedure used to derive the NAWQC is to specify a concentration that, in chronic exposures, results in effects that are greater than 20%, on average.

2. The subchronic tests used to regulate effluents based on their toxicity cannot reliably detect reductions of less than 20% in the test endpoints (Anderson and Norberg-King, 1991). Once again, this is a consequence of the manner in which the EPA regulates effluents rather than a conscious policy decision.
3. The approximate detection limit of field measurement techniques used in regulating aqueous contaminants based on bioassessment is 20%. For example, the community metrics for an exposed benthic macroinvertebrate community must be reduced by more than 20% relative to the best communities within the ecoregion to be considered even slightly impaired in the EPA rapid bioassessment procedure (Plafkin et al., 1989). Measures for other taxa that are more difficult to sample may be even less sensitive. For example, the number of fish species and individuals must be reduced by 33% to receive less than the top score in the EPA rapid bioassessment procedure for fish (Plafkin et al., 1989). Once again, this effects level is a consequence of the manner in which the EPA regulates effluents rather than a conscious policy decision.

The 20% level is also consistent with practice in assessments of terrestrial effects. The lowest-observed-effects concentration (LOEC) for dietary tests of avian reproduction (the most important chronic test endpoint for ecological assessment of terrestrial effects of pesticides and arguably the most applicable test for waste sites) corresponds to approximately a 20% effect on individual response parameters (Office of Pesticide Programs, 1982).

Therefore, a decrement in an ecological assessment endpoint that is less than 20% is generally acceptable based on current EPA regulatory practice and could not be reliably confirmed by field studies. Therefore, it is *de minimis* in practice. To allay concerns about the use of the 20% effects level of protection, statistically significant levels of effects may be considered important as well. Because conventional statistical significance levels usually correspond to biological effects levels greater than 20%, statistical significance is seldom an issue in the interpretation of a particular set of ecological effects data. When using both types of significance criteria, any significant effect should be identified as either biologically significant (>20% effect) or statistically significant (<5% chance the difference from control or reference is due to chance).

This definition of a significant effect (20% or statistically significant decrement) is not recommended for general use and will not be acceptable to regulators or other risk managers at many sites. However, risk assessors and risk managers must bear in mind that, if they choose levels of effect lower than 20%, they will need to increase the level or replication in standard toxicity tests and design much more labor-intensive field studies or accept levels of Type I error greater than 5%.

In addition, some exceptions apply to the use of a 20% level of effect or of statistical significance to define ecological assessment endpoints. Threatened and endangered species are protected from any adverse effects; therefore, neither a 20% effect nor a statistically significant effect can be considered acceptable. Wetlands are protected from any net loss, so a 20% reduction could not be considered acceptable for ecosystems that are so classified. At particular sites there may be other species, communities, or ecosystems that have exceptional importance and therefore require greater protection than is afforded by the 20% level or statistical significance. These exceptions must be identified on a site-by-site basis.

2.6 CONCEPTUAL MODELS

Conceptual models summarize the results of the problem formulation and guide the analytical phase of the risk assessment. They are working hypotheses about how the hazardous agent or action may affect the endpoint entities (Barnthouse and Brown, 1994; EPA, 1998). Conceptual models include descriptions of the source, of the receiving environment, and of the processes by which the receptors come to be exposed directly to the contaminants and indirectly to the effects of the contaminants on other environmental components.

Conceptual models are developed and used iteratively in the risk assessment process. First, following the initial site survey, draft conceptual models should be developed as input to the problem formulation process. These models should include all sources, receptor classes, and routes of exposure that are plausibly of concern. This preliminary conceptual model also serves as the conceptual model for the scoping or screening assessment (depending on information available) performed to support the problem formulation process. During the problem formulation process, the conceptual model is modified to be more relevant to the decision. The model is simplified by eliminating (1) receptors that are not deemed to be suitable assessment endpoints, (2) routes of exposure that are not credible or important, (3) routes of exposure that do not lead to endpoint receptors, and (4) potential sources that are not deemed credible or important. In addition, the problem formulation process makes the conceptual model more specific by identifying particular endpoint species, defining the spatial and temporal scale of the assessment, and making other judgments. The results of the problem formulation process are presented in the conceptual models published in the analysis plan. If a new screening assessment is performed for the analysis plan or for an interim report of a phased assessment, it should be based on this modified conceptual model. The conceptual models reappear in the baseline ecological risk assessment. In most cases, they are the same as in the analysis plan. However, the results of ongoing communications among the parties and the results of the field and laboratory investigations or exposure modeling may result in modification of the conceptual model.

The bases for developing the conceptual models depend on the stage in the remedial process and the amount of existing information. The first conceptual model is based on qualitative evaluation of existing information and expert judgment. It should be conservative in the sense that sources, pathways, and receptors should be deleted only if they are clearly not applicable to the site. Before or during the problem formulation process, a screening assessment should be performed using existing data (see Chapter 5). The results of that screening assessment can be used to eliminate receptors or even entire media for which no contaminants present a potentially significant risk. In addition, the participants in the problem formulation process can apply their professional judgment and managerial authority to modify the draft conceptual model presented by the assessment scientists. For example, the parties may decide that the results of the screening assessment are not based on data of sufficient quality and quantity to justify deleting media or receptors. Some receptors may be eliminated because they are not judged to be sufficiently important or sensitive or not sufficiently related to the remedial decision. If the data gathering is conducted in phases, the screening assessments performed at the end of each phase should be used to modify the conceptual model. Typically, this process involves further reducing the model by eliminating components that were shown by the assessment to be unimportant or even not present. If the assessment is based on predictive exposure modeling, preliminary modeling results may eliminate certain pathways, because they are incomplete (e.g., the contaminated groundwater plume will never intersect the surface) or are unimportant (i.e., will contribute minimally to the total exposure). Those pathways may be eliminated from the conceptual model.

The development of conceptual models is a valuable heuristic exercise. It forces the assessors to clearly work out the implications of what is known about the site and raises issues about which information is lacking. Alternative models may be developed to demonstrate the implications of alternative assumptions about the site. Once developed, a conceptual model can be highly useful in communicating the assessors' understanding of and assumptions about the system to the risk manager, stakeholders, and communications media.

2.6.1 Conceptual Models of Alternative Baseline Scenarios

A conceptual model should be developed for each distinct scenario. For waste sites, scenarios include the current case in which wastes are being released in some manner, future scenarios involving continued routine releases, future scenarios involving increased releases such as tanks rupturing or drums corroding through, and future scenarios involving a change in ecological conditions.

At old waste sites with rapid transport of contaminants and habitat quality unlikely to increase (e.g., most areas of the Oak Ridge Reservation), the current state represents the maximum baseline risk, and ecological risks will decline in the future and need not be assessed. However, separate ecological risk assessments should be performed if these risks could increase in the future. The increased threat could occur if waste containment has not yet failed, if contaminant transport has not brought waste leachates to the surface, or if succession or other changed ecological conditions could bring more susceptible species onto the site. Often, the prediction of future risks does not require a different conceptual model. For example, if range expansion is hypothesized to bring a more susceptible species to the site than the current representative species for a trophic group (e.g., river otters in place of mink) or a more protected species (e.g., bald eagles in place of osprey), the model need not change except to add the future endpoint species to the list of current endpoint species. However, other future scenarios, such as development of a forest ecosystem on a currently bare or mowed site, require a separate model of future conditions.

2.6.2 Components of a Conceptual Model

A conceptual model should be presented in both graphic and narrative form. The graphic form may be pictorial, but pictorial representations are typically costly to produce and often ambiguous. Therefore, diagrams (i.e., flowcharts) are generally recommended. The diagrams should include (1) sources, (2) routes of transport from sources to contaminated media, (3) routes of exposure of receptors to media, (4) endpoint receptors, and (5) output of contaminants to other units (Figure 2.1). This flow of contaminants lends itself to representation as a flowchart. However, in some cases factors and processes other than contaminants and their flows may be included. First, significant indirect effects should be included. For example, if sediment contaminants affect the abundance of benthic invertebrates, and the fish community is an endpoint entity, then effects of loss of food on benthic-feeding fish should be included. Second, effects of physical disturbance, or other aspects of the site other than contaminants, that significantly influence the endpoint entities should be included. For example, if the fish community in a particular stream reach is degraded due to channelization as well as potential contaminant effects, inclusion of that physical effect in the conceptual model could clarify the assessment. Guidance

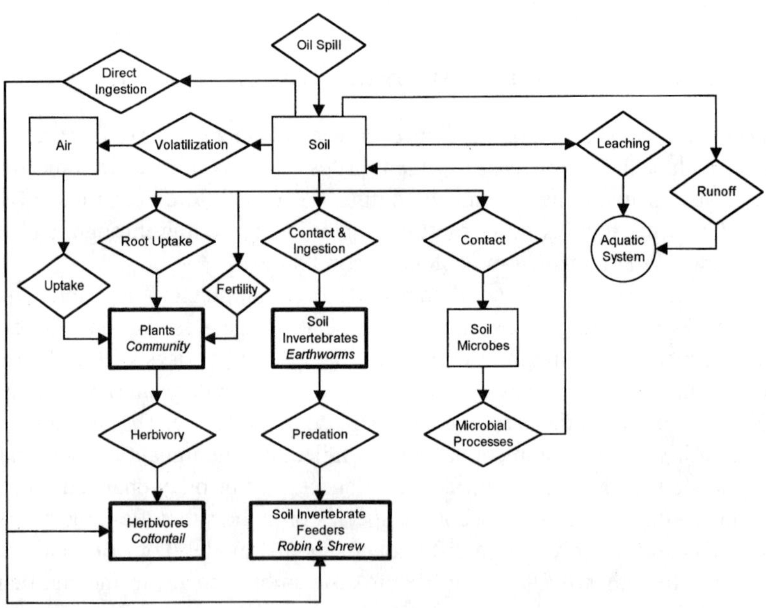

FIGURE 2.1 An example of a conceptual model for a contaminated site. It represents a terrestrial site contaminated by an oil spill. States are rectangles, processes are diamonds, and other conceptual models are circles. Endpoint compartments are indicated by heavy borders, and endpoint entities are in italic font.

has been published for developing complex conceptual models that involve stressors other than chemicals (Suter, 1999a).

The EPA recommends that the narrative should be presented as risk hypotheses (EPA, 1998). That is, it should take the form of a series of statements such as "toxic exposure of herbivorous mammals is hypothesized to occur by ingestion of contaminated soil and plants in the area of the spill." This approach may become stilted and difficult to relate to complex diagrams. Alternatively, the narrative may follow the logical flow of the diagram, providing sufficient detail concerning the components and processes to ensure that it can be understood by an educated layperson. This connection between the narrative and diagram can be aided by numbering the links in the diagram and then using corresponding numbers in the narrative. In any case, the narrative conceptual model should not duplicate information that is present in other sections of the document in which it occurs, such as the site description.

In addition to explaining the components and the processes that link them, the narrative conceptual model should explain the underlying logic of the model, including the following.

- The narrative should describe the spatial bounds on the assessment and any subdivision of the unit into reaches or other subunits.
- If receptors or routes of exposure are left out due to lack of information or knowledge, that omission should be acknowledged in the description of the conceptual model and included in the analysis of uncertainty.
- If receptors or routes of exposure are left out because of judgments of the parties to limit the scope of the assessment to critical pathways and receptors, those judgments should be acknowledged and explained. However, those choices should not be treated as uncertainties, because they are risk management decisions.
- If the chosen receptors are representative of a class of receptors, that relationship should be explained.

2.6.3 Unit Types and Default Conceptual Models

As discussed above, we divide units into four classes: source, watershed, groundwater, and wildlife units. For each of these classes of unit, a generic conceptual model is presented, including a flow diagram of the routes of transport and exposure and generic assessment endpoints (Figures 2.2 through 2.5). The conceptual model should describe the composition of each compartment, inputs to the compartment, and outputs from the compartment. Developers of conceptual models may begin the development process with these generic models. The discussions below explain how to modify those models to make them site specific.

2.6.4 Sources

All conceptual models for contaminated sites begin with sources. On source units, the wastes deposited in pits, trenches, ponds, tanks, etc. are treated as the ultimate sources (Figure 2.2). Each distinct type of ultimate source should be identified by

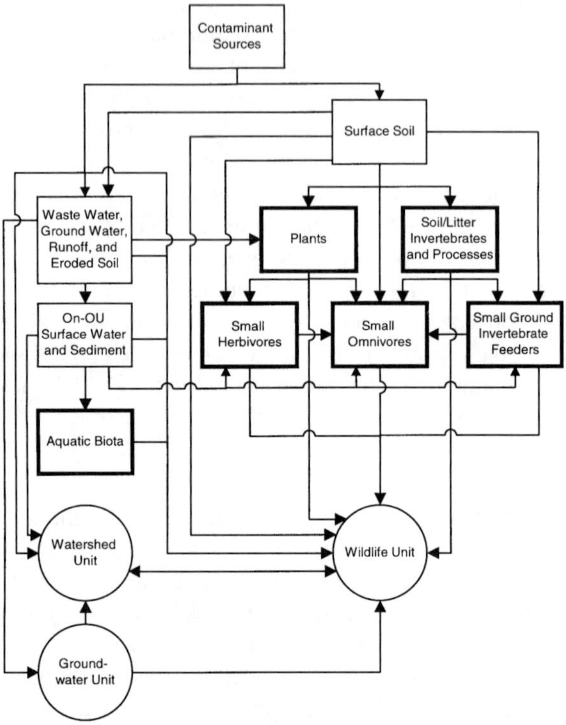

FIGURE 2.2 A generic conceptual model for source units. Endpoint compartments are indicated by heavy borders. Output to other units is indicated by round boxes.

a separate box in the model. Types of sources should be distinguished when they contain wastes that are distinctly different in form or composition or when the wastes are disposed of in different manners (e.g., ponds vs. tanks) or in situations that would result in different modes of transport (e.g., flood plains vs. uplands). Because of the ambiguous status of aqueous waste sumps and ponds (discussed above), it is important to clearly explain the nature of all such bodies, including the purpose for which they were created and their current ecological condition.

Watershed and terrestrial integrator units may have no ultimate sources, but they have proximate sources, the contaminated abiotic media: surface water, shallow groundwater, sediments, and soils. The inputs are in the form of fluxes of contaminated surface water, groundwater, eroded soil, or suspended sediments. These inputs should be identified in terms of their nature and source. In addition, some contaminant inputs are not from the wastes being assessed. For example, a stream may receive contaminants from permitted outfalls as well as leaching of buried wastes. Finally, although watersheds and terrestrial integrator units are only secondarily contaminated by the identified ultimate sources, they may be found to be contaminated by past spills of transported waste or other actions.

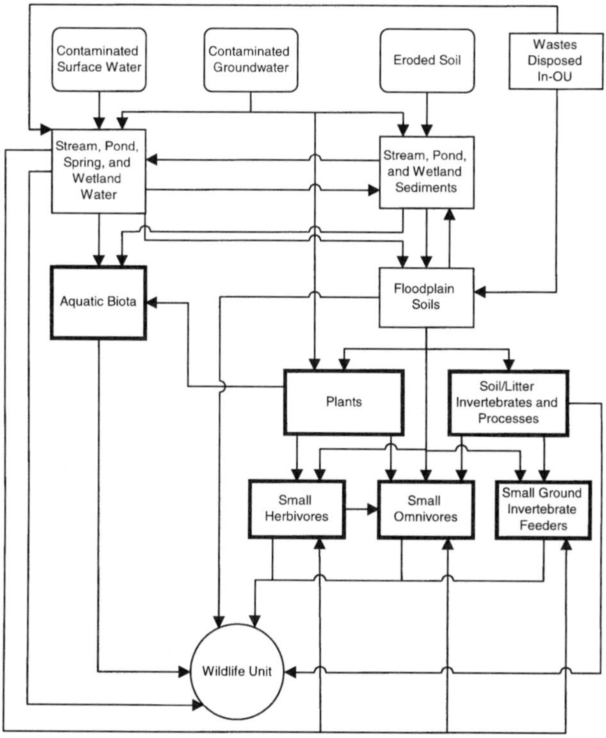

FIGURE 2.3 A generic conceptual model for watershed units. Endpoint compartments are indicated by heavy borders. Output to other units are indicated by round boxes. Rounded rectangular boxes indicate inputs of contaminated materials from other units.

2.6.5 Routes of Transport

The conceptual model should identify the routes by which contaminants in the sources are transferred to the ambient media to which organisms may be exposed. The description of the routes should be specific. For example, the transport from sources to surface water should be identified as occurring in leachate emerging at seeps, in leachate mixed with groundwater entering streams through gaining reaches, or in overland flow, either dissolved or on eroded soil. Anthropogenic pathways must be considered, including storm sewers, underground utility lines, dewatering, soil movement (e.g., site grading), and sedimentation basins. The conceptual model should identify the specific features rather than the generic compartments shown in the generic models. For example, rather than simply "surface water," water in specific streams or ponds should be identified.

The routes of transport for ecological conceptual models do not normally include deep groundwater transport, because it does not contribute to surface water contamination, and because wildlife seldom drink well water. However, a generic conceptual

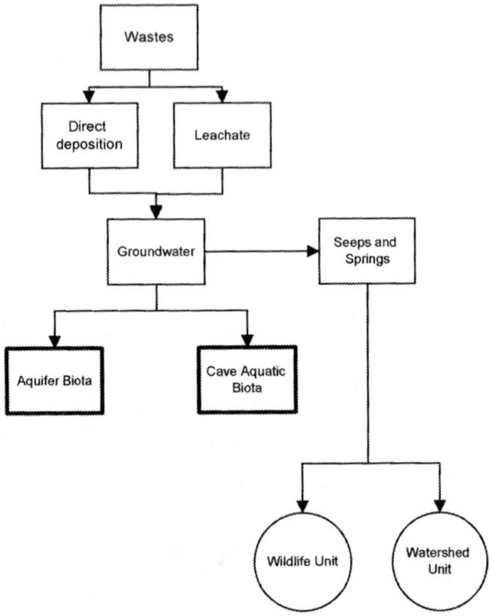

FIGURE 2.4 A generic conceptual model for groundwater units. Endpoint compartments are indicated by heavy borders. Outputs to other units are indicated by round boxes.

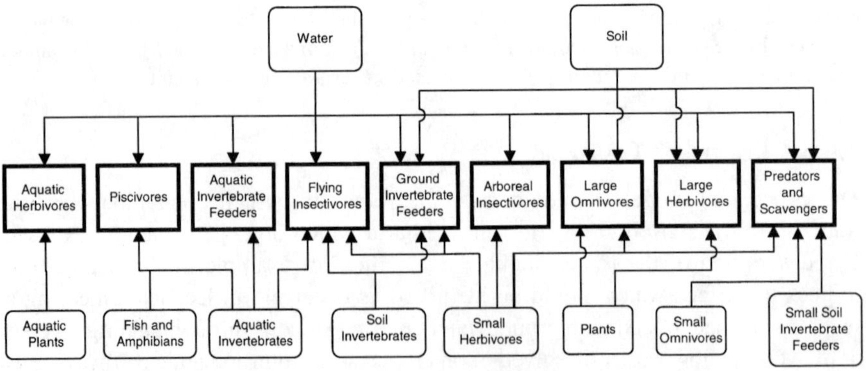

FIGURE 2.5 A generic conceptual model for wildlife units. Endpoint compartments are indicated by heavy borders. Rounded rectangular boxes indicate inputs of contaminated materials from other units.

model for groundwater is included (Figure 2.4) largely because of concern for the biota of caves. Caves commonly support communities of rare species, and, unlike aquifer communities, they may be of concern to regulators and resource managers.

Except for movement into downstream units, the conceptual models described here do not include fate processes that remove contaminants from the system (i.e., degradation and sequestration). That is because these conceptual models are intended to illustrate how ecological receptors come to be exposed rather than illustrating the fate of the contaminants. We must note, however, that the EPA Environmental Response Team recommends including processes of contaminant removal in the conceptual model (Sprenger and Charters, 1997).

2.6.6 Exposure Media

The conceptual model should identify the media that are known to be significantly contaminated, are currently hypothesized to be significantly contaminated, or are predicted to be significantly contaminated in the future and to which endpoint organisms may be exposed. If possible, judgments concerning the significance of contamination should be based on the screening of measured contaminant concentrations against ecotoxicological benchmarks and background concentrations (see Chapter 5). Alternatively, chemical concentrations estimated by modeling may be screened in the same way. In the absence of measured or modeled concentrations, expert judgment should be conservatively applied. An environmental medium should be included in the model if any chemical in the medium is retained by the screening process, or any chemical is judged to be potentially present at significant concentrations.

In some cases, a contaminated medium is also the waste (i.e., the source of the contaminant chemicals). This is true, for example, of the coal ash in an ash pond on the Oak Ridge Reservation. The ash functions as soil, which supports a plant community where small mammals receive toxic exposures to arsenic and heavy metals. Aqueous waste sumps that support aquatic ecosystems may also be treated as receptors as well as sources to watershed units. In such cases, for clarity, the source box should identify the source of the waste rather than the waste *in situ*. For example, in the coal ash case, if the ash in the ash pond is treated as an exposure medium, the ash sluice should be treated as the source.

2.6.7 Routes of Exposure

The conceptual model should identify the routes of exposure that are assumed to result in uptake of chemicals from contaminated biotic and abiotic media. The routes of exposure are limited to those that are deemed to be important for the endpoint receptors. The following points, which are discussed in greater detail in Chapter 3, should be considered.

- Fish, aquatic invertebrates, and aquatic plants are assumed to be exposed to contaminants in water. Conventionally, the EPA and most risk assessors have assumed that dietary exposures are negligible, and that assumption is likely to be true for most chemicals. For example, the NAWQC for Protection of Aquatic Life are based on toxicity tests in which organisms are unfed or fed clean food (Stephan et al., 1985). This is reasonable given the relatively high rate of exposure of organisms to chemicals in the water that passes their respiratory surfaces and the fact that most chemicals are not highly bioaccumulative and do not biomagnify.

- Although dietary exposures of fish and aquatic invertebrates are not routinely included, they may be important for a few long-lived and biophilic chemicals such as methyl mercury, PCBs, and dioxins, and may be important for a wider variety of chemicals than is currently recognized. However, appropriate and defensible dietary exposure models are not available for chemicals in streams such as occur on most waste sites, and toxicity information based on dietary exposure of aquatic organisms is uncommon and poorly standardized. Fish body burdens integrate dietary and direct aqueous exposures, but toxicity information relative to body burdens is poorly standardized and is not available for many chemicals. Therefore, aquatic dietary exposures should be included if the assessors have reason to believe that it is a significant route and have a method for assessing risks due to that route.
- Wildlife exposure routes usually include ingestion of food, drinking water, and incidental soil ingestion. Soil ingestion may be excluded for species that have little exposure to soil (e.g., ospreys).
- Dermal exposure of wildlife is not normally included. Unlike humans, birds and mammalian wildlife are covered with feathers and fur. These coverings exclude most dermal exposures. However, they create another route of exposure, grooming and preening. Amphibians are likely to experience significant dermal uptake, but neither exposure models nor toxicity data are available to address this route and receptor for terrestrial exposures. Aqueous dermal exposures for amphibians are equivalent to respiratory exposure of fish, in that they are assumed to be due to direct uptake of dissolved chemicals through a respiratory epithelium (skin or gills).
- Respiratory exposure of wildlife is not normally included. Few waste sites have significant concentrations of contaminant chemicals in the air. It is possible that burrowing animals may be significantly exposed to volatile chemicals emitted from soils, but there is no direct evidence for significant occurrence of that route, and the existing models are crude.
- Plants, soil invertebrates, and soil microbes are assumed to be directly exposed to whole soil.
- In cases where shallow groundwater is contaminated, plants are exposed to that water, particularly near springs and seeps.
- Benthic invertebrates are exposed to sediment pore water and whole sediment. Although the graphic version of the conceptual model need not depict this distinction, the narrative should include it. Although the EPA sediment quality criteria are based on exposure to the aqueous phase of sediments (i.e., pore water), the evidence is strong that some benthic invertebrates are significantly exposed to a variety of chemicals by ingestion of sediment particles. Further, pore water concentrations cannot be reliably modeled for chemicals other than neutral organic compounds. Therefore, it is important to characterize risks due to both modes of exposure.

2.6.8 Receptors

The receptors presented in the conceptual model should be those that have been proposed to be or designated as assessment endpoint receptors (organisms, populations, communities, or ecosystems). The conceptual model need not redefine the endpoint receptors, but early conceptual models may need to define receptors if endpoints have not been selected. In addition, the conceptual model should provide relevant information, such as location, size, type, and assumed functional properties of wetlands and other endpoint ecosystems and dependence of the receptors on physical and biological features of the site that may be at risk.

2.6.9 Indirect Exposure and Effects

The generic conceptual models include indirect routes of exposure but not indirect effects. An endpoint may be affected indirectly through toxic effects on lower trophic groups, by toxic effects on groups that provide physical habitat, or by other mechanisms. The importance of explicitly including indirect effects depends on the nature of the ecological relationship that causes the indirect effect and the relative sensitivity of the groups involved. For example, it is assumed for most chemicals that aquatic invertebrates and fish are more sensitive than the algal community on which they depend. Therefore, while that trophic relationship should be acknowledged in the conceptual model, it should be made clear that the indirect effects on fish and invertebrates of direct toxicity to algae are not included (if that assumption is assumed to hold). The indirect effect that is most often of concern in ecological risk assessments is the reduction in food for fish due to toxic effects on invertebrates. Planktonic crustaceans and benthic insects are often more sensitive to contaminants than fish, and benthic invertebrates are more exposed to contaminated sediments than are fish.

When indirect effects are included in the conceptual model, it is important to distinguish them from the transfer of contaminants. The generic conceptual models are based on chemical transport and transfer. If indirect effects are hypothesized instead of or in addition to transfer relationships, they should be distinguished by a different sort of arrow in the graphical model or by labels that specify the process causing the indirect effect, and they should be explained and justified in the narrative. If numerous or diverse indirect effects are included in the model, it is desirable to include the processes that link the compartments of the system as distinct boxes in the conceptual model (Suter, Gillett, and Norton, 1994; Suter, 1999a). This practice ensures that the cause–effect linkages are clear to assessors and reviewers.

2.6.10 Output to Other Units

If a site has multiple units, the assessment of each unit should include a characterization of its contributions of contaminants to other units. This responsibility is obvious with respect to hydrologically transported contaminants. For example, an assessment of a stream should characterize its output of contaminants to the larger stream of which it is a tributary. However, the contribution of source and watershed units to the contaminant burden of wide-ranging wildlife species should also be characterized. Therefore, the conceptual model should show connections to downstream watershed and groundwater units and connections to the terrestrial integrator unit. The routes of transport in the former case would be identified as dissolved chemicals in surface water or groundwater flow and transport of chemicals sorbed to suspended particles. In the latter case the routes of exposure would be consumption of contaminated food, soil, or water by wide-ranging species using the subject unit.

2.6.11 Relationship to Other Conceptual Models

The conceptual models for the ecological risk assessment should be consistent with any other conceptual models developed for the remedial investigation. Commonly, there is an overall conceptual model for the unit and a conceptual model for human health risks.

It is often desirable to present one or more overall conceptual models for the unit as part of the site characterization that typically precedes the ecological and human health risk assessments. Such models depict the sources and routes of transport of contaminants. They may emphasize particular physical aspects of the site, such as surface flow patterns, or the relationship between groundwater transport and geologic stratigraphy. They may be in the form of maps showing, for example, the location of streams and seeps relative to wastes and drainage patterns. The ecological conceptual models should be consistent with these more general conceptual models and should refer to them to provide the reader of the ecological risk assessment a context for the ecological conceptual model. Ideally, the ecological conceptual models should be an extension and elaboration of a generic conceptual model for the site. The generic conceptual model would identify the sources, the routes of transport of contaminants from the sources, the contaminated media, the degradation and sequestration of the contaminants within the unit, and the transport of contaminants out of the unit. The ecological conceptual models as well as the human health conceptual models could then be limited to the components that are particular to ecological and health risks: contaminated media, routes of exposure, and receptors.

The conceptual model for ecological risks must be consistent with the conceptual model for human health risks. That is, they should identify the same contaminant sources, routes of transport of contaminants, and contaminated media. However, the routes of exposure and receptors will be different for the two assessments.

2.6.12 Form of the Conceptual Model

In general, the graphic conceptual model consists of a flow diagram. The boxes designate physical compartments in the environment that contain the contaminants. These may be specific entities such as a stream or a waste tank, or they may be classes of entities such as herbivorous small mammals. The clarity of the diagram may be increased by using different shapes to distinguish sources, media, receptors, and output to other units. The compartments are connected by arrows that designate flows of the contaminants between compartments. In most cases, the nature of these transfers is self-evident. However, if some of the relationships are not contaminant transfers (e.g., loss of habitat structure due to loss of plants) or if for some other reason the diagram would be difficult to interpret, the arrows may be labeled or separate process boxes may be added between the compartment boxes. As mentioned above, numbered arrows or boxes allow the diagram to be clearly keyed to the narrative.

Various conventions may be used to impart additional information in the conceptual model. For example, if different transfers differ significantly in the amount of the contaminant that they convey, arrows of different thicknesses can be used to represent that difference. As another example, receptor boxes that are assessment endpoint entities may be distinguished from those that are not by shading, thickness of box outline, etc. Obviously the assessors must make judgments regarding what information is most important to convey.

If the diagram becomes complex, it may be useful to make it hierarchical. For example, a high-level diagram might contain an "aquatic biota of White Oak Creek" compartment. Then a separate, lower-level diagram would show the components of the White Oak Creek biota and the relationships among them.

Pictorial or map-based versions of the conceptual model are often useful, particularly if geographic relationships are important (e.g., dependence of wetlands on seeps or springs that are downgradient of wastes). Examples may be found in Office of Emergency and Remedial Response (1992a), ASTM (1999), and Environmental Response Team (1995a), and in Figure 2.6. In addition to showing the locations of sources and receptors ecosystems, maplike conceptual models can show flow paths and other likely routes of contaminant movement. However, these graphics are supplementary to the flow diagrams, rather than alternatives. Maps cannot clearly represent the contaminant exposures and other causal relationships.

A software tool for developing conceptual models for contaminated sites has been developed for the DOE (http://tis-nt.eh.doe.gov/oepa/programs/scem.cfm). It is particularly useful in the representation of sources and release mechanisms, but provides much less help with ecologically important exposure processes and indirect effects.

2.7 ANALYSIS PLAN

The final product of the problem formulation is a plan for conducting the assessment, including the sampling, chemical analysis, toxicity testing, measurement of environmental properties, data analysis, and modeling to be used to estimate risks. Under CERCLA, this plan is contained in an RI Work Plan, a Field Sampling Plan, and a Quality Assurance Project Plan (Office of Emergency and Remedial Response, 1988). Since these components must be integrated, they are best developed as a unitary plan which addresses the needs of both human health and ecological risk assessment. The EPA guidelines for ERA refer to this overall plan as the analysis plan (EPA, 1998). The plan should include sufficient background information for the reader to understand the rationale for the proposed activities. It should include specific sampling, analysis, and measurement methods that will be used to characterize effects, sources, and exposure, as well as the models that will be used to relate those measures to each other and to estimate risks. The analysis plan should include procedures for ensuring the quality of the data, including methods for data management. The plan should also specify why the measurements are needed and how they

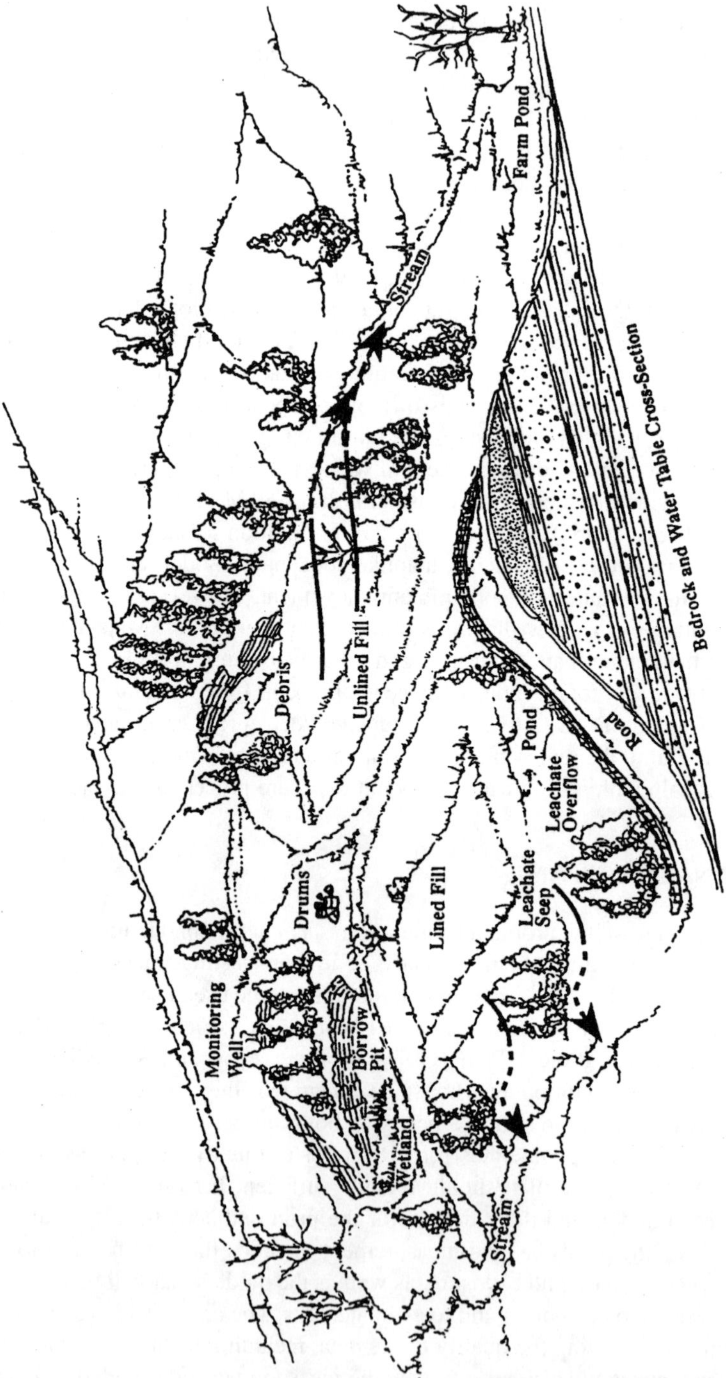

FIGURE 2.6 A pictorial conceptual model of a contaminated site. (From Office of Emergency and Remedial Response, 1992a.)

BOX 2.6
Checklist for Reviewing the Work Plan

- Does the work plan address the objectives of the baseline risk assessment?
- Does the work plan document the current understanding of the site history and the physical setting?
- Have historical data been gathered and assessed?
- Has information on probable background concentrations been obtained?
- Does the work plan provide a conceptual site model for the baseline risk assessment, including a summary of the nature and extent of contamination, exposure pathways of potential concern, and a preliminary assessment of potential risks to human health and the environment?
- Does the work plan document the decisions and evaluations made during the project scoping, including specific sampling and analysis requirements for risk assessment?
- Does the work plan address all data requirements for the baseline risk assessment and explicitly describe the sampling, analysis, and data review tasks? (*Source*: Office of Emergency and Remedial Response, 1992c.)

will be used in the risk assessment. Therefore, the development of the analysis plan requires that the assessors plan the analytical and risk characterization phases of the risk assessment, so that data needs are specified. An EPA checklist for reviewing RI Work Plans in presented in Box 2.6.

To ensure that the planned work will fulfill the objectives of the risk assessment, the EPA recommends that the analysis plan be organized in terms of the ecological risk assessment paradigm (Office of Emergency and Remedial Response, 1992b). This plan should begin with a problem formulation that describes the source, site, and receiving environment and presents the conceptual model of how the three components are related. The sections on analysis of exposure and analysis of effects should describe what observations or samples will be taken, how and when samples will be taken, and how they will be processed and analyzed or tested. The statistical summarization and analysis of the data should also be specified. In the risk characterization section, the plan should explain how the data will be used to estimate risks. It should explain how each type of effects data will be related to exposure to estimate risks and how the results from multiple lines of evidence will be weighed to reach a conclusion. Finally, quality assurance procedures should be presented in an appendix or addendum. It should address issues like chain of custody, blanks and duplicates, holding times, analytical standards, standards for positive and negative controls in toxicity tests, and limits of detection.

If the risk assessment is to be performed iteratively, that process must be part of the plan. That is, each tier of assessment must be described, as well as the relationships between them. In particular, rules should be identified for determining whether the assessment may terminate or additional tiers are needed, and for determining what additional data should be collected in subsequent tiers.

2.7.1 Measures of Exposure, Effects, and Environmental Conditions

The interface between the environmental sampling, analysis, and testing activities and the analyses of exposure and effects is the specific numerical output of the former activities, which are termed measures of effects and measures of exposure (EPA, 1998). Measures of effects (formerly known as measurement endpoints) are statistical or arithmetic summaries of observations used to estimate the effects of exposure on the assessment endpoint (Suter, 1989, 1993a; EPA, 1998). They include test endpoints such as median lethal dose (LD50) values or dose–response functions and summaries of field measurements, such as catch per unit effort or mean density. The distinction between assessment endpoints and measures of effects is important, because the endpoint for a set of measurements should not simply be adopted as the endpoint of the assessment. The property measured is at best an estimate of the property to be protected (e.g., a mean aboveground standing biomass), and it is often a related effect that must be extrapolated to the assessment endpoint (e.g., an EC50 for lettuce seedling growth). If a test endpoint were simply adopted as the assessment endpoint, that use could create the impression that a remedial decision is based on protecting a test species. For example, it could appear that millions of dollars will be spent to prevent reduced reproduction by the zooplanktonic species *Ceriodaphnia dubia*, when in fact the *C. dubia* 7-day test is used to estimate the response of sensitive aquatic species, and the actual goal of the remediation is to restore the natural species richness of the aquatic community. Considerations in choosing measures of effects are listed in Box 2.7. Note that one of the first two considerations must be met. The others are desirable. These considerations are discussed with respect to specific types of measures of effects in Chapter 4.

Note that, unlike some other discussions of measures of effects, the list in Box 2.7 does not include sensitivity as a desired characteristic. This is because consideration of sensitivity should occur in the selection of assessment endpoints (above). The measures of effect should be chosen to represent the assessment endpoint and its sensitivity. In the example above, responses of the sensitive zooplankter *C. dubia* were chosen because the assessment endpoint was defined in terms of protection of the most sensitive members of the aquatic community. However, if the endpoint were some property of the fish community, it would be inappropriate to use *C. dubia* as a measure of effect for direct toxic effects, because it is more sensitive, on average, than fish.

Often, more than one measure of effect is chosen for a particular assessment endpoint, primarily because more than one line of evidence may be used in the risk characterization. For example, inferences about risks to aquatic communities may be based on single chemical toxicity tests, ambient media toxicity tests, or biological surveys (Chapter 4). In addition, more than one measure of effects may be employed for an individual line of evidence. For example, all of the available and acceptable aquatic toxicity data for a contaminant chemical may be used in the single chemical toxicity line of evidence (Chapter 6).

Measures of exposure must also be specified during the problem formulation. Most commonly in studies of contaminated sites, measures of exposure are in the

BOX 2.7

Considerations in Selecting Measures of Effects

Correspondence to an assessment endpoint — The most useful measures of effects are those that are direct measures of the assessment endpoint entity and property.

Quantifiable relationship to an assessment endpoint — Measures of effects that do not correspond to an assessment endpoint should bear a quantifiable relationship to it. Such a relationship is known as an extrapolation model.

Existing data — Measures of effects for which existing data are available have the obvious advantage of being immediately available at no cost.

Readily measured — Measures of effects that can be rapidly obtained with little effort are more desirable than those that are not.

Appropriate scale — Measures of effects should have appropriate spatial and temporal scale. For example, if exposures are episodic, then measures of effects should not have durations much longer than the episodes.

Appropriate to the exposure pathway — Measures of effects should address species and life stages that are exposed by the same routes and to the same extent as the assessment endpoint. If the assessment endpoint is defined as a community or taxon, highly exposed members should be chosen for measurement.

Appropriate to the mode of action — When the contaminants of concern are known to have a specific mode of action in the endpoint organisms, the measures of effects should include the responses that characterize that mode of action.

Diagnostic — Measures of effects that are characteristic of a particular contaminant or class of contaminants can aid the assessor in attributing causation to effects in the field.

Low variability — Low natural variability increases the likelihood of detecting an effect.

Broadly applicable — Broad applicability increases the likelihood that reference sites can be identified and that comparisons to assessments of other sites can be performed.

Standard — Standard tests and measurement methods are preferred when available.

form of summaries of concentrations of contaminants. These include the mean concentration of a chemical or the hectares of soil with concentrations greater than some prescribed value. However, alternative expressions are available (Chapter 3). As with measures of effects, policy considerations as well as technical constraints may influence the choice of measures of exposure. For example, the best estimate of the exposure of soil biota may be concentrations in an aqueous extract, but regulators often prefer the conservatism of total concentrations (e.g., a Soxhlet extraction). Specification of the measures of exposure should include the media, constituents, limits of detection, and enough information about the needed spatial and temporal coverage and desired level of precision to allow the statistical design of the sampling and analysis plan.

The plan must include adequate measures of effects and corresponding measures of exposure for each assessment endpoint. It is useful to prepare a tabulation of assessment endpoints, the measures of effects that will be generated for each of them, and the measures of exposure that are applicable to each measure of effects. In some cases, an assessment endpoint may have been selected for which no good measures of effects or exposure are available. For example, few good toxicity data exist for reptiles, there are no standard protocols for reptile toxicity tests, and field surveys of snakes are not practical as a means of estimating effects because snakes are relatively rare and difficult to sample quantitatively. Therefore, if a threatened species of snake was chosen as an assessment endpoint, the analysis plan is the place to make it clear to the decision makers that the assessment of that species will be highly uncertain, because, unless a research program is included, it will be necessary to assume that snakes are represented by birds.

In addition, the analysis plan should specify what measurements or other characterizations of the environment and receptors must be performed. These activities address characteristics that are needed to fully describe or estimate the risks but are not measures of exposure or effects. Examples include flows of streams, rooting depth of plants, and dietary preferences of endpoint species. These are parameters of exposure or effects models rather than the actual exposure or effects metrics.

2.7.2 Sampling and Analysis Plan

Most of the volume of a typical analysis plan for a contaminated site is devoted to detailed description of the planned environmental sampling and analysis. This text should be based on the results of the DQO process or other interactions with the risk manager. In addition, voluminous guidance for sampling and analysis is available from the EPA (for an entrée, see Environmental Response Team, 1995a-c, 1997; or visit the Environmental Response Team's W7eb site, http://204.46.140.12). Guidance concerning sampling and chemical analysis cannot, and need not, be summarized here since (with the exception of some biological sampling) it is not particular to ecological risk assessment. Ecological risk assessors need not know how to sample a well or perform a mass spectral analysis. Rather, ecological risk assessors must understand the issues that will influence the utility of the data to their assessment and must ensure that the plan will meet their data needs.

If the bases for the sampling design have not been established by DQO or equivalent process, it is incumbent upon the assessors to ensure that the sampling design is based on the goals of the assessment. That is, before the sampling design can be developed, one must know what must be detected, within what areas, over what time, and with what precision. The answers to these questions depend in turn on the questions to be answered by the assessment and the models and other methods that will be used to answer them. For that reason, the assessors who participate in sampling design must be prepared to answer the DQO questions (see Box 2.1). Sampling design for ecological risk assessment of contaminated sites is discussed in Stevens et al. (1989a).

A major issue to be considered in sampling design is assurance that the samples are representative. That is, the design must represent site conditions in a manner that is unbiased and relevant to the exposure of the endpoint receptors. Sampling may be performed by random, stratified random, grid, transect, or other designs as described in any text on sampling design. The selection of the type of design and of the location of samples within the selected design should consider the following.

Encompass the whole site/unit — The sampling should fully characterize the site or unit within the site to be assessed. If the extent of contamination is not well defined, sampling should be designed to determine extent, including areas beyond the originally designated site.

Represent the sources — The sampling should separately characterize all potential sources of contamination.

Represent the pathways — The sampling should separately characterize all potential contaminant transport pathways.

Represent the media — The sampling should characterize all media to which endpoint receptors and intermediate receptors are potentially exposed.

Represent the biota — Biota that are to be analyzed as food items or endpoint organisms that are analyzed for an estimate of internal exposure should be sampled to characterize the biota of the unit.

As much as possible, common sampling designs and locations should be used for the sources, pathways, media, and biota. The coordination of samples increases the utility of the data by allowing the assessor to relate the various results. For example, colocated soil and plant samples allow the assessors to model the relationship between plant and soil concentrations. If concentrations are variable in time, samples must be colocated in time as well as space. For example, if contaminants are thought to be flushed into a stream during storm events, samples of leachate and stream water should be sampled for analysis and toxicity testing during such events.

Sampling methods should be selected to provide appropriate data for transport or exposure estimation. For example, an auger would not be appropriate for soil sampling if analyses of specific horizons are needed. Similarly, composite areal samples of soil would be appropriate for estimating exposure of wildlife but not of plants.

Although the ecological risk assessor is not normally involved in the selection of analytical methods, the methods must meet the requirements of the ERA. First, the detection limits must be low enough to permit detection of contaminants at toxic concentrations. In the United States, default detection limits are the method detection limits for the EPA standard chemical analysis methods. However, these detection limits are often well above toxic concentrations. As part of the planning process, risk assessors should provide a list of required detection limits. These should be concentrations that are low enough to ensure that, if the chemical is not detected, significant toxic effects are unlikely to occur. Ecotoxicological screening benchmarks may serve this purpose (Chapter 4). Second, the speciation of a chemical should be determined if it is a contaminant of concern and if different species may occur on the site that have significantly different toxicities. At minimum, arsenic (III vs. V), chromium (III vs. VI), and mercury (organic vs. inorganic) should be speciated, because there are different water quality criteria for the listed pairs of species. In some cases, more detailed speciation is appropriate. For example, in the floodplain of the lower East Fork Poplar Creek in Oak Ridge, the mercury was nearly all inorganic, but the dominant form was sulfide, which is far less toxic than the chloride that the EPA initially assumed was the form of all inorganic mercury.

Quality Assurance and Quality Control (QA/QC) procedures must be included in the analysis plan. They are intended to ensure, or at least to document, the quality of the methods used to collect, prepare, and analyze samples and to report results. They include:

- Detailed documentation of the sampling, preparation, and analysis methods
- Holding conditions and maximum holding times
- Chain of custody forms and procedures
- Calibration methods and documentation
- Methods for preparation of trip, method, and rinsate blanks
- Methods for preparation of matrix spikes
- Methods for preparation of replicates for methods error
- Methods for preparation of performance evaluation samples
- Methods for analyte identification and quantification
- Methods for documentation of results, including analytical and total method error

QA/QC information provides the basic input to the uncertainty analysis. In addition, the absence of adequate QA/QC procedures and documentation can render data unusable. Without adequate QA/QC procedures and documentation, the data cannot be defended if challenged, and therefore the assessment cannot be used if the remedial decision is contentious. The process of reviewing the QA/QC documentation to determine the acceptability of the data is termed *data validation* (EPA, 1990b, 1994a). It involves determining whether the documentation is adequate, whether procedures were followed, whether the results are acceptable (e.g., whether results of duplicate analyses are sufficiently similar), and whether the results make sense. For example, reported concentrations may be far below the detection limit of the analytical method, unstable daughter products may be reported to be more abundant than relatively stable parent compounds or radionuclides, or concentrations of a hydrophobic chemical may be reported to be much higher in solution phase than in unfiltered water. During validation, data that have deficiencies are marked with qualifier codes, which the assessors must use to determine the usability of the data. The standard reference addressing these issues is the EPA guidance on data usability (Office of Emergency and Remedial Response, 1992c). A guide to the EPA QA/QC guidance can be found at http://es.epa.gov/ncerqa/qa/index.html. The QC procedures for sampling and analysis on the Oak Ridge Reservation provide an example of a site-specific procedure (Sampling and Environmental Support Department, 1992).

QA/QC procedures are most clearly specified for chemical sampling and analysis, but the QA plan must extend to all aspects of the assessment. A data management plan must ensure that the data do not become corrupted, that data descriptors (e.g., units and definitions of names for variables) are correct, that meta-data (information describing the data such as sampling date and location) are correctly associated with the data, and that data are archived. QA procedures are also applied to the risk assessment process itself. These include assuring that all steps in the assessment are completed, that appropriate models are used, that all relevant toxicity data are included, and that the implementations of models are correct. For example, results of an exposure model implemented in a spreadsheet should be checked by hand calculations. In addition, the assessors should evaluate the data to verify that the data validation was adequate. The authors have seen cases of incorrect units (including concentrations greater than a million parts per million), short-lived anthropogenic chemicals detected in background samples, and toxicity test results with >100% survival of test organisms. Use of these QA procedures must be documented so that the quality of the assessment can be defended during audits. A general quality assurance procedure for risk assessments performed for the Oak Ridge Reservation is contained in Risk Assessment Program (1997).

2.7.3 Reference Sites and Reference Information

When performing a risk assessment, one must have information concerning the baseline or nominal uncontaminated state for purposes of comparison and normalization. For example, one may need to know the concentrations of contaminant metals in uncontaminated media or the number of fish species in uncontaminated streams. This is referred to as reference information. The sources of reference information must be clearly specified in the analysis plan. Three sources of such information can be recognized.

2.7.3.1 Information Concerning the Precontamination State of the Contaminated Site

In some cases, studies will have been done at the currently contaminated site prior to the contamination. Such historic information may have been developed for permitting or other regulatory purposes, for resource management, for research, for a monitoring program, for environmental impact assessments, or for evaluation by private conservation organizations. Hence, it is important to contact past landowners, local universities, regulatory and resource management agencies, and private conservation organizations to determine whether information is available. However, such information must be used with caution. Information may not be of high enough quality for risk assessment, it may be so old as to be no longer relevant, or it may have been obtained using inappropriate techniques. For example, chemical analyses may not have achieved adequate limits of detection.

2.7.3.2 Model-Derived Information

In some cases, no measurements or observations can supply adequate reference information. For example, there may be no undisturbed or uncontaminated streams in the area. In such cases, it may be possible to use a model to estimate characteristics of the uncontaminated state. In particular, habitat models may be used to estimate the flora and fauna of a site in the absence of contamination or physical disturbance

(U.S. Fish and Wildlife Service, 1987; Wright et al., 1989). The results of such models are seldom sufficiently accurate to determine that a decline in abundance has occurred, but they can be used to indicate that species, which should be present in abundance, are absent or severely depleted.

2.7.3.3 Information Concerning Other Sites

The most common source of reference information is studies of reference sites, that is, sites that resemble the site being assessed except for the presence of the source and contaminants. Conventionally, the reference is a single site. For contaminated streams and rivers, a single upstream site may be used, and for contaminated terrestrial sites, a single location that is clearly beyond the contaminated area is typically chosen. This approach is inexpensive and, if exposure and effects are clear, it may be sufficient. However, a pair of sites may differ for any of a number of reasons other than contamination. For example, upstream and downstream sites differ due to stream gradients. In addition, the use of a single reference site does not provide an adequate estimate of variance in the reference condition, because repeated samples from a single reference site are pseudoreplicates (Hurlbert, 1984). This issue has been treated at length elsewhere (see, for example, Wiens and Parker, 1995). Pseudoreplication is perhaps best explained here by a relevant example.

The San Joaquin kit fox population on the Elk Hills Naval Petroleum Reserve crashed in the early 1980s, and it was suspected that the oil field wastes were responsible. A study was conducted that, among other things, analyzed the elemental content of fur from the foxes to determine whether they had elevated exposures to the wastes (Suter et al., 1992). The sponsor initially insisted that reference information be limited to analyses of fur from foxes trapped on the periphery of the oil field. However, a pilot study using that reference indicated that the concentrations of several elements were statistically significantly higher in the fur from the oil field than in reference fur. A subsequent study, which included areas away from oil fields and an oil field where kit fox populations had not crashed, found that the initial comparison had been misleading. The metal concentrations in fur from Elk Hills oil field foxes was not high relative to these other reference sites. Rather, the fur from foxes at the peripheral reference site was unusually low in metals. Without true replication of reference sites, an incorrect conclusion might have been drawn about the cause of the observed decline.

One solution to this problem is the one employed in the Elk Hills study: obtain information from a number of reference sites sufficient to define the nature of and variance in the reference conditions. Besides the obvious problem of increased cost, the chief problem in this approach is finding enough suitable sites.

Selection of reference sites involves two potentially conflicting goals. First, relevant properties of the reference sites should resemble those of the contaminated site as closely as possible, except for the presence of the contaminants. The relevance of properties of the sites depends on the comparisons to be made. For example, if a contaminated stream reach has a natural channel, but upstream reaches are channelized, that difference is probably not relevant to comparisons of metal concentrations, but it precludes using the upstream reach as a reference for population or community properties. Second, the reference sites should be independent of the contaminated site and of each other. For example, if animals move between the sites, measurements of body burdens, abundances, or any other property involving those mobile organisms cannot be used for comparison, because the sites are not independent. In most cases, the sites that resemble the contaminated site most closely are those that are adjoining or nearby, but those are also sites that are least likely to be independent of the contaminated site.

The problem of deciding whether reference and contaminated sites differ in any relevant way requires knowledge of the physics, chemistry, or biology controlling the properties to be compared. For example, the occurrence of heavy metals in soils is a function of the cation exchange capacity, so that factor must be considered, but that relationship does not depend on the identity of the clay minerals providing the capacity. For ecological comparisons the relevant habitat parameters must be similar, but the parameters that must be similar depend on the taxa being assessed. For example, the fish species composition of a water body is affected by the diversity of habitat types, but is unlikely to be significantly affected by moderate differences in production. However, condition measures such as weight/length ratios are likely to be affected by production. As a result of considerations such as these, the availability of appropriate reference sites may constrain the measures of exposure and effect that are used. It is pointless to measure a property of a contaminated site if it is not possible to obtain relevant reference information against which to compare it. The EPA has suggested parameters that should be included when determining the similarity of sites, but that guidance cannot substitute for site-specific consideration of the relevance of differences in site characteristics to the types of comparisons among sites that will be part of the assessment (Office of Emergency and Remedial Response, 1994b).

In some cases, suitable criteria for similarity may already exist. The most important example is the soil classification systems developed for the United States by the U.S. Soil Conservation Service and elsewhere by other agencies. For the most part, a reference site with the same soil type as a contaminated site is suitable for measures of soil properties such as metal concentrations. However, for other properties such as abundance of soil invertebrates, additional factors such as compaction, organic matter content, and vegetation must be considered. In any case, the soil type is a good starting point in selecting reference soils. Another example would be the use of the same stream characteristics that have been used to establish regional references (see below) to select reference stream reaches for local comparisons.

The magnitude of difference that is necessary to declare a potential reference site to be unacceptably dissimilar from other sites is determined by the magnitude of exposure or effects to be discriminated and by judgment and experience. For example, experienced invertebrate ecologists know approximately how much difference in sediment texture is likely to cause a 25% difference in the species composition of the community. The magnitude of difference to be detected should be determined by interaction with the risk manager, as discussed above.

The independence problem requires that we consider not only factors that introduce common biases in the sites, but also factors that inappropriately reduce intersite variance. For example, it is clearly inappropriate to simply establish reference sites at different distances upstream and use them to estimate reference variance. The Oak Ridge Reservation Biological Monitoring and Abatement Program has addressed this problem by monitoring a suite of reference streams both on and off the reservation. This use of multiple reference streams establishes information equivalent to a regional reference, but more localized. There are technical and policy elements in the selection of a set of reference sites, because it is necessary to decide what variance is relevant and what properties must be independent. For example, all soils on the Oak Ridge Reservation are at least slightly contaminated by mercury due to atmospheric deposition from mercury use at the Y-12 plant and coal combustion at nearby power plants. However, it was decided that low-level contamination was not relevant to determining which sites could be considered background, and therefore the lack of independence of mercury concentrations among reference sites on the reservation was judged to be irrelevant. However, the variance in metal content from one site to another within a soil type was deemed relevant and was determined (Watkins et al., 1993).

2.7.3.4 Information Concerning the Regional Reference

Regional reference information is a special case of reference information from other sites. Regional reference information is derived from a population of sites within a region that encompasses the contaminated site and that is deemed to be acceptably uniform and undisturbed with respect to the properties of interest. One example is the use of the U.S. Geological Survey summaries of elemental analysis of soils from the eastern and western United States as reference soil concentrations for sites in those regions (Shacklette and Boerngen, 1984). Some states have compiled background concentrations for environmental media (Slayton and Montgomery, 1991; Webb, 1992; Toxics Cleanup Program, 1994). Regional reference information is increasingly available for biotic communities (Davis and Simon, 1995; Barbour et al., 1996). For example, Ohio has been divided into ecoregions and reference fish and benthic invertebrate community properties have been defined for each (Yoder and Rankin, 1995). The use of regional reference information has clear advantages. Regional reference information is available from the responsible agency, eliminating the need for reference sampling; it is based on multiple independent sites so it properly accounts for variance; and, when the reference information is generated by a regulatory agency, it is likely to be acceptable to that agency.

Regional reference information has some disadvantages. One is that the data used to establish the regional reference may not be suitable for risk assessment. Detection limits for analyses of reference media may be too high or the quality assurance and quality control may not be up to the standards required by regulators or responsible parties. For example, the results of metal analyses of uncontaminated waters performed prior to the early 1990s were often too high because of inadvertent contamination during sample handling and analysis (Windom et al., 1991). Alternatively, the regional reference may not include the parameters that are needed to estimate the exposure or the assessment endpoint, as specified by the problem formulation. For example, the abundance or production of a particular fish species may be an assessment endpoint for a site, but regional references for fishes are commonly defined in terms of community properties. In addition, the ubiquity of contamination and disturbance may result in reference sites that are degraded. This is the case in Ohio where the 25th percentile of the distribution of reference community indexes is the threshold for acceptability (Yoder and Rankin, 1998). That is, 25% of the best streams in Ohio are declared unacceptable. Finally, the risk managers often prefer site-specific reference information. This preference is particularly likely when the region is large, making the relevance of the reference information questionable.

An alternative to the use of multiple reference sites to provide reliable reference information is the use of gradients. This is particularly useful where there is a gradient of contamination or where there is a biological or physical gradient. One use of gradient analysis is to establish a reference concentration of a contaminant in terms of the asymptotic concentration. This approach is particularly useful for sites where the extent of contamination is unknown. An analogous use is to establish the natural gradient of stream community properties as a reference against which to compare the state of the community in contaminated reaches. For example, fish species richness naturally increases along downstream gradients due to increased habitat quantity and diversity. Therefore, sampling fish on a gradient from well upstream of the site to well downstream can establish both the natural gradient and any deviations from it associated with the site. Gradient analyses eliminate the problem of balancing the need for similarity against the need for independence by eliminating both. That is, if the relevant factors do follow a gradient, then the lack of similarity and the lack of independence are incorporated into the gradient model. Gradients are relatively seldom used in risk assessments of contaminated sites, in part because their existence is not recognized, and in part because the natural gradient may be disrupted by disturbances other than the contaminated site.

2.7.3.5 Positive Reference Information

In addition to the conventional or negative reference information discussed above, positive reference information may be needed. Positive reference information concerns contaminated sites or tests of contaminants that are not the sites or contaminants being assessed. While the purpose of a negative reference is obvious, the uses of positive references are more diverse and difficult. The most common is the upstream reference on a stream that has sources of contamination upstream of the source being assessed. For example, Poplar Creek is contaminated by various sources on the Oak Ridge Reservation that were assessed for possible remediation, but also by upstream sources including coal mines and a small city. Analyses and tests performed on water and sediment samples from Poplar Creek upstream of the reservation provided a measure of the level of contamination and toxicity to which the DOE sources were an addition. Therefore, they provided an indication of the improvement that might be expected if remediation was implemented. Another use of a positive reference is to determine whether a hypothesized cause of observed effects is plausible. For example, Poplar Creek receives mercury from its East Fork. If the effects observed in Poplar Creek were due to that mercury input, then the same or greater effects with the same symptomology should have been observed in lower East Fork, where the mercury is less dilute. The fact that toxic effects were not observed there indicated that mercury was at most a minor contributor to the observed effects. Finally, positive controls should be used with site-specific toxicity tests to demonstrate that the tests, as performed, are sensitive to concentrations of chemicals that are known to cause relevant toxic effects. Cadmium chloride and other well-studied chemicals have been recommended as positive control chemicals, but for tests at contaminated sites, chemicals should be selected that are not only well studied but also are contaminants of potential concern at the site or at least have the same mode of action.

2.7.3.6 Background

Background is a special case of negative reference sites. Background sites are those that are believed to represent an uncontaminated or undisturbed state. In many cases, this ideal cannot be achieved. For example, the atmospheric deposition of anthropogenic mercury is ubiquitous. Therefore, one can identify negative reference sites for mercury with respect to a specific mercury source, but not true background sites. Nevertheless, background concentration is an important concept in risk assessment, because background concentrations are used to determine what concentrations of metals and other naturally occurring chemicals should not be remediated and are not even worth assessing (Chapter 5). Therefore, it is important in the analysis plan to identify sites that will be treated as effective background sites. The considerations that go into choosing a reference site apply here as well. The difference is that background sites are considered clean, whereas other reference sites are simply clean enough to represent an acceptable condition.

2.7.4 Field Verification of the Plan

At sites such as the Oak Ridge Reservation that have been well studied prior to the planning of the ecological risk assessment, it should be possible to design a feasible field sampling and analysis program based on experience. However, for unfamiliar sites, it is necessary to go into the field and confirm that the planned sampling is feasible; that the sampling effort is sufficient; that important locations, communities, and media have not been excluded; and that proposed reference areas are appropriate. It may be found, for example, that the proposed sampling equipment will not work due to media characteristics (e.g., hard vs. soft sediments), that the proposed trapping methods capture few or no animals, that a sediment sample a meter from the bank is called for in a 0.75-m-wide stream, or that the stream is ephemeral. Where needed, such field verification should occur before the plan is approved. Otherwise, the DQOs may not be met and the risk assessment may not be adequate for decision making. Field verification is discussed at length by Sprenger and Charters (1997).

2.7.5 Deviations and Contingencies

Despite the best efforts of all involved, the analysis plan will inevitably be incomplete or inadequate in certain particulars. The species to be sampled may be absent, the contamination may extend further downstream than expected, or a proposed test species may not tolerate the local water. If such events require approval of amendments to the plan, progress may be considerably slowed. Therefore, contingencies should be written into the plan, to the extent that they can be anticipated. In addition, the assessors should negotiate with regulators and responsible parties for latitude to modify their activities in response to unanticipated conditions. The ultimate degree of latitude is the use of the concept of accelerated site characterization (ASTM practice P53-95). Under this concept, the on-site manager conducts a basic sampling and analysis program, interprets results as they become available, refines the conceptual model, and collects additional data as appropriate to meet the goals of the remedial investigation.

3 Analysis of Exposure

The first part of the case will deal with the plaintiff's evidence which tends to show or [sic] designed to show ... that the toxic materials were on the lands of the defendants, that it migrated, that it got into the water in sufficient quantity to constitute a potential hazard .... When the evidence on these issues is presented ... you will be asked to make a decision as to whether the plaintiff has established all of these elements by a preponderance of evidence. If the answer is negative, that is the end of the case.

—Judge D. J. Skinner (1986), instructions to the jury
in the Woburn, MA civil action

Exposure is the contact or co-occurrence of a contaminant with a receptor. The analysis of exposure estimates the magnitude of exposure of the endpoint entities to contaminants, distributed in space and time. For contaminated sites, it is necessary to consider the exposures due to current conditions and future conditions. Under current conditions the endpoint organisms are exposed to the contaminants that have been determined, through chemical or radiochemical analysis, to occur in surface water, sediments, soil, shallow groundwater, air, and food items. This is referred to as the current baseline exposure.

One future condition that must be considered is the future baseline. This is the exposure that will occur if no remedial actions are taken to isolate, remove, or destroy the contaminants. In general, the future baseline is assumed to be the same as or not worse than the current baseline, so it need not be estimated. However, if exposures may increase in the future, resulting in increased risks, a separate estimate of future exposure must be generated. The most common reason for exposures to increase is that contaminants are migrating out of the source leading to increasing concentrations or extent of contamination. Another reason, which is seldom considered, is that succession will occur on the site resulting in improved habitat quality, more species exposed, and more individuals exposed. This second reason is an issue only if the current habitat quality is poor, if succession is allowed or encouraged to occur, and if succession is rapid relative to degradation and dispersal of the contaminants. The future baseline should also be of interest if the exposure is significantly declining. That is because, if natural processes are rapidly reducing exposures, the expense and damage caused by remediation may not be justified (Chapter 9).

Another future condition that must be considered is post-remedial exposure. As part of the feasibility study (Chapter 9), it is necessary to determine that exposures will be acceptable following remediation. In most cases, this is simply a matter of assuring that the remedial goals will be met (Chapter 8). However, if the remedial actions result in changes in physical or biological conditions that may modify exposure processes, a new exposure assessment may be performed.

Analyses of exposure should be carried out in such a way to facilitate risk characterization. That is, the exposure estimates should be appropriate for characterizing risks by parameterizing the exposure variables in the exposure–response models. This requires that the exposure estimates address the same forms or components of the contaminants as the effects assessment and also have concordant dimensions. For example, the estimation of effects on plants may require that concentrations of chemicals in the aqueous phase of the soil be estimated, that concentrations be averaged over the rooting depth of the plants on the site, and that the results be expressed as a median concentration and other percentiles of the empirical distribution of point concentrations. In contrast, the estimation of risks to wildlife due to ingestion may require total concentrations in surface soil, averaged over the foraging range of the species, expressed as the mean and standard deviation.

In all cases, the analysis of exposure must appropriately define the intensity, the temporal dimension, and the spatial dimension of exposure (Suter, Gillett, and Norton, 1994; EPA, 1998). Intensity is usually expressed as concentration in a medium that is in contact with a receptor, but dose and dose rate are also used. Time is usually the duration of contact, but other relevant aspects of time are the frequency of episodic exposures and the timing of exposure (e.g., seasonality). The spatial dimension is usually expressed as area within which an exposure occurs or as linear distance in the case of streams. If contamination is disjunct (i.e., spotty), the spatial pattern may be important. The definition of exposure must be some measure of intensity with respect to space and time. The simplest case is an average concentration over the entire site that does not vary with time.

The degree of detail and conservatism in the analysis of exposure depends on the tier of the assessment. Scoping assessments need only determine qualitatively that an exposure may occur by a prescribed pathway. Screening assessments must quantify exposure but should use conservative assumptions to minimize the likelihood that a hazardous exposure is inadvertently screened out. Definitive assessments should be realistic and therefore should treat the estimation of exposure and uncertainty separately. This may be done by estimating distributions of exposure or by estimating both the most likely exposure and upper-bound exposure.

This chapter begins with a discussion of the component activities that comprise an analysis of exposure. Next, it discusses issues specific to the individual environmental media: water, sediment, soil, and biota. It then discusses analysis of exposure for particular taxa that are exposed to multiple media. The next topic is modeling of uptake of contaminants by biota, primarily for estimation of food chain exposures. Wastes such as petroleum and its derivatives that are clearly a mélange are discussed in a separate section because they are most logically treated as a complex contaminant rather than as a collection of chemicals that are assessed separately. Finally, presentation of the results of an analysis of exposure is discussed.

3.1 COMPONENTS OF ANALYSIS OF EXPOSURE

This section discusses in general terms the activities that comprise the analysis of exposure. Specific issues in sampling, analysis, and modeling are discussed in the following medium-specific and receptor-specific sections.

3.1.1 Sampling and Chemical Analysis of Media

Most of the funds and effort expended on studies of contaminated sites are devoted to the collection and chemical analysis of the abiotic media: soil, water, and sediment. Similarly, most of the guidance for site studies is devoted to media sampling and analysis. These activities should be performed as specified in the analysis plan, and the quality of the data should be verified before they are used in the risk assessment (Chapter 2). The issues to be addressed here are summarization, analysis, and interpretation of those data. These issues are particularly problematical when chemicals are detected in some, but not all, samples (Box 3.1).

BOX 3.1
Handling Nondetects

One perennial problem with analytical data is sets of results that include both reported concentrations (detects) and reported inability to detect the chemical (nondetects). Thus, the low end of the distribution of concentrations is censored. The problem is that nondetects do not signify that the chemical is not present, merely that it is below the method detection limit (MDL). If a chemical is detected in some samples from a site, it is likely that it is also present at low concentrations in the samples reported as nondetects. For screening assessments, this problem can be handled simply and conservatively by substituting the detection limit for the nondetect observations in order to calculate moments of the distribution, or by using the maximum measured value as the estimate of exposure. However, for definitive assessments, such conservatism is undesirable. The most appropriate solution is to estimate the complete distribution by fitting parametric distribution functions (usually lognormal) using procedures such as SAS PROC LIFEREG or UNCENSOR (SAS Institute, 1989; Newman et al., 1989, 1995). Alternatively, a nonparametric technique, the Product Limit Estimator, can be used to give more accurate results when data are not fit well by the parametric functions (Kaplan and Meier, 1958; Schmoyer et al., 1996).

The problem of censoring is exacerbated by the EPA analytical procedures, which were based on the recommendations of the American Chemical Society (Keith, 1994). The MDLs are not actually the lowest concentration that a method can detect, but rather the lowest concentration that the EPA believes can be detected with sufficient accuracy and precision. Therefore, an analytical laboratory may detect a chemical at 7, 9, and 11 μg/l in three samples, but, if the MDL is 10 μg/L, the reported results are <MDL, <MDL, and 11 μg/l. While the two lower concentrations in this example are more uncertain than the highest concentration, those measured values are clearly more accurate than the estimates generated by the methods discussed above. The best procedure from a risk assessment perspective would be to report all detected concentrations with associated uncertainties rather than allowing chemists to censor data that they deem to be too uncertain.

At some sites, chemical concentrations in site media that were measured prior to the remedial investigation are available. The assessors must decide whether they should be used in the assessment. Although more data are generally advantageous, these encountered data may not be useful because of their age, quality, sampling techniques, or design. In general, the utility of encountered data must be determined by expert judgment. Considerations relevant to the age of the data include the rate of degradation of the contaminants, the rate of change in the rate of release of contaminants from the source, and the rate of movement of the contaminated media. Even if concentrations are declining, old data may be useful for screening assessments because they provide conservative estimates of current concentrations. The quality of the data and the acceptability of the sampling methods must be judged in terms of the uncertainty that is introduced relative the uncertainty due to not having the data. For example, metal analyses that were performed without clean techniques may be acceptable if the contaminants of concern occur at such high concentrations that trace contamination of the sample is inconsequential. Another important consideration is the detection limits. Analyses with high detection limits may create misleading results in screening as well as definitive assessments.

In addition to analyses of contaminant concentrations, analyses must be performed of the physical and chemical characteristics of the tested media that influence toxicity. These analyses are particularly important when toxicity tests of the ambient media are performed, because the media may be unsuitable for the test organisms due to basic properties. For water, these include pH, hardness, temperature, dissolved oxygen, total dissolved solids, and total organic carbon. For sediments, they include particle size distribution, total organic carbon, dissolved oxygen, and pH. For soils, the same properties are measured, except that dissolved oxygen is omitted and water content (e.g., field capacity) and major nutrients (e.g., nitrogen, phosphorus, potassium, sulfer) are added. For example, differences in plant growth between contaminated and reference soils may be due to fertility, pH, or texture rather than toxicity. Without information concerning those properties, the case for toxic effects cannot be defended.

Exposure analyses for ambient media toxicity tests require analyses of chemicals of potential ecological concern (COPECs) from samples that are representative of the tested material. Therefore, results of analyses that are performed independently of the test should be used with great caution. Aqueous concentrations are highly variable over space and time. In our experience, storm events or episodic effluent releases may cause concentrations to change significantly over the course of a 7-day static replacement test, potentially making the analysis of only one of the three samples inadequate for exposure characterization. Soil samples are variable over space both vertically and horizontally. Therefore, exposures in a soil toxicity test may not be well characterized by analyses of samples that were collected "nearby" or were collected from a different range of depths. Sediments may be relatively stable in time like soils or may be mobile and therefore temporally variable.

At most sites, abundant analytical data are generated which must be summarized and presented. The data summarization must meet the needs of the risk characterization. Depending on the effects and characterization models, the data may be presented as means and variances, distribution functions, percentiles, or other forms.

Care must be taken in statistical summarization to avoid bias. For example, because many sets of environmental data have skewed distributions that approximate the lognormal distribution, the geometric mean is commonly recommended. However, this results in an anticonservative bias when the value is used in calculations or interpretations that involve mass balance (Parkhurst, 1998). For example, if fish are exposed to varying concentrations in water, the best exposure metric for calculating their body burdens is the arithmetic mean concentration. Use of the geometric mean would improperly minimize the influence of high concentrations on uptake.

In addition, the chemical data must be summarized for presentation to other members of the assessment team, risk managers, and stakeholders. The goal of these presentations should be to make important patterns in the data apparent. The best general approach to displaying relationships between parameters (e.g., stream flow and contaminant concentrations) is the conventional *x–y* scatterplot. Although maps are generally not as good as scatterplots for showing potentially causal relationships, they provide an important means of presenting spatially distributed data. The difficulty comes in converting data that are associated with points to areal representations. The simplest approach is to present the results on a map at the point where the sample was taken. The results may be in numeric form or as a glyph such as a circle with area proportional to the concentration. Alternatively, various geostatistical approaches can be used to associate concentrations with areas. These may be discrete areas (e.g., Theissen polygons), isopleths (e.g., Kriging), or gradients (e.g., polynomial interpolation) (Figure 3.1). Discussions of data presentation for contaminated sites can be found in Stevens et al. (1989b) and Environmental Response Team (1995a). More technical guidance may be found in Goovaerts (1997). This is an area in which a little creative thought can be useful. Good general guidance for data visualization is provided by Tufte (1983, 1990, 1997).

Toxicity normalization provides a means of summarizing exposure data for numerous chemicals in an interpretable form. This is done by converting the concentrations (*C*) to toxic units (TU), which are proportions of a standard test endpoint such as the *Daphnia magna* 48-h EC50.

$$\text{TU} = C/\text{EC50} \tag{3.1}$$

TU may be plotted as the value for each reach, subreach, transect, or other unit (Figure 3.2). The height of the plot is the sum of toxic units (ΣTU) for that location. The advantage of this approach is that it displays the contaminant concentrations in units that are indicative of toxicity rather than simply mass per unit volume. Therefore, one can see which units are most likely to pose significant risks, and which chemicals are likely to be major contributors to the toxicity. The purpose of this analysis is heuristic.

3.1.2 Chemical Analysis of Biota and Biomarkers

Analysis of abiotic media provides a measure of external exposure to contaminants, but not internal exposure or exposure through trophic transfers. These require estimates of uptake from media and transfer between biotic compartments. In the

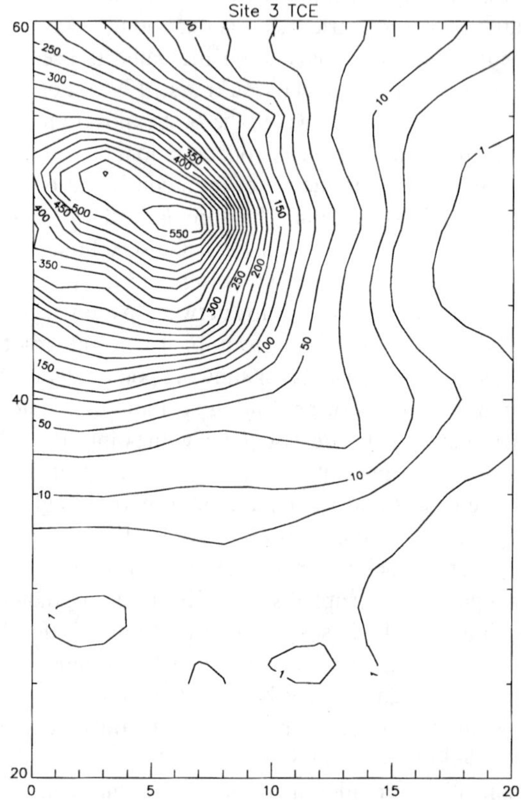

FIGURE 3.1 An example of a map generated by Kriging which, given a spatial array of chemical measurements, defines areas estimated to have chemical concentrations within prescribed ranges. (Provided by Yetta Jager, ORNL)

absence of reliable models of uptake and transfer, internal exposures and trophic transfers can be estimated by collecting and analyzing biota from the contaminated site or from laboratory exposures to contaminated media. This approach has the advantage of avoiding the use of highly variable empirical models or unvalidated mechanistic models. However, analytical chemistry is expensive, and some chemicals are rapidly metabolized or may not accumulate to detectable levels. Similarly, body burden analyses are not feasible for some species such as those designated as threatened and endangered.

Care must be taken to ensure that the body burden analyses are relevant to the assessment. One issue is the treatment of unassimilated material. For example, if soil or sediment oligochaetes are not purged, the analysis may be dominated by chemicals in the gut contents which have not been incorporated. This may either overestimate or underestimate internal exposure of the worms and dietary exposure by vermivores, depending on whether the uptake factor (organism concentration/

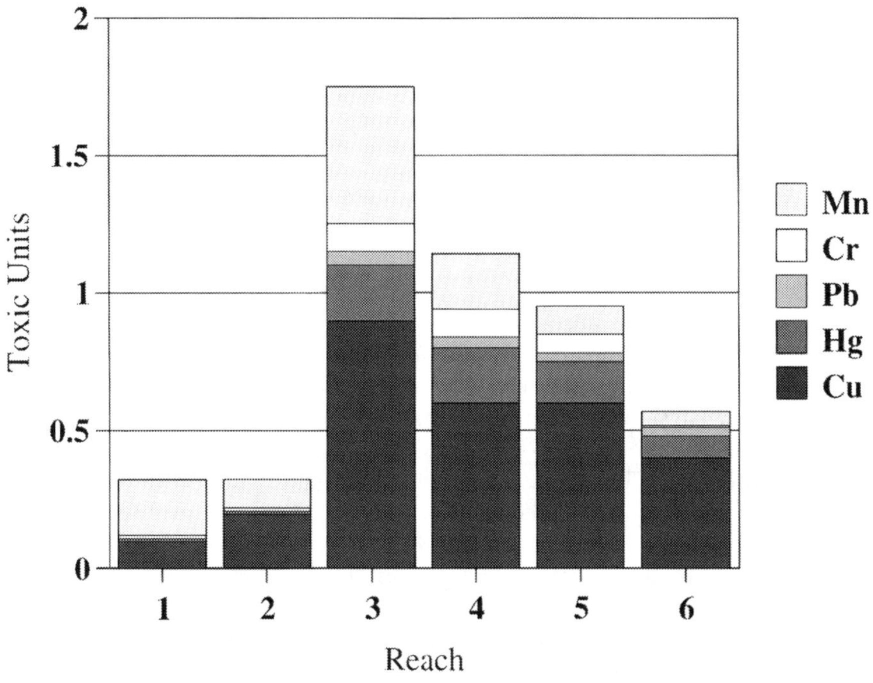

FIGURE 3.2 A heuristic display of the toxicity normalized concentrations of metals in six reaches. The height of each bar is the sum of toxic units.

soil concentration) is less than or greater than 1. However, for chemicals that are rapidly depurated following assimilation, long holding times for purging may result in underestimation of exposure. Although 24 h is the standard holding time to evacuate gut contents, as little as 6 h may be sufficient (Mount et al., 1999). The issue of unassimilated material also arises with contamination of the surfaces of leaves and the fur and feathers and gut contents of wildlife. Decisions concerning this issue should be based on careful consideration of the actual mode of exposure of the endpoint organisms and of the exposure model used in the assessment. For example, if soil ingestion is included as a separate route in the exposure model, care should be taken to avoid incorporating soil into the chemical analyses of endpoint organisms or their food.

Another aspect of ensuring relevance of analyses to the risk assessment is selection of appropriate species, higher taxa (e.g., insects), or assemblages (e.g., benthic invertebrates) for sampling and analysis. This depends on the purpose of the sampling. In general, the purpose is either to estimate the dietary exposure of consumers (i.e., analyzing plants to estimate exposure of herbivores) or to estimate the internal exposure of endpoint organisms. In the first case, sampling should focus on the primary food organisms and on the parts that are consumed. In the latter case, the sampling should focus on the endpoint species or, if that is impractical, on a closely related species with similar habits. If the endpoint entity is a community or higher taxon, then one may choose a representative species, representative set of species, or the entire group. To the extent that they can be identified and are relevant, the species that have the highest level of accumulation should also be selected. When other criteria are satisfied, organisms may be chosen on the basis of practical considerations such as ease of collection and body size. These issues are discussed by Phillips (1978).

A third aspect of ensuring relevant analyses is selection of appropriate components of the organisms for analysis. This requires first deciding whether to analyze the whole organism, or some organ or other component. Second, one must decide which component, if not the whole organism. Once again, the primary consideration is the relationship of the analysis to the mode of exposure. If one is interested in the dietary exposure of a grazing or browsing animal, the leaves of plants would be analyzed; for beavers, the bark and cambium of small branches; for granivores, the seeds. If the analysis is performed to estimate internal exposure of an endpoint receptor, one should perform the analysis that is appropriate for the exposure–response model. This issue is discussed in greater detail below.

It is also necessary to consider the relevance of analyses of mobile organisms. Mobile organisms collected on a site may have spent little time on that site. To the extent that it is consistent with the endpoints of the assessment, organisms that are most associated with the site should be preferred. Those are less mobile organisms and organisms with small home ranges. However, if the organisms of concern are not confined to the site, body burden analyses can still be relevant in that they realistically represent the proportional exposure of those organisms to the site and its contaminants. This rationale is applicable only if the organisms are not significantly exposed to sources of the contaminant outside the site.

In some cases, analysis of organisms from the site is not practical because the site is small or highly disturbed or is not sufficiently consistent because of variability across the site. In those cases, organisms can be exposed to the contaminated site media under controlled conditions. For example, at the Concord, CA, Naval Weapons Station, plants and earthworms were exposed to site soils in the laboratory, and caged clams were exposed to site waters in the field (Jenkins et al., 1995). Similarly, earthworms were exposed in containers of soil at the Baird and McGuire Superfund site in Holbrook, MA (Menzie et al., 1992). While providing consistent bioconcentration data, such studies can also provide information on toxicity.

An alternative to body burden analysis is analysis of biochemical biomarkers such as hepatic mixed-function oxidase enzymes (Huggett et al., 1992). Biomarkers may be detected when the contaminant cannot, and in some cases they may be measured without sacrificing the animal. For example, blood ALAD was used to estimate lead exposure in birds on the contaminated floodplain of the Coeur d'Alene River, and liver lead concentrations were determined in a subsample of birds (Johnson et al., 1999). However, biomarkers tend to be nonspecific, tend to increase nonlinearly with increasing exposure levels (e.g., to decline at high exposures due to inhibited protein synthesis), and tend to vary with extraneous variables such as the animal's breeding cycle or nutritional state. In addition, few reliable exposure–response functions are available to relate biomarker levels to effects on organisms. For these reasons, biomarkers have been used much less than analysis of contaminant burdens in ecological risk assessments. However, one potentially important use is as bioassays (discussed below).

Body burdens and biomarkers of exposure must, in most cases, be related to concentrations in media to which the organisms are exposed, because it is the media that will be remediated. The derivation of such relationships requires sampling and analysis of the exposure media colocated with the sampled biota. A series of such analyses of colocated biological and media samples can be used to develop a site-specific uptake factor or other model (Section 3.10). If the range of sites encompasses the range of contaminant levels, and if the uncertainty in the site-specific factor or model is sufficiently low, the factor or model can then be used to predict body burdens or biomarker levels at locations where media samples, but not biological samples, have been analyzed. Because media and biota concentrations may vary, samples should also be colocated in time. The acceptable interval between samples depends on the rate of variance of the biota and media, but the samples should not be taken in different seasons.

A wide variety of methods is available for the collection of biota samples for residue analyses, with sampling methods generally being medium or taxon specific. Common collection methods for taxa generally of interest in risk assessments are outlined in Appendix A. General guidance on biota sampling is presented in Box 3.2.

BOX 3.2
Rules for Sampling Biota

- Take enough samples to represent the variability at the site adequately.
- Sample endpoint taxa for which internal measures of exposure are useful.
- Sample organisms or parts of organisms that represent the food of assessment endpoint species.
- Take samples of biota and contaminated media at the same locations and at effectively the same time.
- Take samples at reference and contaminated locations or on contamination gradients.
- Because chemical concentrations in organisms may vary seasonally, take samples from all sites at approximately the same time.
- Be aware of the information that is lost when samples are composited.

3.1.3 Bioassay

Bioassays are measures of biological responses that may be used to estimate the concentration or to determine the presence of some chemical or material. Bioassays are seldom used since the development of sensitive analytical chemistry. One valuable use of bioassays is to determine the effective concentration of chemicals with a common mechanism of action. For example, the H4IIE bioassay provides a toxicity-normalized measure of the amount of chlorinated diaromatic hydrocarbons in the food of an organism (Tillit et al., 1991; Giesy et al., 1994). This use is analogous to the use of biomarkers to estimate internal exposure (Section 3.1.2), except that the goal is to estimate external response-normalized concentrations.

It has also been proposed that activity of contaminant-degrading microbes be used as a bioassay for bioavailable contaminant constituents (Alexander, 1995). A weak interpretation of this bioassay is that, if biodegradation has stopped, there is no more bioavailable chemical to cause toxicity. This conclusion requires the assumption that biodegradation has stopped because the residue is unavailable rather than because it is resistant to biodegradation. A stronger interpretation would be that bioavailable concentration is a function of biodegradation rate so that one could estimate exposure from measures of degradation. This idea requires the assumption that the availability of a chemical for degradation by microbes is proportional to availability for uptake by endpoint plants and animals. The use of microbial toxicity tests as measures of bioavailability or ecological effects is beyond the current state of practice.

3.1.4 Biosurvey

Surveys of the organisms inhabiting a site (biosurveys) are used in ecological risk assessments primarily as a means of determining effects (Chapter 4). However, they may play a role in the analysis of exposure. Specifically, they can be used to determine whether a species or taxon is present in contaminated areas, what life-stages are present, their abundance, and how long a migratory or otherwise transient species is present on the site. Without biological surveys, these presence and abundance parameters must be estimated using habitat models (discussed below). Biosurveys of a contaminated site provide estimates of these parameters for the current baseline condition. Biosurveys of uncontaminated reference areas can provide estimates of these parameters for precontamination or postrestoration scenarios. Biosurveys may be conducted in conjunction with the collection of organisms for chemical analysis (Section 3.1.2 and Appendix A), but care must be taken to ensure that sampling designs are adequate for both purposes.

3.1.5 Transport and Fate Models

Transport and fate models play a relatively small role in ecological risk assessment for contaminated sites, because concentrations in media can usually be determined by sampling and analysis. However, transport and fate models are needed when contamination is relatively recent, so that local media are not yet contaminated, or when sampling and analysis are not possible. Most models simulate transport and fate in a single medium. A set of transport and fate models for surface water, groundwater, soil, and air is available from the EPA through its Center for Exposure Assessment Modeling (CEAM) in Athens, GA (www.epa.gov/CEAM). However, for ecological assessments, it may be necessary to estimate concentrations in multiple media. For this purpose multimedia fate models are appropriate (Mackay, 1991; Cowan et al., 1995). Some models have been developed specifically for estimating transport and fate on contaminated sites, but in general they have focused on groundwater and other routes to human exposure (McKone, 1993). Several such models are listed and three are reviewed in Moskowitz et al. (1996). A basic set of models for the transport of petroleum constituents from contaminated terrestrial sites is presented as part of the ASTM Risk Based Corrective Action (RBCA) procedure (ASTM, 1994). When such models are used, ecological risk assessors must carefully review them to ensure that they provide the needed output for estimating ecological exposures.

3.1.6 Exposure Models

After contaminant concentrations have been measured or estimated in ambient media, it is necessary to estimate the actual uptake of contaminants using exposure models. Uptake may be modeled empirically (e.g., uptake factors) or mechanistically (i.e., toxicokinetic models). Empirical and mechanistic approaches have been developed for uptake of organic chemicals in water by aquatic organisms, but uptake from soil has been relatively poorly characterized. In general, development of empirical uptake factors is hindered by the problem of variance in chemical form and bioavailability. Uptake factors developed for soil and sediment are highly variable because of the large variance in the properties controlling bioavailability in those media. Other measures of soil or sediment concentration such as concentrations in pore water or aqueous extracts may be more useful for deriving uptake factors and other models, but are seldom used.

Mechanistic exposure modeling depends on an understanding of contact and uptake mechanisms. For example, it is commonly assumed that plants take up soil contaminants largely or entirely through their roots, but compounds with low solubility and relatively high Henry's law constants are likely to be taken up from the air more than from soil, and material taken up by roots is often poorly transported to aboveground parts (Wild et al., 1992; Bromilow and Chamberlain, 1995). This generalization suggests that models of plant uptake and accumulation of contaminants should be mechanistic and multimedia, but understanding of the processes is poor for most chemicals. On the other hand, studies of mammalian toxicokinetics for human health risk assessment are abundant and should be applied to mechanistic models of wildlife exposure.

3.2 EXPOSURE TO CONTAMINANTS IN SURFACE WATER

In most cases, ecological risk assessments of contaminated waters are based on measurements of aqueous concentrations of chemicals. In such cases, the major issues to be considered by the assessors are the appropriate averaging times for the measurements and the forms of chemicals in water that must be measured or estimated. In some cases, it is necessary to model chemical transport and fate.

Unlike other contaminated media, the concentrations of chemicals in water may be highly variable over relatively short time periods. The resolution of temporal issues in aqueous sampling and data reduction must be based on the variability of concentrations in the stream and the toxicokinetics and toxicodynamics of the chemicals and receptors. Human health risk assessments are nearly always based on the assumption that effects result from long-term average exposures (i.e., years or decades). Hence, sampling and analysis plans based on human health concerns are designed to characterize those averages. In contrast, ecological effects may result from short-term (i.e., less than a week) exposures of small organisms (e.g., zooplanktors or larval fish) who rapidly reach equilibrium with highly mobile chemicals (e.g., metal ions). Hence, the sampling plan should include episodes of high concentration (Chapter 2), and the analysis of exposure should include an analysis of the frequency and duration of such episodes.

The greatest controversies with respect to aqueous exposures for aquatic biota have to do with the forms of metals to analyze in water. Forms include dissolved metal, particulate metal (e.g., associated with suspended clay), and metal complexed with dissolved material (i.e., organic colloids and colloidal hydrous metal oxides). The EPA Office of Water has recommended that assessments of effects of aqueous metals on aquatic biota be based on dissolved metal concentrations as determined by analysis of 0.45-μm-filtered water (Prothro, 1993). However, some states and EPA regions still require that total concentrations be used in ecological risk assessments for the sake of conservatism. In fact, even nominally dissolved concentrations are likely to be conservative in most cases because they include complexed metals as well as dissolved metals. Total concentrations are useful for screening assessments, but the dissolved form is appropriate for definitive risk assessments of aquatic biota for two reasons. First, the form in the exposure assessment should match the form in the effects assessment. Exposure–response models for aquatic biota are usually based on toxicity tests performed in clean water with highly soluble forms of the tested metal. Hence, the best match would be with dissolved concentration estimates. Second, the risk assessments should be based on the form that is best correlated with effects. For exposures of aquatic animals to metals in general, this appears to be the free metal ion (Bergman and Dorward-King, 1997).

It must be noted that use of total concentrations is not always conservative. First, because the high levels of acid-extractable metals may cause analytical interferences, the limits of detection for metals may be greater for total concentration analyses, and therefore toxic concentrations of metals may not be detected. This apparently occurred in the assessment of Bear Creek on the Oak Ridge Reservation where copper was a chemical of concern in filtered samples but was not detected in total samples. Second, when comparing concentrations at contaminated sites to background, if the dissolved concentration is small relative to the total concentration, there may be a significant increase in dissolved concentrations relative to background but no significant increase for total. That is, the particle-associated background concentration may mask a relatively small but toxicologically significant increase in dissolved concentrations. The solution to this problem of dissolved vs. total concentrations that was adopted in Oak Ridge was to determine both total and dissolved concentrations, use the dissolved concentrations to realistically estimate risks, but also present results for total concentrations to satisfy the state and regional regulators. The total metal analyses were performed in any case for the assessment of risks from human and wildlife drinking water.

Speciation should be considered for both inorganic and organic chemicals that have multiple ionization states within the range of realistic ambient conditions. In general, nonionic forms are more toxic because they partition more rapidly from water to biota. Hence, unionized ammonia is more toxic than the ammonium ion, and unionized alcohols and phenols are more toxic than the ionized species. This rule does not apply to metals, particularly those that may have multiple ionic species within the range of ambient conditions. Metal speciation is an expensive addition to the analytical budget. It is justified for metals of ecological concern that (1) may occur at the site as multiple species that have significantly different toxicities or (2) are believed to occur predominantly as a single species which is different from the one assumed by regulators. Assessors should particularly consider speciation of arsenic, chromium, mercury, and selenium.

Forms and species of metals in water can be estimated from measurements of total concentration by applying metal speciation models. A recent workshop recommended use of Model V of WHAM as the best of the available metal speciation models (Tipping, 1994; Bergman and Dorward-King, 1997). However, speciation models are less reliable than analytical chemistry and are not used in routine regulatory practice. The EPA has developed guidance for using a "metals translator" to convert between dissolved and total metal concentrations when only one type of analysis has been performed (Kinerson et al., 1996). The simplest and most defensible translator is the locally derived fraction of total metal that is dissolved in samples of the receiving water. However, standard empirically derived conversion factors or partition coefficients may also be used. The translator may be a function of suspended solids concentration, hardness, or other receiving water properties.

The issue of bioavailability is relevant to organic chemicals as well as metals (Hamelink et al., 1994). Like metals, organic chemicals may bind to dissolved or suspended particles, making them less available for biological uptake. However, unlike metals, there is no guidance from the EPA to use concentrations of organic chemicals in filtered water to represent aqueous exposures. This is in part because the issue of bioavailability has not been considered as important for aqueous organic chemicals, and there has been little pressure on the agency to consider it.

If chemical concentrations cannot be measured, they must be modeled. In many assessments it is sufficient to use a simple dilution model. If a seep, spring, drain, or tributary will add chemical contaminants to a stream, one may assume that the contaminated water and receiving water will be fully mixed within a short distance. The formula is

$$C_d = [(C_c\, F_c) + (C_u\, F_u)] \;/\; (F_c + F_u) \tag{3.2}$$

where

C_d = diluted concentration (mg/l)
C_c = contaminated source concentration (mg/l)
F_c = flow of the contaminated source (l/s)
C_u = upstream concentration (mg/l)
F_u = flow of the receiving stream (l/s).

The use of this formula is appropriate if one is concerned only about risks to biota in a receiving stream near the source, or if performing a screening assessment for a stream. If the assessment must address a sufficient length of stream for processes other than dilution to be significant, more complex fate models are appropriate. The most commonly used fate model for streams is EXAMS, which is available from the EPA CEAM (Burns et al., 1982).

3.3 EXPOSURE TO CONTAMINANTS IN SEDIMENT

Ecological risk assessments of contaminated sediments are typically based on chemical concentrations in bulk sediment or sediment interstitial water. The principal issues to be addressed are the heterogeneity of sediments and sediment contamination and the bioavailability of measured chemical concentrations. It may be useful or necessary to also estimate contaminant uptake and trophic transfers. These and other issues associated with sediment ecological risk assessments are addressed in detail by Ingersoll et al. (1997).

Unlike surface water, sediment contaminant concentrations generally vary spatially (vertically and horizontally) more than temporally. The assessment of exposure must consider the distribution of contamination in relation to the distribution of the receptors. Most sediment-associated organisms are exposed to surface sediment (e.g., the top 5 to 10 cm), rather than deep sediment. For example, the burrowing depth of four orders of sediment-dwelling insects (Charbonneau and Hare, 1998) and three species of oligochaetes (Lazim et al., 1989) varied greatly among taxa and seasons, but seldom exceed 10 cm. Often, ecological assessors cannot specify the sampling depth and must assume that the concentrations reported for the top-most layer of a core or for a surface grab sample represent the exposure of benthic and epibenthic organisms at the sampled location. This may be because the assessor was not involved in the sampling or because the ability to define concentrations in narrowly defined surface layers is limited by the available equipment or by sample size requirements for chemical analyses. Careful consideration of the vertical dimension should be applied as well to species other than benthic invertebrates. For example, most Centrarchid fish form nests in the sediment where the eggs and larvae develop. These early life stages are sensitive to many contaminants. For a screening assessment, one might conservatively assume that they are exposed to epibenthic water that is equivalent to sediment pore water, but the ventilation of the nest by the guarding male renders that assumption unrealistic. If sediments are contaminated in an area that is heavily used by these species and risks are uncertain, a special effort to sample epibenthic water from the area of the nests may be justified.

Because most sediment-associated organisms are relatively immobile, it is not reasonable to assume that benthic organisms average their exposure to sediment contamination over large areas (e.g., entire stream reaches) or depth ranges (e.g., the top 2 ft of sediment). Rather, the median surface sediment concentration is an appropriate measure of the central tendency of sediment exposures for the benthic infauna in a given area (e.g., stream reach), and the maximum detected concentration is an appropriate conservative estimate of this exposure for use in contaminant screening.

Sampling design considerations for spatially heterogeneous media have been discussed at some length elsewhere (EPA, 1997). The design selected will determine the methods used to characterize sediment exposures. For example, random or stratified random sampling allows the assessor to estimate the percentage of the sediment in which the chemical concentrations exceed a particular level of concern. This was the approach used for the Clinch River/Poplar Creek assessment (Jones et al., 1999). If depositional hot spots are identified and sampled, then the characterization is limited to the high end of the distribution of exposure concentrations, with some unquantified lack of certainty. That is, the percentage of all deposited sediments with concentrations exceeding a particular effect level cannot be estimated from biased data alone.

Sediment contaminant concentrations at a given location generally vary little during the life cycle of most benthic infauna, given that most benthic insects can complete one or more life cycles in a year (Merritt and Cummins, 1984). Notable exceptions include lotic systems (e.g., estuaries) in which contaminant sources (e.g., aqueous discharges) and physicochemical characteristics (e.g., salinity, dissolved oxygen, and hydrodynamics) of the overlying water vary on a biologically relevant timescale (Luoma and Fisher, 1997). These changes may alter the partitioning of contaminants to and from the sediment compartment. The resulting variations in benthic infauna exposures should be considered in the collection and evaluation of estuarine sediment data.

Even if the partitioning dynamics are relatively stable, the flux of particles may require the collection of current data to replace existing sediment data. The evaluation of the need for sediment data should account for the frequency and magnitude of scouring events, the rate of sedimentation, and the sources of sediment. For example, surface sediment in lakes and large reservoirs may be scoured very infrequently, but it may be buried by new sediment if the deposition rate is high. In both cases, however, the exposures may not have changed since samples were collected if the source sediment has not changed with respect to contaminant concentration or bioavailability. Data for historic or buried sediments may provide a conservative estimate of current exposures if upgradient remedial actions are known to have reduced the flux of sediment-associated contaminants. Such data may be useful for scoping and screening assessments but are unlikely to be acceptable for definitive assessments.

Two different expressions of sediment contamination are commonly used in ecological risk assessments: concentrations of chemicals in whole sediments and concentrations in sediment interstitial (pore) waters. The use of pore water is based on the assumption that chemicals associated with the solid phase are largely unavailable and, therefore, sediment toxicity can be estimated by measuring or estimating the pore water concentration. This assumption is supported by much empirical data and is widely accepted by the scientific community (NOAA, 1995). The exception is that sediment ingesters may be more exposed to particle-associated chemicals than are nonsediment ingesters. Adams (1987) reviewed the feeding habits of benthic species and concluded that burrowing marine species frequently were sediment ingesters, but that most freshwater species were not sediment ingesters, except for the oligochaetes (aquatic earthworms) and some chironomids. Still, these taxa can constitute most of the benthic assemblage in areas of fine sediment deposition.

Extracting the pore water from sediment samples can be labor intensive, can require large amounts of sediment in order to obtain sufficient sample volume for multiple analyses, and can alter the form and speciation of the chemicals measured.

For example, to quantify the species of arsenic in Clinch River sediment pore water, Ford et al. (1995) performed the extractions under argon gas to prevent the oxidation of arsenic (V) to arsenic (III). The advantage is that measured pore water concentrations can be evaluated using the same techniques and effects data used for surface water. Measuring pore water concentrations is particularly useful for metals and ionic organic chemicals, because the particle–pore water partitioning mechanisms are complex and difficult to model. Of course, the chemical form and speciation issues for surface water exposures also apply to pore water (Section 3.2).

Two ways to adjust bulk sediment concentrations to account for biological availability are (1) estimate the free pore water concentration of nonionic organic chemicals by normalizing the bulk concentration to organic carbon content and (2) measure the fraction of the bulk metal concentration that is bound to reactive sulfide. The EPA has favored the first approach over the direct measurement approach for nonionic organic chemicals (EPA, 1993a). The free chemical concentration in pore water can be estimated directly from the organic carbon–normalized sediment concentration using the equilibrium partitioning approach. Specifically, this approach assumes that hydrophobic interactions with organic carbon control the partitioning of nonionic organic chemicals between particles and pore water:

$$C_{\mathrm{pw}} = C_s / (K_{\mathrm{oc}} \times f_{\mathrm{oc}}) \tag{3.3}$$

where

C_{pw} = concentration in pore water
C_{s} = concentration in solid phase
K_{oc} = the chemical-specific partition coefficient
f_{oc} = the mass fraction of organic carbon (kilograms organic carbon per kilogram sediment)

This estimate is independent of the dissolved organic carbon concentration. Using the pore water chemical concentration to estimate the free pore water chemical concentration requires that the dissolved organic carbon concentration and partition coefficient be known. This is because the proportion of a chemical in pore water that is complexed to dissolved organic carbon can be substantial. However, it is the free, uncomplexed component that is bioavailable and that is in equilibrium with the organic carbon–normalized sediment concentration. Therefore, for highly hydrophobic chemicals and where there is significant dissolved organic carbon complexing, the solid-phase chemical concentration gives a more direct estimate of the bioavailable pore water contaminant concentration than does the pore water concentration (EPA, 1993a).

Where sufficient data are available, a site-specific partitioning model may be developed. For example, in the Elizabeth River in Virginia, the two-phase model did not fit the partitioning data for polycyclic aromatic hydrocarbons (PAHs) well (Mitra and Dickhut, 1999). However, a three-phase model that included dissolved organic matter concentrations did adequately describe the data.

The second approach actually measures the lack of biological availability and is currently considered valid for five divalent metals: cadmium, copper, lead, nickel, and zinc. Acid volatile sulfide (AVS) is a reactive pool of solid-phase sulfide that is available to bind metals and render that portion unavailable and nontoxic to biota (DiToro et al., 1992). The AVS is extracted from sediment using hydrochloric acid. The metal concentration that is simultaneously extracted is termed the simultaneously extracted metal (SEM). Acute toxicity is highly unlikely if the SEM:AVS molar ratio is less than 1. These metals are potentially toxic (available) when the SEM:AVS molar ratio is greater than 1. As Hare et al. (1994) noted, the molar SEM concentration in excess of AVS (SEM – AVS) is a better representation of exposure than the molar ratio. This is because two sediments with the same SEM:AVS molar ratios can have very different excess (i.e., potentially available) SEM concentrations. There are several caveats and limitations associated with the AVS approach (NOAA, 1995):

- It does not predict bioavailability, because the excess SEM may be bound to other ligands.
- It is not meant for evaluating individual metals (SEM must include all five aforementioned metals).
- It is currently considered valid for evaluating acute toxicity only, although progress is being made with respect to chronic toxicity evaluations.
- It does not necessarily apply to bioaccumulation or community-level effects.
- It is invalid if the sediment AVS content is very low (e.g., in fully oxidized sediments).

Another consideration is that AVS concentrations are highly variable in space and time (Luoma and Fisher, 1997). Consequently, the measured SEM concentrations may not be representative of the *in situ* exposures. For example, sediment samples are generally collected to a fixed depth and homogenized, but the SEM:AVS ratio is depth dependent (Hare et al., 1994). Sampling designs should account for variations in exposure with season and depth, to the extent practical.

Although unadjusted bulk sediment concentrations are relatively poor estimators of effective exposure, they are still an important part of the sediment exposure assessment (NOAA, 1995). Collection and analysis methods for bulk sediment concentrations are standardized and these measurements are often required for assessments of hazardous waste sites. Bulk sediment concentrations can be (1) compared with available effects concentrations (see Chapter 4), (2) used to meet minimum screening requirements by comparison to EPA screening values (Office of Emergency and Remedial Response, 1996), (3) used to focus sampling and assessment efforts on the contaminants of greatest concern, and (4) used to estimate exposure of sediment ingesters to highly particle-associated chemicals.

Overlying water concentrations are an alternative type of analysis that is potentially useful (Chapman et al., 1997). Pore water may be in equilibrium with overlying water if fine-grained sediment particles and organic carbon are relatively absent. Contaminated water is the primary exposure pathway in these sediments, because they are poor habitat for sediment ingesters. Sediment may also be a source of contaminants to the surface water. Epibenthic organisms may be primarily exposed to chemicals released into the overlying water in such instances.

Contaminant concentrations in benthic organisms are direct measures of exposure that can be compared with body burden–based effects concentrations for benthic organisms (Chapter 4) and are widely recognized as a potentially useful tool for the assessment of sediments (Office of Water, 1998). Although this approach is still limited by the paucity of effects data, it is an area of ongoing research and development. Body burdens also are used to estimate dietary exposures for predators of benthic organisms. For example, some flying insectivores (e.g., bats and swallows) forage over water and consume adult insects which have aquatic larval stages (e.g., mayflies and midges). These emergent insects can be an important vector for the movement of chemicals out of sediment deposits and into the terrestrial food chain (Larsson, 1984; Currie et al., 1997; Froese et al., 1998).

Tissue concentrations must be estimated in the absence of measured values. Biota–sediment accumulation factors (BSAFs) and models can be derived empirically from colocated sediment and biota samples. They are conceptually analogous to bioaccumulation factors (BAFs) for water (Section 3.10.1). They have the same form, assumptions, and uncertainties. Unfortunately, a compendium of widely applicable BSAFs is not currently available, although efforts to review and compile the relevant literature are ongoing (Bechtel Jacobs Company, 1998a). As with contaminant uptake from water, uptake from sediments can be modeled mechanistically if the assumptions for using accumulation factor models are not met (Section 3.2).

3.4 EXPOSURE TO CONTAMINANTS IN SOIL

Plants, soil invertebrates, and microorganisms are continually exposed to chemicals in soil. Because they are immobile or nearly so at the scale that remedial actions occur, their activity patterns (except for exposure depth) are not relevant for an ecological risk assessment. Wildlife receptors that are directly exposed to soil dermally and by incidental ingestion are discussed in Section 3.9. The major soil exposure issues considered in the planning phase of the risk assessment are: (1) if concentrations in soil or soil water should be measured, (2) the appropriate sampling depth for exposure of the endpoint organisms, and (3) the applicability of existing soil data (Chapter 2). In the exposure assessment and risk characterization, additional issues should be considered: (1) soil and chemical factors that control the transfer of contaminants to biota and (2) the relevance of the type of exposure in published exposure–response relationships to the site of concern. In a retrospective risk assessment for a contaminated site, it is rarely necessary to model chemical transport and fate in soil. An exception is the risk assessment for a future scenario in which buried waste constituents are carried to surface soil during a period of high precipitation or contaminant concentrations are reduced by degradation or other losses.

3.4.1 Sampling, Extraction, and Chemical Analysis

The occurrence of contaminants within the soil matrix may be characterized with respect to either bulk soil or soil pore water. Sampling and analytical methods should be selected that are precise and easily related to measures of effects. Soil chemical analyses, associated estimates of exposure, and measures of exposure used in exposure–effects relationships are presented in Table 3.1.

The most direct and common approach for estimating exposure to soil contaminants is the collection and analysis of the bulk medium. Rigorous extraction tech-

TABLE 3.1
Alternative Methods of Soil Analysis and Associated Methods for Estimation of Exposure and Toxicity

Soil Analyses	Estimate of Exposure	Exposure in Effects Test
Total extractable chemical analysis	Total extractable concentration	Total extractable concentration in test soil
		Concentration added (spiked) to soil
	Solution-phase concentration modeled from total soil concentration	Modeled solution-phase concentration in test soil
		Aqueous concentration in a toxicity test in solution culture
	Total concentration normalized for soil factors that determine bioavailability	Soil concentration normalized for toxicity soil factors that determine bioavailability
Aqueous extract analyses	Concentration in aqueous extract	Aqueous concentration in toxicity tests in solution culture
		Modeled solution-phase concentration associated with test in soil

niques, such as concentrated nitric, sulfuric, perchloric, hydrochloric, or hydrofluoric acids, sometimes with heat, permit the estimation of total concentrations of inorganic elements (Hesse, 1971; Baker and Amacher, 1982; de Pieri et al., 1996). Similar extractions using organic solvents and heat allow the estimation of organic compounds (Hatzinger and Alexander, 1995; Hendriks et al., 1995). These analyses have the reassuring feature of including the full extent of contamination, as well as background concentrations of elements. Nothing is missed or ignored. However, because organisms do not extract chemicals so thoroughly, results of rigorous extractions tend to overestimate exposure. Also, as with total concentrations in water (discussed above), strong extractions of soil may obscure increases in bioavailable forms of chemicals and may raise detection limits by increasing analytical interferences. An advantage of total extractions is that the results can be compared with similarly rigorous analyses of contaminated site soils used in toxicity tests, or, with much less accuracy, to nominal concentrations of chemicals added to test soils.

Total chemical analyses are poorly predictive of toxicity; the fraction of the chemical that is bioavailable is highly variable, depending on soil and contaminant characteristics. Because of variation in soil properties that control the availability of chemicals to organisms, total concentrations in different soils or even the same soil at different times may result in very different levels of effects. Aged organic chemicals (i.e., chemicals that have persisted in soil for months to years) are less bioavailable to earthworms and microorganisms than freshly added forms of the same chemicals (Hatzinger and Alexander, 1995; Ma et al., 1995; Kelsey and Alexander, 1997; Kelsey et al., 1997). This is also the case for metal uptake and toxicity to soil invertebrates, microbes, and plants (Posthuma et al., 1998). The extractability of aged chemicals with some solvents has been observed to be associated with bioavailability to earthworms and to bacteria for degradation (Kelsey et al., 1997).

However, the extent of sequestration and the time required varies markedly among soils (Chung and Alexander, 1998). Therefore, the likelihood that the same solvent would predict bioavailability of many aged organic chemicals in a wide range of soils is low. At this time, all that is clear is that the bioavailable fraction of a chemical freshly added to soil is greater than that of the same chemical that has resided in the same soil for years. Similarly, if a chemical has resided in a soil for years and is taken up by plants, a large fraction of the chemical may be immobilized in living or decaying biotic material or humus (e.g., selenium in decomposing plant tissue; Banuelos et al., 1992). The following approaches to exposure estimation deal with the issue of bioavailability by either estimating the bioavailable component of contaminant concentrations in soil or producing an estimate of exposure that is better correlated with toxicity than the total concentration.

3.4.1.1 Partial Chemical Extraction and Normalization

Soil pore water is assumed to be the bioavailable compartment for plants and other soil endpoint organisms, but the measurement of various elements in pore water can be difficult and imprecise (Sheppard, Thibault, and Smith, 1992). One approach is to measure total concentrations in bulk soil, and then estimate concentrations in soil pore water. As discussed above, this approach has been employed by the EPA in the derivation of proposed sediment quality criteria (EPA, 1993a). Neutral organic compounds are assumed to be in equilibrium between the aqueous phase (pore water) and the organic component of the solid phase. The approach has also been proposed for soils (Lokke, 1994). If it is assumed that exposure of soil organisms is solely to the aqueous phase, the estimated pore water concentrations can then be used with toxicity data based on aqueous toxicity tests (plants in hydroponic solutions, invertebrates on blotter paper, or even aquatic invertebrates in water) to estimate risk. This equilibrium partitioning (EqP) approach remains controversial when applied to sediments and is largely hypothetical for soils. Unlike in sediments, the variation in water content of soils can lead to saturation and other nonequilibrium dynamics. Fewer assumptions would be required if the EqP models were simply used to normalize soil concentrations. That is, responses in soil toxicity tests expressed as a function of estimated pore water concentrations could be used with estimated pore water concentrations from contaminated site soils to generate more accurate estimates of effects.

Empirical methods of normalization may provide better estimates of effective concentrations than simple equilibrium partitioning between aqueous and organic phases of soils. For example, Dutch reference values for various chemicals in soil were derived by normalizing values to a standard soil with 10% organic matter and 25% clay content using linear regression (VROM, 1994). For example, the reference value for cadmium (R_{Cd}) in mg/kg is

$$R_{Cd} = 0.4 + 0.007(c + 3o) \tag{3.4}$$

where c is % clay and o is % organic matter (van Straalen and Denneman, 1989). However, recent studies indicate that the Dutch equations must be expanded to at least include pH (Posthuma et al., 1998). Concentrations of some organic chemicals in soil could be normalized using organic matter alone. If, at a particular site, it could be shown that effective exposure concentrations for chemicals are a function of a set of soil properties, it would be possible to normalize soil concentrations across test soils and site soils.

Another approach is to perform aqueous extractions of soil that are designed to simulate the extraction processes of organisms. That is, the mass of the chemical extracted by an aqueous solution (somewhat less than the total), divided by the mass of the soil, would approximate the bioavailable concentration. For example, the extraction of uranium with an ammonium acetate solution has been shown to correlate well with the soil–biota uptake factors for plants and earthworms in several soils (Sheppard and Evenden, 1992). Appropriate procedures would depend on the organisms for which exposure is being estimated; relatively mild extractions would be appropriate for root uptake, and stronger or sequential extractions would be expected to correlate with uptake by a mammalian gastrointestinal system (Section 3.9). Although many extraction procedures have been proposed, none has been demonstrated to be reliable for a variety of soils and contaminants. For example, although concentrations of DTPA-extracted contaminants from soils sometimes correlate with those taken up by plants (Sadiq, 1985), this estimate of bioavailability has been observed not to be valid for some metals (Sadiq, 1985, 1986; Hooda and Alloway, 1993) or for soils of varying pH (Miles and Parker, 1979). Zinc toxicity to invertebrates, microbes, and a plant were most consistent across soils expressed as a 0.01 *M* $CaCl_2$ extraction (Posthuma et al., 1998). Exposure estimates based on various extractions could be compared with similar estimates of exposure from extraction of soil used in toxicity tests. Alternatively, extractions with dilute aqueous salts or acids could be assumed to approximate the concentrations in soil pore water, with the appropriate correction for dilution. With this extraction method, comparison with aqueous toxicity test results would be possible, as in the EqP approach.

For a site contaminated with multiple inorganic and organic chemicals, the best alternative is often to measure total concentrations of contaminants in soil. Many published studies relate the total concentration of a chemical in soil to toxicity. For some chemicals in some environments, measurements or estimates of concentrations in solution may be useful. Then the numerous studies of the toxicity of plants in nutrient solution can be utilized. However, if the particular contaminants were highly sorbed, the solution concentration would not be in equilibrium with the soil, and the bioavailable fraction would decrease with time. Normalization techniques such as the Dutch reference values are useful only if researchers record the required soil characteristics, and if the normalization is demonstrated to be related to toxicity.

3.4.1.2 Nonaqueous-Phase Liquids

When nonaqueous-phase liquids (NAPLs) are present in soil, the bioavailable fraction of each chemical is not expected to be correlated with the total concentration in soil. Hydrocarbons or other constituents of NAPLs are divided among three principal phases: water, soil solid phase, and NAPL. The most bioavailable fraction is the aqueous portion. When the concentration of a lipophilic chemical in the water phase is close to saturation, and an NAPL is present, the aqueous concentration is independent of the total concentration in soil. Additionally, the NAPL fraction may be available to slowly sorb to and perhaps enter plant roots, earthworm skins, and microorganisms. Measurement of the concentration of the dissolved fraction in soil water would come close to approximating exposure. However, this type of measurement is not practical, since it is difficult to exclude NAPL from extractions of the aqueous phase. Further, petroleum and some other NAPLs may have a greater effect through their influence on the physical properties of soil than through the toxicity of their constituents (Section 3.10). Thus, there is a high degree of uncertainty associated with the exposure of soil organisms to NAPLs, and the uncertainty should be acknowledged.

3.4.2 Soil Depth Profile

The level of exposure of soil organisms to contaminants is defined partly by depth. The importance of selecting the depth of sampling carefully is illustrated in Table 3.2. If rooting depths are improperly estimated for plants, erroneous conclusions could be reached: (1) depth-averaged soil concentrations may not exceed toxicity benchmarks where they should have, or (2) adverse effects observed during a vegetation survey might be associated with different chemical concentrations than those actually in the rooting zone. The depths of exposure of plants, earthworms, and other assessment endpoint organisms residing in soil are thus important to characterize; these are discussed below.

3.4.3 Future Exposures

If an evaluation of a future scenario is included in the risk assessment, a model must be used to estimate the future exposure of endpoint organisms to contaminants in soil. A simple model would be to assume that the concentrations 50 years in the future are equivalent to current concentrations. If no continual source exists, this assumption is probably very conservative for most organic compounds (which biodegrade or volatilize) and somewhat conservative for most inorganic elements (which may leach). Sometimes, chemicals in the subsurface may be mobilized to surface soil. For example, after 4 years in a lysimeter at 0.45 m depth, iodine, technetium, neptunium, and cesium migrated to the surface (Sheppard and Thibault, 1991). A conservative assumption is that the concentrations of chemicals in buried waste trenches in the subsurface will someday be the concentrations in surface soil. A problem with any such conservative assumption is that the uncertainty may be forgotten when the remedial decision must be made. An alternative is to model vertical transport of contamination, although results are likely to be highly uncertain because of the variability in precipitation patterns and the influences of soil and vegetation.

Multimedia models are most useful for soils that have significant inputs from other environmental media. Since most contaminated sites that are candidates for remediation do not have a significant existing air source or other ongoing source of the contaminants of concern, multimedia models are probably not significantly more useful for estimating future soil concentrations at most sites than soil models.

TABLE 3.2
Examples of Chemical Gradients with Depth at a Single Location

Site	Chemical	Concentrations (mg or Bq/kg) at Various Depths[a]	Ref.
Semiarid region far from highways or towns	Lead	0–5 cm: 1400 0–10 cm: 1080 0–15 cm: 770	Sharma and Shupe, 1977
	Arsenic	0–5 cm: 655 0–10 cm: 489 0–15 cm: 351	Sharma and Shupe, 1977
Agricultural sites in Iran that are irrigated with wastewater	Cadmium	0–5 cm: 1.16 0–10 cm: 1.10 0–15 cm: 0.87 0–20 cm: 0.69 0–30 cm: 0.50 0–40 cm: 0.39 0–50 cm: 0.33 0–60 cm: 0.30	Shariatpanahi and Anderson, 1986
Soils east of Rocky Flats, CO	Pu-239 and Pu-240	0–3 cm: 11655 0–6 cm: 10101 0–9 cm: 8498 0–12 cm: 7095 0–18 cm: 5001 0–24 cm: 3806 0–36 cm: 2575 0–48 cm: 1935 0–72 cm: 1291 0–96 cm: 969 72–96 cm: 3.7	Litaor et al., 1994

[a] Concentrations at subsurface depth intervals are averaged with those from the surface to obtain the estimate from 0 to depth.

3.5 EXPOSURE OF TERRESTRIAL PLANTS

Chemicals are taken up by plants directly from soil and from air. Most contaminants are taken up passively from soil water in the transpiration stream, although nutrients such as copper and zinc are actively taken up from solution. Plants are stationary organisms; thus behavior and mobility are not determinants of exposure. The exposure of individual plants to contaminants in soil is controlled by the distribution of roots in the soil profile, physicochemical characteristics of soil, and interactions among chemicals. In addition, physiological differences among plant species are responsible for differential accumulation of contaminants among taxa. The exposure assessment for the plant community can be synonymous with the distribution of exposures of individual plants across the area occupied by the endpoint community. However, if the risk manager desires to protect a certain portion of the plant community, the spatial distribution of exposure is important. The level of detail presented here may not be required to characterize exposure if chemicals are screened out using conservative assumptions (Chapter 5). The information and associated references are more necessary for the definitive risk characterization.

3.5.1 Rooting Depth

The optimal depth interval for sampling soil should be the interval where most feeder roots are found for plant populations or communities that are assessment endpoint entities or are the major food sources of endpoint herbivores. Ideally, the risk assessor should determine this depth by measurement, taking random cores across the plant community and biased cores where threatened and endangered species are located. Roots may be weighed to determine the depth profile where rooting is most dense. Alternatively, one could measure the concentration gradient of the limiting nutrient or water (if it limits plant growth) to estimate the relative uptake of toxicants at different depth intervals. For example, available plant nutrients and root uptake in grassland soils tend to be concentrated in the top 10 cm (Syers and Springett, 1983). The rooting depths of plants vary with species; nutrient and oxygen availability; soil water; soil temperature; presence of pathogens; soil pore size, distribution, and compaction (Foxx et al., 1984); and location of rock–soil interfaces (Parker and van Lear, 1996). Thus, estimates of rooting depth at one site are not applied to another without added uncertainty. However, if measurements are not performed during the sampling phase of a risk assessment, an educated guess based on published studies may be used.

The key to choosing an optimal sampling depth is to estimate the depth that contains the major part of the rooting profile. Given that root densities often decrease exponentially with depth (Parker and van Lear, 1996) and that chemical concentrations may decrease with depth (Table 3.2), averaging soil concentrations over the interval from the surface to the maximum rooting depth would in most cases be a nonconservative error. If samples could be taken at multiple depth intervals, the surface intervals should be weighted more than subsurface intervals in the estimate of exposure (because of the exponential decrease in rooting density with depth, described above). However, in the absence of multiple samples, the sampling depth should be somewhat less than the maximum rooting depth of the plant community if the chemical concentrations decrease with depth.

Information about the normal rooting profile for vegetation in various terrestrial biomes is presented in Table 3.3 (Jackson et al., 1996). The majority of the root density of all biomes is in the top 30 cm of soil. Thus, 30 cm is a good default estimate of the depth of plant root exposure to contaminants in soil. Of course, general information about biomes should be supplemented with site knowledge. A shallower depth may be more appropriate if the site is dominated by grasses. Among different biomes, grasses have 44% of their roots in the top 10 cm of soil (Jackson et al., 1996). Similarly, water availability at a site may alter the relative uptake of contaminants from different depth intervals. In oak stands, the distribution of water uptake among different soil layers has been observed to change during drought (Breda et al., 1995). Also, if one soil depth must serve to estimate exposure to plants, soil invertebrates, and wildlife (incidental ingestion), a compromise depth for estimating all exposures may be necessary.

TABLE 3.3
Rooting Profiles Relevant to Exposure Assessment for Plants in Different Biomes

Biome	% Root Biomass in Upper 30 Cm of Soil
Boreal forest	83
Crops	70
Desert	53
Sclerophyllous shrubs	67
Temperate coniferous forest	52
Temperate deciduous forest	65
Temperate grassland	83
Tropical deciduous forest	70
Tropical evergreen forest	69
Tropical grassland savanna	57
Tundra	93

Source: Modified from Jackson, R. B. et al., *Oecologia*, 108, 489, 1996. With permission.

For single-chemical, exposure–response relationships that are derived from published studies (e.g., toxicity benchmarks, Chapter 4), it is assumed that the plants were exposed to contamination in the depth interval from which the soil was sampled for analysis. In pot studies and tilled field studies, the chemical concentration is typically uniform throughout the soil. Chemical concentrations in untilled field soil that are associated with toxicity are typically measured in the top 10 to 20 cm, but sampling depths range from 5 to 50 cm. Thus, concentrations of chemicals in soil in published studies may not exactly represent the exposure of the plants.

3.5.2 Rhizosphere

As stated above, chemicals taken up by plant roots are in solution. However, plants may influence solubility of chemicals in the rhizosphere (the soil in the vicinity of roots). Also, the rate of degradation of organic chemicals in the rhizosphere may be higher than in bulk soil (Reilley et al., 1996). Thus, the solution concentration to which a plant is exposed may be somewhat different from that in bulk solution. It is not practical for assessors to measure chemical concentrations in soil water in the rhizosphere. However, it is important to be aware of this uncertainty in estimates of exposure of vegetation to contaminants.

3.5.3 Wetlands

Wetlands straddle the line between water and soil exposures. The exposure of most wetland plants to chemical contaminants is assumed to be better represented by the concentration in spring water or groundwater than that in soil. The rationale is that (1) concentrations of chemicals in soil are not necessarily in equilibrium with those in water and (2) roots are more directly exposed to concentrations in the water phase than in bulk soil. Moreover, in wetland water treatment systems, exposures of the wetland plants to chemical contaminants have been found to consist largely of surface water concentrations (Detenbeck et al., 1996). Thus, exposure–effects relationships for wetland plants may be derived from studies of plants grown in inorganic salts solution containing the chemical of potential concern.

The exposure of wetland plants in swamps may not be easily attributed to a single medium. Whether chemical concentrations in surface water or sediment most accurately represent exposure depends in part on whether the media are in equilibrium. For example, in a risk assessment of a swamp in the Florida panhandle region that was exposed to boron, almost all correlations between the boron concentration in leaf tissue of various plants and sediment or water concentrations were significant (Powell et al., 1997). An exception was the herbaceous plant *Typha latifolia*, in which the concentration of boron was related only to the concentration in water. Even in the case of plants other than *Typha*, the concentration in sediment provides no additional information; sediment and water were probably close to equilibrium conditions, or the dual correlations would not have existed. The risk assessor should use the concentration of a contaminant in spring water or surface water as an estimate of exposure of wetland plants, unless conditions at the contaminated site suggest a different route of exposure.

3.5.4 Soil Properties

Soil physicochemical properties determine the fraction of the mass of a chemical in soil that is available for uptake by plants. The interpretation of exposure data in the final risk characterization may rely heavily on the consideration of these characteristics at the contaminated site. Thus, certain evidence from published literature may be ruled to be inapplicable to the site of concern. Soil properties such as pH (Miller et al., 1976; Sims and Kline, 1991; He and Singh, 1994), cation exchange capacity (Bysshe, 1988), and particle size fraction (Miller et al., 1976; Jiang and Singh, 1994) strongly affect the concentrations of inorganic chemicals in soil solution. Metals in certain fractions of soil (e.g., cadmium, zinc, and lead in the carbonate and exchangeable fractions; Xian, 1989) may be more closely related to concentrations in plants than total soil concentrations. Hence, total concentrations of elements in soil may be normalized for these descriptors (Section 3.4.1.1).

Of the soil characteristics that control exposure of plants to nonionic organic chemicals, soil organic matter is primary (Means et al., 1980; Topp et al., 1986; Sheppard et al., 1991). However, many models of the uptake of organic chemicals by plants rely exclusively on chemical characteristics rather than soil characteristics (Section 3.10.2.2). This fact suggests that existing soil–plant uptake models for organic chemicals are unlikely to be accurate across a wide range of soil types. That is, the exposure of plants to organic contaminants is likely to be higher in a sand than in a muck soil, but existing models would predict them to be equal.

Additional factors may control the accumulation of chemicals by plants. In one study of the uptake of radiocesium by plants, Guillitte et al. (1994) speculated that at equivalent bioavailabilities of elements in soil, the intensity and frequency of water flows determine the ultimate concentration in the plant. In this study, factors that controlled these water flows were presence of mycorrhizal fungi, duration of the growing season, and amount of rainfall. In the definitive risk characterization (Chapter 6), it is important that exposure data from existing exposure–response relationships be evaluated for their applicability to the particular contaminated site.

3.5.5 Chemical Form

As stated above, the currently bioavailable fraction of most chemicals at contaminated sites is lower than the total concentration. The fraction is determined in large part by the soil characteristics above. If a metal is added to soil in inorganic salt form (typical for a laboratory toxicity test or accumulation assay), the added concentration of the element is likely to be more bioavailable than the same total concentration of an element in a contaminated field soil. For example, based on calculated soil–plant uptake factors, 10 of 15 metals added to soil as salts were more available than the background fractions (Cataldo and Wildung, 1978). The uptake of arsenic and cobalt by plants from soil at a nuclear waste disposal area (Welcome Dump site, Eldorado Nuclear Limited, Port Hope, Ontario) was lower than that of the elements in salts-amended soil; however, the uptake of uranium was the same under both conditions (Sheppard et al., 1985). Soluble metals added to soil tend to become less available with time. For example, the soluble fraction of elements added in salt form to soil ranged from less than 1% for lead to greater than 40% for tellurium after 13 days of incubation (Cataldo and Wildung, 1978). Finally, the speciation of chemicals may be changed after deposition in soils. For example, hexavalent chromium deposited in cooling tower drift in Oak Ridge was almost completely transformed to the less bioavailable trivalent form. Thus, incubation time may be an important consideration when exposure–effects relationships are derived from published toxicity tests using metal salts (Chapter 4).

3.5.6 Chemical Interactions

Interactions of inorganic chemicals have begun to be studied in the context of their accumulation by plants. A primary source of uncertainty associated with the single-chemical-toxicity line of evidence (typified by the use of toxicity benchmarks for effects assessment, Chapter 4) is the exclusion of interactions among chemicals. These interactions must be researched by the assessor during the definitive risk characterization. For example, lead has been widely observed to increase cadmium uptake, but the results are not so clear for the effect of cadmium on the uptake of lead (Carlson and Bazzaz, 1977; Miller et al., 1977; Carlson and Rolfe, 1979; Burton et al., 1984). The amendment of soil with high concentrations of arsenic (50 mg/kg) increased the uptake of mercury by foliage of Bermuda grass (Weaver et al., 1984). Because the solubility of hydrophobic organic compounds in water is influenced by the presence of other such compounds (Eganhouse and Calder, 1976), it is likely that the uptake of one would be reduced by the presence of others.

3.5.7 Interspecies Differences

The physiological differences that explain differences in accumulation of metals by different plant species are largely unknown. In general, the uptake of inorganic chemicals varies much more among plant than animal species (there are no known instances of hyperaccumulation among animals), although occasionally differences in uptake among plant groups at a site are not large (e.g., uranium and thorium in sagebrush, grasses, and forbs; Ibrahaim and Whicker, 1988). The variability may be particularly large when plants (such as some that accumulate selenium or nickel) are hyperaccumulator species. Lakin (1972) notes that some vegetation in South Dakota and Israel accumulates selenium to levels toxic to grazers, but that vegetation on soils of Puerto Rico and Hawaii that have high selenium levels does not. The existence of hyperaccumulating plants detracts from the expected relationships between the chemical concentration in soil or the plant and toxicity.

Some characteristics of plants that determine the uptake of chemicals have been identified. For example, in a study of radiocesium, the rooting depth of plants was most important (Guillitte et al., 1994). Lipid levels in the plant have also been demonstrated to be important factors that determine the uptake of nonionic organic compounds. Trapp (1995) developed an empirical model of the uptake of organic chemicals from water by plant roots that was based on the lipid and water contents of the root. Similarly, Simonich and Hites (1994) normalized the plant–air partition coefficient to lipid level, based on the large role that the lipid content of foliage played in the uptake. It is reasonable to assume that hydrophobic organic chemicals entering plant foliage from the transpiration stream would be substantially taken into the solid fraction of the foliage if it has a high lipid content.

For the definitive risk characterization, it would be useful to be able to estimate the variability in chemical accumulation among species. However, since the measure of exposure of plants to chemical contaminants is typically a chemical concentration in soil or soil water rather than a concentration in the plants, the variability of the latter is not usually part of the exposure assessment for plants. This is not true for the exposure assessment for wildlife, where the variability in chemical concentrations in food items is an important input to the analysis (see Section 3.9.2). Similarly, if exposure–effects relationships used to evaluate risks to plants were based on concentrations in plant tissues, the variability of this measure would be important.

3.5.8 Air as an Exposure Route

Although this book is not intended to address exposures from active air sources such as incinerators or coal-fired power plants, the air exposure route must be discussed because of the likelihood of certain compounds to volatilize from soil and transfer to plant foliage. High-molecular-weight, nonionic organic compounds are the primary chemicals for which this pathway holds. Indeed, the uptake of organic compounds by plant foliage is often more important than accumulation by the roots. For example, soybean plant tops have been observed to be contaminated by foliar uptake of vapors of pesticides such as DDT, dieldrin, endrin, and heptaclor (Beall and Nash, 1971). Similarly, the major transfer of PCBs from soil to soybean foliage (Fries and Marrow, 1981), radish, and bean plants (Sheppard et al., 1991) has occurred via vapor sorption. 2,3,7,8-Tetrachlorodibenzo-*p*-dioxin (TCDD) is not transported in detectable amounts within the transpiration stream of herbaceous plants; aboveground plant parts are contaminated primarily from air (Trapp and Matthies, 1997). In particular, lower, aboveground plant parts may be significantly exposed if soils are highly contaminated.

Some forms of some inorganic elements are also transferred from soil to plants largely via air. In contrast to other metals, most mercury in aboveground plant tissue is taken up as volatile, elemental mercury through the leaves (Lindberg et al., 1979; Bysshe, 1988; Siegel and Siegel, 1988), with limited accumulation from the soil via the roots and transpiration stream. Similarly, a significant portion of the transfer of iodine from soil to beet foliage has been observed to occur from the air (Sheppard et al., 1993). In addition, volatile alkylselenides are produced from soils amended with selenium (Karlson and Frankenberger, 1989), and these may be taken up by plant foliage.

The atmospheric route of exposure can be ignored if (1) concentrations of the chemical in air and soil are assumed to be in equilibrium and (2) the soil is the only source of the chemical in the vicinity of the plant. Although mercury, for example, may be taken into the plant via the air, a significant correlation between the concentration in soil and plant tissues may exist (Shaw and Panagrahi, 1986). Relationships between concentrations of elements and organic compounds in soil and in plants are discussed in more detail in Section 3.10.2. At or near background levels of organic chemicals, it is reasonable to assume that concentrations in air and soil are close to equilibrium (Trapp and Matthies, 1997).

3.6 EXPOSURE OF SOIL INVERTEBRATES

Most of the information on soil invertebrates presented below relates to earthworms because (1) more information is available on their exposure than on that of soil arthropods, (2) exposure of earthworms to chemicals and toxicity have been related to concentrations in soil, and (3) in the United States earthworms are more often identified as assessment endpoint organisms than any other soil invertebrates. In addition, exposure to earthworms is probably through the skin, while arthropods such as isopods with exoskeletons are probably exposed primarily through food (van Brummelen et al., 1996). The exposure of earthworms to chemicals in soil is determined by several factors. These include concentration of chemical in soil, depth of burrowing, material ingested, activity patterns, soil characteristics, and interactions with other contaminants. Although exposure may be related to the concentration of a chemical in soil water (van Gestel and Ma, 1988; Janssen et al., 1997), more often the measurement that is taken is the total concentration of the chemical in soil. An exposure time of only a few weeks is required for metal concentrations in the earthworm to reach equilibrium with those in soil (Janssen et al., 1997).

3.6.1 DEPTH OF EXPOSURE AND INGESTED MATERIAL

The depth of burrows and the amount of time spent at each depth are factors that determine the exposure of earthworms to chemical contaminants. Earthworm burrows may be categorized into three types (Lee, 1985). First, the primarily vertical refuges of litter-feeding species may extend 3 m or more below the surface in soils of the northeastern United States. Second, burrows of geophagous species that forage for food in the subsurface are generally horizontal with few vertical components. Thus, these earthworms are exposed to contaminants at primarily one soil depth, which may be in the A, B, or C soil horizons. A third type of burrow is an ephemeral, vertical form constructed by earthworms inhabiting surface soil as they rest during cold or dry conditions. We have compiled information on diets and burrowing depths of several earthworm species that are common in both North America and Europe (Table 3.4).

Numerous species of earthworms exist that are not as well studied as those in Table 3.4. For example, one giant species in South Africa, *Microchaetus microchaetus*, burrows to depths of 70 cm (Reinecke, 1983). Another, *Octochaetus multiporus*, in New Zealand, may burrow to 300 to 500 cm (Lee, 1985). Zicsi (1983) identified several deep-burrowing species of earthworms in Hungary and elsewhere in central and southeast Europe that produce casts as deep as 100 or 150 cm.

As stated above, the time spent at each depth, or the annual average earthworm density at each depth is the information needed to determine the optimal soil sampling depth. For example, the litter-feeding species in Table 3.4 spend much more time at the soil surface than their burrow depths would indicate. In a study of the density of *Aporrectodea caliginosa* with depth in an unplowed field in July and August, Pitkanen and Nuutinen (1997) found 104 worms at 0 to 8 cm, 82 worms at

TABLE 3.4
Diet and Burrowing Depth as Determinants of Earthworm Exposure

Earthworm Species	Feeding Preference[a]	Burrowing Depth
Aporrectodea caliginosa	Geophagous (humus, much-decomposed plant remains) Root feeder	≤ 20 cm in New Zealand pasture soils (Stockdill and Cossens, 1966) 2–30 cm in uncultivated fields near Seveso (Martinucci et al., 1983)
Allolobophora chlorotica	Geophagous (humus, much-decomposed plant remains) Root feeders in pastures and leaf feeders in forests	2–10 cm in uncultivated fields near Seveso (Martinucci et al., 1983)
Aporrectodea longa	Mixture of highly decomposed and little-decomposed plant remains	
Lumbricus rubellus	Raw humus (little-decomposed plant remains)	
L. terrestris	Leaf plus small portion of root litter	>80 cm in an unplowed field in southern Finland (Pitkanen and Nuutinen, 1997) 30 cm in uncultivated fields near Seveso (Martinucci et al., 1983) 9.0 cm mean depth in microcosm (Haukka, 1991)
Dendrobaena octaedra	Maybe epigeic (Dymond et al., 1997)	2–10 cm in Aspen and pine forests in the Canadian Rocky Mountains (Dymond et al., 1997)

[a] Information from K. E. Lee's (1985) review of the literature unless stated otherwise.

8 to 15 cm, 50 worms at 15 to 20 cm, and 11 worms at 20 to 30 cm. No earthworms were found in deeper layers. Occasionally, one can infer an appropriate sampling depth from earthworm activity observed in a published study (if soil characteristics are similar). Stockdill and Cossens (1966) observed that *Allolobophora caliginosa* mixed lime and DDT into approximately the top 20 cm of soil. Thus, it may be assumed that the earthworms were fairly active in that portion of the soil profile.

As is evident from Table 3.4, the burrowing depths of various earthworm species are quite variable, making it difficult to choose a default sampling depth. It is advisable for the risk assessor to determine which earthworm species or other invertebrate endpoint species are present at the site before the appropriate sampling depth is chosen. If funding permits only one surface soil interval to be sampled, the 30-cm default depth for exposure of the plant community is reasonable, although many earthworm species do not burrow to that depth, and a few burrow much deeper (Table 3.4). This default depth should not be used, however, to estimate exposure of earthworms that feed on litter and humus (see below). Dietary preferences and seasonal activity patterns should be considered prior to choosing a sampling depth, as described below.

For single-chemical, exposure–effects relationships that are derived from published field studies, it is assumed that the test earthworms were exposed to soil in the depth interval that was sampled for analysis. This is a source of uncertainty because the earthworms may have been concentrated in a depth interval that is different from the sampled soil interval.

The dietary preferences of earthworms are important factors in the exposure assessment for litter feeders. Ireland (1983) notes that since some earthworms show dietary preferences for certain types of leaves, their exposures may not closely resemble concentrations in soil. Indeed, concentrations of PAHs in *Lumbricus rubellus* correlated better with concentrations in fragmented litter and humus than with those in mineral soil or unfragmented litter (van Brummelen et al., 1996). Few empirical studies of the uptake of chemicals from litter by litter-feeding earthworms have been undertaken; thus the relative accumulation of contaminants from litter and burrows in underlying soil is not well understood. Similarly, it is not known whether equilibrium relationships generally exist between chemical concentrations in litter and surface soil, although concentrations of elements in plants are typically correlated with those in soil. Although the exposure of litter-feeding earthworms to chemicals may be represented by the measured concentration of the chemical in litter or the measured body burden of the chemical, the risk assessment for these organisms will be highly uncertain until the associated exposure–response relationships are developed.

The greatest exposure of earthworms to contamination generally occurs during periods of high activity. Ireland (1983) notes that the maximum activity of soil-dwelling earthworms in temperate climates occurs when temperatures are between 4 and 11°C and when soil water content is high. Thus, she observed maximum lead uptake under these conditions. Although it is usually not feasible for soil samples at a contaminated site to be taken at multiple times during the year, any uncertainties related to the timing of sampling relative to temporal variance in contaminant concentrations, soil properties, and earthworm activity may be considered in the risk characterization.

Most arthropods inhabit the near-surface soil, but some are exposed to much deeper soils. Termites, for example, obtain clay-rich soil materials with high water content to build surface mounds and feeding galleries (Lee, 1983). Often the materials come from a 50-cm or lower depth. While most termites feed on aboveground plant tissue, some African termites ingest soil, apparently digesting organic matter (Lee, 1983). Because of the relatively impermeable exoskeletons of arthropods, food may be a more important determinant of uptake than direct exposures to soil. For example, the uptake of PAHs by three isopod species was correlated to levels in humus and fragmented litter rather than fresh litter or mineral soil (van Brummelen et al., 1996). Little is known about the exposure of microarthropods and free-living nematodes that are often highly abundant and diverse components of the soil fauna.

3.6.2 Soil Properties and Chemical Interactions

The exposure of earthworms to inorganic contaminants is dependent on soil chemical properties. Janssen et al. (1997) observed that the accumulation of several inorganic elements by earthworms was controlled by the same soil characteristics that affect the partitioning between the solid phase and pore water. Thus, they concluded that uptake is either from the soil pore water or a related route, but that bioavailability, as measured by the soil–earthworm uptake factor, is predictable using total metal concentrations supplemented with local soil characteristics. Soil factors determining exposure to inorganic chemicals include: pH (Ma, 1982, 1987; Corp and Morgan, 1991), calcium content (Andersen, 1979), cation exchange capacity (Ma, 1982), and organic matter (Corp and Morgan, 1991). Although these soil properties are seldom used for the screening risk assessment (Chapter 5), they should contribute to the evaluation of exposure–response relationships in the definitive risk characterization.

As stated above, the exposures of earthworms to at least some organic chemicals are dependent on the concentrations in soil water (van Gestel and Ma, 1988). The importance of exposure via soil solution applies to other soil organisms, such as the potato cyst nematode (*Globodera rostchiensis*; Houx and Aben, 1993). The accumulation of organic chemicals by earthworms is typically dependent on the fraction of organic matter in the soil, but for some chemicals (e.g., chlorophenols), pH is also important (van Gestel and Ma, 1988). Hendriks et al. (1995) suggest that the soil–earthworm uptake factor (see Section 3.9.3) should be independent of the octanol–water partition coefficient, if the primary compartments in which the organic contaminant resides are in the lipids of earthworms and in the organic matter of soil.

It is advisable for the assessor to obtain information on relevant soil characteristics for contaminants that remain a potential concern to invertebrates following the screening assessment (Chapter 5), so that exposure data may be interpreted well in the definitive risk characterization. Thus, because of differences in soil properties at the site of concern and in studies from which exposure–response relationships are derived, certain evidence from published literature may be ruled to be inapplicable to the risk assessment.

3.7 EXPOSURE OF SOIL MICROBIAL COMMUNITIES

Microorganisms reside in distinct microenvironments of soil. Therefore, they may be exposed to a wide range of local concentrations, including concentrations in soil water, potentially higher concentrations at or near the surface of soil particles, or even droplets of organic liquids to which they may attach, such as petroleum or PCB oil. Because typical assessment endpoints for microorganisms are ecosystem-level microbial processes, the relevant exposure is an average exposure, rather than the exposure of an individual organism. The exposure that is typically used is the concentration in bulk soil, primarily because the available effects tests have used that measure. The concentration in soil water is an alternative measure of exposure, but most endpoints of interest to risk assessors (nitrogen transformation, enzyme activity) have not been tested in liquid culture in the presence of toxicants. Because the assessment endpoints of concern are the microbial processes that influence ecosystem dynamics, the microbes of primary concern are aerobic organisms in surface soil. Thus, the sampling depths that are selected for plants or soil invertebrates are reasonable for microbial processes.

3.8 EXPOSURE OF WILDLIFE

Unlike some other endpoint assemblages, wildlife at contaminated sites are likely to be significantly exposed to contaminants in multiple media. They may drink or swim in contaminated water, ingest contaminated food and soil, breathe contaminated air, or absorb contaminants through dermal contact. To provide an accurate and realistic representation of exposure, exposure models for wildlife must therefore include exposure through multiple media. In addition, because most wildlife are mobile, moving among and within habitats, exposure is not restricted to a single location. They may integrate contamination from several spatially discrete sources. As a consequence, the accurate estimation of wildlife exposure also requires the consideration of habitat requirements and spatial movements.

The contaminant exposure experienced by wildlife can be described using either internal or external measures. Internal measures are chemical concentrations or biomarker levels in tissues of the endpoint species while external measures are concentrations in media or rates of contaminant intake measured or estimated at entry points to the animal (skin, lungs, or digestive system).

3.8.1 Exposure Based on Internal Measures

Most commonly, internal measures of exposure are chemical concentrations in tissues of the endpoint species. These tissues may be target organs for toxic effects (e.g., liver, kidney, or brain), nontarget tissues that simply accumulate contaminants (e.g., bone, hair, or feathers), or whole-animal concentrations. Internal exposure measures have several advantages:

- Integration of all exposure pathways through which the individual may have been exposed
- Averaging of exposure over both time and space
- Possible indication of site-specific contaminant bioavailability (if field-collected data are used)
- Elimination of exposure model error and parameter uncertainty

Additionally, if contaminant concentrations are determined for target organs, then exposure data may be directly related to toxicity. The association of tissue concentration with effects is discussed in more detail in Section 4.1.6.

If internal measures of exposure are to be employed in an assessment and site-specific field data are to be collected, it is important to know something of the toxicokinetics of the chemical of interest. Toxicokinetic data will provide an indication of which tissues should be sampled and analyzed such that effects from exposure can be evaluated. Tissue types most frequently sampled include liver and kidney, as these are the primary organs for metabolism and excretion and are therefore likely to be adversely affected by contaminants. Brain tissues are also frequently analyzed for contaminants that are neurotoxic and accumulate in lipid. Chemical concentrations in eggs are widely used to evaluate the exposure of birds to contaminants that may be transferred through eggs and have adverse effects on reproduction.

Some tissues are analyzed not because they are clearly associated with effects but because they are reservoirs for contaminants. Contaminant reservoirs include bone and fatty tissues. These tissues become reservoirs because of the chemical-specific affinities. For example, because most organochlorine contaminants are hydrophobic and highly fat-soluble, they tend to accumulate in fatty tissues. Similarly, because lead and strontium are analogs for calcium, these inorganic contaminants tend to accumulate in bones. Contaminants in reservoir tissues may be mobilized episodically. Examples include mobilization of organochlorine chemicals from fat during starvation (e.g., hibernation, migration) or mobilization of lead associated with calcium mobilization during pregnancy or egg formation.

Other tissues, such as hair and feathers, are analyzed because they represent contaminant sinks. Contaminants in these tissues cannot be reabsorbed. Chemicals that accumulate in these sink tissues are generally restricted to inorganic contaminants. Advantages of analyzing hair and feathers are that both tissues can be sampled nondestructively (without sacrificing the animal) and that they may be sampled repeatedly from the same individual so that contaminant exposure may be tracked over time. In contrast, most other tissue sampling requires sacrificing the animal and therefore can be performed only once per animal. There is extensive literature concerning chemical concentrations in hair and feathers from various locations around the world (Huckabee et al., 1972; Jenkins, 1979; Chatt and Katz, 1988; Burger, 1993). Burger and Gochfield (1997) provide data relating mercury concentrations in feathers to impaired reproduction among birds in the New York Bight. While hair and feathers provide a good and repeatable measure of contaminant exposure, use of these data in risk assessment has been problematic because there are relatively few data to relate concentrations to effects. Additional research on this topic is needed.

To aid in the interpretation of ecotoxicological significance of chemical concentrations in tissues of wildlife, it is important to collect samples both at the contaminated site of interest and at one or more reference locations (Section 2.7.3). The use of reference sites allows the tissue concentrations measured at the contaminated site to be evaluated in a local context.

While estimation of wildlife exposure using internal measures generally requires field collection of samples followed by laboratory analysis of tissue concentrations, research by Shore (1995) suggests that concentrations in the liver and kidney may be estimated from soil concentrations. Statistically significant semi-log regression models were developed for bioaccumulation of lead and cadmium by small mammals based on data obtained from the literature. While these regression models are based on a small sample size and do not account for variation in bioaccumulation due to age differences or other factors (Shore, 1995), they do suggest an alternative approach for estimating tissue concentrations in wildlife. A recent study by Sample et al. (1998) found that chemical concentrations in whole-body tissues of small mammals may also be estimated from soil concentrations (Section 3.9.5). Statistically significant log–log regression models were developed and validated for 14 inorganic and 2 organic chemicals. While the greatest utility of these models is for estimation of exposure for predators of small mammals, if toxicity data are available in terms of whole-body contaminant concentrations, they may also be used to estimate exposure and effects for small mammals.

Physiologically based pharmacokinetic (PBPK) models are another tool for estimation of chemical concentrations in target tissues. PBPK models are mathematical simulation models that estimate the internal dose of a chemical from external dose estimates by simulating the uptake, internal transport, partitioning among internal compartments, transformation, metabolism, and depuration of chemicals (Suter, 1993a). Application of PBPK models for estimation of chemical concentrations in wildlife is likely to remain limited for the foreseeable future, because these models are data intensive, requiring species-, tissue-, and contaminant-specific compartmental transfer rates and other parameters that are generally not available for wildlife species.

3.8.2 Exposure Based on External Measures

As wildlife move through the environment, they may be externally exposed to contamination via three pathways: oral, dermal, and inhalation. Oral exposure occurs through the consumption of contaminated food, water, or soil. Dermal exposure occurs when contaminants are absorbed directly through the skin. Inhalation exposure occurs when volatile compounds or fine particulates are inspired to the lungs. The total exposure experienced by an individual is the sum of exposure from all three pathways or

$$E_{total} = E_{oral} + E_{dermal} + E_{inhal} \tag{3.5}$$

where

E_{total} = total exposure from all pathways
E_{oral} = oral exposure
E_{dermal} = dermal exposure
E_{inhal} = exposure through inhalation

Dermal exposure can generally be assumed to be negligible for birds and mammals on most hazardous waste sites. While methods are available to assess dermal exposure to humans (EPA, 1992b), data necessary to estimate dermal exposure are generally not available for wildlife (EPA, 1993b). Additionally, feathers and fur of birds and mammals reduce the likelihood of significant dermal exposure by limiting the contact of skin with contaminated media. Therefore, dermal exposure is expected to be negligible relative to other exposure routes in most cases. If contaminants that have a high affinity for dermal uptake are present (e.g., organic solvents and pesticides) and an exposure scenario for an endpoint species is likely to result in significant dermal exposure (e.g., burrowing mammals or swimming amphibians), dermal exposure may be estimated using the model for terrestrial wildlife presented by Hope (1995).

Inhalation of contaminants by wildlife is also likely to be negligible at most hazardous waste sites. This is for two reasons. First, because most contaminated sites are either capped or vegetated, exposure of contaminated surface soils to winds and resulting aerial suspension of contaminated dust particulates are minimized. Second, most volatile organic compounds (VOCs), which are the contaminants most likely to present a risk through inhalation exposure, rapidly volatilize from soil and surface water to air, where they are rapidly diluted and dispersed. Paterson et al. (1990) suggest that organic compounds with soil half-lives of <10 days are generally lost from soil before significant exposure can occur. As a consequence, significant exposure to VOCs through inhalation is unlikely. In situations where significant inhalation exposure of endpoint species is believed to be occurring or is expected to occur, models for vapor or particulate inhalation (Hope, 1995) may be employed. In these cases, EPA (1993b) recommends consulting an inhalation toxicologist.

Because contaminant exposure experienced by wildlife through both the dermal and inhalation pathways is generally negligible, at most sites exposure may be attributed to the oral exposure pathway. Equation 3.5 typically reduces to

$$E_{\text{total}} \approx E_{\text{oral}} \tag{3.6}$$

Oral exposure experienced by wildlife may come from multiple sources. Vertebrates may consume contaminated food (either plant or animal), drink contaminated water, or ingest soil. Soil ingestion may be incidental, while foraging or grooming, or purposeful, to meet nutrient needs. In some cases, waste materials may be also directly ingested (e.g., oil, ethylene glycol in antifreeze). The total oral exposure experienced by an individual is the sum of the exposures attributable to each source and may be described as

$$E_{\text{oral}} = E_{\text{food}} + E_{\text{water}} + E_{\text{soil}} \tag{3.7}$$

where

E_{food} = exposure from food consumption
E_{water} = exposure from water consumption
E_{soil} = exposure from soil consumption

For exposure estimates to be useful in the assessment of risk to wildlife, they should be expressed in terms of a body weight–normalized daily dose or milligrams of contaminant per kilograms body weight per day (mg/kg/day). Exposure estimates expressed in this manner may then be compared with similarly reported toxicological data presented in the literature. Models for the estimation of exposure from oral ingestion have been reported in the literature (EPA, 1993b; Sample and Suter, 1994; Hope, 1995; Freshman and Menzie, 1996; Pastorok et al., 1996; Sample et al., 1997b) and are generally of the form

$$E_j = \sum_{i=1}^{m} (I_i \times C_{ij}) \tag{3.8}$$

where

E_j = oral exposure to contaminant (j) (mg/kg/day)
m = number of ingested media (e.g., food, water, or soil)
I_i = ingestion rate for medium (i) (kg/kg body weight/day or l/kg body weight/day)
C_{ij} = concentration contaminant (j) in medium (i) (mg/kg or mg/l)

E_j represents the daily exposure averaged over the exposure duration, which at most waste sites is likely to be chronic, measured in terms of months or years.

Very few wildlife species consume diets that consist exclusively of one food type. To meet nutrient needs for growth, maintenance, and reproduction, most wildlife species consume varying amounts of multiple food types. Because it is unlikely that all food types consumed contain the same contaminant concentrations, dietary diversity is of one of the most important exposure-modifying factors. To account for differences in contaminant concentrations of different food types, exposure estimates should be weighted by the relative proportion of daily food consumption attributable to each food type and the contaminant concentration in each food type. Wildlife may drink from different water sources and consume soils that differ in contaminant concentrations. These differences should also be accounted for. This may be done by modifying Equation 3.8 as follows:

$$E_j = \sum_{i=1}^{m} \sum_{k=1}^{n} p_{ik}(I_i \times C_{ijk}) \tag{3.9}$$

where

n = number of types of medium (i) consumed
p_{ik} = proportion of type (k) of medium (i) consumed (unitless)
C_{ijk} = concentration of contaminant (j) in type (k) of medium (i) (mg/kg or mg/l)

It is generally assumed that chemical absorption by the wildlife species in the field is 100% (Sprenger and Charters, 1997) or equivalent to that in the toxicity test species used to evaluate exposure (EPA, 1993b). Because chemical forms used in toxicity tests are likely to be more bioavailable (e.g., test based on solutions of metal salts or organic chemicals in vegetable oils) than chemical forms at field sites, this assumption may result in overestimates of exposure. To address this issue, some authors have included absorption factors in wildlife exposure models (Hope, 1995; Pastorok et al., 1996). These factors represent the proportion of the ingested dose that is actually absorbed through the gastrointestinal tract. Few absorption factors are available. Owen (1990) presents values for 39 chemicals. Models for estimation of absorption of organic chemicals by birds and nonruminant mammals are presented by Hope (1995).

If the contaminated site is spatially heterogeneous with respect to either contamination or wildlife use, the exposure model should be modified to include spatial factors. The most important spatial consideration is the movement of wildlife. Animals travel varying distances, on a daily to seasonal basis, to find food, water, and shelter. The area encompassed by these travels is defined as the *home range* (we use the term here to include territories). If the spatial units being assessed are larger than the home range of individuals of an endpoint species and provide the habitat needs of the species, then the previously listed models are adequate. However, endpoint species often have home ranges that are larger than contaminated units, or the contaminated site may not supply all of a species' habitat requirements. In those cases, the wildlife exposure model must be modified.

If a spatial unit has similar habitat quality to the surrounding area but is smaller than the home range, use of the unit may be described as a simple function of site area. That is, one can assume that for wildlife that use the entire contaminated area, exposure is proportional to the ratio of the size of the unit to home range size:

$$E_j = \frac{A}{HR}\left[\sum_{i=1}^{m}\sum_{k=1}^{n} p_{ik}(I_i \times C_{ijk})\right] \tag{3.10}$$

where

A = area (ha) contaminated
HR = home range size (ha) of endpoint species

Note that A is the area of a contaminated unit, not the entire area that has been designated a hazardous waste site. Because boundaries of waste sites are often drawn conservatively, they may contain a considerable uncontaminated area.

Equation 3.10 implies that all of the habitat within a contaminated area is suitable and that use of all portions of the contaminated area is equally likely. Because many waste sites are industrial or highly modified in nature, it is unlikely that all areas within their bounds provide habitat suitable for endpoint species. If it is assumed that use of a waste site is proportional to the amount of suitable habitat available on the site, Equation 3.10 may be modified to read

$$E_j = P_h\left(\frac{A}{HR}\left[\sum_{i=1}^{m}\sum_{k=1}^{n} p_{ik}(I_i \times C_{ijk})\right]\right) \tag{3.11}$$

where

P_h = proportion of suitable habitat in the contaminated area.

One complication is the spatial heterogeneity of contaminants on waste sites. These models are based on the assumption that either contaminants are evenly distributed on the site or wildlife forage randomly with respect to contamination on the portion of the site that constitutes habitat so that they are exposed to mean concentrations. However, if contaminant levels vary and are not independent of habitat quality, these assumptions do not hold. For example, contaminant concentrations might be greatest near the center of a site, but the habitat quality might be highest near the edges. In such cases, it might be necessary to model the proportional contribution of each area with a distinct combination of contaminant level and habitat quality

$$E_j = \sum_{l=1}^{o}\left(\frac{A_l}{HR}\left[\sum_{i=1}^{m}\sum_{k=1}^{n} p_{ik}(I_i \times C_{ijkl})\right]\right) \quad (3.12)$$

where

o = number of distinct contaminated habitat areas
A_l = area (ha) of a distinct contaminated habitat area
C_{ijkl} = concentration of contaminant (j) in type (k) of medium (i) from the lth area (mg/kg or mg/l)

These areas should correspond to the assessment units defined during the problem formulation (Chapter 2).

As can be seen, if the distribution of contamination and habitat quality is complex, this approach to exposure estimation rapidly becomes ungainly. In such cases, it is advisable to implement the exposure in a Geographic Information System (GIS). GIS represents a tool to facilitate evaluation of the spatial co-occurrence of contamination and either habitats or endpoints. Using a GIS, maps displaying the spatial distribution of habitats may be overlaid with maps of contaminant distribution to accurately determine the degree to which habitat is contaminated. Furthermore, if information on the distribution or movements of endpoint species (generated by radiotelemetry or censuses) is available, these data may be combined with the habitat and contaminant data to provide a more accurate visualization of exposure. Examples of the application of GIS to exposure and risk assessments can be found in Clifford et al. (1995), Banton et al. (1996), Henriques and Dixon (1996), and Sample et al. (1996a). Clifford et al. (1995) provide a good description of the use of GIS with spatial statistics to generate unbiased estimates of wildlife tissue concentrations.

A variation of the use of GIS to estimate exposure is presented in Sample et al. (1996a). In this assessment, habitat preferences of 57 wildlife species expected to use the Oak Ridge Reservation were compared with a GIS-based habitat map of the ORR. The presence or absence of suitable habitat within areas of known contamination on the ORR was used to identify species with the highest potential for contaminant exposure and to identify those sites on the ORR that may present exposure sources to the greatest number of species. This analysis was used to aid in the selection of wildlife endpoint species and to prioritize sites for assessment.

3.8.2.1 Exposure-Modifying Factors

Factors other than those described in these models modify contaminant exposure experienced by wildlife endpoint species. These factors include age, sex, season, and behavior patterns.

The models above imply that the endpoint species have uniform body size, metabolism, diet, home range, and habitat requirements. However, these properties may differ between juveniles and adults and between males and females. For example, because they are actively growing, metabolism (and therefore food consumption) is generally greater for juveniles of most endpoint species. Diet composition may also differ dramatically between juveniles and adults of the same species. Similarly, the food requirements of females during reproduction are greater than those for males for many endpoint species. These factors may serve to make certain age classes or a particular sex experience greater contaminant exposure than other segments of the population. Because of their greater exposure, contamination may present a greater risk to these segments of the population. If a particular life stage or sex is likely to be highly exposed, it should be emphasized in the exposure assessment.

Behavior may modify exposure by increasing or decreasing the likelihood of contact with contaminated media. Wildlife behaviors are frequently seasonally variable. Some foods may be available and consumed only at certain times of the year. Similarly, some habitats and certain parts of the home range may be used only in certain seasons. In addition, many species hibernate or migrate; by leaving the area or restricting their activity to certain times of year, their potential exposure may be dramatically reduced. All of these factors should be considered when evaluating contaminant exposure experienced by wildlife, and exposure models should be adjusted accordingly. The simplest approach to modifying the exposure estimates to take into account some of these exposure-modifying factors is to generate multiple exposure estimates. For example, if diet differs by season or by sex, the assessor should calculate exposure estimates for each sex or season. Comparison of exposure estimates generated for differing exposure scenarios aids in identifying the segments of population at greatest risk or times of year when risk is greatest.

3.8.2.2 Probabilistic Exposure Estimation

Contaminant exposure estimates for wildlife are frequently generated using single, conservative values (e.g., upper 95% confidence limits on the mean, maximum observed value) to represent parameters (e.g., contaminant concentration in soil, food, water, or air; ingestion rates; or diet composition) in the exposure model. These single parameter values, known as point estimates, are selected because they are believed to be protective of most individuals and their use simplifies the calculation of an exposure estimate. While the use of conservative assumptions is suitable in a screening assessment, the use of point estimates is not recommended in a definitive assessment. Employing point estimates for the input parameters in the exposure model does not take into account the variation and uncertainty associated with the parameters. Contaminant exposure that endpoint organisms may receive in any given area may therefore be either over- or underestimated. As a consequence, remediation may be recommended for areas where it is unnecessary, or significant risks may be overlooked. Parameterization of the exposure model with point estimates also produces only a point estimate of exposure. This exposure estimate provides no information concerning the distribution of exposures among individuals or the likelihood that individuals within an area will actually experience potentially hazardous exposures. To incorporate the variation or uncertainty in exposure parameters and to provide a better estimate of the potential exposure experienced by wildlife, it is highly recommended that exposure modeling be performed using probabilistic methods such as Monte Carlo simulation (Chapter 7).

3.8.3 Parameters for Estimation of Exposure

To estimate contaminant exposure by terrestrial wildlife using the models described above, species-specific values for the parameters are needed. Because of large within-species variation in values for life history parameters, data specific to the site in question provide the most accurate exposure estimates and should be used whenever available. Because site-specific life history data are seldom available, published values from other areas within the geographic range of an endpoint species must generally be used to estimate exposure.

Summaries of wildlife life history information are available from multiple sources. The "Wildlife Exposure Factors Handbook" (EPA, 1993b) presents life history data for 15 birds, 11 mammals, and 8 reptiles or amphibians. Sample and Suter (1994) present information for 15 species resident on the Oak Ridge Reservation, 4 of which are not discussed in EPA (1993b). Life history parameters for an additional eight mammal and five bird species likely to occur at DOE facilities are summarized in Sample et al. (1997a). A database of wildlife exposure parameters is available from the California Environmental Protection Agency at http://endeavor.des.ucdavis.edu/calecotox/. Additional sources of life history summaries include the *Mammalian Species* series (published by the American Society of Mammologists) and the *Birds of North America* series (published by the American Ornithologists Union and the Philadelphia Academy of Natural Sciences). The *Mammalian Species* series currently addresses over 300 mammal species, while *Birds of North America* series addresses 240. Additional information on the Birds of North America may be obtained from the Internet: http://www.acnatsci.org/bna/. The U.S. Army Corps of Engineers has also been developing species profiles for threatened and endangered species in the southeastern United States. Copies of these reports may also be obtained from the Internet: http://www.wes.army.mil/el/tes/.

3.8.3.1 Body Weight

Body weight is an extremely important parameter in the estimation of exposure. Not only is it a factor in determining the exposure rate, but because metabolism and body weight are related (Davidson et al., 1986), body weights may be used to predict food and water consumption rates. On a per-individual basis, larger animals consume more food or water than do smaller animals. However, because larger animals have lower metabolic rates than smaller ones, smaller animals have higher food and water consumption rates per unit body weight. This means that smaller animals will experience greater oral exposure per unit body weight than will larger animals.

Body weights for selected terrestrial wildlife are reported in EPA (1993b), Sample and Suter (1994), and Sample et al. (1997a). Additional sources include Dunning (1984, 1993), Burt and Grossenheider (1976), Silva and Downing (1995), the *Mammalian Species* series, and the *Birds of North America* series.

3.8.3.2 Food and Water Consumption Rates

Field observations of food, water, or soil consumption rates produce the best data with which to estimate exposure. With very few exceptions, these data are unavailable for most wildlife species. The second best data to use to estimate exposure are media consumption rates derived from laboratory studies of wildlife species. Uncertainties associated with these data may be significant, however, because ambient activity regimes and environmental variables (temperature, humidity, etc.) that influence metabolism (and therefore consumption rates) are difficult to approximate in a laboratory setting.

In the absence of experimental data, food consumption values can be estimated from allometric regression models based on metabolic rate. Nagy (1987) derived equations to estimate food consumption (in kg dry weight) for various groups of birds and mammals:

$$I_{fd} = (0.0687(\mathrm{BW})^{0.822})/\mathrm{BW} \quad \text{placental mammals} \tag{3.13}$$

$$I_{fd} = (0.0306(\mathrm{BW})^{0.564})/\mathrm{BW} \quad \text{rodents} \tag{3.14}$$

$$I_{fd} = (0.0875(\mathrm{BW})^{0.727})/\mathrm{BW} \quad \text{herbivores} \tag{3.15}$$

$$I_{fd} = (0.0514(\mathrm{BW})^{0.673})/\mathrm{BW} \quad \text{marsupials} \tag{3.16}$$

$$I_{fd} = (0.0582(\mathrm{BW})^{0.651})/\mathrm{BW} \quad \text{all birds} \tag{3.17}$$

$$I_{fd} = (0.1410(\mathrm{BW})^{0.850})/\mathrm{BW} \quad \text{passerine birds} \tag{3.18}$$

where

I_{fd} = food ingestion rate (kg food (dry weight) / kg body weight/day)
BW = body weight (kg live weight)

Food ingestion rates estimated using these allometric equations are expressed as kilograms of dry weight per day. Because wildlife do not generally consume dry food (unless maintained in the laboratory), food consumption must be converted to kilograms of fresh weight by adding the water content of the food. Percent water content of wildlife foods are listed in Table 3.5. Additional data may be obtained from the literature (e.g., Holmes, 1976; Redford and Dorea, 1984; Bell, 1990; Odum,

1993). Calculation of food consumption in kilograms of fresh weight is performed as follows.

$$I_{ff} = \sum_{i=1}^{m} \left(P_i \times \frac{I_{fd}}{1 - WC_i} \right) \tag{3.19}$$

where

I_{ff} = total food ingestion rate (kg food (fresh weight) / kg body weight / day)
m = total number of food types in the diet
P_i = proportion of the ith food type in the diet
WC_i = percent water content (by weight) of the ith food type

Food ingestion rates may also be estimated based on the amount of metabolizable energy in foods and metabolic rates. A detailed description of the determination of food ingestion using this method is presented in EPA (1993b).

Water consumption rates can be estimated for mammals and birds from allometric regression models based on body weight (Calder and Braun, 1983):

$$I_w = (0.099(\text{BW})^{0.90})/\text{BW} \qquad \text{mammals} \tag{3.20}$$

and

$$I_w = (0.059(\text{BW})^{0.67})/\text{BW} \qquad \text{birds} \tag{3.21}$$

where

I_w = water ingestion rate (l water/kg body weight /day)
BW = body weight (kg live weight)

3.8.3.3 Inhalation Rates

Allometric equations, based on body mass, have also been developed to estimate inhalation rates of resting mammals (Stahl, 1967) and nonpasserine birds (Lasiewski and Calder, 1971):

$$I_a = (0.54576(\text{BW})^{0.8})/\text{BW} \qquad \text{mammals} \tag{3.22}$$

and

$$I_a = (0.40896(\text{BW})^{0.77})/\text{BW} \qquad \text{nonpasserine birds} \tag{3.23}$$

where

I_a = inhalation rate (m^3 air/kg body weight /day)
BW = body weight (kg live weight).

The applicability of Equation 3.23 to passerines is not known. However, the similarity between the models for mammals and birds suggests that Equation 3.23 is likely to be suitable for passerines.

TABLE 3.5
Percent Water Content of Wildlife Foods

Food Type		Percent Water Content		
		Mean	STD	Range
Aquatic invertebrates	Bivalves (without shell)	82	4.5	
	Crabs (with shell)	74	6.1	
	Shrimp	78	3.3	
	Isopods, amphipods			71–80
	Cladocerans			79–87
Aquatic vertebrates	Bony fishes	75	5.1	
	Pacific herring	68	3.9	
Aquatic plants	Algae	84	4.7	
	Aquatic macrophytes	87	3.1	
	Emergent vegetation			45–80
Terrestrial invertebrates	Earthworms (depurated)	84	1.7	
	Grasshoppers, crickets	69	5.6	
	Beetles (adult)	61	9.8	
Mammals	Mice, voles, rabbits	68	1.6	
Birds	Passerines (with typical fat reserves)			68[a]
	Mallard duck (flesh only)			67[a]
Reptiles and amphibians	Snakes, lizards			66[a]
	frogs, toads	85	4.7	
Terrestrial plants	Monocots: young grass			70–88
	Monocots: mature dry grass			7–10
	Dicots: leaves	85	3.5	
	Dicots: seeds	9.3	3.1	
	Fruit: pulp, skin	77	3.6	

[a] Single values indicate only one value available.

Source: From EPA (1993b).

3.8.3.4 Soil Consumption

In addition to consuming food and water, many terrestrial vertebrates consume soil. Soil consumption may occur inadvertently while foraging (i.e., predators of soil invertebrates ingesting soil adhering to worms, grazing herbivores consuming soil deposited on foliage or adhering to roots) or grooming. In addition, soil may be consumed purposefully to meet nutrient requirements. Diets of many herbivores are deficient in sodium and other trace nutrients (Robbins, 1993). Ungulates, such as white-tailed deer (*Odocoileus virginianus*), have been observed to consume soils with elevated sodium levels, presumably to meet sodium needs (Weeks, 1978). Because soils at waste sites may contain very high contaminant concentrations, direct ingestion of soil is potentially a very significant exposure pathway. For example, field observations indicated that white-tailed deer were consuming coal ash from an ash disposal basin on the Oak Ridge Reservation (DOE, 1995). Because the sodium content of the coal ash was high (approximately four times higher than background soils), it was assumed that deer were consuming ash to meet their sodium needs. Contaminant exposure was estimated by assuming that deer consumed a sufficient volume of ash to alleviate their sodium deficit. Ash consumption to meet sodium requirements was 7.3 times greater than consumption estimated via incidental ingestion (DOE, 1995).

In contrast to food and water consumption, generalized models do not exist with which to estimate soil ingestion by wildlife. Soil ingestion rates, however, have been measured for some wildlife species (Table 3.6).

3.8.3.5 Home Range and Territory Size

Home ranges are the spatial areas occupied by individuals. These areas provide each species with food, water, and shelter and may or may not be defended. Home range size is a critical component in estimating exposure. Species with limited spatial requirements (e.g., small home ranges) may live exclusively within the bounds of a contaminated site and therefore may experience high exposure. Conversely, species with large home ranges may travel among and receive exposure from multiple contaminated sites.

Multiple factors may influence home range or territory size. These factors include habitat quality, prey abundance, and population density. In general, home range size decreases with increasing habitat quality. In contrast, home range generally decreases with increasing population density because antagonistic interactions with neighbors reduce movements.

While species-specific home range or territory size data are reported in the literature for many species, data are not available for all potential endpoint species. In the absence of measured values, home range or territory size may be modeled based on body weight and trophic relationships. For example, McNab (1963) observed that home range size in mammals was a function of body weight:

$$HR = 6.76\ (BW)^{0.63} \tag{3.24}$$

where

HR = home range (acres)
BW = body weight (kg live weight)

Differences in home range requirements were observed between "hunters" (includes species that rely on widely distributed foods, e.g., granivores, frugivores, insectivores, and carnivores) and "croppers" (species that rely on foods that are spatially more concentrated, e.g., grazing and browsing herbivores; McNab, 1963). Home ranges of "hunters" may be as much as four times greater than that of "croppers" of the same body mass. Home ranges for each group may be estimated using the following models:

$$HR_h = 12.6\ (BW)^{0.71} \tag{3.25}$$

and

$$HR_c = 3.02(BW)^{0.69} \tag{3.26}$$

TABLE 3.6
Measured Soil Ingestion Rates for Selected Wildlife Species

Species		Soil Ingestion (% of diet)	Ref.
Reptiles			
Box turtle	*Terrapenne carolina*	4.5	Beyer et al., 1994
Eastern painted turtle	*Chrysemys picta*	5.9	Beyer et al., 1994
Birds			
Blue-winged teal	*Anas discors*	<2.0	Beyer et al., 1994
Ring-necked duck	*Aythya collaris*	<2.0	Beyer et al., 1994
Wood duck	*Aix sponsa*	11	Beyer et al., 1994
Mallard	*Anas platyrhynchos*	3.3	Beyer et al., 1994
Canada goose	*Branta canadensis*	8.2	Beyer et al., 1994
Stilt sandpiper	*Micropalama himantopus*	17	Beyer et al., 1994
Semipalmated sandpiper	*Caladris pusilla*	30	Beyer et al., 1994
Least sandpiper	*C. minutilla*	7.3	Beyer et al., 1994
Western sandpiper	*C. mauri*	18	Beyer et al., 1994
American woodcock	*Scolopax minor*	10.4	Beyer et al., 1994
Wild turkey	*Meleagris gallopavo*	9.3	Beyer et al., 1994
Mammals			
Opossum	*Didelphis virginiana*	9.4	Beyer et al., 1994
Short-tailed shrew	*Blarina brevicauda*	13	Talmage and Walton, 1993
Nine-banded armadillo	*Dasypus novemcinctus*	17	Beyer et al., 1994
Black-tailed jackrabbit	*Lepus californicus*	6.3	Arthur and Gates, 1988
Meadow vole	*Microtus pennsyvanicus*	2.4	Beyer et al., 1994
Cotton rat	*Sigmodon hispidus*	2.8	Garten, 1980
White-footed mouse	*Peromyscus leucopus*	<2.0	Beyer et al., 1994
		1	Talmage and Walton, 1993
Black-tailed prairie dog	*Cynomys ludovicianus*	7.7	Beyer et al., 1994
White-tailed prairie dog	*C. leucurus*	2.7	Beyer et al., 1994
Woodchuck	*Marmota monax*	<2.0	Beyer et al., 1994
Feral hog	*Sus scrofa*	2.3	Beyer et al., 1994
White-tailed deer	*Odocoileus virginianus*	<2.0	Beyer et al., 1994
Mule deer	*O. hemionus*	<2.0	Arthur and Alldredge, 1979
			Beyer et al., 1994
Elk	*Cervus elaphus*	<2.0	Beyer et al., 1994
Moose	*Alces alces*	<2.0	Beyer et al., 1994
Bison	*Bison bison*	6.8	Beyer et al., 1994
Pronghorn	*Antilocapra americana*	5.4	Arthur and Gates, 1988
Domestic dog	*Canis familiaris*	2–4	Calabrese and Stanek, 1995
Red fox	*Vulpes vulpes*	2.8	Beyer et al., 1994
Raccoon	*Procyon lotor*	9.4	Beyer et al., 1994

where

HR_h = home range for hunters (acres)
HR_c = home range for croppers (acres)
(*Note:* 1 acre = 0.4047 ha = 4047 m^2)

More recent research by Harestad and Bunnell (1979) produced the following relationships between body mass and home range in mammals:

$$HR_{herb} = 0.002\ (BW)^{1.02} \quad (3.27)$$

$$HR_{omn} = 0.59\ (BW)^{0.92} \quad (3.28)$$

and

$$HR_{carn} = 0.11\ (BW)^{1.36} \quad (3.29)$$

where

HR_{herb} = home range for herbivores (ha)
HR_{omn} = home range for omnivores (ha)
HR_{carn} = home range for carnivores (ha)
BW = body weight (g)

It is not known whether McNab (1963), Harestad and Bunnell (1979), or some other model produces the best estimates.

A strong positive relationship also exists between body mass and territory or home range size among birds (Schoener, 1968). Predators tend to have larger territories than omnivores or herbivores of the same weight. Territory size also increases more rapidly with body weight among predators than among omnivores or herbivores. Schoener (1968) believes these relationships reflect the higher density of available food for omnivores and herbivores compared with that for predators. Schoener (1968) developed regression models describing the relationship among body size, home range size, and foraging habits, but all parameters needed to implement the models are not presented. A summary of home range or territory sizes for 77 species of land birds (and source references) are listed, however.

3.9 UPTAKE MODELS

To estimate the contaminant exposure that wildlife may experience at any given location, measures of chemical concentrations in the foods that wildlife eat are needed. Direct measurement of chemical concentrations in plant and animal foods is *always* preferred. Because direct measurement of chemicals in site biota incorporates all site-specific parameters that may influence the uptake and transfer of contaminants through the food web and provides the best indication of actual bioavailability and bioaccumulation, incorporation of these data will result in the least uncertainty in the final exposure estimate. In principle, uptake models could also be used to estimate internal exposures of endpoint organisms.

Despite the clear superiority of site-specific data in ecological risk assessment, direct measurement of chemical concentrations in biota is frequently not performed.

Biota may not be collected because of incompatible sampling schedules (i.e., scheduled completion date for assessment dictates that biota sampling occur during winter when some species are unavailable); insufficient time, personnel, or finances to support field sampling; and lack of interest in or understanding of biota sampling among project managers. When direct measurement is not possible, the only alternative is to estimate chemical concentrations in biota using models.

Three general types of models exist for the estimation of contaminant concentrations in biota; however, all types of these models are not available for all biota types. Listed in order of increasing complexity, these models include simple uptake factors, empirical regression models, and mechanistic bioaccumulation models. Availability of models for different types of wildlife foods is discussed below.

Uptake factors are quotients of ratios of chemical concentrations in biota to concentrations in associated abiotic media. Uptake factors are also referred to as transfer coefficients or (particularly in aquatic studies) as bioconcentration factors (BCFs). Multiplication of an uptake factor by the chemical concentration in an abiotic medium produces an estimate of chemical concentrations in a tissue or organism. While uptake factors may be simple to use, uncertainty associated with the estimates that they produce may be quite high. An implicit assumption in the use of uptake factors is that uptake is a simple linear relationship, increasing as the ambient concentration increases, with an intercept of zero. Analyses by Alsop et al. (1996) and Sample et al. (1998, 1999) suggest that this assumption is incorrect and that bioaccumulation (at least for inorganic elements) is nonlinear with respect to soil concentration. Consequently, the use of uptake factors to estimate contaminant concentrations in biota at highly contaminated sites may result in gross overestimates of actual concentrations in biota (Sample et al., 1998, 1999).

Empirical regression models are derived using biota and chemical data from contaminated sites. In general, regression models are preferable to simple uptake factors. First, soil parameters known to influence bioavailability and uptake of contaminants from soil, such as pH, cation exchange capacity (CEC), and organic matter (OM), may be included in multiple-regression models. The resulting models explain more of the variability of the data and are likely to result in improved estimates of tissue concentrations by permitting more site-specific information to be included. Another advantage of regression models is that they can address the nonlinearity of bioaccumulation where the rate of accumulation decreases at higher concentrations of contaminants in soil, whereas uptake factors cannot. While nonlinear regression methods may be used to fit models to these bioaccumulation data, it is easier to log-transform the data and perform simple linear regression analyses. Regression models based on log-transformed data, while linear in log-space are nonlinear in untransformed space (Figure 3.3) and are equivalent to the power model. The regression model for transformed data may be expressed as

$$\ln(\text{biota}) = B_0 + B_1 (\ln \text{soil}) \tag{3.30}$$

where

ln(biota) = estimated natural log-transformed concentration in biota
ln(soil) = measured natural log-transformed concentration in soil

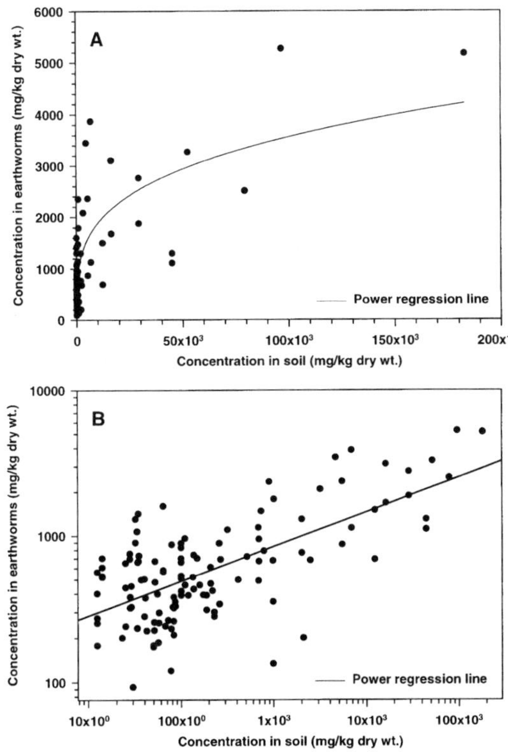

FIGURE 3.3 Scatter plots of untransformed (A) and log-transformed (B) bioaccumulation data for zinc from soil by earthworms. Data from Sample et al., 1999.

B_0 = intercept
B_1 = slope
(*Note:* Either $\log_{10}$ or natural log transformations may be used. While the same slope results from either transformation, intercepts differ.)

In arithmetic space, the transformed model described above is equivalent to the nonlinear power model:

$$\text{biota} = B_0'(\text{soil})^{Be} \tag{3.31}$$

where

biota = estimated untransformed concentration in biota
soil = measured untransformed concentration in soil
B_0' = antilog of intercept (e^{B0})
B_1 = slope

In place of these empirical bioaccumulation models, one may model accumulation mechanistically. Mechanistic models are developed based on knowledge of the biota physiology (e.g., metabolic transformation, transfer among internal compartments, and assimilation efficiencies for foods and contaminants) and the characteristics of the chemical of interest (e.g., solubility and partition coefficients) that determine the rates of the processes. In some models, food web transport and transfer of contaminants are also included (Ram and Gillett, 1993). Although these models may produce accurate estimates of bioaccumulation, they are complicated and data-intensive, requiring many parameters that may not be readily available. As a consequence of these limitations, mechanistic models have been rarely used for site-specific assessments (e.g., MacIntosh et al., 1994; Nichols et al., 1995). However, they are increasingly common in situations where time and resources are not available for site-specific data collection (Hope, 1999).

3.9.1 Aquatic Organism Uptake

Commonly, chemical concentrations are measured in water but not biota. In such cases, it is necessary to model concentrations in aquatic biota if effects are to be estimated based on body burdens in aquatic biota or if risks to wildlife that feed on aquatic organisms must be estimated. The most common approach is to use BCFs derived from conventional laboratory studies or BAFs derived from field studies. In either case, the factor is

$$F = C_o / C_w \tag{3.32}$$

where

F = BCF or BAF
C_o = concentration in the organism (mg/kg)
C_w = concentration in water (mg/l)

BAFs are more realistic than BCFs, because they include dietary exposures and long exposure durations. However, the water concentrations for BAF derivation are often highly variable and poorly defined, and BAFs are influenced by the particular structure and function of the ecosystems from which they were derived. BCFs are more precise and accurate because they are derived in laboratory studies for which aqueous concentrations and conditions are well specified. Therefore, BCFs are often preferable unless the uptake of the chemical of concern is known to have important dietary components, unless the BCF is judged to be of low quality or relevance (e.g., the exposures were short relative to the uptake kinetics of the chemical as indicated by failure to reach an asymptotic concentration), or if the conditions under which the BAF was derived are similar to those at the site being assessed. The EPA water quality criteria documents are useful sources of BAFs and BCFs, but they must be updated with literature reviews.

BCFs and BAFs from the literature are unreliable substitutes for site-specific values. In the Trinity River, maximum BCFs from EPA criteria documents provided estimates of metal concentrations in fish that were within a factor of 10 of measured values in most cases (Parkhurst et al., 1996). However, cadmium concentrations were consistently overestimated, and some predictions were in error by more than a factor of 1000. In the same study, geometric mean BCFs overestimated concentrations of pesticides in nearly all cases, usually by more than a factor of 10.

Quantitative structure–activity relationships (QSARs) are commonly used to estimate uptake of organic chemicals by aquatic organisms. The conventional assumption is that body burdens of nonionic organic chemicals are due to equilibrium partitioning between water and the lipid fraction of the organisms (Veith et al., 1979). The model most commonly endorsed by the EPA is

$$\log \mathrm{BCF} = 0.79 \log K_{ow} - 0.40 \tag{3.33}$$

(Veith and Kosian, 1983; EPA, 1993c). The authors indicate that the estimates of this model have order-of-magnitude confidence limits for most chemicals and larger limits for chemicals with K_{ow} values greater than 6.5. This model works reasonably well for chemicals with short half-lives but underpredicts concentrations of persistent chemicals in field-collected fish (Oliver and Niimi, 1985). Therefore, an analogous model has been developed for bioaccumulation of PCBs and similar chlorinated organic compounds by Oliver and Niimi (1988):

$$\log \mathrm{BAF} = 1.07 \log K_{ow} - 0.21 \tag{3.34}$$

For chemicals that accumulate to a significant extent via diet, BAFs should increase with trophic level. To account for this, the EPA Great Lakes Water Quality Initiative has proposed food chain multipliers (FCMs) (EPA, 1993c,d). BAFs are estimated by multiplying BCFs from Veith and Kosian's model by FCMs which are specific to trophic level and K_{ow} of the chemical. Even for trophic level 4 (piscivorous fish), the FCMs do not exceed a factor of 2 below a K_{ow} of 5, and they are undefined above 6.5. Therefore, FCMs are available and important for only a small range of K_{ow} values. BAFs for chemicals of concern in the Great Lakes are provided by Stephan (1993).

Recently, a set of BCF models for fish has been proposed by the Syracuse Research Corporation and the EPA Office of Pollution Prevention and Toxics (Box 3.3). This set is based on a large literature review and is more broadly applicable than the models discussed above which are limited to chemicals that are neutral or at least "relatively nonpolar." It does not incorporate dietary considerations or lipid content, but the authors argue that they are not needed. Like other QSARs for bioconcentration or bioaccumulation, these equations have large uncertainties (roughly $\pm$ 10×) and should be used for screening purposes or when site-specific analyses are precluded.

Concentrations of chemicals in aquatic plants are rarely measured although they are a significant component of the diet of some fish, turtles, waterfowl, and mammals. Some BCFs may be derived from the literature. Examples include mercury in *Vallisneria spiralis* (Gupta and Chandra, 1998); nickel, chromium, manganese, copper, and zinc in *Elodea canadensis* (Kähkönen et al., 1997; Kähkönen and Manninen, 1998); and cadmium in *Hydrilla verticillata* (Garg et al., 1997). In addition, environmental variables such as salinity may be important determinates of accumulation of heavy metals by macrophytes (e.g., cadmium by the seagrass

BOX 3.3

A Method for Estimating Bioconcentration and Bioacccumulation in Fish

Equations

Nonionic compounds

$\log K_{ow} < 1$	$\log BCF = 0.50$
$\log K_{ow}$ 1 to 7	$\log BCF = 0.77 \log K_{ow} - 0.70 + \Sigma F_i$
$\log K_{ow} > 7$	$\log BCF = -1.37 \log K_{ow} + 14.4 + \Sigma F_i$
$\log K_{ow} > 10.5$	$\log BCF = 0.50$
Aromatic azo compounds	$\log BCF = 1.0$

Ionic compounds (carboxylic acids, sulfonic acids and salts, compounds with nitrogen of +5 valence)

$\log K_{ow} < 5$	$\log BCF = 0.05$
$\log K_{ow}$ 5 to 6	$\log BCF = 0.75$
$\log K_{ow}$ 6 to 7	$\log BCF = 1.75$
$\log K_{ow}$ 7 to 9	$\log BCF = 1.00$
$\log K_{ow} > 9$	$\log BCF = 0.50$
Compounds with $\geq C_{11}$ alkyl	$\log BCF = 1.85$

Tin and mercury compounds

Use the appropriate equation with the appropriate factor (below) or 2.0, whichever is greater.

Factors

The following correction factors are used as the F_i parameters in the equations above.

Compounds with an aromatic *s*-triazine ring (3 compounds)	–0.32
Compounds containing an aromatic alcohol (e.g., phenol) with two or more halogens attached to aromatic ring (17 compounds)	–0.40
Compounds containing an aromatic ring with a *tert*-butyl group in a position ortho to an –OH group (e.g., *tert-butyl ortho*-phenol) (6 compounds)	–0.45
Compounds containing an aromatic ring and aliphatic alcohol in the form of –CH–OH (e.g., benzyl alcohol) (4 compounds)	–0.65
Phosphate ester, O=P(O–R)(O–R)(O–R), where R is carbon (one R can be H) (18 compounds)	–0.78
Ketone with one or more aromatic connections (18 compounds)	–0.84
Nonionic compounds with an alkyl chain containing eight or more –CH2–groups (13 compounds)	–1.00 (log K_{ow} of 4–6) –1.50 (log K_{ow} of 6–10)
Compounds containing a cyclopropyl ester of the form cyclopropyl-C (=O) –O– (e.g., permethrins) (6 compounds)	–1.65
Compounds containing a phenanthrene ring (4 compounds)	+0.48
Multiply halogenated biphenyls and polyaromatics containing only aromatic carbons and halogens (e.g., PCBs) (19 compounds)	+0.62
Organometallic compounds containing tin or mercury (12 compounds)	+1.40

Source: Meylan, W. M. et al., *Environ. Toxicol. Chem.*, 18, 664, 1999.

Potamogeton pectinatus; Greger et al., 1995). However, insufficient information is available to determine the relative importance of water and sediment in the contamination of rooted macrophytes or the possible nonlinearity of contaminant uptake as a function of concentration.

If biota concentrations must be estimated and the assumptions of the QSARs do not hold (e.g., aqueous concentrations are not constant or multiple routes of exposure are important), it may be appropriate to model uptake dynamically. Dynamic models require experimentally derived uptake rates, loss rates, and, for long durations, growth rates or physiological models to estimate those rates. In addition, if effects are estimated from concentrations in specific organs, a model with multiple internal compartments must be used. Dynamic bioaccumulation models have been reviewed in Calabrese and Baldwin (1993). Bioaccumulation has been incorporated into both equilibrium and dynamic models of chemical fate in aquatic ecosystems (Thomann, 1989; Mackay, 1991; Cowan et al., 1995). Aquatic food web models have been applied to assessments of contaminated sites, including the Clinch and Detroit Rivers (MacIntosh et al., 1994; Russell et al., 1999).

3.9.2 Terrestrial Plant Uptake

Although the emphasis of this section is on models for use in estimating the chemical content of vegetation, the actual measurement of these concentrations is advisable to minimize uncertainty in the estimation of exposure. Site-specific data on the bioaccumulation of contaminants in vegetation are often not available, because of constraints in funding or time. However, the field sampling and chemical analysis may ultimately be cost-effective, particularly if wildlife exposures calculated from model outputs point to the need for remediation, and there is some doubt that the models are representative of site conditions. Situations for which concentrations of elements and compounds in plants should generally be measured include:

1. Small sites dominated by only one vegetation type (the chemical analysis is cheap)
2. Sites where roots, fruits, or seeds constitute a significant portion of the diet of the wildlife assessment endpoint population (no good models of uptake exist)
3. Sites where mercury, chromium, or other speciated metals are expected to be important in exposure (existing models do not differentiate uptake based on speciation)
4. Sites that have significant air contamination contributing to the total exposure (models give highly uncertain results, particularly for the portion of the chemical depositing on the plant)
5. Sites where the contamination includes chemicals for which accumulation by plants has rarely been studied (e.g., chemical weapons and elements such as vanadium, molybdenum, cobalt, manganese, and uncommon radionuclides)
6. Bogs, arid soils, or other soils for which few data regarding plant uptake of particular chemicals are available

7. Sites where soil contamination is predominantly in the form of NAPLs (e.g., petroleum), where the measured concentration of the chemical in soil is probably not proportional to that dissolved in soil water

The simplest model for estimating the concentrations of chemicals in plants is the soil–plant uptake factor (also known as a transfer coefficient or BCF), the ratio of the concentration of a chemical in a plant or portion of a plant to that in soil, usually expressed on a dry weight basis. The use of the uptake factor is based on the assumption that within the relatively nontoxic range, concentrations of contaminants in plants increase linearly with increasing soil concentration, with intercepts of zero. For example, Sheppard et al. (1985) found statistically significant uptake factors for the transfer of uranium, cobalt, and arsenic from soil to shoots and roots of Scots pine, although the factors for cobalt and arsenic were lower when derived for a waste site soil than when salts were added to soil. Occasional uptake studies result in linear equations with a *y*-intercept (contaminant concentration in plant at zero concentration in soil) different from zero (e.g., Muramoto, 1989). However, soil–plant uptake factors are more common than nonzero-intercept regressions in the published studies and in risk assessments.

3.9.2.1 Empirical Models for Inorganic Chemicals

Many ecological and human health risk assessments have relied on the soil–plant uptake factors for radionuclides and nonradioactive elements that were compiled by Baes et al. (1984). Baes et al. developed uptake factors for each contaminant in two different types of plant tissue: (1) vegetative tissues of food crops and feed plants (foliage and stems) and (2) reproductive portions of food crops and feed plants (fruits, seeds, tubers) at edible maturity. The "Baes factors," intended for use in human health risk assessments, are based on studies conducted prior to 1984. Baes et al. (1984) provided uptake factors for elements even if data were not available. For example, uptake factors were estimated from trends observed in periods or groups of the periodic table. Alsop et al. (1996) found that the use of Baes factors underpredicted the uptake of lead and zinc by oats at concentrations within background levels in soil and overpredicted metal concentrations in the plants at concentrations above background. Thus, Alsop et al. provided evidence that the relationship between the contaminant concentration in the plant and in soil is not linear.

Over a large range of contaminant concentrations in soil, uptake by plants cannot be reliably predicted by uptake factors that were derived for smaller ranges of concentrations. Often, the uptake factor is observed to decrease as soil concentration increases (as in Alsop et al., 1996, described above). This behavior is not surprising, since the uptake of ions in solution by plants has been observed to be more efficient at lower concentrations than at higher concentrations (Cataldo and Wildung, 1978). Thus, above a certain soil contamination level, the concentration of metals in plants probably reaches a plateau or toxicity is observed, and uptake factors may overestimate accumulation at high soil concentrations and underestimate accumulation at low concentrations. For some elements, such as copper and zinc, the soil–plant contaminant uptake factor has been inversely correlated with the soil concentration (Baes et al., 1984), suggesting that the use of an uptake factor might overestimate plant concentrations at high contaminant concentrations in soil. Copper and zinc are plant nutrients, and their uptake would be expected to be regulated by the plant.

The uncertainty associated with soil–plant contaminant uptake factors can be represented if distributions of factors are developed: for multiple species at a single site, for a single species at multiple sites, or for multiple species at multiple sites. The first type would require some site-specific measurement, and the second would require much existing data; however, the greatest uncertainty would be associated with the use of a distribution of uptake factors of multiple species at multiple sites to represent few species at a single site. Risk assessors may use these distributions as inputs to Monte Carlo simulations of wildlife exposure, although better models exist (regressions are described below). The range of the distribution for a particular chemical in a wide range of species at multiple sites can be four orders of magnitude or more; a distribution of plant uptake factors for selenium derived from the literature is depicted in Figure 3.4. This distribution spans about 3.5 orders of magnitude, with almost half of the uptake factors exceeding 1.

For inorganic chemicals we recommend alternative models to uptake factors. For several elements, including arsenic, cadmium, copper, lead, mercury, nickel, selenium, and zinc, the concentration of the chemical in aboveground vegetation (dry weight) is predictable using a log–log linear relationship (Bechtel Jacobs Company, 1998b). We developed these regressions using data from published and unpublished field studies of the uptake of chemicals from soil by plants. Concentrations

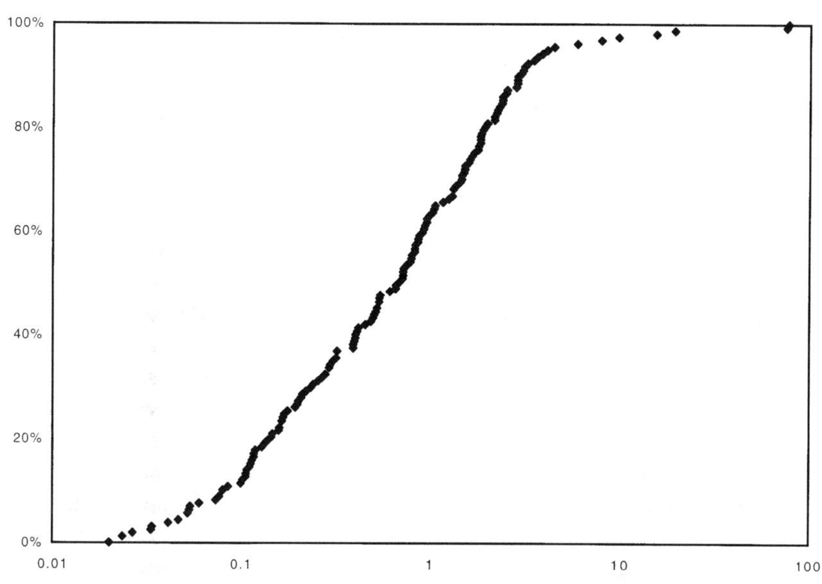

FIGURE 3.4 Cumulative distribution of soil–plant uptake factors for selenium from various published field studies.

TABLE 3.7
Results of Regression of ln (concentration in plant) on ln (concentration in soil)

Chemical	N	$B_0 \pm$ SE	$B_1 \pm$ SE	R^2	*P* Model Fit
Arsenic	122	-1.992 ± 0.431[a]	0.564 ± 0.125[a]	0.145	0.0001
Cadmium	207	-0.476 ± 0.088[a]	0.546 ± 0.042[a]	0.447	0.0001
Copper	180	0.669 ± 0.213[a]	0.394 ± 0.044[a]	0.314	0.0001
Lead	189	-1.328 ± 0.350[a]	0.561 ± 0.072[a]	0.243	0.0001
Mercury	145	-0.996 ± 0.122[a]	0.544 ± 0.037[a]	0.598	0.0001
Nickel	111	-2.224 ± 0.472[a]	0.748 ± 0.093[a]	0.371	0.0001
Selenium	158	-0.678 ± 0.141[a]	1.104 ± 0.067[a]	0.633	0.0001
Zinc	220	1.575 ± 0.279[a]	0.555 ± 0.046[a]	0.402	0.0001

Model: ln (conc. in plant) = $B_0 + B_1$[ln(conc. in soil)]; concentrations are mg/kg and weights are dry.
[a]$p \leq 0.001$.
Source: Bechtel Jacobs Company (1998b).

of chemicals in plants associated with salts added to soil in pots were determined not to be representative of concentrations in the field. Simple regressions derived from log-transformed data for eight inorganic chemicals are presented in Table 3.7.

An example regression of selenium in aboveground plant biomass vs. the concentration in soil is shown in Figure 3.5. The relationship between the chemical concentration in foliage and stems and that in soil is fit better by the log–log relationship than the linear, zero-*y*-intercept relationship that is equivalent to an uptake factor. The upper 95% prediction limit for the regression model is recommended for screening assessments (Bechtel Jacobs Company, 1998b), although 90th percentile uptake factors are also considered to be adequately conservative for most chemicals and soils at most waste sites.

The use of uptake factors or simple log–log regression models does not account for (1) soil properties, such as clay content, OM content, pH, CEC, or particle size; (2) plant properties, such as age, taxonomy, growth form, root depth, and evapotranspiration rate; (3) exposure time; and (4) other site characteristics, such as average temperature during the growing season. Many of the factors that control plant exposure to chemicals (Section 3.4) are pertinent here. For example, Sims and Kline (1991) found significant multiple regression models between nickel, copper, and zinc in wheat and soybean and soil metal concentrations and pH, but not with soil metal concentrations alone. Similarly, the authors found significant regression model fits between several inorganic chemicals in plants and chemical concentrations in soil and pH at various field locations, where pH contributed significantly to the model fit (Bechtel Jacobs Company, 1998b). Regressions for these ln-transformed regression models are presented in Table 3.8. Sheppard and Sheppard (1989)

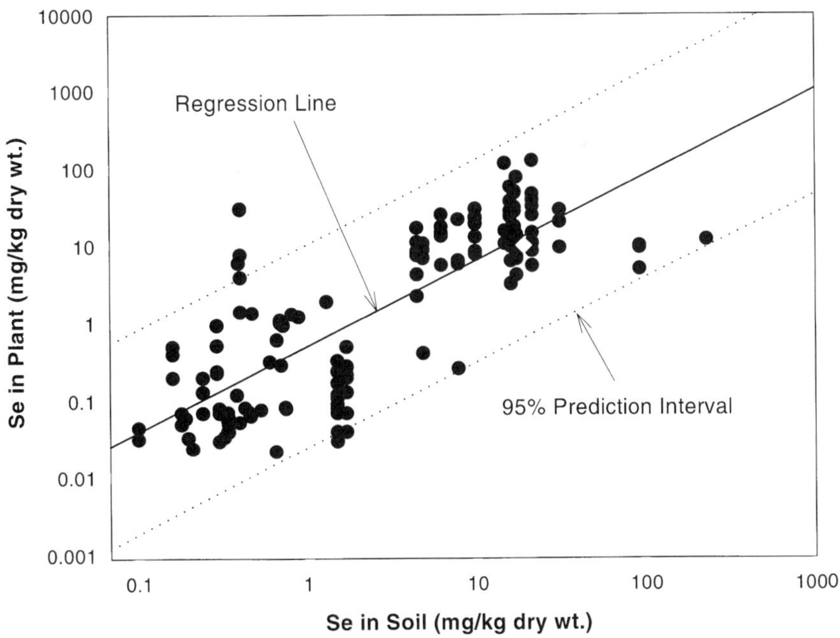

FIGURE 3.5 Scatterplot of plant concentration vs. soil concentration of selenium. The line represents a simple regression model of natural log-transformed data from published studies.

reviewed studies that demonstrated that K_d, the soil-soil water partition coefficient, and the soil–plant uptake factor were negatively correlated.

3.9.2.2 Empirical Models for Organic Chemicals

Soil–plant uptake factors are the models that are generally used to estimate the uptake of organic chemicals by plants; the chemical parameters from which they are derived are described below. Studies of the accumulation of organic chemicals by plants from soil typically focus on the chemical characteristics that determine the soil–plant uptake factor, notably the octanol–water partition coefficient (K_{ow}). In some experiments the major question is whether the dominant pathway of uptake is through air or soil (see Section 3.6.8). Less research is devoted to the role of soil characteristics or exposure time in determining uptake; the assumption is that the former are inconsequential and that the latter is not important because equilibrium is achieved quickly. Similarly, existing published data are probably insufficient to evaluate whether the uptake factor is independent of chemical concentration in soil; thus, it is unclear whether a log–log regression model of the form described for inorganic chemicals above would improve prediction of organic chemical concentration in plants. Accumulation models have been developed for estimating concentrations of organic com-

TABLE 3.8
Results of Regression of ln (concentration in plant) on ln (concentration in soil) and pH

Chemical	N	$B_0 \pm$ SE	$B_1 \pm$ SE	$B_2 \pm$ SE	R^2	*P* Model Fit
Cadmium	170	1.152 ± 0.638^{NS}	0.564 ± 0.047^{c}	-0.270 ± 0.102^{b}	0.42	0.0001
Mercury	82	-4.186 ± 1.144^{c}	0.641 ± 0.062^{c}	0.423 ± 0.186^{a}	0.67	0.0001
Selenium	148	-8.831 ± 0.723^{c}	0.992 ± 0.050^{c}	1.167 ± 0.106^{c}	0.87	0.0001
Zinc	193	2.362 ± 0.440^{c}	0.640 ± 0.057^{c}	-0.214 ± 0.077^{b}	0.49	0.0001

Model: ln(conc. in plant) = $B_0 + B_1$(ln{conc. in soil}) + $B2$(pH); concentrations are mg/kg and weights are dry.

[NS] $p > 0.05$

[a] $0.01 < p \leq 0.05$

[b] $0.001 < p \leq 0.01$

[c] $p \leq 0.001$

Source: Bechtel Jacobs Company (1998b). With permission.

pounds (many of them, pesticides) in plants from concentrations in soil. These are often referred to as partitioning models, because they are typically based on the chemical-specific octanol–water partition coefficients. Fugacity-based multimedia models that incorporate plants have also been constructed (Calamari et al., 1987; Paterson et al., 1994; Hung and Mackay, 1997), but these are not typically used for retrospective assessments, particularly if air sources are not present. Soil–plant uptake factor models for organic chemicals are presented in Table 3.9.

Several aspects of models in Table 3.9 are worthy of emphasis. Most models relate the soil–plant uptake factor to the octanol–water partition coefficient. Indeed, some good relationships have been derived for some chemical and plant combinations. However, it is noteworthy that in some equations, the uptake factor is positively correlated with the octanol–water partition coefficient and in other equations it is negatively correlated. Scheunert et al. (1994) point out that the correlation should be positive if soil water is used, but if the concentration of the compound in total soil is measured, the two-step partitioning process (soil to soil water and soil water to root) determines whether the correlation of uptake with the octanol–water partition coefficient is positive or negative. For this reason, K_{ow} may not be useful as a predictor of uptake by plants if a different chemical, plant species, or soil is used than what was previously tested.

Topp et al. (1986) found that bioconcentration factors for organic contaminants in total plants (roots plus foliage) were best described by the molecular weight of the chemical rather than the octanol–water partition coefficient. Dowdy and McKone (1997) compared the precision and accuracy of the use of the molecular connectivity index (MCI) and the octanol–water partition coefficient to predict uptake of organic chemicals from soil to roots or aboveground plant biomass. The MCI was developed

TABLE 3.9
Equations Used to Estimate Soil–Plant or Soil–Water–Plant Uptake Factors for Organic Chemicals

Equation	Chemical Parameter Required	Soil or Soil Water	Plant Part	Dry or Wet Weight	Chemicals Used in Derivation	Plants Used in Derivation	Ref.
$\log U_p = -0.578 \log K_{ow} + 1.588$	K_{ow}	Soil	Aboveground	Plant-dry; soil-dry	29 organic chemicals, including pesticides, Aroclor 254, TCDD, benzo(*a*)pyrene, and others.	Multiple species from published studies	Travis and Arms, 1988
$\log U_p = 0.233 \log K_{ow} + 0.971$	K_{ow}	Soil water	Root	Plant-wet	5 chlorinated benzenes	Barley (no linear correlation for cress)	Scheunert et al., 1994
$\log U_p = -0.24 \log K_{ow} + 1.47$	K_{ow}	Soil	Whole plant	Plant-wet; soil-dry	5 chlorinated benzenes	Barley	Scheunert et al., 1994
$\log U_p = 0.391 \log K_{ow} - 1.84$	K_{ow}	Soil	Whole plant	Plant-wet; soil-dry	5 chlorinated benzenes	Cress	Scheunert et al., 1994
$\log (U_p - 0.82) = 0.77 \log K_{ow} - 1.52$	K_{ow}	Soil water	Root	Plant-wet	*O*-Methylcarbamoyloximes and phenylureas	Barley	Briggs et al., 1982
$\log U_p = -2385 \log M + 5.943$	M	Soil	Whole plant	Plant-wet; soil-dry	14 pesticides, chlorinated benzenes; benzene and pentachlorophenol did not fit	Barley	Topp et al., 1986
$\log U_p = -0.204\ \text{MCI} + 0.589$	MCI	Soil	Aboveground	Plant-dry; soil-dry	30 organic chemicals, including pesticides, TCDD, and other chlorinated organics	Multiple species from published studies	Dowdy and McKone, 1997

U_p = the soil–plant uptake factor (conc. of chemical in plant / conc. in soil); K_{ow} = the octanol–water partition coefficient; M = the molecular weight of the chemical; MCI = the polar-corrected normal path first-order molecular connectivity index.

by Kier and Hall (1986) and is correlated with molecular size, branching, molar volume, and surface area. For aboveground vegetation, the MCI was more precise than K_{ow} in predicting chemical accumulation. For uptake from soil solution to roots, the two quantities were equally precise, and both had low accuracy. Although most of the models in Table 3.9 permit the calculation of a soil–plant uptake factor that is greater than 1, this level of bioaccumulation of organic chemicals in plants is rare for most chemicals. It should be also noted that the equations in Table 3.9 are not uniformly based on dry or wet weights of plants or soil. A common mistake in risk assessments is to combine an uptake model based on dry weight with a wildlife exposure that assumes that foods are expressed as wet weight.

Clearly, the uptake models for organic chemicals require validation for a wide range of soils, plants, and chemicals. Alsop et al. (1996) compared concentrations of octachlorodibenzodioxin (OCDD) measured in oat plants with those predicted by the model by Travis and Arms (1988). The model, which was based on chemicals with log K_{ow} values between 1.75 and 6.15, underestimated the uptake of OCDD, with a log K_{ow} of 9.05, by a factor of almost 200 (Alsop et al., 1996). It is not known whether there was an air source of OCDD. For most contaminated sites, the models of Dowdy and McKone (1997) probably give the most accurate estimates of plant concentration for the broadest range of chemicals, soils, and plant species, although the results may be highly uncertain.

Even if the primary uptake pathway of some nonionic organic chemicals is not through the soil (see Section 3.5.8), correlations between soil and plant concentrations of organic compounds may exist where soil is the primary source of atmospheric chemical in the vicinity of the plant.

3.9.2.3 Mechanistic Models

Mechanistic models of plant uptake of chemical contaminants have been developed for few chemicals, soils, and plant species in few environments. Mechanistic models may be dynamic, so as to incorporate the variable of time (which is not considered in the empirical accumulation models above). However, these models typically require individual plant parameters that are difficult and time-consuming to measure, and the models have usually been developed in the laboratory without field validation. An example is a model of Lindstrom et al. (1991), which is tested using the uptake of bromacil by soybean (*Glycine max*) from solution in Boersma et al. (1991). The plant consists of one root, three stem, and three leaf compartments, with each compartment subdivided into xylem, phloem, and storage (Boersma et al., 1991). Mechanistic models for plant uptake of chemicals are not recommended for use in ecological risk assessments at this time.

3.9.2.4 Surface Contamination

The assessor should be reminded that the equations in Tables 3.7 to 3.9 do not include any contributions from atmospheric deposition; thus these models do not provide an estimate of the contamination on the plant surface. In most retrospective risk assessments for waste sites, there is little interest in estimating current deposition of chemicals from air to plants. If such a model for estimating surface deposition is required, a dynamic model may be found in van de Water (1995). Vegetation-specific parameters required as inputs for the model include vegetated area, biomass per area, and fraction of chemical intercepted by foliage.

Rain splash may add contamination to the plant surface that is not accounted for in models of plant uptake. Rain splash models can incorporate environmental parameters that risk assessors do not typically measure, such as soil particle size, foliage height, canopy cover and surface area of vegetation, rainfall intensity, and kinetic energy of the storm (Dreicer et al., 1984). Also, rain splash studies are designed to simulate tilled agriculture, so they are likely to overestimate rates for natural vegetation. An estimate of the contamination from rain splash is not necessary if wildlife exposure models assume a reasonable rate of soil ingestion.

3.9.2.5 Plant Tissue Type

The extent of accumulation of contaminants in plants often differs among tissue types. Metals are generally found at higher concentrations in foliage than in fruits and seeds (Parr et al., 1983; Greenleaf-Jenkins and Zasoski, 1986; Sadana and Singh, 1987; Jiang and Singh, 1994). Thus, the use of the uptake models described above probably provides conservative estimates of uptake by most reproductive tissues. In contrast, zinc has been observed to concentrate in wheat grain (Sadana and Singh, 1987), and nickel may be remobilized from senescing tissues and accumulated in seeds (Cataldo and Wildung, 1978). In wheat, oats, soybean, and root crops, similar concentrations of metals have been found in storage tissues and foliage (Parr et al., 1983). Chromium, lead, and selenium often concentrate in roots and are transported slowly to other parts of the plant (Bysshe, 1988). Silver and chromium tend to accumulate in lower stems (Cataldo and Wildung, 1978). Uranium and arsenic concentrated in roots rather than shoots of Scots pine (Sheppard et al., 1985).

Interestingly, the variability in the concentration of mercury in plant tissues was reduced when the concentration was expressed on a wet weight basis (Shaw and Panigrahi, 1986). Differences in water and lipid content in plant tissues may partly explain differences in contaminant uptake. For example, Trapp (1995) generalized the model of Briggs et al. (1982) (Table 3.9) to all plant species, using the water and lipid content of roots and stems. The relative accumulation of organic chemicals in various plant tissues depends on the route of uptake; volatile chemicals should concentrate more in the leaves, and soluble chemicals in plant roots (see Section 3.5.8). Little is known about the differences in the accumulation of chemicals by different plant taxa; this uncertainty is discussed in Section 3.5.7.

3.9.3 Earthworm Uptake

Earthworms are a significant contaminant transfer pathway considered in many ecological risk assessments. This is because they are more highly exposed to contaminants in soil through both ingestion and dermal contact than are other soil and litter invertebrates (Davis and French, 1969; Ma, 1994), and because they may represent a significant fraction of the diet of some vertebrates (e.g., shrews, moles, woodcock).

Uptake factors for earthworms may be based on either chemical concentrations in soil or chemical concentrations in soil pore water. Uptake of inorganic contaminants is generally estimated using soil-to-earthworm uptake models while uptake of organic contaminants is more frequently estimated based on soil pore water. Soil–earthworm uptake factors are available in the literature for some chemicals (Table 3.10) but may be highly variable because of soil properties and the form of the contaminant. While it can be assumed that chemical concentrations in soil and earthworms from field studies are at equilibrium, this assumption should be demontrated for laboratory studies. Previous research has shown that steady state was reached for a variety of organochlorine chemicals in 10 days (Belfroid et al., 1995).

An alternative to soil–earthworm models is soil pore water–earthworm uptake models. They are consistent with soil–worm models if the solid, aqueous, and biotic phases of the soil are all in equilibrium. This approach has been used for sediments by the EPA and others (DiToro et al., 1991). It is well supported for sediment invertebrates and is supported by some evidence for earthworms and possibly other soil invertebrates (van Gestel and Ma, 1988; Connell, 1990; Lokke, 1994; Ma et al., 1998). However, it has been suggested that the model may underestimate accumulation of a few chemicals for which the dietary route is dominant (Belfroid et al., 1994). In addition to potentially making extrapolations between soils more accurate than soil–worm partitioning, the pore water model has the advantages of making available for use the large literature on water–biota partitioning factors (BCFs) and the QSARs for water–biota partitioning. However, it adds the burden of estimating soil pore water concentrations. The conventional formula for estimating soil pore water concentrations is

$$C_w = C_s / K_d \tag{3.35}$$

where

K_d = the soil (or sediment)/water partitioning coefficient (l/kg sediment)
C_w = water concentration (mg/l)

Values of K_d are available from the literature for many metals and some organics but are highly variable (Baes et al., 1984). If literature K_d values are used, this model is not expected to be more accurate than the use of soil–earthworm uptake factors. However, K_d values are available for some chemicals for which soil–earthworm uptake factors are not. For nonionic organic compounds,

$$K_d = f_{oc} K_{oc} \tag{3.36}$$

or

$$K_d = f_{om} K_{om} \tag{3.37}$$

or, from Karickhoff (1981),

$$K_d = 0.58 f_{om} K_{oc} \tag{3.38}$$

TABLE 3.10
Summary of Sources of Earthworm Uptake Factors and Empirical Regression Models

Study Location	Analytes with Uptake Factors	Analytes with Regression Models	Ref.
Pennsylvania	Cadmium, copper, lead, nickel, and zinc	Cadmium	Beyer et al., 1982
Maryland		Lead, copper, cadmium, and selenium	Beyer et al., 1987
Finland	Aluminum, cadmium, copper, iron, mercury, manganese, vanadium, and zinc		Braunschweiler, 1995
Wales	Lead	Calcium, cadmium, copper, lead, and zinc	Corp and Morgan, 1991
Warsaw	Cadmium, copper, lead, and zinc		Czarnowska and Jopkiewicz, 1978
Germany	Cadmium, lead, and zinc		Emmerling et al., 1997
Denmark		Selenium	Nielsen and Gissel-Nielsen, 1975
Netherlands	Cadmium, copper, manganese, nickel, lead, and zinc		Hendriks et al., 1995
	Cadmium, chromium, copper, iron, manganese, nickel, lead, and zinc	Lead and zinc	Ma, 1982
	11 PAHs		Ma et al., 1998
		Cadmium, copper, lead, and zinc	Ma et al., 1983
Seveso, Italy	TCDD		Martinucci et al., 1983
Models fit to data from multiple locations		Cadmium, copper, nickel, lead, and zinc	Neuhauser et al., 1995
Montana	Arsenic, cadmium, copper, and zinc		Pascoe et al., 1996
Illinois	Cadmium, chromium, copper, nickel, and lead		Pietz et al., 1984
Models fit to data from multiple locations	Arsenic, cadmium, chromium, copper, mercury, lead, manganese, nickel, selenium, zinc, PCB, and TCDD	Arsenic, cadmium, chromium, copper, mercury, lead, manganese, nickel, selenium, zinc, PCB, and TCDD	Sample et al., 1999
Reading, U.K.	Cadmium, copper, lead, and zinc	Cadmium, copper, lead, and zinc	Spurgeon and Hopkin, 1996a
Tennessee	Cadmium, lead, zinc		Van Hook, 1974

where

f_{oc} = fraction organic carbon in the soil (unitless)
K_{oc} = water–soil organic carbon partitioning coefficient (kg/kg or l/kg)
f_{om} = fraction organic matter in the soil (unitless)
K_{om} = water–soil organic matter partitioning coefficient (kg/kg or l/kg)

This formula adjusts for the organic content of soil (expressed as either organic matter or organic carbon content), which is the major source of variance among soils in the uptake of neutral organic chemicals. This normalization makes this model more accurate than soil–earthworm uptake factors for neutral organic chemicals. For ionic organic chemicals, van Gestel et al. (1991) recommend correcting the coefficient (K_{oc} or K_{om}) by dividing by the fraction nondissociated (f_{nd}), which is estimated from

$$f_{nd} = 1/(1 + 10^{pH\text{-}pKa}) \qquad (3.39)$$

where

pK_a = the negative log of the dissociation constant

When K_{oc} and K_{om} are both unavailable, they may be estimated from QSARs. Models that have been used by the EPA were developed by DiToro et al. (1991):

$$\log_{10}(K_{oc}) = 0.983 \log_{10}(K_{ow}) + 0.00028 \qquad (3.40)$$

and by Karickhoff (1981):

$$\log_{10}(K_{oc}) = 0.989 \log_{10}(K_{ow}) + 0.346 \qquad (3.41)$$

where

K_{ow} = octanol–water partitioning coefficient (unitless)

Van Gestel et al. (1991) provide a formula for K_{om}:

$$\log_{10}(K_{om}) = 0.89 \log_{10}(K_{ow}) - 0.32 \qquad (3.42)$$

Values for K_{ow} are available in the literature for most organic chemicals, or they can be calculated from QSARs.

From these formulas, the earthworm concentration can be calculated as

$$C_v = K_{bw} C_w \qquad (3.43)$$

where

C_v = concentration in earthworms (mg/kg fresh weight)
K_{bw} = biota–water partitioning coefficient (l/kg organism)

K_{bw} values for chemicals in earthworms may be assumed equivalent to BCFs for aquatic invertebrates from the literature. Alternatively, QSARs can be used to estimate this factor. The model developed by Connell and Markwell (1990) for uptake by earthworms of 32 "lipophilic" organic chemicals (log K_{ow} 1.0 to 6.5) is

$$\log K_{bw} = \log K_{ow} - 0.6 \ (n = 60, r = 0.91) \tag{3.44}$$

It has been suggested that for lipophilic compounds, earthworm accumulation should also be a function of lipid content of the worms (Connell and Markwell, 1990). Worm lipid content (f_{lip}) and f_{om} were used with considerable success to normalize worm and soil concentrations when calculating BCFs for low-molecular-weight PAHs in field-collected earthworms (Ma et al., 1998). However, it is not clear how much of the reduction in variance is due to each of the factors. Worm lipid content is not a component of the standard sediment model and makes no contribution to predictive accuracy in practice because the site-specific lipid content of worms is unknown in nearly all cases and would vary in an unquantified manner seasonally and among species. However, based on a study of marine sediment oligochaetes (Markwell et al., 1989), Menzie et al. (1992) recommend a model for earthworms in soil that contains soil organic content and worm lipid content but not K_{ow} or any other property of the chemical:

$$K_{vs} = L\,(0.66 f_{oc})^{-1} \tag{3.45}$$

where

K_{vs} = worm–soil partitioning coefficient
L = proportion lipid in worms (unitless)

L was estimated by Menzie et al. (1992) to be 0.02, but Connell and Markwell (1990) used 0.0084 for their calculations.

This model predicts that all chemicals have equal concentrations in earthworms at a site, which was not far from true for the set of contaminants of concern at the site where it was applied. There, the mean BCFs for four DDT residues and total chlordane ranged from 0.10 to 0.35, and the estimated mean BCF for all chemicals was 0.25. The lipid-based model is not recommended because (1) the addition of L adds nothing without information that is seldom available (lipid content of field organisms), (2) L is likely to be highly variable across time and species, and (3) the deletion of K_{ow} or other chemical-specific information is not well justified. The model is discussed here because it has been widely adopted in the United States for estimating earthworm concentrations.

Finally, Connell (1990) proposed an extremely reduced formula:

$$K_{vs} = 0.44(K_{ow})^{0.05} \tag{3.46}$$

This model shows worm concentrations to be a weak function of K_{ow} but not of any soil or worm property. It would be appropriate only if the site soils were similar to the test soils used in the study from which this formula was derived (Lord et al., 1980).

In addition to uptake factors, empirical regression models have been developed for estimation of bioaccumulation in earthworms. With the exception of models for TCDD and PCBs (Sample et al., 1999), all of these models have been developed for inorganic contaminants. While most regression models were developed based on site-specific data, models by Neuhauser et al. (1995) and Sample et al. (1999) were developed based on published data from multiple locations with widely varying levels of soil contamination. Examples of available models are presented in Table 3.10. It is known that uptake of contaminants by earthworms may be influenced by soil parameters such as pH, organic matter, and exchange capacity (Beyer et al., 1987; Ma, 1987; Corp and Morgan, 1991), and these parameters have been incorporated into multiple regression models. However, model validation analyses by Sample et al. (1999) indicate that inclusion of these additional soil parameters at best marginally improves estimates over those obtained from simple regression models.

While mechanistic bioaccumulation models specific for earthworms are not currently available, generalized two- or three-compartment models (e.g., Goldstein and Elwood, 1971; Kowal, 1971; Ma et al., 1998) may be employed, if data are available for parameterization.

For any model used, the risk assessor must verify the units of the estimate from the source literature. While earthworm concentration values estimated from the soil–earthworm uptake factors and the empirical regression models reported above are expressed in dry weight, those based on soil pore water uptake factors are expressed in terms of fresh weight. Estimates generated by all methods reported above are for depurated worms. Because wildlife do not consume dry food, all dry weight estimates must be converted to mg/kg of wet weight before they are employed in exposure estimation

$$C_{\text{wet}} = C_{\text{dry}} * P_{\text{dry}} \tag{3.47}$$

where

C_{wet} = wet weight concentration
C_{dry} = dry weight concentration
P_{wet} = proportion dry matter content of worm or other tissue

Water content of earthworms is reported to range from 82 to 84% (EPA, 1993b). Concentrations may also be reported for undepurated worms, but there is no basis for correcting those values because of the variability in mass of ingested material.

3.9.4 Terrestrial Arthropod Uptake

Few models are available for the estimation of bioaccumulation by terrestrial arthropods. Examples of uptake or transfer factors from the literature are presented in Table 3.11. While no empirical regression models are currently available for arthropods, some mechanistic models have been developed. Examples of some simple mechanistic models for arthropods are presented in Kowal (1971), Van Hook and Yates (1975), Webster and Crossley (1978), and Janssen et al. (1991). Because these models are parameterized for only a few species and contaminants, application of these models for risk assessment purposes is problematic.

TABLE 3.11
Summary of Sources of Uptake Factors for Terrestrial Arthropods

Study Location	Taxa	Analytes with Uptake Factors	Type of Transfer	Ref.
Montana	Grasshoppers (Orthoptera, Acrididae)	Arsenic, cadmium, copper, zinc	Soil to invertebrate	Pascoe et al., 1996
Virginia	Ground-dwelling insects	Barium	Soil to invertebrate	Hope et al., 1996
Uptake factors based on data from published studies	Sowbugs (Isopoda), insect larvae, caterpillars, insects, spiders	Cadmium, DDT	Soil to invertebrate	Jongbloed et al., 1996
Austria	Sowbugs (Isopoda)	Copper	Diet to invertebrate	Wieser et al., 1976
Great Britain	detritivores (collembola, isopods, and earthworms), herbivores (flies and beetles), and carnivores (beetles and spiders)	Cadmium, copper	Diet to invertebrate	Hunter and Johnson, 1982
India	Hyacinth weevils (Coleoptera, Curcilionidae)	Cadmium, mercury, manganese, zinc	Diet to invertebrate	Jamil and Hussain, 1992
Tennessee	Sowbugs (Isopoda), termites (Isoptera), grasshoppers (Orthoptera, Acrididae), harvestmen (Phalangida), centipedes (Myriapoda), ground beetles (Carabidae)	Mercury	Diet to invertebrate	Talmage and Walton, 1993
Missouri	Litter invertebrates, spiders, other invertebrate predators	Cadmium, copper, lead, zinc	Diet to invertebrate	Watson et al., 1976

3.9.5 Terrestrial Vertebrate Uptake

To estimate the risks that contaminants present to wildlife that prey on terrestrial vertebrates, uptake models for birds, mammals, reptiles, and amphibians are needed. Uptake factors and models for reptiles and amphibians are not known and those for birds and mammals are few. Examples of uptake factors for estimation of chemical concentrations in whole biota based on food or soil are summarized in Table 3.12.

Uptake factors have also been developed to estimate chemical concentrations in animal tissues based on concentrations in food. While these factors are generally applied as part of exposure assessments for human health, they may be of use in ecological risk assessments. Baes et al. (1984) present diet-to-muscle and diet-to-

TABLE 3.12
Summary of Sources of Uptake Factors for Birds and Mammals

Study Location	Taxa	Analytes with Uptake Factors	Type of Transfer	Ref.
Montana	Small mammals	Arsenic, cadmium, copper, zinc	Soil to biota	Pascoe et al., 1996
Virginia	White-footed mouse (*Peromyscus leucopus*), hispid cotton rat (*Sigmodon hispidus*)	Barium	Soil to biota	Hope et al., 1996
Uptake factors based on data from published studies	Small mammals	14 inorganic and 2 organic chemicals	Soil to biota	Sample et al., 1998
Switzerland	Chickens (*Gallus* spp.)	7 dioxin and 10 dibenzofuran congeners	Soil to egg	Schuler et al., 1997
Uptake factors based on data from published studies	Birds, mammals	Cadmium, DDT	Food to biota	Jongbloed et al., 1996
Great Britain	Short-tailed vole (*Microtus agrestis*), woodmouse (*Apodemus sylvaticus*), common shrew (*Sorex araneus*)	Cadmium, copper	Food to biota	Hunter and Johnson, 1982
Great Britain	Short-tailed vole (*Microtus agrestis*), common shrew (*Sorex araneus*)	Lead, zinc	Food to biota	Andrews et al., 1989 a,b

milk uptake factors for inorganic contaminants. Additional factors for transfer of radionuclides to meat, milk, and eggs are reported in IAEA (1994).

Uptake of lipophilic organic contaminants may be estimated based on chemical-specific K_{ow} values. Garten and Trabalka (1983) present diet-to-fat uptake factors for 93 organic chemicals for sheep, poultry, small birds, rodents, dogs, cows, swine, and primates. Because these uptake factors produce estimates in terms of mg/kg fat, the resulting estimates must be converted from mg/kg fat to mg/kg body weight using species-specific lipid content data.

Only one source of empirical regression models for estimation of whole-body bioaccumulation by terrestrial vertebrates is known. Sample et al. (1998) developed and validated empirical regression models for 16 chemicals in small mammals. Separate models were developed for trophic groups (e.g., insectivore, herbivore, or omnivore) as data permitted. In comparison to the uptake factors generated from the same data, these regression models generally provided more accurate estimates of whole-body small mammal chemical concentrations.

Several mechanistic models of uptake of contaminants by terrestrial vertebrates have been developed. Models are available for only a few contaminants (e.g., PCBs, dioxins, mercury). Examples include a bioenergetic model for accumulation of PCBs by tree swallow nestlings (Nichols et al., 1995), a pharmacokinetic model linked to a mass-based food web model for estimation of PCB bioaccumulation by piscivorous wildlife (Ram and Gillett, 1993), and a probabilistic food web model for the estimation of body burdens of PCBs and mercury in great blue heron adults and eggs (Macintosh et al., 1994).

3.10 PETROLEUM AND OTHER CHEMICAL MIXTURES

Few contaminated sites are contaminated with only one chemical. Therefore, nearly all exposure assessments must deal with mixtures. However, most sites involve incidental mixtures of chemicals that are individually identified, analyzed, and modeled. Other sites are contaminated with mixtures such as crude petroleum, petroleum derived fuels, and PCB product mixtures (e.g., Aroclor®) that are generally treated as a single material. Analysis of exposure to these complex materials can be divided into a set of distinct questions. (1) How can chemical mixtures be analyzed so as to generate useful fate and exposure properties? (2) How can media be analyzed to provide useful estimates of exposure? (3) Might analyses of biota be used to estimate dietary exposure or internal exposure? (4) Might bioassays substitute for chemical analyses? (5) How might exposure and uptake be modeled from current concentrations, and how might future exposures be modeled?

The preliminary characterization of the released material should define as fully as possible the material released (e.g., North Slope Crude or PCB transformer oil). That characterization should be followed by analyses of the material in the contaminated media. Both the composition and concentration of the material should be determined. While the composition of the released material may be known, the weathering process may be quite rapid, so that by the time an assessment is conducted the composition may have changed considerably. The problem is deciding how to analyze the mixture in a way that both adequately characterizes the mixture and provides a basis for characterizing the fate, transport, and toxicological properties of the mixture. Three approaches are proposed: analysis of the whole material, analysis of chemical classes, and analysis of individual chemicals (Table 3.13).

The simplest approach is to analyze the whole material by determining total petroleum hydrocarbons (TPH), oil and grease, total PCB, or some equivalent metric of gross contamination. Such measures typically provide little basis for performing a risk assessment because they provide little information about the composition and therefore the properties that determine potential fate and toxicity of the material. However, these gross metrics may be useful for determining the extent of contamination or the locations of the greatest contamination. An exception to this general-

TABLE 3.13
Methods for Analyzing Chemical Mixtures and Characterizing Their Physical, Chemical, and Toxicological Properties

Mixture Analysis	Property Characterization
Whole material	Whole material properties
Chemical classes	Properties of representative chemicals
	Distribution of properties of class
Individual chemicals	Properties of detected chemicals
	Properties of indicator chemicals

ization is cases in which the composition of the whole material is well characterized, and the contamination is recent. Examples include synthetic lubricating oils and marketed PCB mixtures (e.g., Arochlor 1254®).

Rather than attempt to characterize the whole material, one may characterize constituent classes of compounds. Hydrocarbons are divided into aliphatics and aromatics; long- and short-chain aliphatics; one-, two-, three-, and more-than-three-ring aromatics, etc. In addition to hydrocarbons, petroleum contains metals, nitrogen-, sulfur-, and oxygen-containing organics, and other compounds. Analyses of these classes provide considerably more information than TPH or indicator chemicals, but to model transport and transformation of these chemical classes or to determine their toxicity, they must be associated with concentrations of particular chemicals. This can be done by identifying representative compounds for each class. The representative compounds should be selected based on the following criteria:

- They are an abundant member of the class of chemicals in the material being assessed.
- Data are available concerning their environmental fate and effects.
- They have greater-than-average (screening assessments) or average (definitive assessments) toxicity for their class, and greater-than-average (screening assessments) or average (definitive assessments) persistence and bioavailability.

Once representative chemicals are selected, the assessment can be performed by assuming that the entire mass of each class of chemicals is made up of that representative chemical.

If fate or effects data are available or can be estimated for several chemicals in a class, a more sophisticated approach could be used. The statistical distribution of the fate and effects properties of the members of a chemical class can be used in the assessment to represent the distribution of the properties in the entire class. For example, if water solubility values for several short-chain aliphatic hydrocarbons are found in the literature, a distribution fit to those data is an estimate of the distribution of water solubility for the class. Alternatively, if QSARs are available to estimate fate and effects properties, they could be used to estimate the parameter distributions for the class from the physical properties or structural characteristics of the individual chemicals in the class. For example, the water solubility of hydrocarbons can be estimated from their structure (Lyman et al., 1982). Therefore, by specifying the structures of short-chain aliphatic hydrocarbons, one can estimate the solubilities of all members of the class, and the distribution of those individual solubilities is the solubility distribution for the class. If the relative abundances of the class constituents can be estimated, the distribution can be refined by weighting the observations (e.g., the individual solubility estimates).

If the members of a chemical class have the same mechanism of action, concentrations of all of the chemicals may be normalized to an indicator chemical that has well-characterized properties using toxicity equivalency factors (TEFs). The only example from current practice is the normalization of exposure to all halogenated diaromatic hydrocarbons to 2,3,7,8-TCDD (EPA, 1993c; Newsted et al., 1995; Zabel et al., 1995; Van den Berg et al., 1998). By multiplying the concentration of each chemical in the mixture by its TEF (i.e., its toxicity as a proportion of TCDD toxicity) and summing the results, exposure to a mixture of PCBs, furans, and dioxins can be expressed as an equivalent concentration of 2,3,7,8-TCDD (TCDD equivalents). The resulting normalized concentrations are referred to as toxicity equivalents (TEQs). The TEF approach takes advantage of the fact that one member of a mixture is likely to be better characterized than the others.

Finally, a total analysis of the materials can be performed. For a petroleum-contaminated site, the analysis could include the hydrocarbons and other organic and inorganic compounds found in petroleum, as well as organic and inorganic additives (e.g., oxygenates), and the various chemicals that may occur in mixed wastes (e.g., drilling fluid components). This analytical approach offers the greatest flexibility in that all of the various exposure metrics discussed above can be reconstructed from a total analysis. In addition, if a site has been contaminated by chemicals other than the material of concern, a thorough chemical analysis may identify causes of toxicity that would not be revealed by a material-specific analysis. Finally, analytical quantitation limits are often lower for individual chemicals than for whole materials. For example, using EPA methods, quantitation limits for PCB congeners are three orders of magnitude lower than those for Arochlors and fall within the range of ecotoxicological benchmarks (Valoppi et al., 1999). However, it is likely that the majority of individual chemicals in any particular complex material will have unknown toxicity. Therefore, the risk characterization must be based on a subset of the detected chemicals.

One approach to characterizing the properties of the mixture from analysis of individual chemicals is to identify indicator chemicals (ASTM, 1994). These are chemicals that are assumed to account for the major risks from the mixture. That is, if risks from those chemicals are acceptable, then the risks from the whole mixture are acceptable. For risk assessments of petroleum products, ASTM recommends using benzene and benzo(*a*)pyrene, and, in cases where they are present, ethylene dibromide (EDB) and ethylene dichloride (EDC), because of their carcinogenicity. Clearly, other criteria would need to be used to choose indicator chemicals for ecological risk assessments. Similarly, EPA Region IX has a list of 28 PCB congeners that are of concern with respect to ecological risks (Valoppi et al., 1999).

Another sort of indicator chemical approach is the analysis of chemicals that are characteristic of the complex material. For example, the exposure of kangaroo rats to petroleum from a well blowout was confirmed by analyzing liver tissue for a set of PAHs that are characteristic of oil and coal (Kaplan et al., 1996). Vanadium, which is typically concentrated in petroleum, was found to be elevated in San Joaquin kit foxes from oil fields relative to foxes from other sites, suggesting that the foxes were exposed to petroleum in some form (Suter et al., 1992). Chemical fingerprinting, comparison of the distributions of the abundances of PAHs, was used to determine whether hydrocarbons found in biological samples from the vicinity of the *Exxon Valdez* spill were from the spilled oil, from diesel fuel, or analytical artifacts (Bence and Burns, 1995). These approaches are useful for confirming that exposure has occurred, that exposure is associated with a particular source, or that the material is being transported by a particular route. However, they are not adequate exposure metrics for risk assessment because they cannot be related to exposure–response relationships to predict toxicity.

3.11 SUMMARY OF EXPOSURE CHARACTERIZATION

Since exposure characterization is an intermediate analytical stage in ecological risk assessment, its results should primarily be input to the next stage (risk characterization) rather than voluminous text, tables, or figures. The results are presented as an exposure profile (EPA, 1998). It should summarize, for each endpoint receptor, the exposure pathways addressed to ensure that all pathways included in the conceptual model have been analyzed. If not, the conceptual model should be modified to remove pathways that have been determined to be absent or insignificant. Risk assessors should not claim that their conceptualizations of the exposure process are more complete than is in fact the case. The exposure profile should describe how exposure was estimated for each pathway, summarize the results, and present the associated uncertainties. If multiple lines of evidence are used in the risk characterization of an assessment endpoint, proper exposure metrics should be derived for each. In general, a tabular summary is an appropriate format for the exposure characterization. For each line of evidence and endpoint, it would present exposure media, routes, point estimates of exposure intensity (e.g., mean concentration, median dose, maximum concentration), distributions in space and time, and uncertainties (e.g., sampling variance, analytical precision, model uncertainties).

When performing an exposure characterization for a preliminary phase of a multiphased assessment (e.g., development of a definitive assessment plan or feasibility study) or when presenting preliminary results to the public, it is often useful to summarize the temporal and spatial distribution of contaminants graphically or cartographically. However, since presentations of distributions of exposure may be misleading or cause undue concern, results of the exposure characterization should not be emphasized when risk estimates are available for presentation.

4 Analysis of Effects

What is there that is not poison?
All things are poison, and nothing is without poison.
Solely the dose determines that a thing is not a poison.

—Paracelsus, translation by Deichmann et al. (1986)

In the analysis of effects, assessors determine the nature of toxic effects of the contaminants and their magnitude as a function of exposure. Effects data might be available from field monitoring, from toxicity testing of the contaminated media, and from traditional single-chemical laboratory toxicity tests (Table 4.1). The assessor must evaluate and summarize the data concerning effects in such a way that it can be related to the exposure estimates, thereby allowing characterization of the risks to each assessment endpoint during the risk characterization phase.

In the analysis of effects, available effects data must be evaluated to determine which are relevant to each assessment endpoint, and they must be reanalyzed and summarized as appropriate to make them useful for risk characterization. Two issues must be considered. First, what form of each available measure of effect best approximates the assessment endpoint? This issue should have been considered during the problem formulation. However, the availability of unanticipated data and better understanding of the situation after data collection often require reconsideration of this issue.

The second issue in analysis of effects is expression of the effects data in a form that is consistent with expressions of exposure. Integration of exposure and effects defines the nature and magnitude of effects, given the spatial and temporal pattern of exposure levels. Therefore, the relevant spatial and temporal dimensions of effects must be defined and used in the expression of effects. For example, if the exposure is to a material such as unleaded gasoline that persists at toxic levels only briefly in soil, then effects that are induced in that time period must be extracted from the effects data for the chemicals of concern, and the analysis of field-derived data should focus on biological responses such as mass mortalities that could occur rapidly rather than long-term average properties.

The degree of detail and conservatism in the analysis of effects depends on the tier of the assessment (Chapter 1). Scoping assessments need only determine qualitatively that an effect may occur because a receptor is potentially exposed to one or more contaminants. Screening assessments typically define the exposure–effects relationship in terms of a benchmark value, a concentration that is conservatively defined to be a threshold for toxic effects (Chapter 5). Definitive assessments should define the exposure–response relationship (Chapter 6) and should separately estimate the uncertainty concerning that relationship (Chapter 7).

TABLE 4.1
Types of Effects Data Used in Ecological Risk Assessments of Contaminated Sites and Sources of the Data

Type	Source
Single-chemical toxicity	Published scientific literature reporting results of toxicity tests with individual chemicals or materials and summarizations of that literature such as water quality criteria
Ambient media toxicity	Site-specific *in situ* or laboratory toxicity tests of contaminated water, sediment, soil, or food
Biological survey	Site-specific sampling or observation of organisms, populations, or communities in contaminated areas

4.1 SINGLE-CHEMICAL OR SINGLE-MATERIAL TOXICITY TESTS

In ecological risk assessments for contaminated sites, single-chemical or single-material (e.g., gasoline) toxicity data are usually obtained from the literature or from databases rather than generated ad hoc. One source is the EPA ECOTOX database, which contains toxicity data for aquatic biota, wildlife, and terrestrial plants. It is available from the EPA and commercial sources (http://www.epa.gov/medectox). Assessors must select data that are most relevant to the assessment endpoints and that can be used with the exposure estimates. As far as possible, data should be selected to correspond to the assessment endpoint in terms of taxonomy, life stage, response, exposure duration, and exposure conditions. However, because the variance among chemicals is greater than the variance among species and life stages, any toxicity information concerning the chemicals of interest is potentially useful. If no toxicity data are available that can be applied to the assessment endpoints (e.g., no data for fish or no reproductive effects data), or if the test results are not applicable to the site because of differences in media characteristics (e.g., pH or water hardness), tests may be conducted ad hoc. However, most tests performed for specific sites are tests of local contaminated media (Section 4.2) rather than single chemicals. If combined toxic effects of multiple contaminants are thought to be significant, and if appropriate mixtures are not available in currently contaminated media, synthetic mixtures may be created and tested.

Toxicity tests of single chemicals that are obtained from published literature have biases that should be understood by ecological risk assessors. Assessors must be aware of these biases when these data are used to derive toxicity benchmarks or exposure–response models for chemicals. Potential sources of bias in the test data include the following:

- The forms of chemicals used in toxicity tests are likely to be more toxic than the dominant forms at hazardous waste sites. For metals the tested forms are usually soluble salts, and organic chemicals may be kept in aqueous solution by cosolvents. In dietary or oral dosing tests organic chemicals are typically dissolved in readily digested oils.

- Combined toxic effects are not observed in toxicity tests of single chemicals.
- The test species for toxicity testing may not be representative of the sensitivity of species native to the site.
- The standard media used in toxicity tests may not be representative of those at a particular contaminated site. For example, aqueous tests typically use water with moderate pH and hardness with little suspended or dissolved matter, and soil tests typically use agricultural loam soils or artificial soils.
- Laboratory test conditions may not be representative of field conditions (e.g., temperature, use of sieved soil, and maintenance of constant moisture).

4.1.1 Types of Toxicity Tests

Conventionally, toxicity tests determine effects on individual organisms and are divided into two classes, acute and chronic. Acute tests are those that last a small proportion of the life span of the organism (<10%) and involve a severe effect (usually death) on a large proportion of exposed organisms (conventionally, 50%). Acute tests also usually involve well-developed organisms rather than eggs, larvae, or other early life stages. Chronic tests include much or all of the life cycle of the test species and include effects more subtle than death (e.g., reduced growth and fecundity). In these tests, the endpoint is typically based on statistical significance, so the proportion affected may be large or small. In addition, there are many tests that fall between these two types, which are termed subchronic, short-term chronic, etc. They typically have short durations but include sublethal responses. A prominent example is the 7-day fathead minnow test, which includes growth as well as death. This test includes only part of one life stage but uses the larval rather than juvenile stage and statistical significance rather than effects levels to derive the test endpoint (Norberg and Mount, 1985).

In general, tests with longer durations, more life stages, and more responses reported are more useful for risk assessment, because they provide more information and because the exposures at contaminated sites are typically chronic. However, if exposures are acute, then acute tests should be preferred. Examples include exposures of transients such as migratory waterfowl or highly mobile species that may use a site in transit or exposures during episodes of contamination, such as overflow of waste ponds or flushing of contaminants into surface waters by storms.

Following are general recommendations for selecting toxicity tests of single chemicals or materials. Other issues specific to tests of particular media are addressed later.

Standardization — In general, choose standard tests. Standard test protocols have been developed or recommended by governments (Keddy et al., 1995; EPA, 1996a) and standards organizations (OECD, 1998; APHA, 1999; ASTM, 1999). Most extrapolation models for relating test endpoints to assessment endpoints require standard data (Section 4.1.9.1). In addition, results of standard tests are likely to be reliable because of the QA/QC procedures that are part of the standard methods, and because test laboratories are likely to conduct standard tests routinely. However, nonstandard tests should be used when particular site-specific issues cannot be resolved by standard test results.

Duration — Choose tests with appropriate durations. Two factors are relevant. The first is the duration of the exposures in the field. If exposures are episodic, as is often the case for aqueous contamination, then tests should be chosen with durations as great as the longest episodes. The second factor is the kinetics of the chemical. Some chemicals such as chlorine in water or low-molecular-weight narcotics are taken up and cause death or immobilization in a matter of minutes or hours. Others such as dioxins have very slow kinetics and require months or years to cause some effects such as reproductive decrements. In general, longer durations (i.e., chronic tests) are preferred, but these site-specific considerations may override that generality.

Response — Choose tests with appropriate responses. In particular, if an apparent effect of the contaminants has been observed in field studies, tests that include that effect as a measured response should be used. More generally, chosen tests should include responses that are required to estimate the assessment endpoint. Since most tests are of collections of organisms, and assessment endpoints are usually defined at the population or community level, choose responses that are relevant to higher levels of organization including mortality, fecundity, and growth. Physiological and histological responses are generally not useful for estimating risks, because they cannot be related to effects at higher levels. However, if they are characteristic of particular contaminants, they may be useful for diagnosis (Chapter 6).

Consistency — Prefer tests matching ambient media tests performed at the site (Section 4.2). For example, if 7-day fathead minnow larval tests are performed with ambient water, use of the same tests with individual chemicals can help in interpretation of results.

Media — Prefer tests conducted in media with physical and chemical properties similar to the site media.

Organisms — Prefer taxa and life stages that are closely related taxonomically to the endpoint species. If an assessment endpoint is defined in terms of a community, one may either choose tests of species that are closely related to members of the community or use all high quality tests in the hope of representing the distribution of sensitivity in the endpoint community (Section 4.1.9.1). Species, life stages, and responses should also be chosen so that the rate of response is appropriate to the duration of exposure and kinetics of the chemical. In general, responses of small organisms such as zooplankters and larval fish are more rapid because they achieve a toxic body burden more rapidly than larger organisms. Therefore, if exposures are brief and if those small organisms are relevant to the assessment endpoint, tests of small organisms should be preferred over larger organisms that are no more relevant. However, such tests may not be appropriate if, for example, the endpoint is fish kills, or exposures do not occur during the breeding period of fish.

Multiple exposure levels — Studies that employ only a single concentration or dose level plus a control are seldom useful. If the exposure causes no effect, it may be considered a no observed effects level (NOEL), but no information is obtained about levels at which effects occur. Conversely, if the exposure causes a significant effect, it may be considered a lowest observed effects level (LOEL), but the threshold for effects cannot be determined. Studies in which multiple exposure levels were applied allow an exposure–response relationship to be evaluated and NOELs and

LOELs to be determined. Consequently, studies that apply multiple exposure levels are strongly preferred.

Exposure quantification — To interpret the results of toxicity tests correctly and to apply these results in risk assessments, the exposure concentrations or doses should be clearly quantified. Ideally, the test chemical should be measured at each exposure level; measured concentrations are always preferable to nominal concentrations.

Chemical form — Correct estimation of the dose requires that the form of toxicant used in the test be clearly described. For example, in tests of lead, the description of the dosing protocol should specify whether the dose is expressed in terms of the element (e.g., lead) or the applied compound (e.g., lead acetate). Tests of chemicals in the forms occurring on the site should be preferred. This is particularly important for chemicals that may occur in multiple forms under ambient conditions that have widely differing toxicities.

Statistical expressions of results — The traditional toxicity test endpoints for chronic tests, NOELs and LOELs, have low utility for risk assessment (Suter, 1996a). NOELs are the highest exposure levels at which no effects are observed to differ statistically significantly from controls, while LOELs are the lowest exposure levels at which one or more effects are observed to differ statistically significantly from controls. These endpoints do not indicate whether the statistically significant effect is, for example, a large increase in mortality or a small decrease in growth. The level of effect at a NOEL or LOEL is an artifact of the replication and dosing regime employed. Use of the NOEL or LOEL does not indicate how effects increase with increasing exposure, so the effects of slightly exceeding a NOEL or LOEL are not qualitatively or quantitatively distinguishable from those of greatly exceeding it. To estimate risks, it is necessary to estimate the nature and magnitude of effects that are occurring or could occur at the estimated exposure levels. To do this, exposure–response relationships should be developed for chemicals evaluated in ecological risk assessments. Methods for fitting of exposure–response distributions to toxicity data are presented in Crump (1984), Kerr and Meador (1996), Moore and Caux (1997), and Bailer and Oris (1997).

In some cases these criteria may conflict. Hence, assessors must determine their relative importance to the particular site and assessment, and apply them accordingly.

4.1.2 Aquatic Tests

More toxicity tests are available for aquatic biota than any other type of receptor. In general, flow-through tests are preferred over static-renewal tests, which are preferred over static tests. Flow-through tests maintain constant concentrations, whereas concentrations may decline significantly in static tests. However, in a few cases, static tests are appropriate, because exposure is static, as in the spillage of a chemical into a pond. The most abundant type of test endpoint is the 48- or 96-h LC50. However, chronic test results are more generally useful. They include life cycle tests and, for fish, early life-stage tests. Currently, the most popular aquatic test organisms in the United States are fathead minnows (*Pimephales promelas*) and daphnids (*Daphnia* and *Ceriodaphnia* spp.). Test results for algae or other aquatic plants are less often available. Aquatic microcosm and mesocosm test data are rare, and largely limited to pesticides.

4.1.3 Sediment Tests

Selecting representative sediment tests and test results is complicated by the interactions among the multiple phases (i.e., particles, pore water, and overlying water) of the sediment system. Chemicals can be tested either in the presence (spiked-sediment tests) or absence (aqueous) of the solid phase. Test selection depends on the expected mode(s) of exposure, and more than one test type may be appropriate. Spiked-sediment tests consist of the addition of known quantities of the test chemical or material to a natural or synthetic sediment to which the test organism is exposed. Spiked-sediment tests provide an estimate of effects based on all direct modes of exposure, including ingestion, respiration, and absorption. Hence, toxicity to sediment-ingesting organisms may be best approximated by bulk sediment tests. The primary disadvantage is that the exposure–response relationship is somewhat uncertain due to the unquantified effects of the sediment matrix (Ginn and Pastorok, 1992). Aqueous phase tests are most appropriate if interstitial or overlying water is believed to be the primary exposure pathway for the toxicants and receptors at a site.

As noted in Section 4.1.2, aqueous tests and data are more abundant than any other kind. Most of the species tested live in the water column rather than the sediment. Aqueous tests and data are used to evaluate aqueous exposures of benthic species, based on data suggesting that benthic species are not systematically more sensitive than water column species (EPA, 1993a). The types of aqueous tests and factors to consider in selecting a test type are discussed in Section 4.1.2 and apply here as well.

Sediment and water tests are available for marine and freshwater species (Section 4.2.2). Risk assessors should choose tests in media similar to the site media. Unlike aqueous toxicity data, which are relatively abundant for both fresh water and salt water, there are few test data from freshwater sediment tests relative to estuarine sediment tests. Therefore, it is necessary to consider whether to use saltwater toxicity values for assessments of freshwater systems. Klapow and Lewis (1979) applied a statistical test of medians to freshwater and marine acute toxicity data for nine heavy metals and nonchlorinated phenolic compounds. In only one case (cadmium) was there a statistically significant difference in the median response of marine and freshwater organisms. On the other hand, Hutchinson et al. (1998) found potentially important differences. They compared the aqueous toxicity of several heavy metals, pesticides, and organic solvents to freshwater and saltwater invertebrates—83% of the no observed effects concentrations (NOEC) and 33% of the 50% effects concentrations (EC50) for freshwater and saltwater invertebrates were within a factor of 10. Based on the ratios of EC50s, freshwater invertebrates were more sensitive than saltwater invertebrates to four (2-methylnaphthalene, 1-methylnaphthalene, benzene, and chromium) of the 12 evaluated chemicals. Comparison of NOECs indicated that two (copper and cadmium) of the six chemicals for which sufficient data were available to allow comparison were more toxic to freshwater invertebrates than to saltwater invertebrates. The authors emphasized that the results should be considered preliminary because of the limited amount of appropriate data. The bottom line is that cautiously using data from tests of saltwater sediments to evaluate chemicals in freshwater sediments is probably better than having no data at all in the preliminary stages of an assessment. There is precedent for this in the use of effects range–low values from estuarine and marine sediments (Long et al., 1995) as ecotox thresholds for both marine and freshwater sediments (Office of Emergency and Remedial Response, 1996).

The physical and chemical properties of the test media are particularly important for evaluating chemical toxicity in the sediment system. Characteristics of the sediment (e.g., organic carbon content and grain size distribution) and water (e.g., dissolved organic carbon, hardness, and pH) can significantly alter the speciation and bioavailability of the tested material. Again, tests in media similar to the site media should be preferred. Regression models could be derived to account for confounding matrix factors (e.g., grain size or organic carbon content) (Lamberson et al., 1992). However, such models are species- and matrix factor-specific and would need to be developed on a case-by-case basis. This is not practical for most hazardous waste site assessments, especially for adjustments of multiple variables. The test method also can affect exposure. For example, chemical concentrations and bioavailability can be altered by the overlying water turnover rate, the water-to-sediment ratio, and the oxygenation of the overlying water (Ginn and Pastorok, 1992). Issues associated with sediment toxicity testing are discussed in detail elsewhere (Burton, 1992).

4.1.4 Soil Tests

The available body of soil toxicity tests is relatively small and poorly standardized. For example, few organic chemicals other than pesticides are represented. Soil toxicity test data for inorganic chemicals and some organic compounds are available for plants (mainly crops), soil invertebrates (primarily earthworms), and soil microorganisms (usually expressed as changes in rates of carbon mineralization, nitrification, nitrogen fixation, or other processes).

Tests in both soil and soil solution may be useful for assessing risks from soil contaminants. The relevance of published tests in soil to the assessment of risks to soil organisms seems self-evident, but, unless the properties of the test soil are similar to those of the site soil, the toxicity observed in the test soil concentration may be poorly correlated with effects at the site. For example, Zelles et al. (1986) found effects of chemicals on microbial processes to be highly dependent on soil type. Moreover, it is usually desirable for the assessor to exclude data from tests in quartz sand or vermiculite, unless toxicity of chemicals mixed with these materials is demonstrated to be similar to that in natural soils. Tests conducted in solution have potentially more consistent results than those conducted in soil. Toxicity observed in inorganic salts solution may be related to concentrations in soil extracts, estimated pore water concentrations, or springs where wetland plant communities are located. It has even been proposed that aquatic toxicity test results could be used to estimate the effects of exposure of plants and animals to contaminants in soil solution (van de Meent and Toet, 1992; Lokke, 1994), although we do not recommend this practice.

The risk assessor should be aware that bioavailability in soil from the contaminated site may be substantially different from the bioavailability in published soil tests. As stated in Section 3.4.1, aged organic chemicals are typically less available and less toxic to biota than organic chemicals freshly added to soil in published toxicity tests (Alexander, 1995); thus, the toxicity at the contaminated site may be overestimated if a published toxicity test of a chemical freshly added to soil is emphasized too heavily in the assessment. The risk assessor can make adjustments to observed toxic concentrations to account for differences in soils or chemical speciation. The variance in toxicity among natural soils may be reduced by normalizing the test soil concentrations to match normalized site soil concentrations (Section 3.4.1.1). Or free metal activities in soil solution may be estimated, potentially improving the precision of toxic thresholds for plants, soil invertebrates, or microbial processes (Sauvé et al., 1998). The assessor may be more liberal in including tests in screening assessments (e.g., in the derivation of screening benchmarks) than in definitive assessments. In definitive assessments, soil type and chemical speciation should be factors in decisions about the acceptability of data.

Tests should be chosen for risk assessments based on a relationship to the assessment endpoint. For example, if the assessment endpoint is production of the plant community, tests relating to plant growth or yield or mycorrhizal biomass may be sufficiently relevant to the endpoint, but tests of DNA damage would probably not be. Tests of litter-feeding earthworms may not be representative of those that ingest soil, and vice versa. Similarly, it is not always clear that microbial communities that have become altered in their tolerance of contaminants (pollution-induced community tolerance, PICT; Rutgers et al., 1998) are indicators of a decrease in the rate of a valued microbial process (Efroymson and Suter, 1999). Microcosm tests of the soil community and processes such as decomposition incorporate indirect effects of chemical addition as well as direct toxic effects (Sheppard and Evenden, 1994; Bogomolov et al., 1996; Parmelee et al., 1997; Salminen and Sulkava, 1997; Weeks, 1998). In addition, in microcosms, the responses of communities may be observed directly rather than deduced from effects on single populations of invertebrates.

The assessment endpoint may include a defined level of effects such as a 20% reduction in some endpoint property. However, such a decrease in the rates of some microbial processes such as litter decomposition may be desirable (or acceptable) in particular ecosystems (Efroymson and Suter, 1999); thus, an appropriate level of effects is sometimes unclear. Moreover, the desired level of effect is seldom obtainable from soil toxicity test results in the literature. Frequently, the EC50 is reported, but lower-level or lower-percentile effects are not. Often the lowest observed adverse effects level is a 50% effects level or higher, and lower concentrations (other than the reference) were not tested. No good models for estimating an EC20 from an EC50 exist for plants, earthworms, or other soil organisms. For example, the shape of the dose–response curve may be affected by whether the chemical is an essential element or whether detoxification occurs. It is advisable for the assessor either to use a safety factor or retain the uncertainty associated with the single-chemical toxicity test line of evidence during the risk characterization (Section 4.1.9.2).

4.1.5 Dietary and Oral Tests

Dietary and oral toxicity tests are those in which test animals are exposed to toxicants orally in food, water, or another carrier, with the organ of uptake being the gastrointestinal tract. These tests are employed primarily with birds and mammals and are rarely applied to aquatic organisms.

For dietary tests, the toxicant is mixed with food or water and test animals are allowed to feed *ad libitum*. The amount of food consumed daily should be recorded so that the daily dose can be estimated. A potential problem with dietary tests is that animals may not experience consistent exposure throughout the course of the study. For example, as animals become sick, they are likely to consume less food and water. They may also eat less or refuse to eat if the toxicant imparts an unpleasant taste to the food or water or if the toxic effects induce aversion.

In oral tests, animals receive periodic (usually daily) toxicant doses by gavage (i.e., esophageal or stomach tube) or by capsules. The chemical is generally mixed with a carrier (e.g., water, mineral oil, acetone solution, etc.) to facilitate dosing. Oral tests assure consistent daily doses of test chemicals including those that are repellent or aversive.

The choice of carrier used for oral or dietary tests has been shown to influence uptake by binding with the toxicant or otherwise influencing its absorption. For example, Stavric and Klassen (1994) report that the uptake of benzo(*a*)pyrene by rats is reduced by both food and water but facilitated by oil. Similarly, uptake of inorganic chemicals varies dramatically between tests with food and with water as carriers. Chemicals are generally taken up more readily from water than from food.

Results of most dietary toxicity tests are presented as toxicant concentrations (mg/kg) in food or water. These data can then be converted into doses (mg toxicant/kg body weight/day) by multiplying the concentrations in food or water by food ingestion rates and body weights either reported in the literature or presented in the study (e.g., Sample et al., 1996a). Fairbrother and Kapustka (1996) argue that uncertainty in food consumption rates, particularly in response to toxicity, precludes the accurate estimation of dose, and therefore concentration data should not be converted to dose. They are correct in indicating that the conversion is a significant source of uncertainty. However, toxicity data expressed as concentrations cannot be readily compared to multimedia contaminant exposure estimates (Section 3.10). Therefore, the conversion of concentration to dose is recommended, unless only one source of exposure is significant. Conversion of results from most oral toxicity tests is not needed as the results are generally expressed as dose in mg/kg/day or equivalent metrics.

Standard methods for performing avian and mammalian oral toxicity tests have been developed and are generally applied for testing of drugs, pesticides, and other chemicals. While standard test methods specifically developed for wildlife at hazardous waste sites do not exist, existing standard laboratory tests may be modified and applied. These tests vary from acute tests to subacute dietary tests to developmental and reproductive tests. A summary of selected standard oral test methods is presented in Table 4.2.

4.1.6 Body Burden–Effect Relationships

Single-chemical toxicity tests may be used to develop exposure–response relationships based on internal exposure measures (body burdens), rather than on external exposures (media concentrations or administered doses). In theory, this approach

TABLE 4.2

Selected Standard Oral Toxicity Methods for Birds and Mammals

Taxon	Test Type	Test Species	Duration	Exposure Route(s)	Test Endpoint(s)	Ref.
Mammal	Acute	Not stated	Single dose, 14 day post	Gavage	Mortality	ASTM, 1999
		Rats	Single dose, 7 day post	Gavage	Mortality	ASTM, 1999
	Subacute dietary	Rats	90 day	Capsule, gavage, in diet, in water	Mortality, organ pathology, behavior	ASTM, 1999
	Developmental	Rats, rabbits	Day 6–15 of gestation (rats) Day 6–18 of gestation (rabbits)	Capsule, gavage	Fertility, fetal body weights, number of dead fetuses, number of malformed fetuses	ASTM, 1999
Bird	Subacute dietary	Northern bobwhite, Japanese quail, mallard, ring-necked pheasant	5 day exposure, 3 day post	In diet	Mortality, but other effects can also be considered	ASTM, 1999
	Reproduction	Northern bobwhite, mallard	10 week	In diet	Adult mortality, eggs laid, egg fertility, egg hatchability, eggshell thickness, weight and survival of young	ASTM, 1999, EPA, 1991b

offers considerable advantages. Chemicals cause toxic effects in the organism, so measures of internal exposure should be more predictive of effects than measures of external exposures (McCarty and Mackay, 1993). Estimation of effects from body burdens potentially bypasses all of the variance among sites, species, and individuals associated with the physical, chemical, physiological, and behavioral processes that control intake, uptake, and retention of chemicals. The body burden approach is particularly relevant to chemicals that may be accumulated by aquatic biota through food intake as well as direct exposure to the chemical in water.

In theory, all chemicals acting by the same mechanism of action should be effective at the same molar concentration at the site of action, or the same concentration adjusted for relative potency. If all internal compartments (e.g., muscle, fat, blood plasma) are in equilibrium and have roughly the same relative size across individuals and species, the absolute or adjusted whole-body effective concentration should be the same for all chemicals with the same mechanism of action. Finally, if all individual molecules of chemicals with the same mechanism of action have the same potency, then effective molar concentrations should be constant. These assumptions underlie the compilation of estimated critical body residues for eight groups of chemicals in fish presented in Table 4.3. These thresholds may be used to estimate whether measured body burdens of organic chemicals with known mechanisms of action are likely to be associated with acute or chronic effects. Like all toxicity benchmarks, these should be used with caution, and the original sources should be consulted before using these values to estimate risks. For example, body burdens of 2,3,7,8-TCDD varied 122-fold at the time of death in fathead minnows (Adams, 1986). This variation was apparently due to an interaction between concentration and duration in determining lethality.

If the mechanism of action is unknown or not included in Table 4.3, one may assume that the toxicity of a chemical is at least as great as that of chemicals acting by baseline narcosis. Baseline narcosis is a nonspecific mechanism of action based, apparently, on nonspecific binding to cell membranes and subsequent disruption of membrane function. Since all organic chemicals have at least that level of toxicity, body residues of any organic chemical of 0.8 mmol/kg (the upper limit for chronic narcosis; Table 4.3) or greater is clearly indicative of chronic toxicity in fish. However, since chemicals may have more powerful specific modes of action, concentrations less than 0.2 mmol/kg (the lower limit for chronic narcosis; Table 4.3) cannot be assumed to be nontoxic.

Interpretation of body burdens of metals is more problematic (McCarty and Mackay, 1993). Because of the nutrient role of many metals and the numerous processes that control metal uptake, depuration, distribution, and sequestration, effective concentrations are highly variable, even when measured at the presumed primary site of action for most metals, the gills (McCarty and Mackay, 1993; Bergman and Dorward-King, 1997). However, exposure–response relationships for metal body burdens may be used as a line of evidence in risk assessments. These relationships are no less reliable than simple concentration–response relationships for metal concentrations in water.

There are no standard benchmarks for effects on fish of internal exposures. The body burdens associated with effects in published reports of toxicity tests and field

TABLE 4.3
Summary of Modes of Toxic Action and Associated Critical Body Residue Estimates in Fish

Chemical and effect	Estimated residue (mmol/kg)
Narcosis	
Acute (summary)	2–8
Chronic (summary)	0.2–0.8
Acute (octanol, MS222)	1.68 or 6.32[a]
Polar narcosis	
Acute (summary)	0.6–1.9
Acute (2,3,4,5-tetrachloroaniline)	0.7–1.8
Chronic (summary)	0.2–0.7 (chronic/acute = 0.1–0.3)
Chronic (2,4,5-trichlorophenol)	0.2
Acute (aniline, phenol, 2-chloroaniline, 2,4-dimethylphenol)	0.68 or 1.76
Respiratory uncoupler	
Acute (pentachlorophenol)	0.3
Acute (2,4-dinitrophenol)	0.0015 or 0.2
Chronic (pentachlorophenol, 2,4-dinitrophenol)	0.09–0.00015 (chronic/acute = 0.1–0.3)
Chronic (pentachlorophenol)	0.094 0.08
Acute (pentachlorophenol, 2, 4-dinitrophenol)	0.11 or 0.20
AChE inhibitor	
Acute (malathion and carbaryl, chlorpyrifos)	0.5 and 2.7
Acute (chlorpyrifos)	2.2
Acute (aminocarb)	0.05 and 2
Acute (parathion in blood)	0.13–0.2
Chronic (chlorpyrifos)	0.003
Acute (malathion, carbaryl)	0.16 or 0.38
Membrane irritant	
Acute (benzaldehyde)	0.16 2.1 or 13.2
Acute (acrolein)	0.0014 or 0.94
CNS convulsant[b]	
Acute (fenvalerate, permethrin, cypermethrin)	0.002–0.017 0.000048–0.0013
Acute (endrin in blood)	0.0007
Acute (endrin)	0.0018–0.0026 0.005
Chronic (fenvalerate, permethrin)	0.0005 and 0.015
Respiratory blockers	
Acute (rotenone)	0.0006–0.003 0.008 0.0009 or 0.0028
Dioxin (TCDD)-like	
Lethal (TCDD)	0.000003–0.00004
Growth/survival (TCDD)	0.0000003–0.0000008
Early life stages, lethal (TCDD)	0.00000015–0.0000014
Early life stages, NOAEL (TCDD)	0.0000001–0.0000002

Note: The rainbow trout used in this study weighed 600 to 1000 g; the other data presented are largely for small fish, sometimes early life stages, that typically weighed less than 1 g. Most estimates were converted from mass-based data.

[a] The two values represent residues estimated by two different methods.

[b] Includes three subgroups characterized by strychnine; fenvalerate and cypermethrin; endosulfan and endrin.

Source: McCarty, L. S. and Mackay, D., *Environ. Sci. Technol*., 27, 1719, © 1993. With permission of the American Chemical Society.

studies and body burdens reported for uncontaminated sites should be presented in the toxicity profiles. In addition to the values in Table 4.3, body burdens associated with effects are presented in many of the EPA water quality criteria documents. To be consistent with EPA practices in calculating chronic values (CVs), thresholds for toxic effects can be expressed as geometric means of body burdens measured at the NOEC and lowest observed effects concentration (LOEC). However, other expressions that are more clearly related to effects may also be used. Effective body burdens for a variety of chemicals in sediments are presented in the Environmental Residue–Effect Database (http://www.wes.army.mil/el/ered/index.html). A compilation of body burden and effects data for aquatic toxicity tests is presented in Jarvinen and Ankley (1999).

The use of chemical concentrations in plant tissues to estimate effects may be advantageous. Measurement of tissue concentrations permits the assessor to ignore the very large differences in bioavailability of chemicals in different soils as well interspecies differences. For example, phytotoxicity of metals in soils of low organic matter is not a good predictor of the toxicity of metals in sludge-amended soils. Chang et al. (1992) reviewed the literature and developed empirical models relating concentrations of copper, nickel, and zinc in crop foliage to growth retardation.

Although body burden–effects data are usually obtained from the literature, as discussed above, it is also possible to generate them at the site. As part of the biological surveys (Section 4.3), animals or plants may be collected, examined for signs of toxic effects, and subjected to chemical analysis. A function relating body burdens to the severity or frequency of observed effects may be developed, or a maximum body burden associated with no observable effects may be established. This approach is potentially more reliable than the use of literature values, but must be used with care. For mobile species, the time that the collected individuals have spent on the contaminated site must be considered. In addition, it must be realized that the most sensitive individuals and species may have been eliminated from the site by toxic effects, leaving only resistant organisms. These two phenomena may interact. That is, the loss of individuals to toxicity may result in immigration of relatively uncontaminated individuals and eventually to the evolution of resistant local populations.

An assessment of the Seal Beach Naval Weapons Station used body burdens in a somewhat unconventional way that could be helpful elsewhere. Because of the concern that persistent organic chemicals were reducing tern reproduction, the assessors collected tern eggs that failed to hatch and analyzed them for the chemicals of concern (Ohlendorf, 1998). If those chemicals were responsible for reproductive failure, one would expect that they would have concentrations that were elevated relative to reference populations, and they would be similar to those found in controlled studies that demonstrated reproductive effects. In this case, the analysis of biological materials was used to investigate the cause of apparent effects rather than to estimate the exposure of the population.

4.1.7 Criteria and Standards

Criteria and standards are concentrations of contaminants in water or other media that are intended to constitute the lower bounds of regulatory acceptability given certain conditions. The only national criteria in the United States are the acute and chronic National Ambient Water Quality Criteria (NAWQC). (Criteria have been proposed for sediments by the EPA but not adopted.) The acute NAWQCs are calculated by the EPA as half the final acute value, which is the fifth percentile of the distribution of 48- to 96-h LC50 values or equivalent median effective concentration EC50 values for each criterion chemical (Stephan et al., 1985). The acute NAWQCs are intended to correspond to concentrations that would cause less than 50% mortality in 5% of exposed populations in a relatively brief exposure. The chronic NAWQCs are final acute values divided by the final acute–chronic ratio, which is the geometric mean of quotients of at least three LC50/CV ratios from tests of organisms from different families of aquatic organisms (Stephan et al., 1985). Chronic NAWQCs are intended to prevent significant toxic effects in most chronic exposures. Some are based on protection of humans or other piscivorous organisms rather than protection of aquatic organisms (i.e., final residue values). Those criteria are not appropriate for protecting aquatic life and are, in general, poor estimators of threshold effects levels for piscivorous wildlife.

NAWQCs may be applicable regulatory standards, but they often are not good risk estimators for particular sites. If they are applied to a site, assessors should consider deriving site-specific criteria using the water–effect ratio. This is a factor for adjusting criteria to site water that may be derived using an EPA procedure (EPA, 1994c; Office of Science and Technology, 1994). It requires performing toxicity tests in site waters, and, optionally, with site species. The time and expense required to calculate site-specific criteria are most likely to be worthwhile if the water chemistry at a site differs significantly from conventional laboratory test waters and if risk managers insist on using criteria as the basis for remedial decisions. Otherwise, the effort is likely to be better expended on tests of ambient waters.

Many nations other than the United States have criteria or standards for water and other media, and these comments may not apply to them. The utility of those standards should be considered where they are potentially applicable.

4.1.8 Screening Benchmarks

Screening benchmarks are concentrations of chemicals that are believed to constitute thresholds for potential toxic effects on some category of receptors exposed to the chemical in some medium. Since they are used for screening chemicals, they should be somewhat conservative so that chemicals that do in fact cause effects at a particular site are not screened out of the assessment. It is more important to ensure that hazardous chemicals are retained than to avoid retention of chemicals that are not hazardous. However, excessive conservatism decreases the value of screening assessments, because effort is wasted on nonhazardous chemicals that might better be expended on the truly hazardous ones. Because of this deliberate conservatism, it is important to avoid adoption of screening benchmarks as remedial goals without some additional assessment to determine that they are appropriate to the site.

There is little consensus about the best methods for deriving screening benchmarks. The following alternatives are based on regulatory practice, and therefore are likely to be acceptable. Other alternatives, which were developed to demonstrate potentially more scientifically defensible approaches, are discussed in Suter (1996b).

4.1.8.1 Criteria and Standards as Screening Benchmarks

Water quality criteria or standards are commonly used as screening benchmarks because exceedence of one of these values constitutes cause for concern. Also, NAWQCs have been recommended for the purpose of screening by the EPA (Office of Emergency and Remedial Response, 1996). However, it is not clear that they are sufficiently conservative, since they are assumed to be sufficiently close to the true threshold of effects to justify regulatory action.

For particular chemicals, the chronic NAWQC may not be an adequate screening benchmark for reasons explained elsewhere (Suter, 1996b). These concerns are supported by the recent finding that nickel concentrations in a waste-contaminated stream on the Oak Ridge Reservation that were below chronic NAWQC were nonetheless toxic to daphnids (Kszos et al., 1992). When used for regulation of effluents, their intended purpose, these criteria achieve additional conservatism by being applied to short exposure durations. That conservatism does not operate at contaminated sites.

4.1.8.2 Tier II Values

If NAWQC are not available for a chemical, the Tier II method described in the EPA "Proposed Water Quality Guidance for the Great Lakes System" or a slight variation used at the Oak Ridge National Laboratory (ORNL) may be applied (EPA, 1993c; Suter and Tsao, 1996). Tier II values were developed so that aquatic life criteria could be established with fewer data than are required for the NAWQC. The Tier II values are concentrations that would be expected to be higher than NAWQC in no more than 20% of cases, if sufficient test data were obtained to calculate the NAWQC. The Tier II values equivalent to the final acute value and final chronic value (Section 4.1.7) are the secondary acute values (SAV) and secondary chronic values (SCV), respectively. The sources of data for the Tier II values, and the procedure and factors used to calculate the SAVs and SCVs are presented by EPA (1993c) and Suter and Tsao (1996). The ORNL methods differ from those in the Great Lakes guidance in not requiring that a daphnid EC50 be included in the data set, since that requirement severely restricts the number of benchmarks that can be calculated and does not increase confidence. Tier II values have been recommended by the EPA for use as screening benchmarks for chemicals for which there are no water quality criteria (Office of Emergency and Remedial Response, 1996).

4.1.8.3 Thresholds for Statistical Significance

Test endpoints based on statistical significance are commonly used as screening benchmarks. The endpoint used varies among media and receptors.

Lowest chronic values — CVs are geometric means of the highest concentration not causing a statistically significant effect (NOEC) and the lowest concentration causing a statistically significant effect (LOEC). They were formerly known as maximum acceptable toxicant concentrations (MATCs). They are used to calculate the chronic NAWQC, and are presented in place of chronic criteria by the EPA when chronic criteria cannot be calculated (EPA, 1985a). CVs are not controversial because they are not the result of any mathematical or statistical analysis beyond their derivation as test endpoints. However, they are not conservative. They have not been used for receptors other than aquatic communities.

Wildlife NOAELs — Screening benchmarks for wildlife are conventionally based on no observed adverse effects levels (NOAELs) from chronic or subchronic toxicity tests with mammals or birds. The major variables in derivation of wildlife benchmarks are the test endpoints used and whether allometric scaling or safety factors are used. The ORNL wildlife benchmarks use reproductive effects as endpoints, allometric equations for interspecies extrapolations, and factors to allow for shortcomings in the test design (Sample et al., 1996b).

4.1.8.4 Test Endpoints with Safety Factors

Some states and EPA regions base screening benchmarks on test endpoints divided by safety factors. For example, the EPA Region IV has used the lowest chronic values for fish or invertebrates divided by 10 or lowest acute LC50 values divided by 100 to calculate aquatic screening benchmarks for chemicals with no NAWQC (unpublished table, U.S. EPA Region IV, Atlanta, GA). These factors do not have the scientific basis of the factors used to derive the Tier II values (above) or the factors proposed by Calabrese and Baldwin (1993, 1994); see Section 4.1.9.1. However, the use of factors of 10, 100, or 1000 have a long history in the EPA (Dourson and Stara, 1983; Nabholz et al., 1997), and such factors can be easily applied to any test endpoint.

4.1.8.5 Distributions of Effects Levels

Sets of screening benchmarks for sediments and soils have been derived from distributions of effects or no-effects levels. An estimate of the threshold effects concentration for a particular chemical is derived from a percentile of the distribution of reported effects or no-effects concentrations. These concentrations vary due to variance in the physical and chemical properties of soils or sediments, variance among the measured responses, and variance in the sensitivities of the organisms. Therefore, the benchmarks derived in this way may be thought to protect some proportion of combinations of species, responses, and media. The following are examples of this approach.

Screening level concentration (SLC) for sediments — The SLC approach is used to estimate the highest concentration of a particular contaminant in sediment that can be tolerated by approximately 95% of benthic infauna (Neff et al., 1988). A species SLC is the 90th percentile of the frequency distribution of contaminant concentrations over at least ten sites where the species is present. Species SLCs are plotted as a frequency distribution to determine the contaminant concentration above which 95% of the species SLCs occur. That lower 5th percentile concentration is the SLC.

Effects range–low and effects range–median for sediments — The National Oceanic and Atmospheric Administration (NOAA) uses data from studies of contaminated sediments from coastal marine and estuarine sites in the United States to derive benchmark values. NOAA uses three methods: (1) equilibrium partitioning

(Section 4.1.8.6), (2) spiked sediment toxicity tests, and (3) field surveys to develop exposure–response relationships (Long et al., 1995). Then chemical concentrations observed or estimated to be associated with biological effects are ranked, and the lower 10th percentile (effects range–low, ER-L) and the median (effects range–median, ER-M) concentrations are identified.

Threshold effects levels and probable effects levels for sediments — The Florida Department of Environmental Protection (FDEP) uses the data from Long et al. (1995) (the NOAA approach above) and incorporates chemical concentrations observed or predicted to be associated with no adverse biological effects (MacDonald, 1994). Specifically, the threshold effects level (TEL) is the geometric mean of the 15th percentile of the effects concentrations and the 50th percentile of the no effects concentrations. The probable effects level (PEL) is the geometric mean of the 50th percentile of the effects concentrations and the 85th percentile of the no-effects concentrations.

Oak Ridge National Laboratory benchmarks for soil — Benchmarks for toxicity to plants (Efroymson et al., 1997), soil invertebrates (Efroymson, Will, and Suter, 1997), and microbial processes (Efroymson, Will, and Suter, 1997) have been developed from distributions of effects data. Like the NOAA ER-L, the benchmark is the 10th percentile of the distribution of various toxic effects thresholds for various organisms in various soils. If fewer than ten LOECs for a chemical exist, the lowest LOEC is used as the benchmark. The soil benchmarks are based on toxicity tests and, unlike the NOAA ER-L, do not include field survey data.

4.1.8.6 Other Methods Used for Sediment Benchmarks

Because samples of benthic invertebrates can be associated with a corresponding sample of contaminated sediment, sediment benchmarks have been developed based on the chemical concentrations in whole sediment that are associated to varying degrees with adverse effects on benthic organisms. Those field-derived data may be used alone or mixed with laboratory tests to derive effects distributions (above) or may be analyzed by other means as discussed here (MacDonald et al., 1994). In addition, aquatic benchmarks may be converted into sediment benchmarks and field-contaminated sediments may be tested in the laboratory. Some types of benchmarks that are based on studies of sediments are briefly described below. Examples of each are described in Table 4.4.

Apparent effects thresholds — These benchmarks are sediment chemical concentrations above which statistically significant biological effects always occur in a field study. They are site specific and they may be underprotective, given that biological effects are observed at much lower chemical concentrations. These are generally used for ionic and polar organic chemicals when other, better values are not available.

Screening level concentrations — These benchmarks are derived from synoptic data on sediment chemical concentrations and benthic invertebrate distributions. They are estimates of the highest concentration that can be tolerated by a specified percentage of benthic species. Examples include the Ontario Ministry of the Environment lowest and severe effect levels.

TABLE 4.4
Example Benchmarks for Sediment-Associated Biota

Example Benchmarks	Description[a]	Source
ER-L	The 10th percentile of estuarine sediment concentrations reported to be associated with some level of toxic effects; possible-effects benchmarks	Long et al., 1995
ER-M	The 50th percentile of estuarine sediment concentrations reported to be associated with some level of toxic effects; probable-effects benchmarks	Long et al., 1995
TEL	The geometric mean of the 15th percentile of reported concentrations, which were associated with some level of effects, and the 50th percentile of reported concentrations, which were associated with no adverse effects (all data are for marine and estuarine sediments); possible-effects benchmarks	MacDonald et al., 1996
PEL	The geometric mean of the 50th percentile of reported concentrations, which were associated with some level of effects and the 50th percentile of reported concentrations, which were associated with no adverse effects (all data are for marine and estuarine sediments); possible-effects benchmarks	MacDonald et al., 1996
Ontario Ministry of the Environment Lowest Effect Level	Concentrations estimated to constitute thresholds for toxic effects in Ontario sediments; for most chemicals this is the concentration that can be tolerated by approximately 95% of benthic invertebrates; possible-effects benchmarks	Persaud et al., 1993
Ontario Ministry of the Environment Severe Effect Level	Concentrations estimated to constitute thresholds for severe toxic effects in Ontario sediments; for most chemicals, the concentration that can be tolerated by approximately 5% of benthic invertebrates; probable-effects benchmarks	Persaud et al., 1993
National Sediment Quality Criteria	Proposed sediment quality criteria based on toxicity in water expressed as chronic water quality criteria (recalculated after adding some benthic species) and partitioning of the contaminant between organic matter (1% of sediment) and pore water (in the absence of site-specific data, organic matter content is assumed to be 1% by weight); probable-effects benchmarks	(EPA, 1993g-k)

ORNL Equilibrium Partitioning Benchmarks	Benchmarks derived in the same manner as sediment quality criteria except that the expression of aqueous toxicity is one of five benchmarks: the chronic NAWQC, the SCV, the LCV for daphnids, the LCV for fish, or the LCV for nondaphnid invertebrates (in the absence of site-specific data, organic matter content is assumed to be 1% by weight); the SCV-based value is a possible-effects benchmark; all others are probable-effects benchmarks	(Jones et al., 1997)
Assessment and Remediation of Contaminated Sediments Program's ER-Ls and TELs	Sediment effect concentrations based on the toxicity to *Hyalella azteca* and *Chironomus riparius* associated with contaminants in sediment samples collected from predominantly freshwater sites; possible-effects benchmarks, below which adverse effects to these organisms are not expected	(EPA, 1996b)
Assessment and Remediation of Contaminated Sediments Program's ER-Ms, TELs, and AETs	Probable-effects benchmarks, above which adverse effects to *H. azteca* and *C. riparius* are likely to occur; the majority of the data are for freshwater sediments	(EPA, 1996b)
Apparent Effect Threshold (AET)	A concentration above which toxic effects occurred at all sites in Puget Sound; probable-effects benchmarks	(Ginn and Pastorok, 1992)

[a] Possible-effects benchmarks are conservative estimates of concentrations at which toxicity may occur. Probable-effects benchmarks are concentrations at which toxicity is likely.

Equilibrium partitioning benchmarks — These benchmarks are bulk sediment concentrations derived from aqueous benchmark concentrations based on the tendency of nonionic organic chemicals to partition between the sediment pore water and sediment organic carbon. The fundamental assumptions are that pore water is the principal exposure route for most benthic organisms and that the sensitivities of benthic species are similar to those of the species tested to derive the aqueous benchmarks, which are predominantly water column species. Examples include the proposed EPA sediment quality criteria and ORNL benchmarks derived from five types of benchmarks for aquatic biota (Jones et al., 1997).

Benchmarks from tests of field-contaminated sediments — Benchmarks may be derived by testing ambient sediments in the laboratory using a standard species and protocol to identify concentrations that cause effects. The best example is the sediment effects concentrations (EPA, 1996b).

Each of the example benchmarks described in Table 4.4 is classified as either a possible-effects or probable-effects benchmark. Possible-effects benchmarks are conservative estimates of concentrations at which toxicity may occur, e.g., the 10th percentile of the sediment concentrations reported to be toxic. Probable-effects benchmarks are concentrations at which toxicity is likely, e.g., the 50th percentile of the sediment concentrations reported to be toxic. Recognition of the relative degrees of conservatism associated with each benchmark allows for a more thorough and informed use of the screening values.

4.1.8.7 Summary of Screening Benchmarks

Currently, the development of screening benchmarks is inconsistent across media. The large and relatively consistent body of data for aquatic animals has led to the development of more than a dozen alternative types of benchmarks. Similarly, there are several alternative benchmarks for sediments, but they have been developed for fewer chemicals. Wildlife benchmarks are nearly always based on NOAEL values, so there is usually only one type of benchmark. However, there is considerable variance in what effects are included. Finally, benchmarks for plants, invertebrates, and microbes in soil are highly inconsistent.

ORNL has produced a large set of ecological screening benchmark values (http://www.hsrd.ornl.gov/ecorisk/ecorisk.html). The EPA has published a set of screening benchmarks (termed *ecotox thresholds*) (Office of Emergency and Remedial Response, 1996). Those for water are based on chronic NAWQC values and SCVs. Those for sediments are, in order of preference, conservatively adjusted, draft sediment quality criteria (i.e., the lower limit of the 95% confidence interval); comparable values based on secondary chronic values; and the ER-Ls for marine and estuarine sediments. Other sets of values have been produced by EPA regions, states, and by agencies outside the United States. The authors have deliberately not included any of these benchmark values in this book because they change so rapidly and their acceptability to local decision makers is so inconsistent. Although benchmarks have been compared with each other and with background, there has been no systematic attempt to validate them (Suter, 1996b). The validity of the various sediment benchmarks has been a subject of particular controversy (Long and MacDonald, 1998; O'Connor, 1999).

Given the lack of validation or even a common definition of validity, no single type of benchmark can be demonstrated to be consistently reliable. At ORNL, the authors used a battery of benchmarks for water and sediments to decrease the likelihood of falsely screening out a contaminant (Chapter 5). Alternatively, when there are multiple benchmarks for a chemical and none is clearly superior, "consensus" benchmark values may be simply derived by averaging. Swartz (1999) derived a threshold effects concentration for total PAHs (290 μg/g organic carbon) as the arithmetic mean of five diverse benchmarks. He found that it was a reasonable threshold value for PAH effects in independent data sets from PAH-contaminated sites.

4.1.9 Single-Chemical Test Endpoints and Definitive Assessment

Single-chemical toxicity test endpoints can play two roles in definitive assessments. If biological surveys or ambient media toxicity tests are performed, single-chemical test results may be used to support the conclusion that toxic effects are or are not occurring, to determine what contaminants are responsible, and to help establish remedial goals. If more realistic effects data are not collected, single-chemical test results must be used to estimate risks. In either case, the test endpoints must be appropriately selected and used in extrapolation models to provide useful descriptions of the relationship between exposure and effects on the assessment endpoints.

4.1.9.1 Extrapolation Approaches

Most of the work of effects analysis is devoted to determining the relationship between exposure and effects for each chemical or material of concern. Derivation of exposure–response models from laboratory toxicity tests requires analysis of the test data to derive a test endpoint and extrapolation from the test endpoint to the assessment endpoint. The extrapolation may be performed in various ways, briefly discussed in the following subsections (OECD, 1992a; Suter, 1993a, 1998a).

Classification and Selection — It may be assumed that the endpoint species, life stages, and responses are equal to those in the most sensitive reported test or in the test that is most similar in terms of taxonomy or other factors. This process of classification and selection of test endpoints is the simplest and most commonly used extrapolation method. Sufficient similarity must be judged on the basis of some classification system. For example, plants are often classified by growth form, and the EPA classifies freshwater fish as warm water and cold water species (Stephan et al., 1985). However, species are most commonly classified taxonomically. Studies based on correlations of the LC50s of species at different taxonomic distances indicate that, for both freshwater and marine fishes and arthropods, species within genera and genera within families tended to be relatively similar, which suggests that they can be treated as equivalent, given testing variance (Suter et al., 1983; Suter and Rosen, 1988; Suter, 1993a). The same conclusion was reached by the same method for terrestrial vascular plants (Fletcher et al., 1990).

Safety factors — A test endpoint can be divided by 10, 100, or 1000 to estimate a safe level as in the EPA review of new industrial chemicals (Zeeman, 1995). This method is also easily and commonly used, but it has little scientific basis, and it results in a number that is no longer clearly associated with a particular effect. It is not particularly useful in definitive assessments, because it does not serve to estimate an effect and cannot indicate that a chemical is the cause of an observed effect.

Species sensitivity distributions — A percentile of the distribution of test endpoint values for various species can be used to represent a concentration or dose that would be protective of that percentage of the exposed community. For example, if the distribution of 96-h LC50 values for fish exposed to a chemical is normally distributed (m_t,s_t), then half of the fish species in the field would be expected to experience mass mortality after exposure to concentration m_t for 96 h. This approach is becoming increasingly popular. This approach is based on the species sensitivity distributions (SSDs) that were developed by the EPA for deriving water quality criteria (Stephan et al., 1985). It has been used by European nations to derive environmental criteria and has been recommended as a standard ecological risk assessment technique (Suter, 1993a; Aquatic Risk Assessment and Mitigation Dialog Group, 1994; Parkhurst et al., 1996). The chief limitations on this method are the requirement that enough species have been tested to define the SSD and that they be representative of the receiving community. The EPA requires at least eight species from eight different families and that they be distributed across taxa in a prescribed manner (Stephan et al., 1985). Relatively few chemicals have enough chronic toxicity data to establish the chronic SSD. Another potential problem is that, if the media or the test conditions are variable and influential, the distributions will include extraneous variance.

Regression models — Regressions of one taxon on another, one life stage on another, one test duration on another, one level of organization on another, etc. can be used to extrapolate among taxa, life stages, durations, or levels of organization. This approach is extremely flexible and quantitatively rigorous but is seldom used. For example, when the SSD cannot be estimated for a chemical because there is only one test datum for the chemical, a test species to higher taxon or community regression can be used to estimate the same endpoint. Regression models for aquatic extrapolations are presented below (Section 4.1.9.2) and in Table 4.5. More extensive discussions and examples of these methods can be found in Suter et al. (1983, 1987). Barnthouse and Suter (1986), Sloof et al. (1986), Holcombe et al. (1988), Suter and Rosen (1988), and Calabrese and Baldwin (1994).

Factors derived from regression models — Because factors are more easily employed than even simple regression models, they have been much more popular. Sloof et al. (1986) used the prediction intervals around regression models to derive uncertainty factors. Calabrese and Baldwin (1993) applied this approach to previously developed extrapolation models (Suter et al., 1983, 1987; Barnthouse and Suter, 1986; Suter and Rosen, 1988). Results for acute–chronic extrapolations for defined chronic responses and intertaxa extrapolations are shown in Tables 4.6 and 4.7, respectively. The reader should note that this method retains only the highly conservative 90, 95, or 99% upper-bound estimate of effects levels and not the best estimate.

The intertaxa extrapolations require some explanation. Suter et al. (1983) developed an approach for extrapolating between any test species and reference species that involved aggregation of species within taxonomic hierarchies. By using a large data set of aquatic acute toxicity data, congeneric species were regressed against each other; then congeneric species were aggregated and genera within common

TABLE 4.5
Linear Equations for Extrapolating from Standard Fish Test Species to All Freshwater or Marine Fish (units are log μg/l).

Test Species	Slope	Intercept	*n*	mean *X*	F_1	F_2	PI[a]
Pimephales promelas	1.01	–0.30	354	2.77	0.45	0.0006	1.31
Lepomis macrochirus	0.96	0.17	500	2.52	0.49	0.0005	1.37
Oncorhynchus mykiss	0.99	0.29	480	2.42	0.38	0.0004	1.20
Cyprinodon variegatus	0.97	0.03	51	1.25	0.58	0.0085	1.49

[a] PI, the 95% prediction interval at the mean, is log mean Y ± the number in this column.
Source: Suter, G. W. II, *Ecological Risk Assessment*, Lewis Publishers, Boca Raton, FL, 1993. With permission.

families were regressed against each other; and then confamilial species were aggregated and families within the same order were regressed against each other. This process continued up to a regression of the phylum vertebrata against the arthropoda. The increasing prediction intervals on these regressions as the taxonomic distance increased was used to demonstrate that toxicological similarity is related to taxonomic similarity. Calabrese and Baldwin (1993) used a later version of the regressions for fish taxa to reduce the regressions and prediction intervals to 95 and 99% uncertainty factors for each taxonomic relationship by calculating confidence

TABLE 4.6
Uncertainty Factors for Extrapolations from Acute Lethality to Specific Chronic Effects in Fish

			Confidence Interval		
X Variable	*Y* Variable	n	90%	95%	99%
LC50	Hatch EC25	31	26	50	198
LC50	Parent mortality EC25	28	18	32	106
LC50	Larval mortality EC25	89	18	31	93
LC50	Eggs EC25	42	32	64	228
LC50[a]	Fecundity EC25	26	26	50	206
LC50[a]	Weight[b] EC25	37	28	53	188
LC50[a]	Weight/egg EC25	14	91	246	2247
Mean			34.1	75.1	466.6
Weighted Mean			27.1	54.7	264.9

[a] Regression analysis from Suter et al. (1987).
[b] Decrease in weight of fish at end of larval stage.
Source: Calabrese, E. J. and Baldwin, L. A., *Performing Ecological Risk Assessments*, Lewis Publishers, Boca Raton, FL, 1993. With permission.

TABLE 4.7
Taxonomic Extrapolation: Means and Weighted Means Calculated for the 95% and 99% Prediction Intervals (PIs) for Uncertainty Factors Calculated from Hierarchical Regressions[a]

X Variable	*Y* Variable	*n*	Uncertainty Factor 95% PI	Uncertainty Factor 99% PI
Taxonomic Extrapolation: Species within Genera				
Salmo clarkii	*S. gairdneri*	18	9	13
S. clarkii	*S. salar*	6	6	10
S. clarkii	*S. trutta*	8	6	8
S. gairdneri	*S. salar*	10	7	11
S. gairdneri	*S. trutta*	15	4	5
S. salar	*S. trutta*	7	5	8
Ictalurus melas	*I. Punctatus*	12	5	7
Lepomis cyanellus	*L. macrochirus*	14	6	9
Fundulus heteroclitus	*F. majalis*	12	6	8
Mean			6.1	10.1
Weighted mean			6.0	7.4
Taxonomic Extrapolation: Genera within Families				
Oncorynchus	*Salmo*	56	5	6
Oncorynchus	*Salvelinus*	13	4	5
Salmo	*Salvelinus*	56	5	7
Carassius	*Cyprinus*	8	4	6
Carassius	*Pimephales*	19	7	9
Cyprinus	*Pimephales*	10	7	10
Lepomis	*Micropterus*	30	8	11
Lepomis	*Pomoxis*	8	9	13
Cyprinodon	*Fundulus*	12	6	8
Mean			6.1	8.3
Weighted Mean			5.8	7.7
Taxonomic Extrapolation: Families within Orders				
Centrarchidae	Percidae	47	10	14
Centrarchidae	Cichlidae	6	4	6
Percidae	Cichlidae	5	13	24
Percidae	Esocidae	11	9	13
Atherinidae	Cyprinodontidae	32	7	9
Mugilidae	Labridae	12	55	78
Cyprinodontidae	Poecillidae	12	3	5
Mean			14.4	21.3
Weighted mean			12.6	17.9
Taxonomic Extrapolation: Orders within Classes				
Salmoniformes	Cypriniformes	225	20	27
Salmoniformes	Siluriformes	203	39	51
Salmoniformes	Perciformes	443	12	16
Cypriniformes	Siluriformes	111	11	15
Cypriniformes	Perciformes	219	32	43

TABLE 4.7 (continued)
Taxonomic Extrapolation: Means and Weighted Means Calculated for the 95% and 99% Prediction Intervals (PIs) for Uncertainty Factors Calculated from Hierarchical Regressions[a]

X Variable	*Y* Variable	*n*	Uncertainty Factor 95% PI	Uncertainty Factor 99% PI
Siluriformes	Perciformes	190	63	83
Anguiliformes	Tetraodontiformes	12	13	18
Anguiliformes	Perciformes	34	25	34
Anguiliformes	Gasterosteiformes	8	16	24
Anguiliformes	Atheriniformes	46	9	12
Atheriniformes	Cypriniformes	7	501[b]	786[b]
Atheriniformes	Tetraodontiformes	46	13	17
Atheriniformes	Perciformes	148	25	33
Atheriniformes	Gasterosteiformes	36	20	27
Gasterosteiformes	Tetraodontiformes	8	20	30
Gasterosteiformes	Perciformes	33	32	43
Perciformes	Tetraodontiformes	34	25	34
Mean			23.5	31.7
Weighted mean			26.0	34.5

Uncertainty factors calculated by Calabrese and Baldwin (1994); used with permission.

[a] Values in this table are similar to but differ from those in Barnthouse et al. (1990) due to differences in the algorithm used, particularly the use of ordinary least squares regression by Calabrese and Baldwin (1994).

[b] Not included in calculations.

intervals on the set of prediction intervals for pairs of orders of fish (Table 4.8). Calabrese and Baldwin (1994) later suggested that these generic factors were applicable to taxa other than fish, including humans. For example, when extrapolating between a mouse test and equivalent effects on a mammalian carnivore (order Carnivora), one would divide the mouse test endpoint by 64.8 to be 95% certain of including the carnivore species 95% of the time (Table 4.8).

Allometric scaling — The type of quantitative extrapolation model used most commonly by human and wildlife pharmacologists and toxicologists is allometric scaling. These models are based on the assumption that all members of a taxon have the same response to a chemical, but they differ in the size and in processes that are related to size. The most commonly used allometric model is a power function of weight, $E_x = aW^b$ (E_x is the effect at some weight W). This form has been adopted by toxicologists because various physiological processes, including metabolism and excretion of drugs and other chemicals, are approximated by that functional form (Peters, 1983; Davidson et al., 1986). Recently, the EPA has used the 3/4 power for piscivorous wildlife (EPA, 1993e), and others have followed its lead (Sample et al., 1996b). Although allometric scaling may be applied to aquatic species, it is primarily used for wildlife extrapolations, and is discussed at length in the wildlife section, below.

TABLE 4.8

Upper 95% Uncertainty Factors Calculated for the 95% and 99% Prediction Intervals in Table 4.4

	Prediction Interval	
Level of Taxonomic Extrapolation	**95%**	**99%**
Species within genera	10.0	16.3
Genera within families	11.7	16.9
Families within orders	99.5	145.0
Orders within classes	64.8	87.5

Source: Calabrese, E. J. and Baldwin, L. A., *Performing Ecological Risk Assessments*, Lewis Publishers, Boca Raton, FL, 1993. With permission.

Mathematical models — Toxicodynamic models can be used to estimate effects on organisms from physiological responses, and population or ecosystem models can be used to estimate effects on populations or ecosystems from organism responses. This approach to extrapolation is probably the least commonly used and the most technically demanding, but is potentially the most powerful. Toxicodynamic models are virtually unknown in ecological risk assessment practice. Potentially relevant population and ecosystem models are described in Bartell et al. (1992) and in Suter (1993a).

4.1.9.2 Extrapolations for Specific Endpoints

Different extrapolation approaches are used with different classes of endpoints. These differences are based on the constraints of available data as well as differences in the intellectual traditions of the different groups of toxicologists. In addition, the following recommendations are based on the judgment of the authors concerning the best practices from among those that are currently employed and accepted.

Aquatic biota

If, as is often the case, the endpoint property for the aquatic biota is species richness or diversity (Chapter 2), SSDs are an obvious choice of extrapolation model. Based on modeling results, continuous exposure to concentrations equal to the CV for a species can cause extinction of that species (Barnthouse et al., 1990). Therefore, the proportion of species for which the CV is exceeded by long-term exposures can be assumed to approximate the proportion of species lost from the community. In addition, because toxicity data are relatively abundant for aquatic organisms, it is often feasible to derive such distributions for individual chemicals. As discussed above, this approach is widely accepted, because it is used for the derivation of water quality criteria. If the distributions are to be used to estimate levels of effects, a logistic or other function should be fit to them. The choice of function makes relatively little difference (OECD, 1992a). Distributions may be fit and percentiles calculated by any statistical software. However, convenient software is available for this purpose, including calculation of both HC5 and its lower, one-tailed 95% confidence limit for lognormal, log-logistic, and log-triangular distributions (Aldenberg, 1993; Cadmus Group, 1996). If used to support risk estimates based on site-specific data, an empirical distribution is simpler and more appropriate than a mathematical function. If responses are known to be a function of water chemistry, the individual test endpoints should be corrected before defining the distribution.

One important issue to be resolved is the inclusiveness of the distributions. The EPA includes multicellular aquatic animals (Stephan et al., 1985), but others include algae and other plants as well (Wagner and Lokke, 1991; Aldenberg and Slob, 1993). Although inclusiveness seems desirable, it strains the assumption that the species are drawn from a single unimodal distribution. It also makes the inferences that can be drawn less specific. The authors recommend separating plants and animals in all cases because of their great differences in chemical sensitivity, and, when data allow, separating vertebrates from invertebrates (Figure 4.1).

If there are not enough data to generate an SSD, a number of alternatives present themselves. If a test endpoint for a standard test species is available, the distribution of the endpoint for all fish species can be estimated from the equations like those in Table 4.5 that regress all fish species against a standard test species for multiple chemicals (Barnthouse and Suter, 1986; Suter et al., 1987; Holcombe et al., 1988; Suter and Rosen, 1988). The equations estimate the mean of log LC50 for saltwater

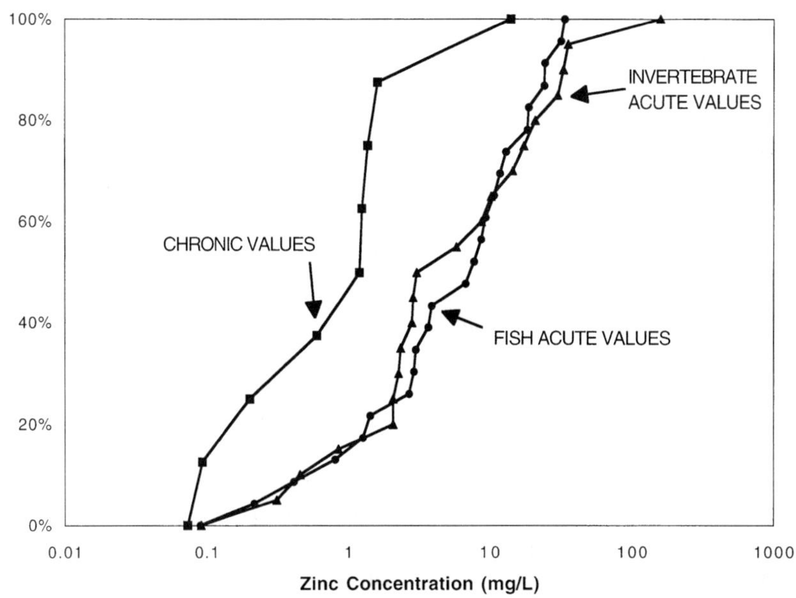

FIGURE 4.1 Empirical cumulative species sensitivity distributions for acute toxicity to fish, acute toxicity to aquatic invertebrates, and chronic toxicity to fish and invertebrates combined for zinc.

fish from *Cyprinodon variegatus* LC50 or for freshwater fish from the standard freshwater species. The 95% prediction interval at the mean is log mean $Y \pm$ PI. The PI is estimated from the variance in LC50 for other species (Y) at a given LC50 for a standard test species (X_0):

$$\mathrm{var}(Y \mid X_0) = F_1 + F_2(X_0 - X)^2 \tag{4.1}$$

Since the second term of the variance is relatively small, the PI at the mean is a reasonable estimate of the PI for all Y. That is, 95% of fish responses would be expected to fall within approximately ±1.3 log units or approximately a factor of 20 of the lognormal mean fish response estimated from the equations.

An alternative approach was developed for the calculation of Tier II water quality values (Section 4.1.8.2). If there is only one acute value (LC50 or EC50), that value is divided by 20.5 if it is a daphnid and 242 if it is not (EPA, 1993c; Suter and Tsao, 1996). Equivalent factors are available for other numbers of acute values in Appendix B of Suter and Tsao (1996). Given the way in which the water quality criteria are derived, the values obtained with these factors should protect 95% of aquatic invertebrate and fish species with 80% confidence. No method is provided for estimating the expected effects level or the distribution of effects levels, so this method is best used to develop screening benchmarks (above).

If the endpoint is a property of a particular population rather than a community, the extrapolations using SSDs are performed or interpreted a little differently. SSDs are still useful, because they can be interpreted as probability distributions for effects on an individual species. Alternatively, one can use the appropriate intertaxa regression models (Barnthouse et al., 1990; Suter, 1993a) or the uncertainty factors derived from them (Table 4.7). That is, if one wanted to predict the toxicity of a chemical to brook trout (a salmonid) from test data for fathead minnow (a cyprinid), one could divide by 20 to be 95% certain of not underestimating the sensitivity of brook trout (or any other salmonid). If the desired taxonomic regression is not available, the appropriate generic factor (which would be 26 in this case of an interorder extrapolation) would be applied. These two approaches for estimating effects on particular species or taxa (SSDs or taxonomic regressions) have different weaknesses, and it is not clear which works better in practice. However, the taxonomic regressions and the factors derived from them require test data for only one species, so they are more generally useful. The factors are quite conservative and may estimate effects levels that are below background. For estimation of probabilities of effects, one should use the original regression models to estimate means and variances (see Table 7.4 in Suter, 1993a).

Acute–chronic extrapolations may be made with regression models or factors. Acute–chronic regression models are presented in Suter (1993a), and factors derived from them are presented in Table 4.6. These factors are based on including the CV or EC25 with 95 or 99% confidence. Alternatively, CVs can be estimated with 80% confidence of not overestimating their value using a factor of 17.9 (Host et al., 1991). Calabrese and Baldwin (1993) recommend generic 95 and 99% uncertainty factors of 50 and 200 for acute–chronic extrapolations, based on the weighted means in Table 4.6. Any of these factors is adequate if one is trying to conservatively estimate a chronically toxic concentration of a chemical to support an assessment based primarily on other lines of evidence. The authors would not recommend using a chronic effects level estimated from acute toxicity data for anything else. If compelled by circumstances to estimate risks to aquatic organisms using only an LC50, one should use one of the regression equations in Suter (1993a) or Sloof et al. (1986) and include the model uncertainty in the analysis rather than using conservative factors.

In some cases, multiple extrapolations are required including those between taxa and life stages. Such multiple extrapolations may be dealt with by chains of factors or by chains of regression models (Barnthouse et al., 1990; Calabrese and Baldwin, 1993; Suter, 1993a). However, estimation of risks by these methods are recommended only if site-specific data cannot be obtained. Therefore, they are not explained in detail here.

Benthic invertebrates

Species or community sensitivity distributions can be derived for toxicity of individual chemicals to benthic invertebrates based on published test results. In the case of exposure of benthic invertebrates to sediment pore water, the effects distributions are the same as the species sensitivity distributions discussed above for aquatic biota. The use of aqueous data to evaluate effects on benthic species is based on data suggesting that benthic species are not systematically more or less sensitive than water column species (EPA, 1993a).

In the case of exposure of benthic invertebrates to chemicals in whole sediment, the effects distributions are for species/sediment combinations and community/sediment combinations. This is necessary because it is not possible to adequately control for the effect of sediment characteristics, including co-contaminants in field-collected sediments, on toxicity. The most prominent examples of effects distributions for benthic invertebrates are those used to derive screening benchmarks for sediment-associated biota (Long et al., 1995; MacDonald et al., 1996). The effects in those distributions include taxa richness, diversity, density, mortality, growth, respiration, behavior, and suborganismal effects. As a result, those distributions only indicate an unspecified level of an unspecified effect. This is adequate for screening purposes, but not for definitive risk characterization. For definitive assessments, such nonspecific distributions can be parsed into distributions of thresholds for specific effects. For example, Jones et al. (1999) developed distributions of community-level effects and lethality from the sediment toxicity data presented in McDonald et al. (1994) and Long and Morgan (1991). As in Figure 4.1, these are cumulative empirical distribution functions.

Wildlife

Literature toxicity data exist for relatively few wildlife species. Common avian toxicity test species include mallard ducks, doves, quail, and chickens. Common mammalian test species include rodents (e.g., mice, rats, and guinea pigs), dogs, and mink. Because toxicity data are frequently not available for wildlife species that may be considered in an ecological risk assessment, extrapolation from test species to endpoint species is required. Interspecies extrapolation of toxicity data for wildlife is generally made using one of four approaches: classification, uncertainty factors, allometric scaling models, or physiologically based pharmacokinetic (PBPK) models.

Classification — This extrapolation approach is by far the simplest. All species in a taxon or other class are assumed equally sensitive to chemicals, and the literature-derived toxicity value is therefore assumed to be directly relevant to each member. The most common classes are birds and mammals. This approach ignores observed interspecies toxicity differences.

Uncertainty factors — Uncertainty factors consist of one or more values, usually multiples of 10, by which the literature-derived toxicity value is divided to estimate a toxicity value for a wildlife species. Uncertainty factors are widely used for the development of toxicity values for human health risk assessment (Dourson and Stara, 1983) and have also been applied for wildlife risk assessment (e.g., Banton et al., 1996; Sample et al., 1996b; Hoff and Henningson, 1998). Uncertainty factors have been applied to account for a wide range of extrapolations (e.g., interspecies, acute-to-chronic, laboratory-to-field, LOAEL-to-NOAEL). While the key advantage to the use of uncertainty factors is their simplicity, application of large uncertainty factors (e.g., those resulting from the multiplication of many individual factors) lead to overly conservative toxicity values. Extensive reviews of the application of uncertainty factors in ecological risk assessments are provided by Fairbrother and Kapustka (1996) and Chapman et al. (1998).

An extrapolation model based on uncertainty factors for estimating wildlife toxicity values has been proposed by Hoff and Henningson (1998):

$$D_w = D_t \,/\, (\mathrm{UF}_a \times \mathrm{UF}_b \times \mathrm{UF}_c \times \mathrm{UF}_d) \tag{4.2}$$

where D_w represents the estimated critical chronic dose for an endpoint wildlife species, and D_t is the literature-derived toxicity value for the test species. UF_a accounts for intertaxon variability and can range from 1 if the test and wildlife species are the same to 5 if the test and wildlife species are in the same class but in different orders. Uncertainty in study duration is represented by UF_b, which ranges from 1 to 15 for the range from chronic to acute. UF_c accounts for the type of toxicity data available and ranges from 0.75 for NOELs to 15 for severe or lethal effects (>>ED50). Finally, UF_d addresses other modifying factors (e.g., species sensitivity, laboratory-to-field extrapolation, intraspecific variability) and may range from 0.5 to 2. Hoff and Henningson (1998) recommend reporting quantitative risk results only if total UF < 100. For total UF > 100, only qualitative (e.g., presence–absence, low, medium, high) estimates of risk should be reported. As with other uses of multiplicative factors, this proposed extrapolation model includes inappropriate error propagation and subjective factors. However, it is similar to current practice in human health risk assessment.

Allometric Scaling — The allometric scaling approach is based on the observation that many morphological, physiological, biochemical, pharmacological, and toxicological attributes of animals vary with some function of an animal's body weight (Davidson et al., 1986). These functions are best described using the allometric power function: $A = a(\mathrm{BW})^b$, where A is the biological attribute, a is the intercept, BW is the animal's body weight, and b is the allometric scaling factor. Reviews of the theory and application of allometric scaling are provided by Fairbrother and Kapustka (1996), Davidson et al. (1986), and Peters (1983).

Allometric scaling has been commonly applied for the estimation of toxic doses to humans based on animal studies. Initial research by Freireich et al. (1966) indicated that scaling for cancer chemotherapy drugs varied in relation to body surface area or $BW^{0.66}$. This 0.66 scaling factor was adopted by the EPA (1986a) for human health risk assessments and was employed for avian and mammalian wildlife in earlier versions of the ORNL wildlife benchmarks (Opresko et al., 1993). The data from Freireich et al. (1966) have subsequently been reanalyzed several times by other authors (Travis and White, 1988; Goddard and Krewski, 1992; Travis and Morris, 1992; Watanabe et al., 1992), resulting in scaling factors that were on average closer to 0.75 than 0.66, but consistent with either value. The 0.75 scaling factor suggests that toxicity varies with metabolic rate, which also scales at $BW^{0.75}$ (Davidson et al., 1986). In the EPA (1992c), the 0.75 scaling factor was adopted for human health risk purposes. More recently, the EPA has investigated the use of 0.75 scaling factor for piscivorous wildlife (EPA, 1993e), and the 1996 revision of the ORNL wildlife benchmarks uses this factor (Sample et al., 1996b). Use of either the 0.66 or 0.75 scaling factor is conservative for humans and mammalian wildlife in that large species such as deer are estimated to be more sensitive than the small rodents that are typically used in mammalian toxicity testing, while small wild species are estimated to be approximately equal in sensitivity to test species.

Little attention has been paid to allometric models for avian toxicology. However, use of the same models for birds as mammals with the same exponents was supported by allometric models of avian physiology (Peters, 1983) and pharmacology (Pokras et al., 1993). In fact, Pokras et al. (1993) present models for the extrapolation of effective doses of drugs from mammals to birds based on a common exponent of 0.75 but with a higher *a* value (see Equation 4.2) for birds. In contrast, Mineau et al. (1996) performed allometric regression analyses on 37 pesticides with between 6 and 33 species of birds. They found that for 78% of chemicals the exponent was greater than 1 with a range of 0.63 to 1.55 and a mean of 1.1. However, because scaling factors for the majority of the chemicals evaluated were not significantly different from 1, Sample et al. (1996) considered a scaling factor of 1 to be most appropriate for interspecies extrapolation among birds.

Allometric scaling is simple to apply, and it has a stronger scientific basis than uncertainty factors. If a toxicity value (e.g., a NOAEL) and the body weights of both the test and endpoint species are known and an appropriate scaling factor (b) is selected, the toxicity value for the wildlife species may be calculated (Sample et al., 1997a):

$$NOAEL_w = NOAEL_t\left(\frac{\mathrm{bw}_t}{\mathrm{bw}_w}\right)^{1-b} \tag{4.3}$$

Drawbacks to allometric scaling include the limited number and type of chemicals upon which current models are based (i.e., mammalian values are based primarily on drugs and avian values are based primarily on organophosphate and carbamate insecticides) and the fact that both avian and mammalian models are based only on acute toxicity data. Because allometric scaling factors can vary widely among different chemicals (Mineau et al., 1996) and because the toxic mode of action varies for acute and chronic exposure to the same chemical, the current practice of applying the same scaling factors for all chemicals and types of exposures may produce inaccurate estimates (Fairbrother and Kapustka, 1996). Sample and Arenal (1999) provide 138 chemical-specific scaling factors for acute toxicity to birds and 94 for mammals. In the absence of a scaling factor for the chemical of interest or for a similar chemical, they recommend generic factors of 1.2 for birds and 0.94 for mammals. Factors must be developed for chronic exposures and for effects other than lethality.

PBPK models — A final approach that may be used for wildlife extrapolation is PBPK models. These are compartmental models that combine contaminant-specific characteristics with species-specific anatomically defined characteristics of organs (e.g., organ, interorgan, and blood volumes, blood flow, and blood interconnections between organs) to estimate the dose of a chemical that reaches a given organ per unit time (Menzel, 1987). Knowledge of how species-specific characteristics vary among species may be used to extrapolate toxic effects among species (Dedrick and Bischoff, 1980). The primary problem with PBPK models is that they are very data intensive and much of the physiological data needed for wildlife species are unavailable or poorly developed (Suter, 1993a; Fairbrother and Kapustka, 1996). However, despite limited data, Fairbrother and Kapustka (1996) suggest the use of even simple PBPK models may significantly increase accuracy of interspecies extrapolations.

These various potential methods for extrapolation of toxic effects between avian species and mammalian species have traditionally been applied independently and not in a systematic manner. In a recent review, Fairbrother and Kapustka (1996) suggested that less reliance be placed on a single approach (e.g., allometric models) for all species and chemicals and that multiple approaches be applied to the problem of wildlife extrapolation.

Plants

Little guidance is available regarding interspecies toxicity extrapolations for plants. In the single major study of this topic, Fletcher et al. (1990) compared the EC50 for sets of plant taxa treated with the same herbicide. For each of 16 chemicals, the EC50 values of between 7 and 36 plant species were compared. The variation in sensitivity ranged from 3.5-fold for linuron to 316-fold for picloram. Out of almost 300 chemical–plant species combinations, 59% of the EC50 values varied by less than a factor of 5 from other EC50 values for the same chemical. Plants that were closely related, taxonomically, had similar sensitivities to the same chemical (Fletcher et al., 1990). No trends in the relative sensitivity of various species, genera, or families to different chemicals were observed. In this study interspecies variability in toxicity was much higher than the variability associated with the extrapolation from greenhouse to field (Fletcher et al., 1990).

Thus, it is not possible at this time to estimate with a high level of confidence the relative sensitivities of two plant populations to a single chemical, if the toxicity has not been tested. However, the finding that taxonomy is more important than test location (Fletcher et al., 1990) is instructive. If the assessor is concerned about risks to a particular plant population (e.g., threatened or endangered species, a dominant forest tree, or a revegetation species), the published toxicity of the chemical to the most closely related species should be used as an estimate of the effect on the endpoint species.

More often, the assessment endpoint entity is the plant community or a generalized plant population. For these cases the risk assessor can construct a species sensitivity distribution for plant toxicity, based on existing tests with single chemicals. For example, in the document in which ORNL plant benchmarks are derived, Efroymson et al. (1997) present distributions of published sensitivities of plant species to chemical concentrations in soil. An example is the cumulative distribution of toxicity of various plants to zinc in soil (Figure 4.2). Because the variance of toxicity among soils is significant and cannot be factored out, soil type is included as a source of variability in plant sensitivity distributions. Hence, the points in Figure 4.2 are species–soil type combinations, and the distribution is the distribution of effective concentrations across species and soils. As a result, single plant species that are tested under different conditions have different LOECs in the distribution. Although LOECs for corn and cowpea are relatively clustered, those for soybean are not. The assessor must continually be aware of the sources of the variance in any effects distribution. An untested species in a particular soil may be assumed to be a random draw from the distribution, or the distribution may represent the proportion of species in a plant community that is likely to be affected by a particular concentration of a chemical given uncertainty concerning the influence of soil type.

A further contribution to the distribution is variance in the level of response. Although tests of herbicides use median lethal concentrations as test endpoints (Fletcher et al., 1990), the test endpoints used by Efroymson et al. (1997) are highly inconsistent. In an attempt to increase consistency among the responses used in the distributions, factors were applied to test endpoints with severe effects.

Soil invertebrates

The community is the typical assessment endpoint entity for soil invertebrates. As with plants and other ecological receptors, the risk assessor can construct a species sensitivity distribution for toxicity to soil invertebrates, based on published tests with single chemicals in soil. For example, in the document in which ORNL benchmarks for soil invertebrates are derived, Efroymson, Will, and Suter (1997) present distributions of sensitivities of earthworms to concentrations of inorganic elements and organic compounds in soil. As in Figure 4.2 they are distributions of species–soil combinations. The distribution for the community endpoint can include tests of invertebrates other than earthworms, if they are available. Only a few microcosm studies or other toxicity tests exist in which invertebrate species including non-earthworm species were tested under identical conditions (e.g., Streit, 1984; Parmelee et al., 1993; Kammenga et al., 1994). The results are not sufficient to support generalizations about the relative sensitivity of different taxonomic groups.

4.1.10 Quantitative Structure–Activity Relationships

At many sites, ecotoxicological data will not be available for some of the contaminant chemicals. In such cases, it is highly desirable to conduct toxicity tests of those chemicals or of the contaminated media. However, if testing is not possible, models of toxicity termed quantitative structure–activity relationships (QSARs) may be applicable.

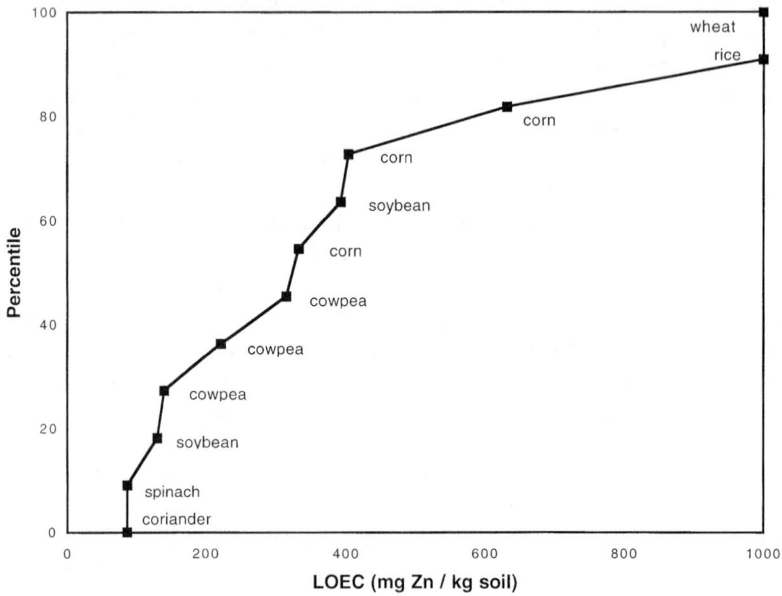

FIGURE 4.2 Cumulative distribution of LOECs for plants exposed to zinc in soil. Effects consist of significant changes in the mass of whole plants or plant tissues.

The most commonly used QSARs are those for the baseline narcosis mode of action which is relatively predictable (Veith et al., 1983; Hansch and Leo, 1995). Narcosis is the least toxic mode of action for organic chemicals, and it is expressed by all xenobiotic organic chemicals, even those that have more specific toxic modes of action. The QSARs for chronic narcosis of fish were generated using subchronic tests, which do not include some important responses such as sexual maturation or fecundity (Call et al., 1985). As a result, narcosis QSARs may underestimate toxicity. Therefore, one may infer that toxic effects are likely to occur if the values generated by this narcosis model are exceeded. However, lack of exceedence should not be used to infer that no effects will occur.

There are other significant limitations on the use of QSARs. First, QSARs are not available for many taxa (e.g., birds). Second, except for pesticides and herbicides, mechanisms of action are poorly defined for invertebrates and plants, and are not well defined for any nonmammalian taxon. Third, chronic effects may result from a different mechanism of action than the mechanism inducing acute lethality, the response that has been used to establish nearly all ecotoxicological QSARs. Therefore, QSARs should be used with great caution. QSARs for nonpesticide chemicals have been compiled in OPPT (1994), and the EPA Assessment Tools for Evaluation of Risk (ASTER) includes aquatic toxicity values plus aquatic biodegradation and physicochemical properties estimated using QSARs (http://www.epa.gov/medatwrk/databases/aster.html).

4.1.11 Mechanistic Models of Effects

Mechanistic models of effects are an alternative that must be mentioned for the sake of completeness. Although they offer considerable promise, they are not sufficiently developed for routine use in ecological risk assessments. For example, a mechanistic model of the acute toxicity of certain metals to fish has been proposed, based on interference of toxic metals with the uptake of sodium or calcium (Bergman and Dorward-King, 1997). However, the model is not yet a predictive tool and does not account for chronic effects which may have multiple mechanisms (Bergman and Dorward-King, 1997). Another example is the modeling of narcotic effects on fish based on partitioning of contaminants among internal compartments that vary in size (Lassiter and Hallam, 1990).

4.1.12 Ecotoxicity Profiles

Ecotoxicity profiles are summaries of the available information concerning the ecotoxicological properties of a chemical or material. They should be developed for chemicals or materials that are chemicals of potential ecological concern (COPECs). However, they should not simply be a dump of everything that is known about the ecotoxicology of the COPEC. They should be limited to information relevant to the contaminated media and to the endpoint entities and responses. Topics to be considered in a toxicity profile should include the relative toxicity of different forms of the chemical, its relative toxicity to different taxa and life stages, its modes and mechanisms of action, the temporal dynamics of its toxicity, and its interactions with other chemicals on the site. Toxicity should be related to all relevant exposure metrics, potentially including media concentrations, body burdens, biomarker levels, doses, and exposure duration.

Organizations that routinely engage in ecological risk assessments of contaminated sites often develop standard ecotoxicity profiles (Sample et al., 1997b). These off-the-shelf profiles are labor saving and also are useful for planning the assessment during the problem formulation process (Chapter 2). Reviews such as those produced by R. Eisler for the National Biological Service may also be useful during problem formulation (www.pwrc.usgs.gov/new/chrback.htm). However, such reviews do not serve the purpose of ecotoxicity profiles because they are lengthy and not organized in terms of the needs of risk assessment. For example, if species sensitivity distributions are to be used (Section 4.1.9.1), toxicity data should be reduced to a set of consistent endpoints, redundancies should be eliminated by calculating species mean values, and the data should be presented in ascending numerical order.

4.2 CONTAMINATED MEDIA TOXICITY TESTS

The toxicity of media can be tested in at least three ways. The most direct approach is to cage, pen, or plant organisms on the site along a gradient of contamination or at contaminated and reference sites. This approach, termed field testing or *in situ* testing, is relatively easy for immobile organisms such as plants and more difficult for organisms that are mobile and forage for food. Field testing may be highly realistic, in that the organisms are subject to realistic conditions and variation in exposure. However, such studies are subject to the effects of variation among sites in conditions other than contamination and to loss of the study due to vandalism, predation, or extreme conditions. In addition, cage effects may modify the sensitivity of the organisms. A good example of this technique is the use of caged mussels or clams to measure uptake of contaminants and associated effects on survival and growth (Jenkins et al., 1995; Salazar and Salazar, 1998). A much less used technique is bringing contaminated biota into the laboratory to test them. This technique is appropriate if the contaminant is persistent and bioaccumulated, or if it is known to cause persistent injury. For example, herring eggs from areas exposed to spilled oil and from unexposed areas were brought into the laboratory, and their hatching rates and frequencies of abnormalities recorded (Pearson et al., 1995). By far the most common approach is to bring contaminated and reference media into the laboratory for toxicity testing. This is a very active area of ecotoxicology, and test methods have been developed for ambient waters, sediments, soils, and biota. Methods specifically recommended for use at contaminated sites in the United States and Canada may be found in Office of Emergency and Remedial Response (1994a) and Keddy et al. (1995).

For nine reasons, testing the contaminated media from the site is generally more useful than testing individual chemicals in laboratory media.

1. The bioavailability of the contaminants is realistically represented. Because of sorption, the formation of complexes, and other processes that reduce the availability of a chemical for uptake by organisms, the toxic effects of a particular concentration of a chemical may be highly variable. In particular, standard single chemical toxicity tests are conducted under conditions that tend to maximize bioavailability so toxicity values from the literature may be highly conservative. Media toxicity tests greatly reduce or eliminate this source of uncertainty by conserving the bioavailability of the contaminants to which organisms are exposed on the site.
2. The form of the contaminants is realistic. The toxicity of chemicals depends on their form including the ionization states and co-ions for metals and other ionizable chemicals. Typically, the form of contaminants at a site is unknown. Even when it is known, the predominant form found at the site may not be one for which toxicity data are available. Media toxicity tests greatly reduce or eliminate this source of uncertainty by conserving the forms of the contaminants to which organisms are exposed on the site.
3. Combined toxic effects are elicited. Few sites are contaminated by only one chemical, and the toxic interactions of chemicals are seldom well known. In addition, the interactions depend on the form of the chemicals, which is itself problematical. Media toxicity tests greatly reduce or eliminate this source of uncertainty by retaining the combination of contaminants in the forms and proportions that occur at the site.
4. The effects of contaminants for which few or no relevant test data are available are included. Ecotoxicological testing has focused on pesticides and metals, not the industrial chemicals found at many contaminated sites.

Even for metals and pesticides, the taxa and responses of interest may not have been tested. Media toxicity tests greatly reduce or eliminate this source of uncertainty by including all contaminants to which organisms are exposed on the site in a test that has been chosen to represent the endpoint response.
5. The type of effects may be determined. The specific effects of the mixture may not be predictable from available knowledge of the effects of the components. The test can be designed to determine the occurrence of effects that are relevant to the assessment endpoint.
6. The spatial distribution of toxicity can be determined. The extent of the area to be assessed or remediated and the priority to be assigned to different sources or receptor ecosystems can be more appropriately determined on the basis of the distribution of toxicity than of chemical concentrations.
7. Remedial goals may be determined. In some cases, toxicity provides a better basis for defining media and areas to be remediated than chemical concentrations (Section 8.2).
8. The potential for achieving the level of anthropogenic effects specified in the assessment endpoint can be determined. In some cases, because of upstream or background contamination, it may be uncertain whether site remediation will significantly improve the ecological condition of the receiving system. Demonstrated toxicity of upstream water and sediments can provide a better basis for this determination than chemical concentrations.
9. The efficacy of remedial actions can be determined (Chapter 10). In many cases, toxicity provides a better basis for defining whether additional remediation other than chemical concentrations is needed.

For reasons such as these, media toxicity testing has been recommended by the EPA for use at contaminated sites (Office of Emergency and Remedial Response, 1994d). However, the qualifiers in the statements above point to the following limitations in media toxicity testing:

1. The medium may be modified by collection and preparation for toxicity testing. This has been a particular concern for testing of sediments, which may lose their physical structure and oxidation state during collection, sieving, and storage. Soils and water may be modified as well.
2. The forms and concentrations of chemicals may be modified by sample collection and processing. These changes may result from the changes in the medium just discussed or from direct effects such as loss of volatile chemicals to air or loss of chemicals from solution due to sorption to the walls of the sampling and testing containers.
3. The samples may be unrepresentative. This problem also occurs in sampling for chemical analysis, but may be more severe for media testing, because typically fewer samples are tested than are analyzed.
4. Most media toxicity tests have short durations, and few response parameters are recorded relative to conventional chronic toxicity tests.

5. The cause of the toxicity is unknown. Toxicity may be due to one or more contaminants in the tested medium, so it may not be clear what remedial actions are needed. In some cases, apparent toxicity may be due to extraneous factors such as chemical or physical properties of the medium or disease. For example, it has been necessary to UV-sterilize water from the Oak Ridge Reservation for fathead minnow larval tests because of an unidentified pathogen.
6. Apparent toxicity may be due to the choice of inappropriate reference locations. For example, relatively rapid growth may be due to high nutrient levels in reference media rather than toxicity in site media.

These limitations do not negate the considerable advantages of media toxicity testing. The first three can be avoided to a considerable extent by care in the collection and handling of samples and in the conduct of the tests. The fourth point requires analysis and interpretation of the results during the risk characterization, as with other test data. The fifth problem requires that additional tests be done to identify which components of the contaminant mixture are responsible, a process called toxicity identification evaluation (TIE) (EPA, 1991a, 1993f). In TIE, the toxic components of a mixture are identified by removing components of a mixture and testing the residue, fractionating the mixture and testing the fractions, or adding components of the mixture to background medium and testing the artificially contaminated medium. TIE is a powerful tool that is commonly applied to effluents but is seldom used at contaminated sites. An example from Oak Ridge is the use of TIE to demonstrate that low concentrations of nickel were responsible for the toxicity to *Ceriodaphnia* observed in Bear Creek (Kszos et al., 1992). Extension of the TIE process to include other properties of tested media could solve the sixth problem. TIE methods have been developed for water and can be extended to sediments by using pore water. TIEs may be performed for soil by spiking background soil. For example, toxicity of soils from the Lehigh Gap, PA, to isopods (*Porcellio scaber*) were correlated with concentrations of several metals, but tests of soils spiked with individual metals showed that zinc was responsible (Beyer and Storm, 1995).

As part of the analysis of effects, it is important to consider whether some qualitative or quantitative extrapolation model needs to be applied to the media toxicity tests to make them relevant to the assessment endpoint. The types of extrapolation models used with single chemical toxicity tests are potentially useful for these tests as well (Section 4.1.9).

Both control and reference media should be tested along with the contaminated media. Control media are laboratory media that are known to be appropriate for the test species. That is, control media support the maximal rates of survival, growth, and reproduction of the test species. The characteristics of control media are usually prescribed in standard test protocols. Reference media are media that come from the vicinity of the site, and are physically and chemically similar to the test media except that they do not contain the site contaminants. Reference media would include waters and sediments collected upstream of the site or soils from the same soil series as the contaminated soils, but not contaminated by the site. If upstream reference media are contaminated by an upstream source or if local soils are contaminated by a source other than the site (e.g., historic use of an arsenical pesticide or atmospheric deposition from a smelter), it may be desirable to obtain reference samples outside the range of those sources (a clean reference as opposed to the local reference). The control tests determine whether the test was conducted properly using healthy organisms. The local reference tests provide the basis for determining how much toxicity the site adds to proximate media. If a separate clean reference is used, it provides the basis for determining whether the differences from controls are due to contaminants or to properties of the media such as pH or texture. For example, water from Poplar Creek on the Oak Ridge Reservation was toxic to Japanese medaka embryos, water immediately upstream was a little less toxic, and water several kilometers upstream, above a municipal waste water treatment plant, was not toxic (equivalent to controls).

As in any form of toxicology, the best evidence for toxic effects is provided by demonstration of an exposure–response relationship. This can be done by testing samples collected on a contamination gradient or by testing a dilution series. An obvious example of the former is sampling and testing waters in a gradient downstream of a site. Often, particularly for soils, contamination gradients do not occur on the site. In such cases, an exposure series can be created by diluting the contaminated medium with clean reference media. It is obviously important to ensure that factors such as nutrient levels or texture do not confound the toxic effects by carefully matching the dilution medium to the test medium. Finally, an exposure–response series can be established by spiking site or reference media with the chemicals of concern. In such studies, it is important to match the forms of the test chemicals to those of the site contaminants. For soils and sediments, it is also appropriate to age the media to establish more realistic bioavailability (Heiger-Bernays et al., 1997). In addition to establishing that toxicity is responsible for observed effects, an exposure–response relationship can be used to establish well-defined remedial goals by establishing what level of the site's contaminant mixture has acceptably low levels of effects (Chapter 9).

Conventionally, media toxicity test data are analyzed using hypothesis testing statistics. That is, responses of each tested medium are determined to be either statistically significantly different from those in reference or control media, or not. If an exposure–response relationship can be established by gradients or dilution series, a function may be fit to the data and used to estimate exposure levels that cause prescribed levels of effects (LC50, EC10, etc.). If, as is nearly always the case, there is a mixture of contaminants, then exposure must be expressed as concentration of a representative chemical or some metric of aggregate concentration (Chapter 3). If there are no exposure gradients, assessors should at least report the level of effects relative to reference or controls in the tests of the individual samples and the associated variance among replicates (e.g., mortality = 0.22 ± 0.12).

In some cases, the site medium may be unsuitable for the test organisms to survive, grow, or reproduce. In those cases, one has the option of adjusting the medium or changing the test species. Adjusting the pH, hardness, or other physicochemical property is problematical because of the potential effects on the form, bioavailability, or toxicity of the contaminants. Such adjustments should not be performed when the medium properties are unsuitable because of properties of the waste. For example, leachate from the S-3 ponds in Oak Ridge was highly acidic, as well as having high metal levels, so it would have been inappropriate to adjust the pH of test waters from the receiving stream. However, in cases where the medium is naturally unsuitable for the test species, and the adjustment will not affect the chemical state of the contaminants, adjustments may be appropriate. The alternative, use of a test species that is appropriate to the medium, is conceptually more appealing. Particularly if the chosen species is resident to the site, use of a species characteristic of the type of ecosystem that is being assessed increases the apparent relevance of the test results. However, a standard test species that is adapted to the site medium may not be available and it is often difficult to develop testing procedures for a nonstandard species.

Tests for specific media are discussed below and test protocols are summarized in accompanying tables. In general, the tests performed at contaminated sites follow standard protocols from the EPA and ASTM (or other organizations outside the United States). Standard tests have the advantage that they are reasonably reliable, can be performed at reasonable cost by many laboratories, are acceptable to most regulators, and have known sensitivities. The most common deviation from standard protocols is the substitution of local species. However, some thought should be applied to determining whether nonstandard tests may be more appropriate. For example, neither acute lethality tests or the "subchronic" 7-day larval toxicity tests can detect effects on fecundity or early development of fish. Any chemical that has a primary mode of action that involves disruption of endocrine control of the formation of gametes or development of embryos would not be adequately tested by those methods. An early life-stage test or reproductive test would be more appropriate. Another example would be chemicals with very slow uptake kinetics, which would not be adequately tested unless that test were long enough for equilibrium between the test organisms and the medium to be attained. In such cases, tests longer than standard durations may be needed. Finally, the relationship of the test species and responses to the assessment endpoints must be considered. For example, if some property of a reptile or amphibian population is the assessment endpoint, none of the standard tests is suitable.

4.2.1 Aqueous Toxicity Tests

Standard toxicity tests have been developed for determining the acceptability of aqueous effluents and are widely used in effluent permitting in the United States. These tests are unique in the extent to which they have been validated against biosurvey data (Dickson et al., 1992; Grothe et al., 1996). In a number of studies, the 7-day fathead minnow and *Ceriodaphnia dubia* tests have been found to be predictive of reductions in the species richness of aquatic communities. As a result of this intensive development and validation, these tests are widely used at waste sites, and many laboratories are available to conduct them. Those and other aquatic toxicity tests that are appropriate for testing contaminated waters are listed in Table 4.9.

4.2.2 Sediment Toxicity Tests

Sediment toxicity tests are in a less advanced stage of development than are aqueous toxicity tests (Office of Emergency and Remedial Response, 1994a). Relatively few

TABLE 4.9
Standard Procedures Used to Test the Toxicity of Ambient Water

Species	Life Stage	Response	Duration (days)	Medium[a]	Ref.
Sea urchin	Eggs and sperm	Fertilization	0.3	SW	Weber et al., 1988; ASTM, 1999
Daphnia sp.	Juvenile	Immobilization	2	FW	Weber, 1991; ASTM, 1999
Bivalve mollusk	Larvae	Mortality, shell development	2	SW	Weber, 1991; ASTM, 1999
Fish[b]	Juvenile	Mortality	4	FW/SW	Weber, 1991; ASTM, 1999
Algae (*Selenastrum capricornutum*)	Cell culture	Growth	4	FW	Weber et al., 1989; ASTM, 1999
Ceriodaphnia dubia or *Mysidopsis bahia*	Juveniles–adults	Immobilization, fecundity	7	FW/SW	Weber et al., 1988, 1989; ASTM, 1999
Fish[b]	Larvae	Mortality, growth	7	FW/SW	Weber et al., 1988, 1989; ASTM, 1999
	Embryo-larvae	Mortality, deformities	7	FW/SW	Weber et al., 1988, 1989
Algae (*Champia parvula*)	Culture	Sexual fecundity	7	SW	Weber et al., 1988; ASTM, 1999)

Note: The tests listed are those that have been recommended by the EPA for contaminated sites (Office of Emergency and Remedial Response, 1994a).

[a] FW = fresh water, SW = salt water, FW/SW = protocols are available for species from both media.
[b] The standard freshwater fish in the United States is the fathead minnow (*Pimephales promelas*) and for salt water is the sheepshead minnow (*Cyprinidon variegatus*) or inland silverside (*Menidia beryllina*).

protocols have been standardized (Table 4.10) and they have not been thoroughly validated against biosurvey data. However, short-term tests with marine and estuarine amphipods and oligochaetes have been extensively used to evaluate the relative toxicity of coastal sediments (Long et al., 1995; MacDonald, 1996). Standard protocols for freshwater sediments are limited to those for the amphipod *Hyalella azteca* and the midge *Chironomus tentans*. The selection of test organism depends, in part, on their sensitivity to the site contaminants and tolerance to ecological conditions (e.g., salinity and grain size; ASTM, 1999; Office of Emergency and Remedial

TABLE 4.10
Standard Procedures Used to Test the Toxicity of Ambient Sediments

Species	Life Stage	Response	Duration (days)	Medium[a]	Reference
Chironomus tentans	Larvae	Mortality and growth	10	FW	Office of Emergency and Remedial Response, 1994a
Hyalella azteca	N/A	Mortality and growth	10	FW/ME[b]	Office of Emergency and Remedial Response, 1994a
Amphipod, marine sp.[c]	N/A	Mortality, emergence, and reburial	10	ME	ASTM, 1999
Polychaetes[d]	Recently emerged juveniles and young adults	Mortality	10	ME	ASTM, 1999
Polychaetes[d]	Recently emerged juveniles	Mortality and growth	20-28	ME	ASTM, 1999

Note: The tests listed include those that have been recommended by the EPA for contaminated sites (Office of Emergency and Remedial Response 1994a).

[a] FW = freshwater sediment, ME = marine or estuarine sediment, FW/ME = protocols are available for species from both media.

[b] *H. azteca* can be tested in estuarine sediments up to 15% salinity.

[c] The standard marine or estuarine amphipods in the United States are *Rhepoxynius abronius*, *Eohaustorius estuarius*, *Ampelisca abdita*, *Grandidierella japonica*, and *Leptocheirus plumulosus*.

[d] The standard marine or estuarine polychaetous annelids in the United States are *Neanthes arenaceodentata* and *N. virens*.

Response, 1994a). For example, *H. azteca* can be used in tests of estuarine (≤ 15% salinity) but not marine sediments, whereas *Rhepoxynius abronius* can only be used for tests of marine sediments.

Seasonal changes in sediment chemistry (e.g., redox potential) may modify the bioavailability and toxicity of sediment-associated contaminants. For example, toxicity and bioconcentration tests of clams (*Mya arenaria*) were performed with a mixture of metals in water designed to simulate the interstitial water of Narragansett Bay (Eisler, 1995). The mixture was lethal at simulated summer temperatures, but not winter temperatures. It is advisable to conduct seasonal tests and to consider environmental characteristics that may modify toxicity when developing sampling and monitoring plans.

Sediment toxicity tests are generally conducted using whole-sediment samples and benthic infauna or epibenthic fauna. Sediment interstitial water can be tested using the standard aquatic toxicity test methods discussed above. The pore water is extracted from the bulk sediment sample and, typically, used in tests with invertebrates. This approach can help identify the mechanisms of exposure for benthic infauna (i.e., respiration of contaminated pore water, ingestion of contaminated sediments, or both). This knowledge can be used to plan further sampling and interpret the results of other analyses (e.g., benthic invertebrate surveys). The disadvantage is that the extraction and testing processes can alter the form and bioavailability of the contaminants. Consequently, tests of interstitial water are not recommended for most assessments, but may be advisable when the mechanism of exposure may change the conclusions or remedial goals.

4.2.3 Soil Toxicity Tests

Testing of contaminated soils in the United States is largely limited to seedlings of vascular plants and earthworms. Some of the standardized soil tests (i.e., those that are recommended by a government entity such as the EPA or a standardization organization such as the ASTM) are presented in Table 4.11. A variety of additional tests have been developed for single chemicals in soil that might be adapted for use with contaminated site soils (Linder et al., 1992; Van Straalen and Van Gestel, 1993; Donker et al., 1994; Heiger-Bernays et al., 1997). In particular, recent research has expanded the range of soil invertebrates used in toxicity testing (Donkin and Dusenbery, 1993; van Gestel and van Straalen, 1994; Kammenga et al., 1996). Many new but not yet standard methodologies for testing invertebrates were developed within a European Union project entitled Development, Improvement and Standardization of Test Systems for Assessing Sublethal Effects of Chemicals on Fauna in the Soil Ecosystem (Laskowski et al., 1998). Tests presented in a summary volume include the enchytraeid *Cognettia sphagnetorum*, *Eisenia fetida*, *Aporrectodea caliginosa*;

TABLE 4.11

Standard Procedures Used to Test the Toxicity of Ambient Soils

Type of organism	Life Stage	Response	Duration (days)	Medium	Ref.
Earthworm[a]	Adult	Mortality	14	Soil	Greene et al., 1988
Earthworm	Adult	Reproduction	35	Soil	ISO, 1991
Plant[a]	Seed	Germination	5	Soil	Greene et al., 1988; Linder et al., 1992, 1990
Plant[a]	Seed	Root elongation	5	Elutriate	Greene et al., 1988; Linder et al., 1990, 1992
Plant	Seedling	Mortality and vegetative vigor	20–90	Soil	Linder et al., 1992
	Seedling	Weight	45	Soil	Linder et al., 1992
	Life cycle	Reproduction and growth	28–44	Elutriate	Linder et al., 1992

[a] Recommended by the EPA for use at contaminated sites (Office of Emergency and Remedial Response, 1994a).

the oribatid mite *Platynothrus peltifer*; collembolans *Isotoma viridis*, *Folsomia candida*, and *Folsomia fimetaria*; the Staphylinid beetle *Philonthus cognatus*; the centipede *Lithobius mutabilis*; the millipede *Brchydesmus superus*; the isopod *Porcellio scaber*; the nematodes *Plectus acuminatus* and *Heterocephalobus pauciannulatus* (competition); and predation of the predatory mite *Hypoaspis aculeifer* on the collembolan *Folsomia fimetaria* (Lokke and van Gestel, 1998). In addition a test of a soil alga has recently been developed (Hammel et al., 1998).

Because these tests are performed in the soil from the contaminated site, there is generally less need to consider normalization of the concentrations of the chemical in soil than when using literature values. However, care must be taken to match reference soils to contaminated soils in terms of chemistry, texture, and nutrient status. Particularly for growth and reproduction endpoints, tests may be highly sensitive to soil properties. Therefore, it is desirable to test soils from multiple reference locations to estimate the natural variance. Normalization of concentrations can reduce that variance. For example, soil metal concentrations were pH-normalized to reduce variation among locations at a metal mining and milling site in Anaconda, MT (Kapustka et al., 1995). If site and reference soils have low organic matter or inorganic nutrient levels, it may be necessary to amend or fertilize them to support the test organisms, to achieve reasonable growth in the reference soils, and to bring site and reference soils to the same levels. For the Bear Creek Valley remedial investigation at Oak Ridge, only upland reference locations were available for a contaminated floodplain. Thus, plants did not germinate or grow well in the reference soil, and the ambient toxicity test was not as large a contributor to the final risk characterization as it would have been if an appropriate reference site had been selected.

As with single-chemical toxicity tests (Section 4.1.4), it is important to choose test organisms and toxicity endpoints that are relevant to the assessment endpoints. If possible, invertebrate species should be representative of assessment endpoints in function, taxonomy, trophic level, life history strategy, and route of exposure to toxicants (Spurgeon and Hopkin, 1996b). Tests of earthworms are relevant to the soil invertebrate community in most nonarid ecosystems. *Eisenia fetida*, the most common test organism in soil toxicology, has about average sensitivity to toxicants among earthworms (Laskowski et al., 1998). It is advisable to include more than one test endpoint if relevant to the assessment endpoint. For example, Spurgeon and Hopkin (1996b) found effects of a contaminated soil on growth of juvenile earthworms but not cocoon production of adults. In contrast, tests of earthworms would not be as relevant as arthropod tests to the most important invertebrate species in arid environments.

Field testing with contaminated soils at the site is useful, but seldom performed. An example of field testing is the placement of worms for 7 days in contaminated soil in plastic buckets buried at the locations where the soil was collected (Menzie et al., 1992). Through this field testing Menzie et al. (1992) found that most toxic soils occurred as veins through site drainage areas. Carabid beetles have been tested in field pens on pesticide-contaminated soils (Heimbach et al., 1994). Nwosu et al. (1991) describe a seed germination test using cucumber (*Cucumis sativus*) and wheat (*Triticum aestivum*) in soil contaminated with Surflan (3,5-dinitro-N^4,N^4-dipropylsulfanylamide). Field tests may be performed for processes as well as organisms. The process of decomposition in a pasture contaminated with timber preservative was tested by burying cotton strips and measuring tensile strength (Yeates et al., 1994). The researchers found that decomposition was impaired in areas with high soil copper, chromium, or arsenic contents. Similarly, bags of leaf litter placed on metal-contaminated soils showed reduced decomposition (Strojan, 1978). Several field testing methods were also proposed by Linder et al. (1992). Field tests incorporate environmental variables and may be used to confirm that analogous laboratory tests represent toxicity at waste sites. In the laboratory, ambient media toxicity tests do not simulate some environmental conditions, such as changing soil moisture with precipitation events, volatilization of the contaminant, the plant rhizosphere community, transport of chemicals through macropores, and aerial deposition.

Most phytotoxicity tests of contaminated soil use crop species. For CERCLA purposes, lettuce is listed as the standard species of the seed germination and root elongation assay, although the use of other species is permitted (Greene et al., 1988, Kapustka, 1997). Other common toxicity test species include alfalfa, beet, clover, corn, cucumber, foxtail millet, mustard, oat, perennial ryegrass, pinto bean, soybean, sorghum, radish, and wheat (Kapustka, 1997). Advantages are that these species have been demonstrated to grow well in the laboratory and to have replicable toxic effects. The potential for litigation is also a driver for using standard, established tests (Kapustka, 1997).

Because ecological risk assessors are interested in the toxicity to native species, it would be advantageous to test these. For example, unrelated plant species have different root morphology, development patterns, and carbon allocation patterns that may make results of a root elongation test more or less relevant to assessment endpoint species (Kapustka, 1997). Similarly, Ernst (1998) argues that exposure durations in phytotoxicity tests should vary with species. LeJeune et al. (1996a) tested hybrid poplar as a surrogate for *Salix* spp. and *Populus* spp. at the Clark Fork River, MT. Examples of the use of native species as test organisms include using common yarrow (*Achillea millefolium*) and yellow nutsedge (*Cyperus esculentus*) to test soils from the Upper Pecos Mine site in northern New Mexico and the Naval Weapons Station at Concord, CA, respectively (LeJeune et al., 1996b; Jenkins et al., 1995). Kapustka (1997) offers advice to the assessor who plans to use nonstandard test species. A few of his rules are: choose nominal performance standards (e.g., percent germination), characterize statistical variability using reference soils, and use one or more species in addition to the nonstandard species. Laskowski et al. (1998) also provide recommendations regarding testing invertebrate taxa that are native to the site, given that their history, age, and physical condition is unknown: (1) maintain collected organisms in the laboratory for at least 2 weeks prior to testing, if culturing is not possible, and (2) make sure that test organisms in control treatments survive beyond the end of the test.

The soil for toxicity tests should be sampled to an appropriate depth for the endpoint plant population or community. Appropriate exposure depths (i.e., rooting depths of plants) are discussed in Section 3.4.3.1. In one of the Clark Fork River studies described above, a shallower depth interval was used for tests with hybrid poplar than with alfalfa, lettuce, and wheat (LeJeune et al., 1996a).

In the final risk characterization, ambient toxicity test results should be correlated with measured soil concentrations of chemicals. For example, at the Anaconda mine site in Montana, concentrations of arsenic, copper, zinc, and, to a lesser degree, lead and cadmium were positively correlated with phytotoxicity (Kapustka et al., 1995). Similarly, metal concentrations in soil were correlated with earthworm growth and mortality in Concord, CA (Jenkins et al., 1995). These results may be used to determine the reliability of the single-chemical toxicity line of evidence (Section 4.1.4). For example, zinc was at least ten times more toxic to *Eisenia foetida* in artificial soil than in contaminated soils collected from the field (Spurgeon and Hopkin, 1995).

Ambient toxicity tests may also be used to prioritize sites for remediation (Chapter 9). For example, Kapustka et al. (1995) reported phytotoxicity scores for locations at the Anaconda site in Montana. Scores were calculated to combine results for three test species and six response parameters for soils from a particular location.

4.2.4 Ambient Media Tests with Wildlife

Toxicity tests for wildlife may be separated into two categories: laboratory tests in which the test species are exposed to contaminated media under a controlled, laboratory setting and *in situ* tests where test animals are exposed to contaminants in the field. Laboratory tests may be for oral, dermal, or inhalation exposures. In general, oral exposure tests are most applicable for risk assessment purposes (see Section 3.9). However, if the dermal or inhalation exposure pathways are considered critical at a particular site, toxicity tests for these pathways may be performed. Standard dermal and inhalation test methods are presented in ASTM (1999). Additional discussion of ambient media toxicity tests for wildlife is presented by Linder et al. (1992).

Oral Toxicity Tests

Contaminated food, soil, or water from the site of interest may be collected and fed to test animals. These tests provide an indication of the toxicity of chemical mixtures present at the site. Standard methods for ambient media toxicity tests for wildlife do not exist. Methods for these tests may be developed ad hoc or by modification of standard test methods (Section 4.1.5). While ambient media toxicity tests with wildlife are not widely performed, examples do exist and some are summarized in Table 4.12.

In Situ *Toxicity Tests*

In *in situ* toxicity tests, animals are introduced or otherwise exposed at the contaminated site. The primary advantage of *in situ* tests over laboratory tests is that they allow ambient conditions to influence toxic effects and therefore may provide a more realistic measure of actual toxicity at the site. However, due to the great mobility of most wildlife species, *in situ* wildlife tests are problematic for most species and generally only suitable for species with small home ranges. Two potential *in situ* approaches are mesocosms or nest boxes.

Mesocosms are enclosed outdoor experimental systems in which effects of contaminants on biota can be studied. For example, Dieter et al. (1995) placed

TABLE 4.12

Examples of Ambient Media Toxicity Tests to Evaluate Effects of Environmental Contaminants on Wildlife

Test Species	Reason for Test	Contaminants of Concern	Test Media	Toxicity Test Endpoint	Ref.
Mink	Determine causes of mortality of ranch mink fed fish from the Great Lakes	Organochlorine pesticides and PCBs	Diets in which different fish species collected from different lakes were incorporated	Reproduction	Aulerich et al., 1971, 1973
	Determine the hazard that consumption of Great Lakes fish may present to wild mink populations	Organochlorine pesticides, PCBs, and dioxins	Diets in which carp collected from Saginaw Bay, MI were incorporated	Mortality, reproduction, hematology, liver pathology, bioaccumulation	Heaton et al., 1995a,b; Tillitt et al., 1996
	Determine the hazard that consumption of fish downstream of a United States DOE facility may present to wild mink populations	PCBs, mercury	Diets in which fish collected from Clinch River and Poplar Creek, TN were incorporated	Reproduction	Halbrook et al., 1997, 1998
Least shrew	Determine if bioavailability and accumulation of heavy metals in sewage sludge present a risk to secondary consumers	Cadmium, copper, lead, zinc	Diets in which earthworms collected from a sewage sludge-treated site were incorporated.	Growth, bioaccumulation	Brueske and Barrett, 1991
Mallard Ferret	Determine toxicity of weathered *Exxon Valdez* crude oil to surrogates for seabirds and sea otters	Weathered crude oil	Weathered *Exxon Valdez* crude oil by capsules, gavage, or incorporated into diets	Mortality, food avoidance, organ pathology	Stubblefield et al., 1995a
Mallard	Determine toxicity of weathered *Exxon Valdez* crude oil to surrogate for seabirds	Weathered crude oil	Weathered *Exxon Valdez* crude oil incorporated into diets	Reproduction	Stubblefield et al., 1995b

mallard ducklings in littoral mesocosms to evaluate effects of aerial application of the organophosphate insecticide phorate on waterfowl in prairie wetlands. In another study, Barrett (1968) evaluated the effects of the carbamate insecticide carbaryl on plants, arthropods, and small mammals within 1-acre oldfield enclosures. Additional mesocosm methods for terrestrial systems are currently being developed. Sheffield (1995) presents methods for the design and construction of terrestrial mesocosms for use at contaminated sites.

An alternative to mesocosms is the use of nest boxes. Nest boxes consist of wooden, plastic, or metal structures attached to posts or trees within or around a contaminated site. Because nest sites are frequently a limiting resource for cavity-nesting birds, free-living individuals are likely to be attracted to nest boxes placed on the site. Once the birds become established, potential effects of contamination on behavior, diet, and reproduction can be studied. A wide variety of bird species with varying foraging strategies may use nest boxes. They include starlings, bluebirds, tree swallows, wood ducks, barn owls, and American kestrels (EPA, 1989b). Guidance for the use of nest boxes for the studies of starlings at contaminated sites is presented in EPA (1989b). Nest boxes have previously been employed to evaluate risks to birds from aerial application of insecticides to agricultural fields or forests (Robinson et al., 1988; Pascual, 1994; Craft and Craft, 1996), PCBs and heavy metals at a Superfund site (Arenal and Halbrook, 1997), and lead along a highway (Grue et al., 1986).

Despite the availability of methods to perform ambient media toxicity tests for wildlife and their great utility in reducing uncertainty and strengthening risk conclusions, these tests are generally conducted for only the largest, most complex sites. There are several reasons for the lack of broad application of these tests. First, the care, housing, and feeding of adequate numbers of test animals required by these toxicity tests may be quite expensive. Obtaining sufficient contaminated food materials to maintain test animals for the duration of the study may also be difficult. For example, the mink toxicity study performed as part of the ORNL Clinch River ERA (Halbrook et al., 1997, 1999) required over 2000 kg of contaminated fish to maintain 50 mink for 6.5 months (DOE, 1996b). Space required to house sufficient test animals can also limit the application of these tests. A final limitation to the wider use of these tests is time. Many wildlife species reproduce only once per year and have generation times measured in years. These long time requirements make tests difficult to perform within the time constraints of most ERAs.

4.3 BIOLOGICAL SURVEYS

Biological surveys of effects include a variety of techniques for enumeration and characterization of biological populations, communities, and ecosystems. In the simplest case, the measure of effect for the biological survey is an estimate of the assessment endpoint. In such cases, the effects analysis consists of summarizing the data in such a way as to reveal the relationship of effects to exposure. Examples would include plotting the species richness of the soil microinvertebrate assemblage on exposure axes such as kilometers from a source, total petroleum hydrocarbons, or concentrations of a particular chemical. The EPA has recommended the use of biological surveys for ecological risk assessment of contaminated sites when feasible and appropriate (Office of Emergency and Remedial Response, 1994b; Sprenger and Charters, 1997).

A frequent problem in the use of biological surveys is that the entities and properties measured bear an undefined relationship to the assessment endpoints. They are often referred to as indicators or surrogates without defining what they indicate or for what they are surrogates. If the measures of effect do not directly estimate the assessment endpoint, then the relationship between them must be clearly characterized. For example, if data are available for stream macroinvertebrates and the assessment endpoint is some property of the fish community, then the relationship between them must be characterized in terms of the trophic dependence of fish on invertebrates, the relative sensitivity of fish and invertebrates, the similarity of their exposure, and other relevant properties. Clearly, this difficulty should be avoided in the problem formulation by selecting measures of effects that correspond as nearly as possible to the assessment endpoint (Chapter 2).

The following points should be considered when deciding whether biological surveys are appropriate for analysis of effects at a site, and, if so, what should be surveyed and how.

Scale — Highly mobile organisms and the populations and communities that include them are seldom appropriate for biological surveys. For example, a survey of breeding birds was conducted on the East Fork Poplar Creek floodplain in Oak Ridge, but it contributed nothing to the results of the ecological risk assessment. Territorial birds are highly mobile and are nearly always space limited, so all sites that contain physically suitable habitat are quickly occupied whatever the longevity or reproductive success of the resident birds may be.

Interpretation — To interpret the variation observed in results of biological surveys, the properties measured must be stable and consistent across similar sites, in the absence of contamination or disturbance, relative to the magnitude of effects that is considered significant. For example, population densities of microtine rodents are notoriously variable across time and space, varying by orders of magnitude in the absence of any anthropogenic effects. In contrast, properties of benthic invertebrate communities and stream fish communities are relatively stable and are commonly used to detect anthropogenic effects by comparing exposed communities to reference ones.

Difficulty — Clearly, if biological surveys are highly costly and time consuming or are likely to fail due to the difficulty of proper execution or if the necessary conditions for success are unlikely to occur, they are inappropriate. For example, determining the reproductive success of kingfisher populations has proved to be quite difficult while the reproductive success of birds that nest colonially and in the open is relatively simple (Halbrook, Brewer, and Buehler, 1999). Similarly, the fish communities of wadable steams can be easily quantified with great accuracy, but the abundances of fish populations and communities of large bodies of water cannot be quantified with sufficient accuracy or precision for many assessments.

Appropriateness — Techniques employed must be suitable for the species or community, season, and habitat of interest and should produce results that meet the objectives of the risk assessment.

Technical expertise — In some cases the assessors do not have the expertise or experience needed to perform certain surveys. In some cases, the need for technical expertise can be greatly reduced by a simple change in the survey techniques or endpoint. For example, technicians who can identify benthic invertebrates to species are in short supply, but little information is lost if only families are identified, and individuals with very little training can sort invertebrates into higher taxa without knowing their names.

Consequences of the survey — In some cases the biological survey may cause unacceptable injury to the site. The destructive sampling of rare species is an obvious example.

Data availability — Data not generated by the assessment program should be used if pertinent and of adequate quality. However, care must be taken to analyze and interpret them appropriately. For example, fish survey data have been collected by the Tennessee Valley Authority for the purpose of comparing the quality of their reservoirs. Those data were used to determine that the Oak Ridge Reservation had not altered the fish community of Watts Bar Reservoir relative to other reservoirs in the system, but they could not be used to infer risks at the scale of embayments, where the contamination was highest.

4.3.1 Aquatic Biological Surveys

Aquatic biota surveyed for waste site assessments may include periphyton, plankton, fish, and benthic macorinvertebrates (Office of Emergency and Remedial Response, 1994b). The choice of assemblage and sampling method depends on the endpoints and habitat characteristics (Appendix A). Care should be taken to ensure that the survey locations capture the variation in exposure while recognizing the scale of the system relative to the habitat requirements and mobility of the surveyed organisms.

Habitat quality information is critical to the ability to discriminate between contaminant effects and natural variability. The information must be accounted for in the survey design and should be quantified to the extent possible for all sites. The relevant habitat factors depend on the types of organisms being surveyed. For example, photosynthetically active radiation is important for algal and periphyton surveys, cover type and stream structure are important for fish surveys, and water chemistry (e.g., pH, hardness, and conductivity) is important for all assemblages.

The common use of biosurveys in aquatic systems, particularly streams, presents both advantages and disadvantages for risk assessors. The primary advantages are that methods are well established, and the expertise to perform surveys is commonly available. In addition, in states such as Ohio where bioassessment programs are under way, community types are already classified, reference conditions are established, and criteria for injury are defined (Yoder and Rankin, 1995, 1998). The primary disadvantage is that the metrics and indexes chosen for monitoring or regulatory programs may not be appropriate for a contaminated site risk assessment and remediation. In particular, the biotic indexes that are commonly used are designed to discriminate sites with common sorts or disturbances, particularly organic enrichment (Karr and Chu, 1997). They are relatively insensitive to toxic effects (Dickson et al., 1992; Hartwell et al., 1995). Biological surveys for risk assessments of contaminated sites should focus on metrics that are sensitive to the contaminants of concern and are sufficiently valued to support a remedial action.

4.3.1.1 Periphyton

Algae and other aquatic plants are much less often included in biosurveys than fish or invertebrates. However, they have been recommended by the EPA, which further recommends species composition, richness and relative abundance, biomass, diversity, evenness, and similarity as periphyton survey metrics (Office of Emergency and Remedial Response, 1994b). Except for estuarine species, aquatic macrophytes are more likely to be considered noxious weeds than valued endpoint entities. Periphyton have long been used as indicators of stream quality because they are ubiquitous, constitute the base of most lotic food chains, are in direct contact with the water, are sessile, are sensitive to a wide range of stressors, respond quickly to changes in water quality, and are easily sampled (Rosen, 1995). Periphyton have the practical advantages of being easily associated with a site and being easily collected by scraping from natural or artificial substrates. However, it is often difficult to demonstrate to decision makers that a change in algal community properties is adverse, particularly when disturbances increase algal production due to increased light or nutrient levels.

Periphyton samples can be collected from artificial substrates (e.g., frosted glass slides), which are placed in the water for a set period of time (e.g., 2 to 4 weeks) and then removed for analysis, or from natural substrates. The principal advantages of artificial substrates are ease of use, repeatability of measurements, and reduced variability of taxonomic composition and relative abundance. However, they are selective for particular species and the results may not be representative of the entire periphyton community. Aloi (1990) reviewed several approaches used for periphyton sampling and recommends that natural substrates should be preferred for studies of biomass, species composition, or primary productivity, and artificial substrates should only be used when experimental design dictates. However, both approaches have their supporters and detractors in the regulatory community (Rosen, 1995), and the choice of natural or artificial substrates should be made in cooperation with the local agencies of concern.

Periphyton communities vary widely in response to microhabitats even within the scale of individual sampling units, especially on rocks or other natural substrates. One can reduce variability by compositing multiple samples collected from a single type of habitat within a stream reach. For the assessment of water quality, collecting only from riffles and runs in streams is generally sufficient and periphyton communities in these habitats (particularly in those with current velocities of 10 to 20 cm/s) are less variable than in pools and edge habitat (Rosen, 1995). Also, sampling the soft substrates found in depositional areas is more difficult and time-consuming than sampling hard substrates (Warren-Hicks et al., 1989).

The measures of effect may be structural or functional. Structural measures are of two categories, measures of taxonomic composition and measures of standing crop (biomass). Common measures of taxonomic composition are species richness and relative abundance. Census of the sampled periphyton is usually not possible, but enough cells must be counted to ensure that uncommon species are included (Rosen, 1995). Taxonomic identification should be at least to genus for soft algae and to species for diatoms (Rosen, 1995). Because diatoms are common, abundant, and relatively easily identified by their silicaceous frustules, they are often counted to the exclusion of soft algae. Standing crop is measured as *chlorophyll a*, ash-free dry weight, cell counts, and cell volume. Each method has limitations, but all are generally acceptable for comparisons between exposed and reference sites. Indexes of structural characteristics include diversity indexes and indexes of similarity between sites. Although useful in conjunction with other structural measurements, such indexes should not be used as the only measure of periphyton structure.

The functional measure used for periphyton is primary productivity. The most common and widely accepted methods for estimation of primary productivity are based on the production of oxygen (O_2 method) or the uptake of radioactive carbon (^{14}C method) (Rosen, 1995). Choose the method that best fits the budgetary, logistical, and quality requirements of the assessment. The O_2 method is inexpensive, relatively simple to perform, and is readily used in the field. O_2 measurements must be converted to units of carbon, which introduces additional error. The advent of microelectrode technology has simplified and improved the measurement of O_2 production. The ^{14}C method is more expensive, more complicated to perform, and much less amenable to field use than is the O_2 method. However, it is a direct measure of primary productivity and is more sensitive than the O_2 method (Rosen, 1995).

Physicochemical parameters to be measured and controlled for in selecting reference sites include substrate composition, current velocity, temperature, photosynthetically active radiation, dissolved oxygen, conductivity, alkalinity, hardness, and nutrients (Warren-Hicks et al., 1989).

4.3.1.2 Plankton

Plankton are the microscopic algae (phytoplankton) and invertebrates (zooplankton) suspended in the water column with little to no ability to resist the current (APHA, 1999). Plankton are traditionally used as indicators of water quality in lentic systems. They are ubiquitous, are in direct contact with the water, are sensitive to a variety of stressors, respond quickly to changes in water quality, and have a direct impact on water quality (APHA, 1999).

Plankton may be collected from discrete depths or be integrated over a range of depths or horizontal distances, depending on the expected distribution of the stressor(s). Method selection depends on the target organisms, target depths, and desired sample quality (Appendix A). Measurements include species richness, relative abundance, and community indexes (e.g., diversity and similarity). Phytoplankton are often used as the sole representative of the plankton community (APHA, 1999). They are sufficiently diverse to permit the evaluation of a variety of stressors. It is especially important to collect physical data and water samples for analyses (temperature, photosynthetically active radiation, dissolved oxygen, conductivity, alkalinity, hardness, contaminants, and nutrients) in conjunction with plankton samples; otherwise, it may be difficult to associate exposure with effects in large open bodies of water (APHA, 1999).

4.3.1.3 Fish

Biological surveys commonly include fish because the value of fish is generally acknowledged and fish respond to a variety of aqueous contaminants. In addition, fish have practical advantages; their environmental requirements are well known, they integrate effects at lower trophic levels, and identification is relatively simple. Collection methods include electrofishing, nets, and traps (Appendix A). Method selection depends on the habitat characteristics and study design. Relevant habitat characteristics include cover type, stream structure, flow rate, pH, hardness, alkalinity, conductivity, and temperature.

Streams are typically sampled using electrofishing or seining methods. High-quality estimates of species presence and abundance can be obtained by blocking the upstream and downstream approaches with nets and then repeatedly sampling the reach. The resulting attributes are expressed per unit area, rather than per unit effort, which is less precise for this method. Electrofishing, seines, or hoop nets are used for large streams and rivers, whereas electrofishing, gill nets, fyke nets, and subsurface trawls are best used for lakes and marine environments (Warren-Hicks et al., 1989). Boat-mounted electrofishing units are used in large rivers and lakes. The inability to restrict fish movement in open bodies of water results in relative measures of fish community metrics (i.e., numbers per unit effort). Stationary nets are highly selective; the results should not be compared with results obtained with other sampling techniques.

Most commonly, fish community properties or indexes that combine several properties are used as measures of effect. The properties may include the number of species, the number of trophic groups, the abundances of species or trophic groups, the biomass of species or the community, and size distributions. The indexes may include conventional diversity indexes or arithmetic combinations of heterogeneous variables, most notably the Index of Biotic Integrity (IBI) and its derivatives (Karr et al., 1986). These indexes are preferred by many state agencies because they are used in water quality management programs (Simon and Lyons, 1995). They have many disadvantages as effects measures in risk assessment, which can be largely mitigated by disaggregating the index to its component metrics (Suter, 1993a). Properties of individual fish populations are less commonly used as endpoints in surveys, but they would be appropriate where game, commercial, rare, or otherwise particularly valued species are present. Appropriate population properties include abundance, size distribution, and production. The only commonly used properties of individual fish are frequencies of gross pathologies and anomalies. These are easily noted while counting and measuring fish from a community survey and are often of concern to the public and risk managers. The EPA recommends species richness and relative abundance as fish survey metrics (Office of Emergency and Remedial Response, 1994b).

Because fish are mobile, attention must be paid to the range of movement relative to the scale of contamination. For this reason, fish surveys are used more in streams, where movement is relatively limited, than in lakes or estuaries. Where movement is a problem, it may be desirable to focus on species such as sunfish that are relatively sessile rather than community properties, which may be influenced by highly mobile or schooling species such as shad.

4.3.1.4 Benthic Invertebrates

Benthic macroinvertebrate communities are commonly surveyed for hazardous waste site assessments, because they are ubiquitous, important components of lentic and lotic food chains, in direct contact with the water or sediment, relatively immobile, and sensitive to a wide range of stressors.

Benthic invertebrates in lotic systems are frequently collected from cobble substrates in riffles and runs. The techniques are well established and the results can be compared with many other similarly sampled sites (DeShon, 1995). These riffle communities are exposed to water-borne contaminants but not sediment-associated contaminants. Benthic invertebrates in riffles are exposed primarily via respiration of contaminated water, whereas benthic invertebrates in sediment depositional areas are often immersed in and ingest the contaminated sediment. Respiration of overlying water may still be an important pathway for sediment-dwelling organisms, especially for those that ventilate their burrows (e.g., *Hexagenia* mayflies), but not to the exclusion of sediment-associated pathways, which include respiration of sediment pore water.

Kerans et al. (1992) analyzed the results of paired riffle and pool surveys for multiple streams in Tennessee. In many instances the results for both riffles and pools correctly classified the streams regarding human impacts (based on a fish community index). However, when the classifications differed between riffles and pools, the results for pools were nearly always "correct" (i.e., consistent with the classification based on the fish community index). The authors concluded that sampling only one habitat type, either riffles or pools, generates biased results (Kerans et al., 1992).

The authors recommend that benthic invertebrate communities in sediment depositional areas be surveyed in addition to the riffle communities if any of the contaminants of concern are likely to be particle associated. This will be the case for most contaminants at hazardous waste sites, because highly soluble chemicals tend to be rapidly dissipated. The exception is streams in which sediment depositional areas constitute a relatively small fraction of the habitat. For example, the benthic invertebrate communities in sediment depositional areas of Upper East Fork Poplar Creek in Oak Ridge, TN were not surveyed because such areas constituted less than 5% of the total available habitat (DOE, 1998). In this case a preliminary stream survey was conducted to measure the size, distribution, and total area of deposited fine sediments. This proved to be a very useful tool for selecting assessment endpoints and exposure pathways for detailed analysis and should be considered for assessments of systems dominated by one type of habitat, either cobble and gravel or soft sediments.

Survey methods vary in rigor from qualitative (e.g., sampling all habitats with a D-frame net) to semiquantitative (e.g., sampling for a specified time or distance with a kick net), to quantitative (e.g., sampling 0.1 m^2 with a Surber sampler) (Appendix A). Kerans et al. (1992) compared the results of quantitative (Surber and Hess samplers) and qualitative (sampling all habitats with D-frame net and hand picking) surveys for multiple streams in Tennessee. The quantitative surveys consisted of three to eight replicate samples per site, whereas the qualitative survey consisted of a single composite sample with collection time limited to 2 h. The qualitative surveys failed to detect human impacts that were detected by the quantitative surveys, probably due to the lack of replication. Thus, the assessor should select methods and survey designs that are quantitative and replicated within sites. A preliminary site evaluation may be limited to qualitative and semiquantitative surveys to establish the presence or absence of invertebrates and provide qualitative taxa richness and semiquantitative abundance estimates. However, a definitive risk assessment should include quantitative, replicated estimates of community metrics. Definitive assessments also include qualitative surveys of all habitats when the quantitative samples are collected using artificial substrates (DeShon, 1995), because artificial substrates are selective and may not be representative of rare taxa or the actual taxa richness at a site.

Data from benthic invertebrate surveys typically consist of counts of individuals of species or higher taxa and, in some cases, biomass. From these data one may calculate species richness (or taxonomic richness if some taxa are not identified to the species level) or other diversity metrics such as evenness, total numbers, or biomass, or may simply use the numbers or biomass of individual taxa. They may be aggregated into multimetric indexes such as the Ohio Invertebrate Community Index (DeShon, 1995). The total abundance of Ephemeroptera, Plecoptera, and Trichoptera (EPT taxa) is also a common metric. However, it is based on the sensitivity of these taxa to organic loading and siltation and may not be relevant to site contaminants. For example, the nominally sensitive ephemeropteran *Hexagenia limbata* was so abundant and so contaminated with mercury and PCB in Poplar Creek that it posed a risk to its predators (Baron et al., 1999). In addition, some waters are unsuitable for EPT taxa, even in the absence of contamination. The EPA recommends biomass, species richness, density, diversity, and relative abundance as benthic invertebrate survey metrics (Office of Emergency and Remedial Response, 1994b). Functional measures are seldom used, but may be assumed to be related to these structural measures (Clements, 1997).

Kerans and Karr (1994) evaluated 18 attributes of benthic invertebrate communities as indicators of biological condition in streams. Four attributes were classified as "excellent" in that they distinguished sites, correlated strongly with water quality characteristics, and were concordant with fish community results: total taxa richness, mayfly taxa richness, caddisfly taxa richness, and dominance. Eight other attributes were concordant with the fish community results, but not standard water quality indicators: stonefly taxa richness, total abundance, and proportions of predators, grazers, *Corbicula*, oligochaetes, omnivores, and filterers. The authors conclude that both sets of attributes should be used because they appear to be indicative of different human impacts (Kerans and Karr, 1994). This is a reasonable approach for hazardous waste sites, provided consideration is given to the possible explanations for the status of each attribute (see Kerans and Karr, 1994, for examples). However, emphasis should be placed on metrics that are related to the assessment endpoint. Carlisle and Clements (1999) found taxa richness measures to be the most sensitive and statistically powerful metrics for evaluating metal pollution in Rocky Mountain streams. Abundance attributes were generally found to be insensitive to metal pollution or highly variable. Of particular note is the species richness of mayflies, which are generally considered to be sensitive to metals (Clements, 1997).

Population or organism properties are seldom considered in benthic invertebrate surveys. In some cases, however, abundance of a particular sensitive and valued species may be an endpoint. Abundance of the widgeon clam (*Pitar morrhuana*) was such an endpoint at Quonset Point, RI and was estimated by surveys (Eisler, 1995).

Assessment of risks from contaminated sediments is complicated by the influence on invertebrates of sediment properties other than contamination. It is important to determine the texture, organic matter content, depth of overlying water, and any other properties that might influence the benthic invertebrate community for each sampled location. Elevated ammonium concentrations are particularly common and likely to result in toxicity that is unrelated to site contaminants. Even at a highly contaminated site, these habitat variables are likely to explain a larger fraction of the variance in invertebrate community properties than contaminant concentrations (Jones et al., 1999).

Spatial variability, rather than temporal variability, is the primary concern for sediment contaminants and sediment characteristics. This is especially true in lentic and slow-moving lotic systems (e.g., large streams and rivers) with relatively stable sediments. Samples for sediment analysis should be collected as close to the biological survey sampling points as practical. Ideally, subsamples of the sediment included in each benthic survey sample, including replicates, should be analyzed for contaminants and sediment characteristics. This is rarely practical for contaminant analyses, but sediment characterization is relatively simple and inexpensive. Recommended sediment quality characteristics include grain size (percent sand, silt, clay), organic carbon content, ammonia, and pH. Quantitative measurements should be preferred over subjective and qualitative measurements. For example, grain size characteristics (percent sand, silt, clay, etc.) should be used rather than subjective designations such as sandy or mucky. Objective measurements result in consistent and comparable data. This allows the assessor to better compare results within and among studies. It also expands the risk characterization techniques available to the assessor. For example, the benthic invertebrate assessment for the Clinch River included multiple regression analyses of the benthic survey data with both contaminant and habitat characteristics as explanatory variables (Jones et al., 1999).

In addition, water quality may influence benthic communities and can vary significantly through time. Water samples should be taken such that representative exposures can be estimated. Hence, water samples may not be collected concurrently with biosurveys. Recommended water quality characteristics include pH, hardness, alkalinity, conductivity, and flow rate.

4.3.2 Soil Biological Surveys

Biological survey techniques are used less frequently for contaminated soils than waters or sediments, even though there are fewer inherent difficulties in obtaining soil samples. Ecological risk assessments rarely include surveys of soil invertebrates, microorganisms, or soil processes. An exception is the assessment of a site in Massachusetts where the leaf litter community was quantified (Menzie et al., 1992).

Examples of surveys at contaminated sites are presented in Table 4.13. Most of these have been performed at metal-contaminated sites, primarily those contaminated by smelters. Methods for surveying soil biota include: (1) collecting samples of soil and extracting taxa in the laboratory; (2) extracting organisms in the field, e.g., with mustard solution; or (3) trapping organisms using pitfall traps. The second method is the least quantitative, as it is likely to extract organisms to variable depths. Although most surveys have focused on invertebrates, microbial community properties, element transformations, and litter accumulation have also been surveyed (Jackson and Watson, 1977; Tyler, 1984; Beyer and Storm, 1995). The variation in quantities of soil biota from location to location is highly dependent on soil characteristics (Nuutinen et al., 1998). Thus, the risk assessor must be certain that adequate reference locations exist.

4.3.3 Wildlife Surveys

Many methods are available for the collection of field data for wildlife populations. These methods may produce data that are useful in ecological risk assessments and may help elucidate the presence, nature, and magnitude of effects. Types of data for wildlife that can be generated from field surveys include presence–absence, abundance, age structure, and food habits. Through comparison of these data between the contaminated site and one or more reference sites, effects attributable to contaminant exposure may be differentiated from population fluctuations or habitat alterations that result from other causes. It is highly recommended that multiple reference sites be selected, each site being as comparable to the contaminated site as possible (Section 2.7.3). It should be pointed out that, although differences observed between the contaminated site and uncontaminated reference sites may show the presence and nature of an apparent effect, they do not indicate the cause. Additional data on the toxicity and biological effects of contaminants found at the site and on habitat properties are needed (Section 7.5). Additional discussions of wildlife survey methods and their use in ecological risk assessment are presented in Appendix A.

4.3.4 Plant Surveys

Because vegetation provides the habitat for all inhabitants of terrestrial communities, it is important to survey and map vegetation on contaminated sites, even before the problem formulation (Section 2.3). In addition, if plant populations or communities are assessment endpoints, biological surveys may be an appropriate line of evidence for estimating risks. Because plants are immobile, they are clearly associated with a localized environment, and they are easily sampled. However, few risk assessors have experience in using plant survey data in ecological risk assessments. Guidance has been provided by the EPA (Environmental Response Team, 1994d, 1996). The agency recommends density, coverage, and frequency metrics as measures of effects for plant populations and communities. For example, LeJeune et al. (1996a) and Galbraith et al. (1995) took transect measurements of percent cover of tree, shrub, forb, and grass species to aid in the estimation of risks to the plant community in

TABLE 4.13
Examples of Surveys of Soil Invertebrates and Microorganisms at Contaminated Sites

Soil Endpoint	Source of Pollution	Soil Contaminants Inversely Correlated or Associated with Endpoint	Ref.
Total macroinvertebrate and oligochaete species abundances	Aqueous discharge to a wetland	Cadmium, copper, lead, zinc, arsenic, selenium	Jenkins et al., 1995
Fungal biomass, density of sporophore-producing macrofungi, species composition	Smelter	Copper	Ruhling et al., 1984
Density of bacteria, actinomycetes, fungi, nematodes, and earthworms	Lead smelter	Lead, arsenic, cadmium, copper	Bisessar, 1982
Density of bacteria, acidophilic *Mortierella* species, actinomycetes, fungi	Zinc smelter	Zinc	Jordan and Lechevalier, 1975
Density of orders (and some families) of arthropods and total density of arthropods	Zinc smelter	Not stated	Strojan, 1976
Microbial biomass carbon, respiration rate, ATP content	Brass mill	Copper, zinc	Baath et al., 1991
Density of spiders, harvestmen, slugs, beetles, ants	Brass mill	Copper, lead	Bengtsson and Rundgren, 1984
Density of nematodes and microarthropods and groups within these categories, total fungal hyphal lengths	Urban environment	Lead, copper, nickel	Pouyat et al., 1994
Density and biomass of earthworms	Urban environment	Cadmium, magnesium, lead, copper, zinc	Pizl and Josens, 1995
Density of field crickets (*Gryllus pennsylvanicus*)	PCB-contaminated landfill	Not PCBs	Paine et al., 1993
Density of earthworms	Pig waste in pastures	Not stated	van Rhee, 1975

the Clark Fork River floodplain and Anaconda site in Montana. Similarly, surveys of vascular plants, mosses, and lichens showed severe effects in zinc-contaminated areas of the Lehigh Gap, PA (Beyer and Storm, 1995).

Since plants are valued and ecologically important because of their primary production, measures of plant growth or production may be particularly useful for sites with contaminated soils or shallow, contaminated groundwater. Tree coring is recommended by the EPA as a means to measure effects of contaminants on tree growth (Environmental Response Team, 1994e). The width of annual growth rings may indicate the effects of contaminants, but, because of the confounding effects of drought, frost, and other environmental factors, the interpretation should be performed by an experienced dendrochronologist. When vegetation is herbaceous, the EPA recommends that growth be determined by repeated clipping and weighing of the aboveground plant parts (Environmental Response Team, 1994b).

It should be emphasized that few risk assessment schedules permit the repeated sampling of vegetation over long periods of time. The usefulness of a vegetation survey depends on whether observed effects can be related to measures at reference sites or reference (precontamination) points in time. Although one tree ring sample provides a time series (and each tree is its own control), the discernment of effects from herbaceous plant clippings generally requires multiple temporal samples. Thus, detrimental effects on production of the forest understory, oldfields, or grasslands are not usually evident from a single vegetation survey.

If unhealthy plants or unvegetated areas are observed, the following question should be asked to determine the usefulness of the survey: Can factors other than contaminants explain the brown foliage or other adverse response? These factors could include seasonal patterns, nutrient deficiency, insect herbivory, salt from winter applications to roads, acid rain, ozone, drought, grazing pressure, fire, or changes in hydrological patterns associated with the development of adjacent land. For example, when adverse impacts on forest trees were observed within the Bear Creek Watershed on the Oak Ridge Reservation, it was unclear whether dead trees were the result of contamination or of altered hydrology associated with logging a neighboring area. Occasionally, specific toxic symptoms may be associated with particular contaminants. For example, "crinkle leaf" of cotton is associated with manganese toxicity, and an accumulation of purple pigment in soybean leaves can signal cadmium toxicity (Foy et al., 1978). However, these symptoms do not necessarily apply to other species, and most symptoms of toxicity such as stunted growth and chlorosis are common to many toxicants and nutrient deficiencies.

Basic soil data should be obtained during the vegetation survey. These characteristics would include major plant nutrients, pH, organic matter content, particle size distribution, bulk density, and perhaps salinity. One or more of these factors might explain differences in plant parameters at different locations.

4.4 BIOMARKERS OF EFFECTS

Biomarkers of effects are biochemical, physiological, or cellular properties of an organism that are indicative of toxic effects. Their use has been inhibited by the fact that few of them are clearly related to the overt effects that constitute assessment endpoints at most contaminated sites. Although it has been proposed that remedial goals be based on elimination of any detectable biomarker response (Depledge and Fossi, 1994), regulators do not normally require remediation on the basis of enzyme induction, even for humans.

Biomarkers of effects may play a supporting role in ecological risk assessments. In particular, biomarkers that are characteristic of a particular chemical, class of chemicals, or mode of action can support the inference that apparent effects are caused by particular contaminants. Even damage that is not particularly diagnostic can be useful if it can be even qualitatively related to population-level responses. For example, histological damage to the gonads of largemouth bass in Poplar Creek embayment in Oak Ridge supported the inference that the low abundance and species richness of fish was due to toxic effects rather than to habitat properties. When biomarkers are used to support inferences concerning causation, it is important to associate their levels or frequencies with contaminant concentrations. The EPA recommends biomarkers for use in assessment of risks to plants on contaminated sites, including chlorophyll and peroxidase activities, but does not specify how they are to be interpreted (Environmental Response Team, 1994a, c, 1996).

Gross pathologies such as tumors, lesions, and skeletal deformities have played a more important role in ecological risk assessments than biomarkers. They are a common source of public concern, particularly where they occur in sport or commercial fish. Frequencies of gross pathologies are easily determined when fish are collected for chemical analysis or for biological surveys. Pathologies that are characteristic of chemicals or chemical classes can also contribute to attributing causation to both the pathologies themselves and any population or community effects.

4.5 INDIRECT EFFECTS

Ecological risk assessments have been consistent with human health risk assessments in emphasizing direct toxic effects of contaminants. However, because nonhuman organisms are much more subject than humans to indirect effects such as habitat modification and reductions in the abundance of food species, indirect effects should receive more attention. The term *indirect effects* refers to effects that result when a contaminant directly affects a population, community, or ecosystem process (not necessarily assessment endpoints), and that direct effect becomes a new stressor with respect to an assessment endpoint entity. Indirect effects of chemical contaminants include the effects on trophic and competitive relationships, such as reduced abundance due to toxic effects on food species. In addition, indirect effects due to habitat alteration should be considered. For example, the toxicity of chemicals to earthworms may result in soil compaction, which could inhibit seed germination and result in other adverse effects to plants. Decomposition of organic contaminants in soil, surface water, or sediment could be associated with depletion of oxygen and reduced availability of nitrogen, adversely affecting endpoint species and processes. In contrast, after decomposition of petroleum is largely completed, plant production may actually be greater due to effects on soil structure, nitrogen availability, or other factors (McKay and Singleton, 1974; Bossert and Bartha, 1984). Like direct toxic effects, habitat-mediated effects of contaminants may depend on the magnitude of exposure. For example, at high exposures, petroleum and other nonaqueous phase liquids may fill soil pores that would otherwise be habitat for microorganisms and mesofauna.

These indirect effects should have been identified in the conceptual model and their relationship to the contaminants should be quantified as far as possible in this component of the assessment. Biological surveys of contaminated areas can potentially reveal indirect effects, but, because the exposures are uncontrolled and unreplicated, indirect effects are difficult to distinguish in such studies. When the results of microcosm, mesocosm, or field tests are available, they can be used to empirically estimate the indirect effects, or, for less selective chemicals, the combined direct and indirect effects. Alternatively, simple assumptions can be made such as an *x*% loss of wetlands will result in an *x*% reduction in the abundance of species that depend on that community for any of their life stages. A somewhat more sophisticated alternative is to define the direct toxic effects as changes in the parameters of habitat suitability models which can then be used to estimate changes in abundance (Rand and Newman, 1998). A potentially more rigorous approach is to use ecosystem models to estimate the consequences for all endpoint taxa of toxic effects on all modeled components of the exposed ecosystem (O'Neill et al., 1982; Emlen, 1989; Bartell et al., 1992; Suter, 1993a). As more and more indirect effects are considered, the uncertainties associated with exposure and effects models and measurements are compounded.

4.6 EXPOSURE–RESPONSE PROFILE

The output of the analysis of effects is the exposure–response profile. For each assessment endpoint and each line of evidence, the exposure–response profile presents the results of an analysis of the relationship between the exposure to site contaminants and the effects on the endpoint.

For the individual chemicals of potential ecological concern, the exposure–response profile is their ecotoxicological profile (Section 4.1.12). It should indicate how the effects increase with increasing duration or frequency and with dose or concentration of exposure. It should also, to the extent that such information is available and relevant, indicate the effects of environmental variables such as soil organic matter content and pH on toxic effects. It should indicate the mechanism of action of the chemical and the variation in sensitivity among taxa, life stages, and processes. It should be limited to information that is relevant to the site and assessment endpoint.

For ambient media toxicity tests, the exposure–response profile should summarize the results in terms of the spatial and temporal distribution, the nature and magnitude, and the consistency of toxicity. If a TIE was performed, its results should be presented in terms of the chemicals that were responsible for toxicity, thresholds or other exposure–response relationships, and interactions among contaminants. Otherwise, if chemical concentrations were determined for the tested media, exposure–response relationships can be presented in terms of exposure concentrations of individual chemicals that are apparently responsible for toxicity. If multiple chemicals significantly contribute to toxicity, a representative chemical or an aggregate concentration such as the ΣTU may be used (Section 3.1.1). If good chemical data are not available, surrogates for exposure such as distance downstream from the source may be used. If more than one test is performed on a contaminated medium, the relative sensitivities of the tests should be explained as far as possible in terms of the relative sensitivities of the species and life stages involved, the nature and duration of the exposure in the test system, or other relevant factors.

For biological survey data, results should be organized in terms of relationships between exposure and effect. As with the media toxicity test, the exposure metric may be concentrations of a chemical of concern, concentrations of a representative chemical, or some other metric. Because of the importance of habitat properties to population and community properties, it is also desirable to include habitat variables in the exposure–response model (Jones et al., 1999). However, biological survey data are unreplicated and, particularly for mobile organisms, are likely to be available for few sites. Therefore, conventional exposure–response relationships are desirable but often impractical. Often, effects must be related to broad categories of exposure and to habitat variables in a qualitative manner.

5 Risk Characterization in Screening Assessments

All organisms with complex nervous systems are faced with the moment-by-moment question that is posed by life: What shall I do next?

—S. Savage-Rumbaugh and R. Lewin (1994)

Screening ecological risk assessments (SERAs) are a critical component of the remedial process, particularly on large sites with diverse wastes. The primary purpose of screening risk assessments is to narrow the scope of subsequent assessment activities by focusing on those aspects of the site that constitute credible potential risks. SERAs are performed by a process of elimination. Beginning with a site description and the full list of chemicals that are suspected to constitute site contaminants, one can potentially eliminate:

- Particular chemicals or classes of chemicals as chemicals of potential ecological concern
- Particular media as sources of contaminant exposure
- Particular ecological receptors as credible assessment endpoints
- Ecological risks as a consideration in the remedial action

A secondary purpose of screening risk assessments is to identify situations that call for emergency responses. That is, a screening assessment may identify ongoing exposures that are causing severe and clearly unacceptable ecological effects or potential sources of exposure that are likely to cause severe and clearly unacceptable ecological effects in the immediate future. In such cases, the usual remedial schedule is bypassed to perform a removal action or other appropriate response. No guidance is provided for such decisions because there are no generally applicable rules for defining an ecological emergency. They must be identified ad hoc.

Finally, screening assessments serve to identify data gaps. Media or classes of chemicals that have not been analyzed, for which analyses are of unacceptably low quality or quantity, or for which the spatial or temporal distribution has been inadequately characterized should be identified during SERAs. This information serves as input to the assessment planning process for any subsequent phases of the remedial process or becomes a component of the uncertainty section of the baseline ecological risk assessment (BERA).

Screening assessments are performed at three stages in the assessment process. First, when an operable unit (OU) is initially investigated, existing information is collected, and a screening assessment is performed to guide the development of the analysis plan (Section 2.7). The SERA is used to help focus the work plan on those elements of the OU that require investigation and assessment. Second, in a phased assessment a screening assessment is performed after the preliminary phase to guide the development of the subsequent phase by focusing investigations on remaining uncertainties concerning credible potential risks. Finally, as a preliminary stage to the definitive assessment, a screening assessment is performed to narrow the focus of the assessment on those contaminants, media, and receptors that require detailed assessment.

Note that screening assessments are final assessments only when they indicate that no potential hazards to ecological receptors exist. Otherwise, SERAs should prompt the parties to the remedial process to consider the need for additional data. Whether or not additional data are collected, a screening assessment that indicates that a site is potentially hazardous must be followed by a more definitive baseline ecological risk assessment which provides estimates of the risks and suggests whether remedial actions are needed.

5.1 SCREENING CHEMICALS

At many sites, concentrations in environmental media will be reported for more than 100 chemicals, most of which are reported as undetected at some defined limit of detection. The assessor must decide which of these constitute chemicals of potential ecological concern (COPECs). That is, which detected chemicals constitute a potential ecological hazard and which of the undetected chemicals may pose a hazard at concentrations below the reported detection limits. The concern about undetected chemicals results from the possibility that the detection limit may be higher than concentrations that cause toxic effects. This screening is done for each medium by applying one or more of the following criteria:

1. If the chemical is not detected and the analytical method is acceptable, the chemical may be excluded.
2. If the wastes deposited at the site are well specified, chemicals that are not constituents of the waste may be excluded.
3. If the concentration of a chemical in a medium is not greater than background concentrations, the chemical may be excluded.
4. If the application of physicochemical principles indicates that a chemical cannot be present in a medium in significant concentrations, that chemical may be excluded.
5. If the chemical concentration is below levels that constitute a potential toxicological hazard, the chemical may be excluded.

Other criteria may be used, particularly if they are mandated by regulators or other risk managers. For example, the California EPA specifies that chemicals be retained if they have ARARs, if they are difficult to treat, or if they are highly toxic or highly bioaccumulative even if they occur at very low concentrations (Polisini et al., 1998). In the "Risk Assessment Guidance for Superfund" (1989a), the EPA has specified that chemicals found in less than 5% of samples may be excluded. These criteria are not recommended here, because they are not risk based.

In the United States the list of chemicals to be screened is assumed to include the EPA Target Compound List and Target Analyte List (EPA, 1985b, 1991c). Chemicals known to be associated with the site contaminants but not on those lists, particularly radionuclides, should be included as well.

Specific methods for applying these criteria are presented in the following subsections. The order of presentation is logically arbitrary. That is, the screening methods can be applied in any order, and the order used in any particular assessment can be based on convenience. In addition, it is not necessary to use all of the five screening criteria in the SERA. Some criteria may be inappropriate to a particular medium or unit, and others may not be applicable because of lack of information.

5.1.1 Screening against Background

Waste sites should not be remediated to achieve concentrations below background; therefore, baseline risk assessments do not normally estimate risks from chemicals that occur at background concentrations. Chemicals that occur at background concentrations may be naturally occurring, may be the result of regional contamination (e.g., atmospheric deposition of cesium-137 from nuclear weapons testing or mercury from incinerators or coal combustion), or may be local contaminants that have been added in such small amounts that their concentrations in a medium have not been raised above the range of background concentrations. Screening against background requires that two issues be addressed. First, what locations constitute background for a particular site? Second, given a set of measurements of chemical concentrations at background locations, what parameter of that distribution constitutes the upper limit of background concentrations?

5.1.1.1 Selection of Background Data

Background sites should be devoid of contamination from the wastes or any other local source. For example, water from a location upstream of a unit cannot be considered background if there are outfalls or waste sites upstream of that location. To ensure that there is no local contamination, a careful survey of watersheds for potential background water or sediment sites should be performed, and for terrestrial sites, the history of land use must be determined. For example, although Norris Reservoir, upstream of Oak Ridge, is quite clean relative to Watts Bar Reservoir, the reservoir receiving contaminants from Oak Ridge, the occurrence of a chloralkali plant in the Norris Reservoir watershed has eliminated it as a background site for mercury. In theory, if a local source releases a small and well-characterized set of contaminants, a location that is contaminated by it could be used as a background site for other chemicals. However, wastes are seldom sufficiently well defined. Background can be defined at multiple scales: regional, local, and unit-specific. Each scale has its advantages and disadvantages.

National or regional background concentrations may be available from existing sources such as U.S. Geological Survey or state publications (Shacklette and Boerngen, 1984; Slayton and Montgomery, 1991; Toxics Cleanup Program, 1994). National or regional background concentrations are advantageous in that they provide a broad perspective. It is not sensible to remove or treat soils at a site for a metal concentration that is higher than local background but well within the range of concentrations of that metal at uncontaminated sites across the region (LaGoy and Schulz, 1993). However, one must be careful when using national or regional background values to ensure that the concentrations were measured in a manner that is comparable with the measurements at the waste site. For example, dissolved-phase aqueous concentrations should not be compared with total concentrations of metals. The EPA has suggested that nationwide background concentrations of metals could be used for comparison with site levels (EPA, 1986b). However, because the use of national or regional background concentrations is often less conservative than the use of local or unit-specific concentrations, the former is seldom favored by EPA or state regulators. An appropriate compromise might be to screen against local or unit-specific background concentrations, but to use the regional background as supporting evidence.

Local background measurements are the most generally useful. Local backgrounds are concentrations of chemicals in environmental samples collected to represent an entire site or a geologically homogeneous portion thereof. Examples include the background soils data for the Oak Ridge Reservation (ORR) (Watkins et al., 1993) and the background soil and groundwater data for the Portsmouth Gaseous Diffusion Plant in Piketon, OH (Geraghty and Miller, 1994). These studies systematically collected soils from all geologic units on the sites. They provided high-quality data that were agreed by all parties to represent local background and its variance.

Unit-specific background values are those that are collected for an individual unit (Chapter 2). Examples include water collected upstream of a unit or soil collected beyond the perimeter of a unit. The use of this type of background has two major disadvantages. First, unit background levels are seldom available for the initial screening assessment. Second, because samples are typically collected in the vicinity of the unit, there may be some danger of undetected contamination. In addition, because unit-specific background measurements are often poorly replicated in space or time, their variance is often poorly specified. However, because the natural variance in background concentrations is lower in the vicinity of an individual unit than across a site or region, use of unit-specific background values is more likely (compared with background estimates on a larger scale) to suggest that concentrations on the unit are above background.

Screening metal concentrations in groundwater presents a peculiar problem. Because these waters are of concern to ecological risk assessors when they enter surface waters, it would seem reasonable to screen them against surface water background concentrations. However, metal concentrations in groundwater are commonly much higher than surface water, even in the absence of contamination. Therefore, site groundwater should be screened against background groundwater, even for assessments of risks to surface water biota. Background groundwater may be obtained upgradient in the same aquifer or in another aquifer from the same geologic formation.

Given the disadvantages of both regional and unit-specific backgrounds, local background values should be used when they are available or can be obtained.

Lacking those values, unit-specific background should be used, with care taken to ensure that background variance is specified and that samples from background and contaminated locations are comparable. For example, because aqueous concentrations of naturally occurring chemicals are sensitive to hydrologic conditions, background samples should be taken at the same time as contaminated samples. Regional and national background values can be used to check the reasonableness of local and unit-specific background values.

5.1.1.2 Quantitative Methods for Comparison to Background

Various methods may be used for comparison of site concentrations to background concentrations. Because concentrations of chemicals at uncontaminated reference sites are variable, a definition of the upper limit of background must be specified. Some EPA regions have specified that chemicals should be retained unless the maximum concentration of the chemical in a medium on the unit is less than twice the mean background concentration (Akin, 1991). Other possibilities include the maximum observed value at reference sites, a percentile of the distribution of reference concentrations, or tolerance limits on a percentile. If there are multiple reference sites, one might use the site with the highest concentrations as best representing the upper limits of background or statistically combine the sites to generate a distribution of concentrations across uncontaminated sites. Some state agencies such as the Ohio EPA (1991) specify statistical procedures for defining the upper limit for background. Procedures may be developed for particular sites. There is no specific guidance for defining the limits of background in SERAs.

5.1.1.3 Treatment of Background in Multimedia Exposures

Wildlife are exposed to contaminants in food, water, and soil. If concentrations of a chemical in all media are at background levels, the chemical can be screened out. However, if concentrations in one or more of the media are above background, the chemical cannot be eliminated from consideration in any of the media with respect to that wildlife endpoint. That is because all sources of the chemical contribute to the total exposure.

5.1.1.4 What Chemicals Can Be Screened against Background?

Screening concentrations of inorganic chemicals and naturally occurring radionuclides against background is generally acceptable. However, elimination of anthropogenic radionuclides and organic chemicals as COPECs by screening against background is often controversial, even when they are detected at background sites that receive no input other than regional atmospheric deposition. For example, some polycyclic aromatic hydrocarbons (PAHs) were detected in the Oak Ridge background soils, and there is no reason to believe that these were due to local contamination (Watkins et al., 1993). PAHs have also been detected in Poplar Creek sediments both upstream of and within the ORR and are believed to be extracted from coal, which is abundant in those sediments because of the extensive coal mining and processing in the watershed. However, the EPA regional office would not officially accept that there were background concentrations of these chemicals. Similarly, cesium-137 from nuclear weapons testing is ubiquitous, and polychlorinated biphenyls (PCBs) are found in all rivers with any urban or industrial development. Even when these observations cannot be used in the screening assessments, they can be used to qualify conclusions concerning risk and remediation.

5.1.1.5 When Is a Concentration Not Comparable to Background?

If there is reason to believe that a chemical occurs in a form that is more toxic or more bioavailable than at background sites, it may be a COPEC even at concentrations that are within the range of background values. An example from Oak Ridge is the highly acidic and metal-laden leachate from the S-3 waste ponds which entered Bear Creek. Because metals are more bioavailable in acidic than neutral waters, metal concentrations in a stream downgradient of the ponds that were within the range of background waters were not screened out on that basis. Considerations include the major physicochemical properties of the waste, such as pH, hardness, and concentrations of chelating agents relative to properties of the ambient media, and the species of the chemical in the waste relative to the common ambient species.

5.1.1.6 Screening Biota Contamination Using Background Concentrations for Abiotic Media

It is possible to use background values for abiotic media to screen biota. For example, if all metals in soil are at background concentrations, it can be assumed that plant and earthworm metal concentrations are also at background. Similarly, if concentrations in both water and sediment are at background levels, concentrations in aquatic biota can be assumed to be at background.

5.1.1.7 Screening Future Exposure Concentrations against Background

If exposure concentrations may increase in the future, current concentrations should not be used to exclude chemicals from the baseline assessment, because future exposure scenarios must also be addressed. If the increased future exposures would result from movement of a contaminated ambient medium such as soil or groundwater, then concentrations measured in those media should be screened against background. For example, if a groundwater plume would intersect the surface in the future, concentrations in the plume should be screened against background. If the increased future exposures would result from changes in a source such as the failure of a tank, the contaminant concentrations predicted to occur in ambient media often may not be screened against background concentrations. That is because regulators often argue that the modeled future concentrations are, by definition, additions to background. Therefore, even if the predicted concentrations are within the range of local concentrations, they are not "really background."

5.1.2 Screening against Detection Limits

Chemicals that are not detected in any sample of a medium can be screened out if the detection limits are acceptable to the risk manager. For example, EPA Region

IV has indicated that the Contract Laboratory Program (CLP) Practical Quantification Limits (PQLs) can be used for this purpose (Akin, 1991). It should be noted that this screening criterion is not risk based. It is entirely possible that undetected chemicals pose a significant risk to some receptor and may account for effects seen in media toxicity tests or biological surveys. The use of this criterion is based entirely on a policy that responsible parties should not be required to achieve lower detection limits than are provided by standard EPA methods. An alternative is to use the limit of detection as the exposure concentration in the screening assessment.

Care should be taken when eliminating undetected chemicals that are known to bioaccumulate to concentrations in biota that are higher than in inorganic media. In particular, mercury and persistent lipophilic organic compounds such as PCBs and chlordane may occur in significant amounts in aquatic biota when they cannot be detected in water or even sediment. If there are known sources of these chemicals, they should not be eliminated as COPECs until biota have been analyzed.

5.1.3 Screening against Waste Constituents

If the waste constituents are well specified either because they were well documented at the time of disposal or because they are still accessible to sampling and analysis (e.g., leaking tanks), chemicals that are not waste constituents should be screened out. However, this is not possible at many sites because of imperfect recordkeeping, loss of records, poor control of waste disposal, disposal of ill-defined wastes, or occurrence of wastes in forms that do not permit definitive sampling and analysis.

5.1.4 Screening against Physicochemical Properties

Chemicals can be excluded as COPECs if their presence in significant amounts in a medium can be excluded by physicochemical principles. For example, volatile organic compounds (VOCs) were excluded from the risk assessment for Lower Watts Bar Reservoir, TN because any VOCs in Oak Ridge emissions would be dissipated by the time the contaminated waters reached the mouth of the Clinch River. Similarly, atmospheric routes of exposure have been eliminated from ERAs at most units because significant atmospheric concentrations are implausible given the nature and concentrations of the soil and water contaminants.

5.1.5 Screening against Ecotoxicological Benchmarks

Chemicals that occur at concentrations that are safe for ecological receptors can be excluded as COPECs. Exposure concentrations that are deemed to be safe are referred to as ecotoxicological screening benchmarks, or simply benchmarks (Section 4.1.2). If benchmarks for a chemical exceed its conservatively defined exposure concentration, the chemical is screened out. Although this is the typical approach to screening for toxic hazards, some risk managers require that the benchmark exceed the exposure level by some factor (e.g., 2 or 10). These safety factors are used because the benchmarks or exposure levels are not conservative or not sufficiently conservative. The screening method developed by Parkhurst et al. (1996) uses a factor of 3. A much more elaborate set of factors can be found in Kester et al. (1998).

The calculation of exposure concentrations to be compared with the benchmarks depends on the characteristics of the receptor. In general, a concentration should be used that represents a reasonable maximum exposure given the characteristics of the medium and receptor. The fundamental distinction that must be made is between receptors that average their exposure over space or time and those that have essentially constant exposure.

- Terrestrial wildlife are like humans in that they move across a site potentially consuming soil, vegetation, or animal foods from locations that vary in their degree of contamination. Therefore, mean concentrations over space provide reasonable estimates of average exposure levels. For the conservative estimate to be used in the screening assessment, the 95% upper confidence limit (UCL) on the mean is as appropriate as in human health assessments (EPA, 1989c).
- Fish and other aquatic organisms in flowing waters average their exposures over time. Similarly, wildlife average drinking water concentrations over time. Therefore, the 95% UCL on the mean is commonly used as a reasonably conservative estimate of chronic aqueous exposure concentrations. If aqueous concentrations are known to be highly variable in time and if periods of high concentration that persist for extended periods can be identified, the averaging period should correspond to those periods.
- Wildlife that feed on aquatic biota average their dietary exposure across those prey organisms. Therefore, they average their exposure over space and time (i.e., over their feeding range and over time as their prey respond to variance in water quality). The 95% UCL on that mean is a reasonably conservative estimate of exposure concentrations.
- Soil and sediment concentrations are relatively constant over time, and plants, invertebrates, and microbes are immobile or nearly immobile. Therefore, there is effectively no averaging of concentrations over space or time. The reasonable maximum exposure for those media and receptors is the maximum observed concentration, if a reasonable number of samples have been analyzed. That is, some organisms occupy that maximally contaminated soil or sediment or would occupy it if it were not toxic. Therefore, exceedence of ecotoxicological benchmarks at any location implies a potential risk to some receptors. Alternatively, an upper percentile of the distribution of concentrations (e.g., 90th percentile) could be used. Such percentiles would be more consistent than maxima because they are less dependent on sample size.

Screening against wildlife benchmarks requires specification of individual wildlife species. Even if endpoint species have not yet been selected for the unit through the assessment planning process, species should be selected for screening. The chosen species should include potentially sensitive representatives of trophic groups and vertebrate classes that are potentially exposed to contaminants on the site.

If no appropriate benchmark exists for a chemical that cannot be screened out by comparison to background, then an effort should be made to find or develop a benchmark for that chemical. However, in some cases, there are no appropriate toxicity data available for a chemical/receptor combination. In such cases, the chemical cannot be eliminated, and its toxicity cannot be addressed. The media in which such chemicals occur should not be eliminated from further assessment. In particular, they provide an argument for media toxicity testing.

The occurrence of multiple contaminants complicates screening. One common approach is to assume that this issue is covered by the conservatism of the screening process. That is, any chemical that significantly contributes to toxicity on a site will be retained by the conservative screening process. An alternative is to explicitly model combined toxicity. The only approach that is appropriate for screening assessments is to assume concentration additivity. The assumption of additive toxicity is accurate for chemicals with the same mechanism of action, and is conservative in most cases for mixed mechanisms of action (Konemann, 1981; Alabaster and Lloyd, 1982; OECD, 1992a). The assessor should calculate

$$\mathrm{HI} = \Sigma(\mathrm{SEC}_i / \mathrm{SBC}_i) \tag{5.1}$$

where HI is the hazard index, SEC is the screening exposure concentration, and SBC is the screening benchmark concentrations for the chemicals. If the sum is greater than one, then one might retain as COPECs those chemicals that contribute more than 10%, 1%, or some other percentile of the HI. Alternatively, one may include chemicals with individual quotients greater than some value. Parkhurst et al. (1996) recommend a minimum quotient of 0.3.

As discussed in previous chapters, petroleum and other complex and poorly specified materials present a particular problem. Currently, the most common approach for sites contaminated with petroleum and its products is to screen total petroleum hydrocarbon (TPH) concentrations against TPH benchmarks. However, a representative chemical approach could be used to screen against benchmark concentrations. For example, when screening a mixture of PAHs at a site, assessors cannot obtain benchmarks or good toxicity data from which to derive benchmarks for many constituents. In that case, it is appropriate for the sake of screening to use a representative PAH for which a benchmark is available and which the assessor is confident is more toxic than the average of the constituents of the mixture.

5.1.6 Exposure Concentrations

The following issues must also be considered when deriving exposure concentrations for screening assessments.

- The screening methods described here presume that measured chemical concentrations are available to define exposure. Use of measured concentrations implies that concentrations are unlikely to increase in the future. Where concentrations may increase in the future due to movement of a contaminated groundwater plume, failure of waste containment, or other processes, future concentrations must be estimated and used in place of measured concentrations for the future scenarios. For screening assessments, simple models and assumptions such as exposure of aquatic biota to undiluted groundwater are appropriate.

- For large units, it is appropriate to screen contaminants within subunits such as stream reaches rather than in the entire unit to avoid diluting out a significant contaminant exposure. The division of a unit into areas or reaches should be done during the development of the conceptual model and should take into consideration differences in contaminant sources and differences in habitat (Section 2.2).
- Some benchmarks are defined in terms of specific forms or species of chemicals. When forms are not specified in the data available for the screening assessment, the most toxic form should be assumed unless there are compelling reasons to believe that other forms predominate. For example, it has been generally recognized and confirmed in studies at Oak Ridge that the hexavalent form of chromium is converted to trivalent chromium in humid soils, sediments, and waters; therefore, EPA Region IV has recommended against assuming that chromium VI is present in significant amounts on the ORR.
- Measurements of chemicals in ambient media often include a mixture of detected concentrations and nondetects with associated detection limits. For screening of soil and sediment, the maximum value is still available in such cases. However, 95% UCLs on the mean concentration cannot be derived directly. If time and resources permit, these values should be estimated using a maximum likelihood estimator or a product limit estimator. Otherwise, the 95% UCL can be calculated using the detection limits as if they were observed values. If the chemical was not detected in any sample and the analytical techniques did not achieve detection limits that were agreed to be adequate by the risk manager, the reported limit of detection should be screened in place of the maximum or 95% UCL value.

5.2 SCREENING MEDIA

If the screening of chemicals does not reveal any COPECs in a particular medium and if the data set is considered adequate, that medium may be eliminated from further consideration in the risk assessment. However, if toxicity has been found in appropriate tests of the medium or if biological surveys suggest that the biotic community inhabiting or using that medium appears to be altered, then the assessors and risk manager must consider what inadequacies in the existing data are likely to account for the discrepancy between the lines of evidence and must perform appropriate investigations or reanalysis of the data to resolve the discrepancy.

5.3 SCREENING RECEPTORS

If all media to which an endpoint receptor is exposed are eliminated from consideration, then that receptor is eliminated as well. For wildlife that are exposed to contaminants in water, food, and soil, this means that all three media must be eliminated. Aquatic biota can be eliminated if both water and sediment have been eliminated. Plants and soil heterotrophs can be eliminated if soil has been eliminated. Any evidence of significant exposure to contaminants or injury of the receptor would prevent its elimination from the assessment.

5.4 SCREENING SITES

A unit can be eliminated from a risk assessment if all endpoint receptors for that type of unit have been eliminated. However, it must be noted that even when there are no significant risks due to contaminant exposures on the unit, the risk assessment must address fluxes of contaminants that may cause ecological risks in other units or incidental use of the site by wildlife, which may cause risks to wide-ranging wildlife populations.

5.5 DATA COLLECTION AND EVALUATION

Initial SERAs typically use all available data. In many cases, it is necessary to use data that are not adequately quality assured or documented to perform a full data evaluation. In such cases, the data should be evaluated as far as possible to eliminate multiple reports of the same measurement, unit-conversion errors, and other spurious data. Such screening assessments also need to consider the relevance of historical data to current conditions. Issues to consider in deciding whether data are too old to be useful include the following:

- If contamination is due to a persistent and reasonably stable source that has been operating since before the date of the historic data, the data are likely to be relevant.
- If the source is not persistent and stable and the chemical is not persistent (e.g., it degrades or volatilizes), the data are unlikely to be relevant.
- If the ambient medium is unstable or highly variable, then historic data are less likely to be relevant. Examples include highly variable aqueous dilution volumes and scouring of sediments.
- Human actions, particularly those taken to stabilize the wastes or partially remediate the site, may make historic data irrelevant.

5.6 DATA ADEQUACY AND UNCERTAINTIES

SERAs performed at the intermediate stages of a phased assessment process or as the initial step in the definitive assessment should have adequate data quality and quantity because the data set used should be the result of a proper assessment plan. However, the initial SERA performed for the assessment work plan is likely to have few data for some media, and the data quality may be questionable. The proper response to questionable data adequacy is to perform the screening assessment with the available data and describe the inadequacies of the data and the resulting uncertainties concerning the results. Highly uncertain screening results for a medium could then constitute an argument for making analyses of some or all of the Target

Analyte List and Target Compound List chemicals a part of the assessment plan for the next phase of the assessment.

5.7 PRESENTATION OF A SCREENING ASSESSMENT

The documentation of a screening assessment should be brief. The results can be presented in two tables. The first table lists the chemicals that were retained with reasons for retaining any chemicals despite their passage of the screening benchmarks and background. The second table lists those chemicals that were rejected with the reason for rejection of each. These results may be presented for each medium, for each unit being assessed, and for each endpoint entity exposed to a medium. The assessment that generated these results should be presented in the format of the framework for ecological risk assessment (EPA, 1998). However, the description and narrative should be minimal. Although it is important to do the screening assessment correctly, it is also important to ensure that the production of a screening assessment does not become a major impediment to completion of a definitive assessment and remediation.

The following information is important to support the screening results:

- The rationale for the list of chemicals that was screened
- The sources of the site, the reference, and the background contaminant concentrations
- Justification of the use of any preexisting concentration data
- The methods used to derive any site-specific background concentrations
- Criteria used to determine the adequacy of the concentration data set
- The sources of existing screening benchmark values
- The methods used to derive any new screening benchmark values

6 Risk Characterization in Definitive Assessments

A wise man proportions his belief to the evidence. In such conclusions as are founded on an infallible experience, he expects the event with the last degree of assurance.... In other cases, he proceeds with more caution: he weighs the opposite experiments: he considers which side is supported by the greater number of experiments: to that side he inclines, with doubt and hesitation; and when at last he fixes his judgement, the evidence exceeds not what we properly call probability.

—D. Hume, *An Enquiry Concerning Human Understanding*

Risk characterization for definitive risk assessments consists of integration of the available information about exposure and effects, analysis of uncertainty, weighing of evidence, and presentation of conclusions in a form that is appropriate to the risk manager and stakeholders. The integration of exposure and effects information should be carried out for each line of evidence independently so that the implications of each are explicitly presented. This makes the logic of the assessment clear and allows independent weighing of the evidence. For each line of evidence, it is necessary to evaluate the relationship of the measures of effect to the assessment endpoint, the quality of the data, and the relationship of the exposure metrics in the exposure-response data to the exposure metrics for the site. The actual characterization for ecological risk assessment is then performed by weight of evidence (Suter, 1993a; EPA, 1998). That is, rather than simply modeling risks, ecological risk assessors examine all available data from chemical analyses, toxicity tests, biological surveys, and biomarkers to estimate the likelihood that significant effects are occurring or will occur and describe the nature, magnitude, and extent of effects on the designated assessment endpoints.

The approach presented in this chapter is based on the assumption that significant effects (i.e., effects that could prompt remediation) have been identified in the problem formulation (Chapter 2). The risk characterization determines whether risks are significant for each endpoint and spatial unit and then estimates the magnitude and extent of effects and the associated uncertainties for the significant risks. This approach simplifies the risk characterization, particularly if there are multiple lines of evidence. However, if significant risks have not been defined a priori, the risk characterization must proceed directly to the estimation of magnitudes and extents of effects for all endpoints and units.

6.1 SINGLE-CHEMICAL TOXICITY

This line of evidence uses concentrations of individual chemicals in environmental media (either measured or modeled) to estimate exposure and uses results of toxicity test endpoints for those individual chemicals to estimate effects (Figure 6.1). They are combined in two steps. First, the chemicals are screened against ecotoxicological benchmarks, against background exposures, and, where possible against characteristics of the source to determine which are chemicals of potential ecological concern (COPECs). This may have been done previously in screening assessments for earlier phases in the remedial process such as the RI work plan, but the process should be repeated for each new assessment. Methods for screening assessments are presented in Chapter 5. The results of the screening assessment should be presented in the definitive assessment as a table listing all of the chemicals that exceeded benchmarks, indicating which are COPECs, and indicating the reasons for acceptance or rejection. The integration of exposure with single-chemical toxicity data is minimally expressed as a hazard quotient (HQ), which is the quotient of an ambient exposure concentration (AEC) divided by a toxicologically effective concentration (TEC):

$$HQ = AEC/TEC. \tag{6.1}$$

For wildlife, doses are typically used in place of concentrations (Chapter 3). The TEC may be a test endpoint, a test endpoint corrected by a factor or other extrapolation model, or a regulatory criterion or other benchmark value. This calculation

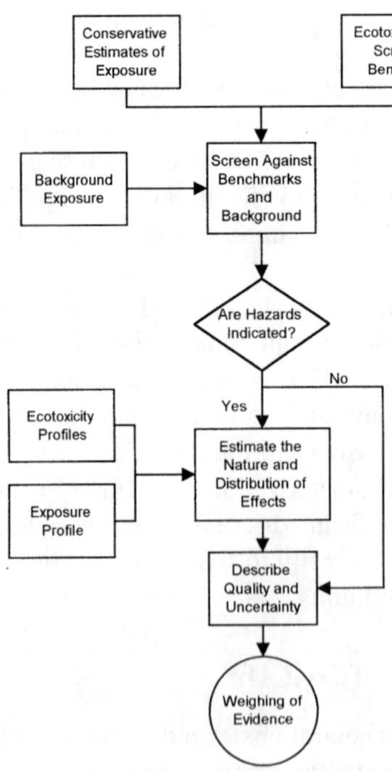

FIGURE 6.1 Risk characterization based on chemical analyses and single-chemical toxicity.

of HQs is simply a generalization of the type of analysis used for risk characterization in screening assessments (Section 5.1.5). In that case, conservative AEC values are used, and an HQ greater than 1 is treated as evidence that the chemical is worthy of concern. For definitive assessments, more realistic exposure estimates are used for the AEC, and effects are expressed as test endpoints that are closely related to the assessment endpoint, or the effects threshold is estimated using an extrapolation model (Section 4.1.9). In addition, in the definitive assessment one must be concerned about the magnitude of the quotient and not simply whether it exceeds 1. Large quotients suggest large effects or at least indicate that the uncertainty concerning the occurrence of the endpoint effect is low.

If numerous chemicals occur at potentially toxic concentrations, it is useful to calculate an index of total toxicity, the sum of toxic units (ΣTU) (Section 3.1.1). This permits the assessor and reviewers to compare the COPECs with each other and examine their distributions across reaches or areas within a site. Since the relative importance of COPECs is a function of their potential toxicity rather than their concentration, toxicity-normalized concentrations or toxic units (TUs) are calculated. This is a common technique for dealing with exposures to multiple chemicals by expressing concentration relative to a standard test endpoint (Finney, 1971). TUs are quotients of the concentration of a chemical in a medium divided by the standard test endpoint concentration for that chemical. A TU is similar to an HQ and a ΣTU is similar to a hazard index (HI; Section 5.1.5 and below) except that, because TUs are used for comparative purposes rather than to draw conclusions, a common test endpoint is used rather than conservative benchmarks or most relevant test endpoints. The expression of concentration and the test endpoint vary among media; for water they are the mean or upper 95% confidence limit concentrations and the 48-h EC50 for *Daphnia sp.* (the most common aquatic test endpoint). If the TU for a chemical equals 1, the interpretation is that the aquatic community in that reach is exposed to a conservatively estimated average concentration sufficient to kill or immobilize *Daphnia* within 48 h. The chemicals that constitute a potentially significant component of toxicity (i.e., TUs > 0.01) should be plotted for each reach or area for water, sediment, soil, and wildlife intake (e.g., Figure 3.1). The choice of a cutoff for inclusion is based on the fact that acute values are used in calculating the TUs, and chronic effects can occur at concentrations as much as two orders of magnitude below acute values. Other values may be used if specific circumstances warrant. The height of the plot at each subreach is the sum of toxic units (ΣTU) for that medium and subreach (Figure 3.1). This value can be conservatively interpreted as the total toxicity-normalized concentration and therefore as a relative indication of the toxicity of the medium in that subreach.

If multiple chemicals appear to be significantly contributing to toxicity, it is highly desirable to perform toxicity tests of the contaminated media to determine the nature of interactions (Section 4.2), or, if that is not possible, to perform tests of the mixture with laboratory media using conventional test methods (Section 4.1). If that is not possible, then one must consider how to estimate the combined toxic effects. The first step is to determine whether data concerning the toxicity of the mixture in the endpoint taxa are available. If the mixture is simple (i.e., few chemicals contribute to the toxicity), then one may find data in the literature indicating their combined effects. For example, mercury and selenium have been shown to be antagonistic in a variety of receptors. If those two metals are contaminants of concern, one should seek test data for that combination in a species similar to endpoint species and use a joint toxicity model to estimate effects at the observed concentrations. If the mixture is complex, an assessor should seek tests of the whole material. For example, one may use results of toxicity tests performed with petroleum, gasoline, or a PCB formulation to estimate risks of those materials on the site. If appropriate toxicity data are not available for the whole material, one may choose representative chemicals for the material or fractions of the material. These options are discussed in Section 3.11.

If there are no data on the toxicity of the mixture, one must infer the nature of the combined toxic effect and model the combined effects from the individual toxicities. Concentration addition is most commonly assumed. If the mechanisms of uptake and toxicity of a set of chemicals in a mixture are the same, but they differ in potency, one may calculate an HI which is the equivalent of the HQ for mixtures:

$$HI = \Sigma(AEC_i \ / \ TEC_i). \tag{6.2}$$

The HI is equivalent to the ΣTU except that, rather than indicating relative toxicity based on a common test endpoint (e.g., *Daphnia* EC50), it indicates the risk of the mixture to an assessment endpoint. The HI for definitive assessments differs from the HI for screening assessments (Section 5.1.5) in that realistic exposure and effects values should be used rather than deliberately conservative ones. As discussed above for HQ calculations, the TEC for each chemical is derived using a test endpoint that is representative of the assessment endpoint or by using extrapolation models. If the HI equals 1, then it is estimated that the assessment endpoint effect will occur. This estimate is appropriate if all of the chemicals have the same mechanisms of uptake and toxic action. For heterogeneous chemical mixtures, the addition of normalized concentrations to estimate effects is likely to yield a conservative estimate, because combined toxic effects of chemicals in environmental samples have usually been found to be additive or less than additive, not superadditive (i.e., not synergistic) (Alabaster and Lloyd, 1982). If chemicals act completely independently, a response addition model is appropriate. This model is used for mixtures of carcinogens in humans, but it is unlikely to be applicable to ecological risk assessments. The alternatives to concentration addition and response addition are synergism and antagonism. While various degrees of synergistic and antagonistic interactions are possible, there is no basis for modeling these interactions without tests of the mixtures. Hence, in the absence of specific information on interactions, concentration addition is the most appropriate default assumption.

A test of the use of literature values and assumed additivity to predict effluent toxicity is found in Bervoets et al. (1996). They found that the ΣTU for the four most toxic constituents based on 24- and 48-h LC50s were predictive of 24- and 48-h LC50s in tests of the effluent. However, the ΣTU based on chronic NOECs overestimated the toxicity of the effluent as indicated by the NOEC. The latter result is not surprising given that the NOEC does not correspond to a level of effect and therefore does not provide a consistent toxic unit for addition.

A variation of concentration additivity is the use of toxicity equivalency factors (TEFs; Section 3.11). Consensus TEFs have been published for conversion of concentrations of halogenated diaromatic hydrocarbons to toxicologically equivalent concentrations of 2,3,7,8-TCDD for fish, birds, and mammals (Van den Berg et. al., 1998). This approach has been successful in estimating effects of real contaminant mixtures on wildlife in the field (Sanderson and Van den Berg, 1999). However, attempts to extend it to other mixtures have not been successful to date (Safe, 1998).

One should not put much faith in the results of any model of combined toxic effects that is based on an assumed mode of combined toxicity. Even models based on tests of mixtures may be misleading if the relative proportions of the chemicals, the response measured, the duration of exposure, the species and life stage are not similar to the situation being assessed. The field has been reviewed by Calabrese (1991) and Yang (1994). New EPA guidance is being prepared (Teuschler and Hertzberger, 1995).

For all COPECs for each endpoint, exposures must be compared with the full toxicity profile of the chemical to characterize risk. For example, the distribution of concentrations in water would be compared with the distribution of concentrations of thresholds for chronic toxicity across endpoint species and across species upon which they depend (e.g., prey and habitat-forming species), the nature of the chronic effects would be described, and the exposure durations needed to achieve effects in the laboratory would be compared with temporal dynamics of concentrations in the field. Characteristics of the chemicals that are relevant to risks, such as the influence of metal speciation on toxicity and tendency of the chemical to accumulate in prey species, are also examined.

Inferences about the risk posed by the COPECs should be based on the distribution of concentrations relative to the distribution of effects. Distributions provide a better basis for inference than point estimates because they allow the consideration of variance in concentration over space or time and of sensitivity across species, measures of effects, media properties, or chemical forms. In all cases, risk is a function of the overlap between the exposure and effects distributions, but the interpretation depends on the data that are used. Interpretations of commonly used distributions are explained below for the different classes of endpoints, and the interpretation of distributions in general is discussed in greater detail in Chapter 7.

For all endpoints the risk characterization ultimately depends on weighing of all of the lines of evidence. To facilitate the weight-of-evidence analysis (Section 6.5) and to make the bases clear to the reader, it is useful to summarize the results of the integration of single chemical exposure and effects information for each endpoint in each reach or area where potentially toxic concentrations were found. Table 6.1 presents an approach to performing that summarization in terms of a table of issues to be considered in the risk characterization and the type of results that are relevant.

6.1.1 Aquatic Organisms

Fish, aquatic invertebrates, and aquatic plants are exposed primarily to contaminants in water. Contaminants in water may come from upstream aqueous sources, including

TABLE 6.1
Summary Table for Integration of Single-Chemical Toxicity

Issue	Result of Risk Characterization for Single Chemicals
Taxa affected	List specific species or higher taxa, life stages, and proportion of tested species affected at estimated exposures
Severity of effects	List types and magnitudes of estimated effects at ambient concentrations
Spatial extent	Define the meters of stream, square meters of land, etc. estimated to experience specified effects
Frequency	Define the proportion of time or number of distinct episodes of prescribed effects
Association with source	Describe the spatial and temporal relationships of effects to hypothesized sources
Estimated effect	Summarize the expected nature and extent of effects and credible upper bounds of effects
Confidence in results	Provide rating and supporting comments regarding confidence

waste sites, other anthropogenic sources, and background; exchange of materials between the surface water and contaminated sediments; or exchange of contaminants between the biota and the water column. As discussed in Chapter 3, aquatic biota should be assumed to be exposed to the dissolved fraction of the chemicals (particularly metals) in water, because that is a reasonable estimate of the bioavailable form (Prothro, 1993). However, because many states and EPA regions currently prefer to use total concentrations as conservative estimates of the exposure concentration, it may be necessary to use dissolved-phase concentrations to provide a best estimate of risk and to use total concentrations to provide a conservative estimate to satisfy regulators.

Because water in a reach is likely to be more variable in time than space, due to the rapid replacement of water in flowing systems and the lack of spatial gradients in the ponds that occur on waste sites, the mean chemical concentration in water within a reach or subreach is an appropriate estimate of the chronic exposure experienced by fishes. The upper 95% confidence bound on the mean is commonly used as a conservative estimate for screening assessments (Section 5.1). However, the full distribution of observed concentrations is used to estimate risks.

Some fish and invertebrates spend most of their lives near the sediment, and the eggs and larvae of some species (e.g., most centrarchid fishes) develop at the sediment–water interface. These epibenthic species and life stages may be more highly exposed to contaminants than is suggested by analysis of samples from the water column. If available, water samples collected just above the sediments provide an estimate of this exposure. Alternatively, the estimated or measured sediment pore water concentrations may be used as a conservative estimate of this exposure.

The aqueous toxicity data from the toxicity profiles and the aqueous chemical concentrations should be used to present distributions of exposure and effects. For exposure of fish and other aquatic organisms to chemicals in water, the exposure distributions are distributions of aqueous concentrations over time, and the effects distributions are usually distributions of sensitivities of species to acutely lethal effects (e.g., LC50s) and chronically lethal or sublethal effects (CV). If the water samples were collected in a temporally unbiased design (preferably stratified random or random), overlap of these two distributions indicates the approximate proportion of the time when aqueous concentrations of the chemical are acutely or chronically toxic to a particular proportion of aquatic species. For example, 10% of the time copper concentrations in Reach 4.01 of the Clinch River, TN are at levels chronically toxic to approximately half of aquatic animals (Figure 6.2). Interpretation of this result depends on knowledge of the actual temporal dynamics of the exposures and effects. For example, 10% of a year is 36 days, which would cover the entire life cycle of a planktonic crustacean or the entire embryo-larval stage of a fish, so significant chronic effects are clearly possible. However, if the 36 days of high concentrations is associated with a number of episodes, the exposure durations are reduced. The 7-day duration of the standard EPA subchronic aqueous toxicity tests could be taken as an approximate lower limit for chronic exposures, so the proportion of the year with high copper concentrations could be divided into five equal episodes and still induce significant chronic effects on a large proportion of species. More precise interpretations would require knowledge of the actual duration of episodes of high concentrations and of the rate of induction of effects of copper on sensitive life stages.

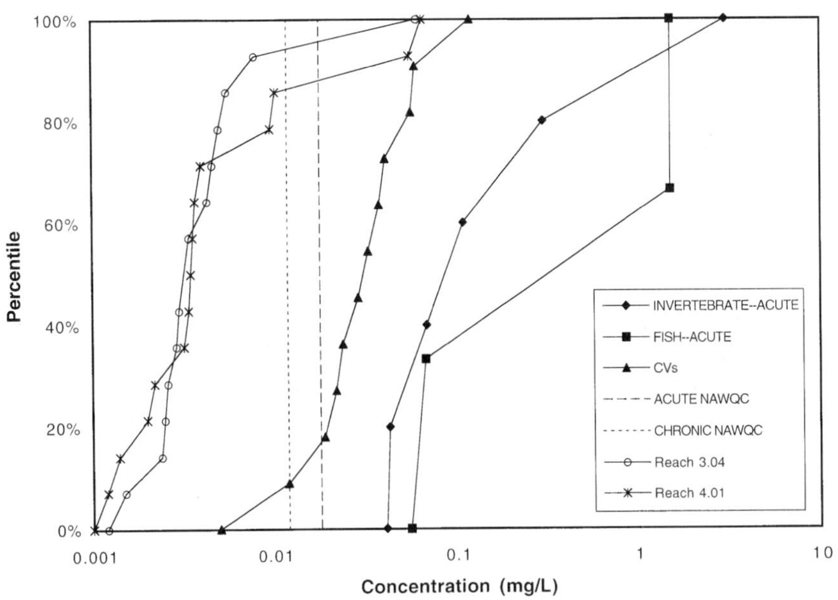

FIGURE 6.2 Empirical distribution functions for acute toxicity (LC50 and EC50 values) and chronic toxicity (chronic values) of copper to fish and aquatic invertebrates and for individual measurements of copper in surface water from two reaches. Vertical lines are acute and chronic National Ambient Water Quality Criteria.

Although the exposure and effects distributions described above are the most common in aquatic ecological risk assessments, numerous others are possible. Exposures may be distributed over space rather than time or may be distributed due to uncertainty rather than either spatial or temporal variance. Rather than distributions of test endpoints across species, effects may be distributed with respect to variance in a concentration–response model or uncertainties in extrapolation models (Section 4.1.9). The risk estimates generated from these joint exposure and effects models must be carefully interpreted.

6.1.2 Benthic Invertebrates

Two different expressions of sediment contamination may be used to characterize risks to benthic invertebrates, whole–sediment concentrations and pore water concentrations. The use of pore water is based on the assumption that chemicals associated with the solid phase are largely unavailable, and therefore sediment toxicity can be estimated by measuring or modeling the pore water concentration. This approach was used by the EPA to calculate proposed sediment quality criteria for organic chemicals. Whole–sediment concentrations do not account for effects of sediment properties on bioavailability. However, they are required by some regulators and may provide a better estimate of risk for highly particle-associated chemicals.

For purposes of screening chemicals, the appropriate estimate of exposure is a concentration that protects the most exposed organisms (Section 5.1). For risk estimation, an appropriate estimate of risks to the community is the percentage of samples exceeding particular effects levels. For each COPEC, the distributions of observed concentrations in whole sediment and pore water are compared with the distributions of effective concentrations in sediment and water. In the case of exposure of benthic invertebrates to sediment pore water, the exposure distributions are interpreted as distributions over space, since sediment composition varies little over the period in which samples were collected, but samples were distributed in space within reaches. The effects distributions are the same as for surface water distributions of species sensitivities in acute and chronic aqueous toxicity tests. Therefore, overlap of the distributions indicates the proportion of locations in the reach where concentrations of the chemical in pore water are acutely or chronically toxic to a particular proportion of species. For example, copper concentrations in sediment pore water from more than 90% of locations in Reach 4.04 are below chronically toxic concentrations for more than 90% of aquatic animal species (Figure 6.3). If the samples are collected by random or some other equal probability sample, these proportions can be interpreted as proportions of the area of the reach. Therefore, an alternate expression of the result is that less than 10% of the reach is estimated to be toxic to as many as 10% of benthic species.

In the case of exposure of benthic invertebrates to chemicals in whole sediment, the exposure distributions are, as with pore water, distributions in space within reaches. If sufficient data are available, three effects distributions are presented for each sediment COPEC: a distribution of concentrations reported to be thresholds for reductions in benthic invertebrate community parameters in various locations, a

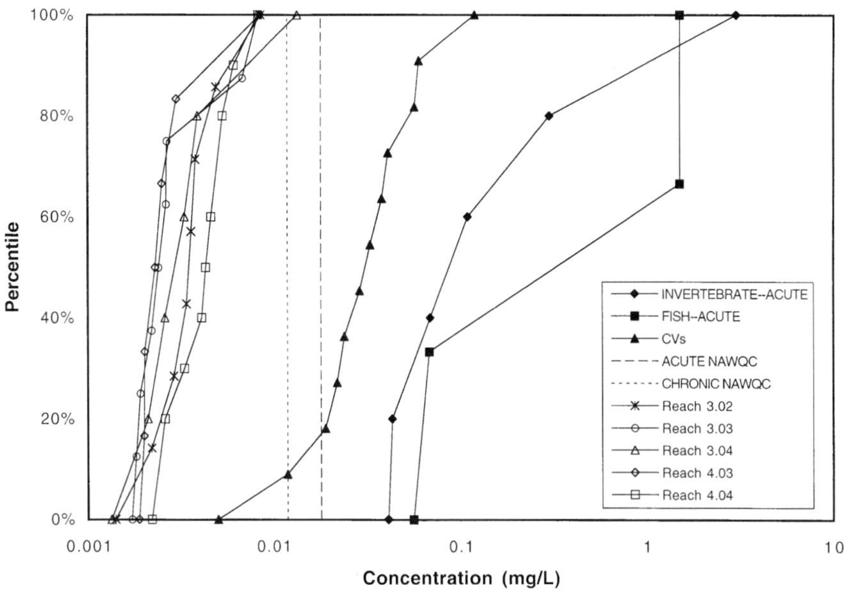

FIGURE 6.3 Empirical distribution functions for acute toxicity and chronic toxicity of copper to fish and aquatic invertebrates and for individual measurements of copper in sediment pore water from five reaches. Vertical lines are acute and chronic National Ambient Water Quality Criteria.

distribution of concentrations reported to be thresholds for lethal effects in toxicity tests of various sediments, and a distribution of concentrations reported to be thresholds for behavioral effects in toxicity tests of various sediments. If one assumes that the effects data set is drawn from studies of a random sample of sediments so that the site sediments can be assumed to be a random draw from the same distribution, and if one assumes that the reported community effects correspond to the community effects defined in the assessment endpoint, then the effects distributions can be treated as distributions of the probability that the chemical causes significant toxic effects on the endpoint at a given concentration. Overlap of the exposure and effects distributions represents the probability of significant alteration in the benthic communities at a given proportion of locations in a reach. For example, copper concentrations in whole sediment from half of locations in Reach 3.02 of Poplar Creek were above the concentration at which there is approximately a 20% likelihood of effects on community composition (Figure 6.4). The other two effects curves are not direct estimates of the endpoint, but they provide independent supporting evidence. Copper concentrations in whole sediment from half the locations in Reach 3.02 of Poplar Creek were above the concentration at which there is approximately a 50% likelihood of behavioral effects on benthic invertebrates and a 15% likelihood of lethal effects.

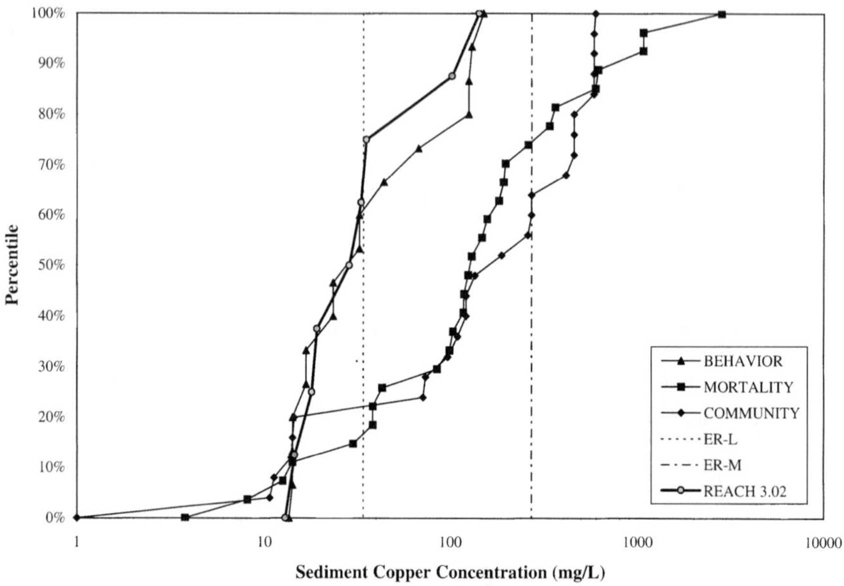

FIGURE 6.4 Empirical distribution functions for toxicity of copper in sediment to sediment-associated organisms and for individual measurements of copper in sediment. Effects are reported thresholds for effects on behavior, survival, and community structure from MacDonald (1994). Vertical lines are National Oceanographic and Atmospheric Administration (NOAA) effects range–low (ER-L) and effects range–median (ER-M) values.

6.1.3 Soil Exposure of Plants, Invertebrates, and Microbial Communities

Exposures to organisms rooted in or inhabiting soil are typically expressed as whole-soil concentrations, although concentrations in soil water are also potential measures of exposure, as well as concentrations normalized for soil characteristics (Section 3.4). For screening purposes, the maximum observed surface soil concentration at each location is an appropriate, conservative estimate of exposure, because soil organisms are essentially immobile (Chapter 5). For the definitive estimation of risks from chemicals in soil, the distribution of observed concentrations of each chemical should be used to estimate exposure. The exposure distributions should be interpreted as distributions over space only, since soil composition should vary little during a single sample collection period, which should be less than a month. The number of concentrations included in the distribution should be the number of locations that comprise an assessment unit area — that is, an area that is expected to be treated as a single unit in remedial decisions.

The distribution of measured concentrations should be compared with the distributions of effective concentrations for plants, invertebrates, and microbial processes. The relevant level of effects should be defined in the problem formulation, although the paucity of data may prevent the use of a consistent level for all toxicity tests used in the distribution. For plants, the effects distributions are distributions of species–soil combinations in published, single-chemical soil toxicity tests. If it is assumed that characteristics of site soils are drawn from the same distribution of soil properties as the tested soils and that test plants have the same sensitivity distribution as plants at the contaminated site, then the threshold concentrations for effects on site plants can be assumed to be drawn from the distributions of threshold concentrations for effects on plants in single-chemical toxicity tests. Therefore, an overlap of concentrations in the distributions indicates the proportion of locations in an area where concentrations of the chemical are expected to be toxic to a fraction of species in the site community (e.g., concentrations at half of the locations are apparently toxic to some of the plant species; concentrations at 20% of the locations are apparently toxic to more than 10% of the plant species; etc.). Assumptions for soil invertebrates are similar except that in ecosystems where earthworms are present, they are assumed to be representative of soil invertebrates. Thus, the distribution of test earthworm species sensitivities may be assumed to be representative of site invertebrates. The distributions of concentrations toxic to microbial processes are not quite equivalent to the other two endpoints because some processes are carried out by individual microbial strains and others are carried out by entire microbial communities. Therefore, we recommend that the distributions of effects data be distributions of process–soil combinations rather than species–soil combinations if processes comprise an endpoint. The overlap of the distributions indicates the proportion of locations in an area where concentrations of the chemical are expected to be toxic to a fraction of the microbial processes carried out in the area.

The definitive risk characterizations for the plant and soil invertebrate communities and microbial processes differ from screening assessments in several ways, even if measurements of single chemicals in soil and associated estimates of toxicity are the only line of evidence in the assessment. First, the role of chemical form and speciation must be considered. If a polycyclic aromatic hydrocarbon (PAH) is a constituent of petroleum, it may be less bioavailable than the chemical in laboratory toxicity tests. If the selenium concentration measured in a soil is mostly selenite, tests of selenate or organoselenium may not be as relevant. Even if these chemicals are equally toxic to plants, the typical measure of exposure, concentration in soil, requires that the assessor consider differences in uptake of the chemical species. It is common for inorganic salts to be more bioavailable than elements aged in contaminated soils. Second, the role of soil type must be considered. Tests in sandy or clay loams may have limited applicability to the estimation of toxicity in a muck soil. Third, differences in sensitivity between tested organisms and those at the site should be considered. The toxicity of chemicals to earthworms may differ substantially from the toxicity to arthropods. In addition to considering these factors in the summary of this single-chemical line of evidence, the assessor should incorporate them in decisions about how much to weigh this line of evidence, relative to media toxicity tests, biological surveys, and others.

An additional factor should also be part of the single-chemical line of evidence in the risk characterization. Even if the sampling methodology (e.g., spatial distribution, timing, soil depth) was agreed to at the DQO meetings, compromises may have been made that were not ideal with respect to the needs of the risk assessment. For example, the number of soil samples taken may have been low, or the assessor may have been forced to use 10-year-old data. The assessor should discuss any limitations in the data in the risk characterization.

6.1.4 Multimedia Exposure of Wildlife

In contrast to the other ecological endpoints that are considered, wildlife exposure is typically estimated for multiple media (e.g., food, soil, and water), with exposure being reported as the sum of intake from each source in units of mg/kg/day (see Sect 3.10). For screening assessments, conservative exposure estimates should be generated using point estimates of exposure parameters (e.g., upper 95% confidence limits for chemical concentrations in media), conservative uptake models, and conservative assumptions (e.g., 100% site use, 100% bioavailability). These conservative estimates are compared with no observed adverse effects levels (NOAELs) to identify COPECs (Chapter 5). For definitive assessments, more realistic estimates of exposure should be generated for the COPECs using distributions of exposure parameters and uncertainty analysis (e.g., Monte Carlo simulation; see Chapter 7). Depending on the nature of the site and the goals of the assessment, exposure simulations are performed for each discrete area at the site of interest at which COPECs were identified or for the range of individuals or populations of an endpoint species. Examples of this approach to wildlife risk characterization are presented in Sample and Suter (1999), Baron et al. (1999), and Sample et al. (1996a).

6.1.4.1 Comparison of Exposure to Toxicity

Comparisons of exposure estimates may be performed in several ways. For screening assessments and even most definitive assessments, HQs are calculated, where

$$HQ = (\text{total dose / NOAEL or LOAEL}) \tag{6.3}$$

NOAEL-based HQs that exceed 1 suggest that adverse effects are possible (i.e., estimated exposure exceeds the highest dose at which no statistically significant effects were observed), while LOAEL-based HQs that exceed 1 suggest that adverse effects are likely (i.e., estimated dose exceeds the lowest dose at which statistically significant effects were observed). The exposure pathway that is driving risk may be identified by comparing HQs calculated for each pathway to the HQ for total exposure (Sample et al., 1996a). Additional discussion of HQs is presented in Chapter 5 and Section 6.1.

If exposure distributions are generated from Monte Carlo simulations, superposition of point estimates of toxicity values on these distributions provides an estimate of the likelihood of individuals experiencing adverse effects. The portion of the exposure distribution that exceeds the given toxicity value is an estimate of the likelihood of effects described by the toxicity value. For example, Figure 6.5 displays distributions for the dose to river otter of mercury within two subreaches of the Clinch River/Poplar Creek system. The LOAEL for mercury to otters crosses the exposure distributions for Subreaches 3.01 and 3.02 at approximately the 15th and 85th percentiles, respectively (Figure 6.5). The interpretation of this figure is

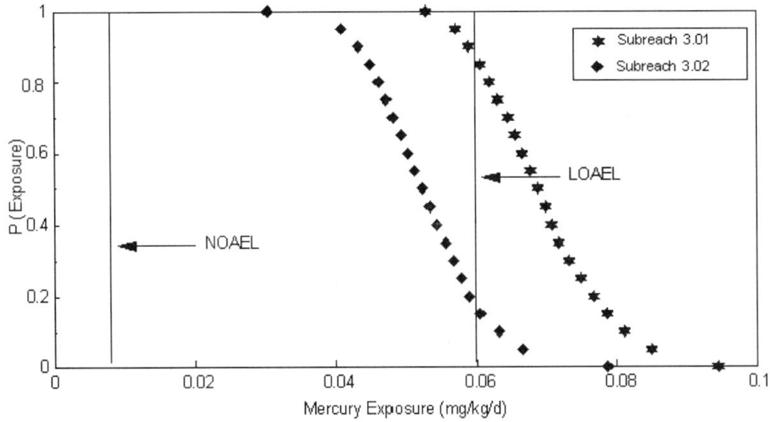

FIGURE 6.5 Probability density of estimated mercury exposures for otters in a subreach of Poplar Creek in relation to the allometrically scaled LOAEL.

that there is a 15% likelihood of effects to individual otters from mercury within Subreach 3.01 and an 85% likelihood of effects within Subreach 3.02. Because the NOAEL for mercury to otters was exceeded by the entire exposure distributions at both subreaches, the potential for effects at both locations is suggested. By superimposing toxicity values for different types and severity of effects (e.g., LD50, growth, hepatotoxicity, etc.), additional information concerning the nature of risks associated with the estimated exposure may be produced.

Monte Carlo-generated exposure distributions may also be compared with effects distributions, if such data are available. Methods for comparison of exposure and effects distributions are discussed in Section 7.7.2.

In addition to comparison to literature toxicity values, exposure estimates from the site of interest should be compared with background. As stated earlier (see Section 5.1.1.3), it is inappropriate to screen out chemicals because they do not present a risk to wildlife in individual media. However, if all media at the background site are combined into a background exposure estimate, this estimate may be used for screening purposes. Exceedence of background exposure estimates by on-site exposure, while not indicative of risk, is supportive of a conclusion of on-site risks. To conclude that risks are potentially significant, on-site exposures must also exceed appropriate measures of effects (NOAELs, LOAELs, or other values). If on-site exposure does not exceed that at the background site, potential for on-site risk is questionable; however, differences in form and bioavailability of contaminants between background and the waste site must be considered. Exposure estimates from background locations should also be compared with NOAELs and LOAELs. If background exposure exceeds NOAELs and LOAELs and the background site is known to be uncontaminated, the suitability of the toxicity values and assumptions used in the exposure assessment (primarily bioavailability of contaminants) should be reevaluated.

The calculation of HQs does not address risks to populations. However, for small sites or units, they may be adequate because the chief problem is to determine whether even one organism will be exposed to a hazardous dose. Figure 6.6 illustrates this point. A contaminant has been dumped or spilled at a point, and the average soil concentration drops off approximately exponentially from that point as a larger area of uncontaminated land is included in the average. Equivalent curves can be plotted for other patterns of contamination. The horizontal dashed lines indicate soil levels that are estimated to be thresholds for effects on birds or mammals. The vertical dashed lines indicate the home range or territory sizes of endpoint species (shrews and woodcock in this case). If the average concentration drops below the effects level before intersecting the home range size, then not even one individual is expected to be affected. In Figure 6.6, no effects are expected on woodcock, but one shrew is estimated to experience reproductive effects and has a small risk of death.

6.1.4.2 Individual vs. Population Effects

Multimedia exposure estimates, whether based on point estimates or Monte Carlo simulation, generally represent exposure of individuals to contaminants. Toxicity data, while it may have population-level implications (e.g., mortality, reproduction), is almost exclusively based on individual-level effects. With the exception of threat-

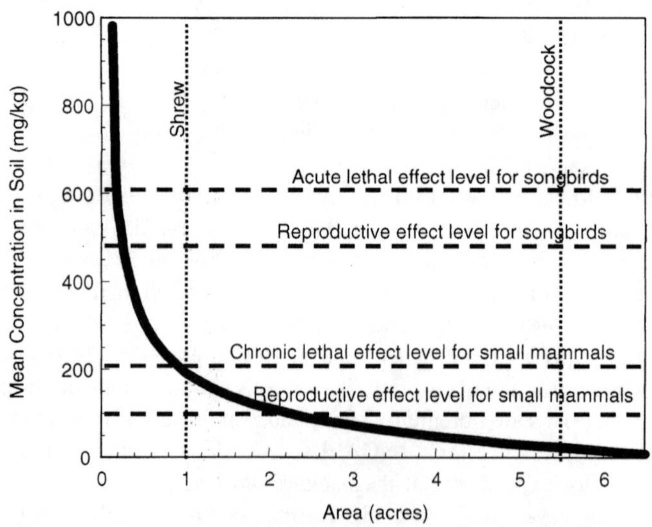

FIGURE 6.6 Mean concentration of a contaminant as a function of averaging area centered on the point of maximum concentration. Vertical lines indicate the home range area for a shrew and a woodcock. Horizontal dashed lines indicate soil concentrations estimated to produce toxic doses associated with specific effects. (Graphic redrawn from an unpublished graphic provided by C. Menzie.)

ened and endangered species, which are protected as individuals, most risk management decisions for wildlife are based on population-level effects. Therefore, for the results of the risk characterization to be of use for risk management, extrapolations from effects on individuals to effects on populations must be made. This extrapolation may be performed in several ways.

First, it may be assumed that there is a distinct population on the site, so that the exposure of the population is represented by the exposure of all of the individuals. All individuals at the site may be assumed to experience equivalent exposure. That is, the exposure distributions represent distributions of total intake rate of the chemical across individuals in the population based on the distributions of observed concentrations in water, soil, and various food items within each modeled area. If it is assumed that the members of a population occurring in each modeled area independently sample the water, soil, and food items over the entire modeled area, then proportions from the resulting exposure distribution represent estimates of the proportion of a population receiving a particular intake rate. This assumption is appropriate for small, nonterritorial organisms with limited home ranges, on large sites, particularly if the site constitutes a distinct habitat that is surrounded by inappropriate habitat. For example, a grassy site surrounded by forest or industrial development might support a distinct population of voles. The risks to that population can be estimated directly from the exposures of the individual organisms. Baron et al. (1999) used this approach to estimate population-level risks to rough-winged swallows along the Clinch River, TN.

Another approach is to assume that a certain number of individuals out of a larger population are exposed to contaminants. The proportion of the local population exposed at levels that exceed toxic thresholds represents the proportion of the population potentially at risk. This was the logic underlying the preliminary assessment for wide-ranging wildlife on the Oak Ridge Reservation (ORR; Sample et al., 1996a). On the ORR, while most habitat for wide-ranging wildlife species exists outside of source units (directly contaminated areas; see Chapter 2), some suitable habitat is present within source units. The proportion of the ORR-wide population potentially at risk from a source unit is estimated by the number of individuals that may use habitat on that unit. The degree to which a source unit is used (and therefore the risk that it may present) is dependent upon the availability of suitable habitat on the unit. An estimate of risks to reservation-wide populations was generated as follows.

1. Individual-based contaminant exposure estimates were generated for each unit using the generalized exposure model (see Sect 3.10). Contaminant data, averaged over the entire unit, were used in the exposure estimate.
2. Contaminant exposure estimates for each unit were compared with LOAELs to determine the magnitude and nature of effects that may result from exposure on the unit. If the exposure estimate was greater than the LOAEL, then individuals on the unit were determined to potentially experience adverse effects.
3. The availability and distribution of habitat on the ORR and within each unit, suitable for each species considered, was determined using a satellite-generated land cover map for the ORR (Washington-Allen et al., 1995).

4. Habitat requirements for the endpoint species of interest were compared with the ORR habitat map to determine the area of suitable habitat on the ORR and within units.
5. The area of suitable habitat on the ORR and within units was multiplied by species-specific population density values (ORR-specific or obtained from the literature) to generate estimates of the ORR-wide population and the numbers of individuals expected to reside within each unit.
6. The number of individuals for a given endpoint species expected to be receiving exposures greater than LOAELs for each measured contaminant was totaled using the unit-specific population estimate from step 5 and the results from step 2. This sum was compared with the ORR-wide population to determine the proportion of the ORR-wide population that was receiving hazardous exposures.

This approach provides a very simple estimate of population-level effects. It is biased because it does not take wildlife movement into account. Wide-ranging species may travel among and use multiple units, therefore receiving exposures different from that estimated for a single unit. In addition, the proportion of a reservation-wide population potentially at risk is limited by the proportion of suitable habitat present in source units. Application of this approach requires knowledge of the types, distribution, and quality of habitats present on the site.

A third approach is to combine the results of Monte Carlo simulation of exposure with literature-derived population density data to evaluate the likelihood and magnitude of population-level effects on wildlife. The number of individuals within a given area likely to experience exposures greater than LOAELs can be estimated using cumulative binomial probability functions (Dowdy and Wearden, 1983). Binomial probability functions are estimated using the following equation:

$$b(y;n;p) = \binom{n}{y} p^y (1-p)^{n-y} \tag{6.4}$$

where

y = the number of individuals experiencing exposures > LOAEL
n = total number of individuals within the watershed
p = probability of experiencing an exposure in excess of the LOAEL
$b(y;n;p)$ = probability of y individuals, out of a total of n, experiencing an exposure > LOAEL, given the probability that exceeding the LOAEL = p

By solving Equation 6.4 for $y = 0$ to $y = n$, a cumulative binomial probability distribution may be generated that can be used to estimate the number of individuals within an area that are likely to experience adverse effects.

This approach was used to estimate the risks that polychlorinated biphenyls (PCBs) and mercury in fish presented to the population of piscivores in watersheds on the ORR (Sample et al., 1996a). Monte Carlo simulations were performed to estimate watershed-wide exposures. It was assumed that wildlife were more likely to forage in areas where food is most abundant. The density or biomass of fish at or near locations where fish bioaccumulation data were collected were assumed to represent measures of food abundance. (Biomass data were preferred but, where unavailable, density data were used.) The relative proportion that each location contributed to overall watershed density or biomass was used to weight the contribution to the watershed-level exposure. The watershed-level exposure was estimated to be the weighted average of the exposure at each location sampled within the watershed. In this way, locations with high fish densities or greater fish biomass contributed more to exposure than locations with lower density or biomass. Because the watersheds were large enough to support multiple individuals, the weighted average exposure estimate was assumed to represent the exposure of all individuals in each watershed. While simplistic, this approach is believed to provide a better estimate of population-level effects than the previously described method. The use of this method, however, requires exposure data from multiple, spatially disjunct areas and data suitable to weight the potential exposure at each area. Sample and Suter (1999) and ODEQ (1998) provide additional application of this method.

Freshman and Menzie (1996) present an additional approach for extrapolating to population-level effects. Their population effects foraging (PEF) model estimates the number of individuals within a local population that may be adversely affected. The PEF model is an individual-based model that allows animals to move randomly over a contaminated site. Movements are limited by species-specific foraging areas and habitat requirements. The model estimates exposures for a series of individuals and sums the number of individuals that receive exposures in excess of toxic thresholds.

6.1.5 Body Burdens of Endpoint Organisms

Body burdens of chemicals are not a common line of evidence in risk assessments, and, when they are used, it is most often in assessments of aquatic toxicity. Although nearly all toxicity data for fishes are expressed in terms of aqueous concentrations, fish body burdens potentially provide an exposure metric that is more strongly correlated with effects (McCarty and Mackay, 1993). The correlation is most likely to be evident for chemicals that bioaccumulate in fish and other biota to concentrations greater than in water. For such chemicals dietary exposure may be more important than direct aqueous exposures, and concentrations that are not detectable in water may result in high body burdens in fish. Three contaminants that accumulate in that manner are mercury, PCBs, and selenium. Since the individual body burden measurements correspond to an exposure level for an individual fish, the maximum value observed in an individual fish is used for screening purposes, and the risk estimate is based on the distribution of individual observations for each measured species. Measurements may be performed on muscle (fillet), carcass (residue after filleting), or whole fish. Since measurements in whole fish are most commonly used in the literature, concentrations of chemicals in whole fish should be either measured directly or reconstructed from fillet and carcass data to estimate exposure (Chapter 3).

Body burdens may also be used to estimate risks to terrestrial wildlife, plants, and, less commonly, soil or sediment invertebrates. The justification of the use of this line of evidence for wildlife is that these organisms may bioaccumulate certain chemicals, just as the fish described above. Body burdens are not often measured in wildlife risk assessments because of the difficulty and ethical considerations inherent in sampling these vertebrates. Moreover, if the target organ is not known, it is unclear whether the chemical analysis of the whole animal, muscle tissue, or a particular organ is correlated with the concentration at the site of toxicity. For example, measurements of a particular metal in liver or kidney of deer is relevant to the risk assessment only if one of these organs is associated with the mechanism of toxicity or if the chemical is at equilibrium in all organs. In addition, for wide-ranging wildlife, the history of the animal (i.e., how much time it has spent on the site) is generally unknown. If concentrations of chemicals are measured in small mammals to estimate risks to predators, these concentrations may also be used to estimate risks to the small mammals.

Concentrations measured in plant tissues may also be used as estimates of exposure in risk assessments. As with wildlife, the concentrations are useful only if the organs (roots, leaves, reproductive structures) in which they were measured are either integral to the mechanism of toxicity or in equilibrium with organs involved in toxicity. An advantage of the use of this measure of exposure over total concentrations of a chemical in soil is that the tissue concentration is a more direct estimate of exposure. However, far fewer studies relate these tissue concentrations to effects. One may assume that for some chemicals, the directness of the estimate of exposure and the availability of published toxicity data cancel each other out in the comparison of the reliability of soil concentrations and plant tissue concentrations as estimates of exposure. A case in which plant tissue concentrations would be more reliable (and therefore the associated evidence would be weighed more heavily) than soil concentrations would be instances where published single-chemical toxicity data for soils were irrelevant, because the contaminants were largely unavailable for uptake. Such an example is the estimate of toxicity to plants in sewage sludge-amended agricultural soils (Chang et al., 1992). As with any measure of single-chemical toxicity, the relevance of toxicity data should be considered. For example, if studies of plants in solution are used to estimate effective concentrations of chemicals in plant roots in soil, the uncertainty associated with using these concentrations to estimate risks should be acknowledged. Plant roots grow differently in solution and in soil.

The estimation of body burdens from concentrations in abiotic media and the subsequent comparison to toxicity thresholds in published studies is rarely advisable. The derivation of tissue concentrations from bioaccumulation models adds a degree of uncertainty to the risk estimation. The only cases where estimates of body burdens may be useful are sites where measurable concentrations in abiotic media have been demonstrated to be poor estimates of exposure — e.g., in the example related to sewage sludge application, above. The measurement of tissue concentrations is generally well worth the expense.

Because few toxicity data are based on body burdens, effects distributions may not be worth using in the risk characterization. The relatively small set of relevant effects data has also inhibited the development of models for species and life stage extrapolations. Therefore, the calculation of a hazard quotient is usually a sufficient integration of exposure and effects. However, as with any measure of single-chemical toxicity, the relevance of the data needs to be carefully reviewed and described.

6.2 AMBIENT MEDIA TOXICITY TESTS

Risk characterization for ambient media toxicity tests begins by determining whether the tests show significant toxicity (Figure 6.7). At Oak Ridge, effects have been considered significant if (1) the hypothesis of no difference between responses in contaminated media and in either reference media or control media is rejected with 95% confidence (i.e., statistical significance), or (2) an effect of 20% or greater in survival, growth, or reproduction relative to either reference media or control media is observed (i.e., biological significance).

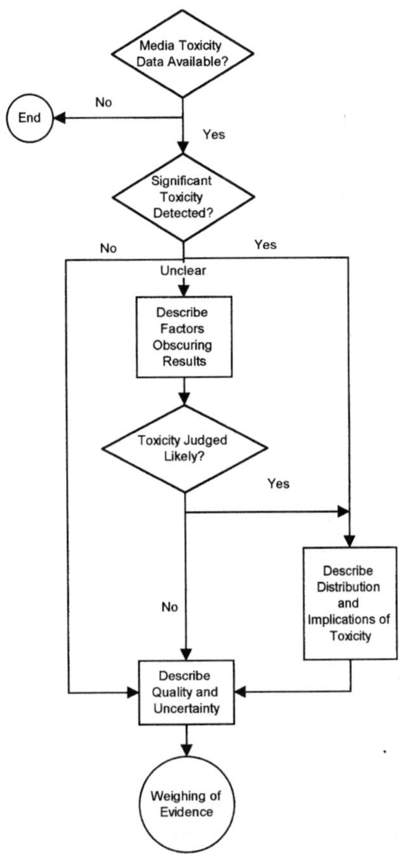

FIGURE 6.7 Risk characterization based on toxicity testing of ambient media.

If no significant toxicity is found, the risk characterization consists of determining the likelihood that the result constitutes a false negative. False negatives could result from not collecting samples from the most contaminated sites or at the times when contaminant levels were highest, from handling the samples in a way that reduced toxicity, or from using tests that are not sufficiently sensitive to detect effects that would cause significant injuries to populations or communities in the field.

If significant toxicity occurs in the tests, the risk characterization should describe the nature and magnitude of the effects and the consistency of effects among tests conducted with different species in the same medium.

Toxicity tests may produce ambiguous results in some cases due to poor performance of organisms in reference or control media (e.g., due to diseases, background contamination, inappropriate reference or control media, or poor execution of the test protocol). In such cases, expert judgment by the assessor in consultation with the individuals who performed the test should be used to arrive at an interpretation of the test results. One particular problem with reference and control media is their nutrient properties. This is particularly apparent with soil toxicity tests where site soils may have very low levels of organic matter and nutrient elements due to site activities that scrape and mix the soil horizons. In contrast, reference soils may have higher nutrient and organic matter levels because they have been vegetated and have not been scraped or mixed, and control soils (e.g., potting mixes) may have much higher levels than all but the best site or reference soils. Such differences should be minimized as far as possible during test design but must still be considered and accounted for in the risk characterization.

If significant toxicity is found at any site, then the relationship of toxicity to exposure must be characterized. The first way to do this is to examine the relationship of toxicity to concentrations of chemicals in the media. The manner in which this is done depends on the amount of data available. If numerous toxicity tests are available, the level of effects or the frequency of tests showing toxic effects could be defined as a function of concentrations of one or more COPECs. For example, if fathead minnow larvae experienced more than 20% mortality in one or more tests, the percent mortality could be plotted against the concentration of each of the COPECs in water samples collected in conjunction with the test. If there is a positive relationship, an appropriate statistical model should be fit to the points. If multiple chemicals with the same mode of action may be jointly contributing to toxicity, the aggregate concentration (e.g., total PAHs or TEF-normalized chlorinated dicyclic compounds; see Section 6.1) could be used as the independent variable. Alternatively, toxic responses might be plotted against HI or ΣTU values (Figure 6.8). Finally, each chemical of concern may be treated as an independent variable in a multiple regression or correlation. For example, phytotoxicity at mine-mill site in Anaconda, MT was correlated with soil arsenic, copper, and zinc (Kapustka et al., 1995). In general, if toxicity is occurring, it should be possible to identify exposure–response relationships. However, there are a number of reasons a true causal relationship between a chemical and a toxic response may not be apparent, even if numerous samples are taken (Box 6.1). Therefore, the lack of an exposure–response relationship does not disprove that one or more of the COPECs caused an apparent toxic effect.

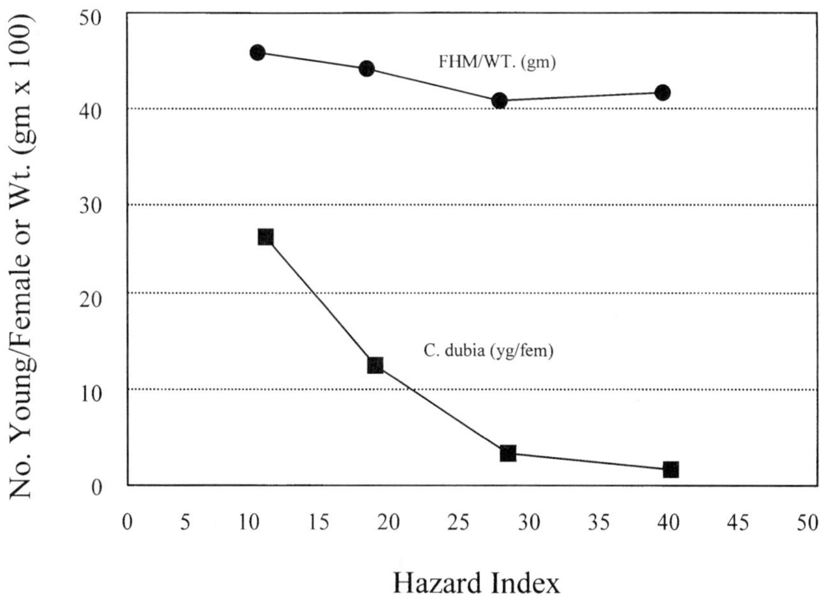

FIGURE 6.8 Results of ambient water toxicity tests (weight of fathead minnows and number of *Ceriodaphnia dubia* young per female) as a function of the chronic hazard index from a risk assessment of the Trinity River. (Redrawn from Parkhurst, B. R. et al., Project 91.-AER-1, Water Environment Research Foundation, Arlington, VA, 1996. With permission.)

An alternative and potentially complementary approach to relating toxicity to exposure is to determine the relationship between the occurrence of toxicity and sources of contaminants (e.g., springs, seeps, tributaries, spills) or of diluents (i.e., relatively clean water or sediments). This may be done by simply creating a table of potential sources of contamination or dilution and indicating for each test whether toxicity increases, decreases, or remains unchanged below that source. The same information may be conveyed graphically. For a stream or river, toxicity may be plotted as a function of reach (if reach boundaries are defined by source locations) or distance downstream (with locations of sources marked on the graph) (Figure 6.9).

Finally, when sources of toxic water have been identified, and tests have been performed on dilution series of those waters, the transport and fate of toxicity can be modeled like that of individual chemicals (DiToro et al., 1991). Such models of toxicity can be used to explain ecological degradation observed in streams and to apportion causation among sources.

To facilitate the weight-of-evidence analysis and to make the bases clear to the reader, it may be useful to summarize the results of this integration for each reach or area where significant toxicity was found using Table 6.2.

BOX 6.1

Why Contaminant Concentrations in Ambient Media May Not Be Correlated with Toxicity of Those Media

Variation in bioavailability
- Due to variance in medium characteristics
- Due to variance in contaminant age among locations (contaminants added to soil and sediments may become less bioavailable over time due to sequestration)
- Due to variance in transformation or sequestration rates among locations

Variation in the form of the chemical (e.g., ionization state)

Variation in concentration over time or space (i.e., samples for analysis may not be the same as those tested)
- Spatial heterogeneity
- Temporal variability (e.g., aqueous toxicity tests last for several days, but typically water from only 1 day is analyzed)

Variation in composition of the waste
- Due to interactive toxicity and variance in the relative proportions of COPECs
- Due to variance in concentrations of waste components that are not COPECs but that influence the toxicity of the COPECs

Variation in co-occurring chemicals
- Due to variance in upstream or other off-site contaminants
- Due to variance in background concentrations of chemicals

Inadequate detection limits (correlation will be apparent if chemicals are not detected when there is low or no toxicity)

Variation in toxicity tests due to variation in test performance

Variation in toxicity test due to variance in medium characteristics (e.g., hardness, organic matter content, and pH)

6.3 BIOLOGICAL SURVEYS

If biological survey data are available for an endpoint species or community, then the first question to be answered is whether the data suggest that significant effects are occurring (Figure 6.10). For some groups, notably fish and benthic invertebrates, there is a good chance that reference streams have been identified and survey data are available for comparison. For most other endpoint groups, references must be established ad hoc, and the lack of temporal or spatial replication may make inference tenuous. For some taxa such as most birds, traditional survey data are not useful for estimating risks from wastes because mobility, territoriality, or other factors obscure demographic effects. However, survey results may be more reliable if efforts are made to control extraneous variance. For example, the reproductive success of birds can be estimated by setting out nest boxes on contaminated and reference sites.

Care must be taken to consider the sensitivity of field survey data to toxic effects relative to other lines of evidence. Some biological surveys are very sensitive

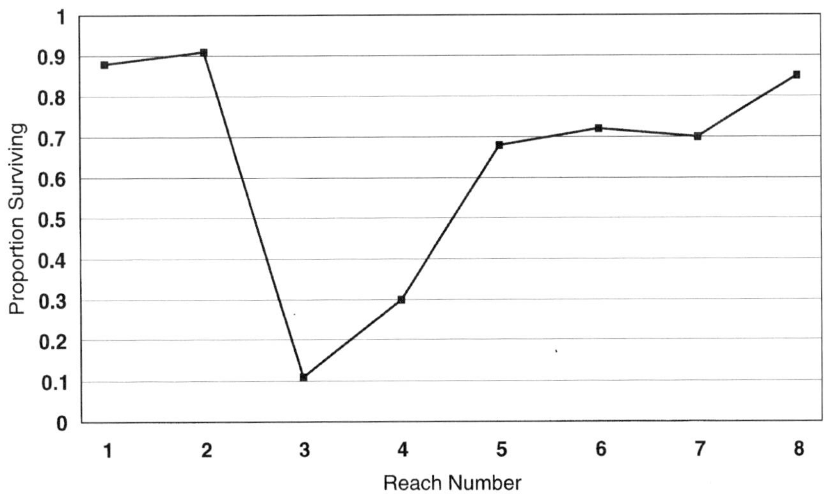

FIGURE 6.9 Mean proportional reduction in survival of fathead minnow larvae in ambient water relative to controls in numbered reaches. The source occurs in Reach 2.

TABLE 6.2
Summary Table for Integration of Results from Ambient Media Toxicity Tests

Issue	Result
Species affected	List species and life stages affected in the tests
Severity of effects	List types and magnitudes of effects
Spatial extent of effects	Delineate meters of stream, square meters of land, etc. for which media samples were toxic
Frequency of effects	Calculate the proportion of time or the number of distinct toxic episodes per unit of time
Association with sources	Define spatial and temporal relationships to potential sources
Association with exposure	Define relationships to contaminant concentrations or other measures of exposure
Estimated effect	Summarize the nature and extent of estimated toxic effects on the assessment endpoint and credible upper bounds
Confidence in results	Provide a rating and supporting comments

(e.g., surveys of nesting success of colonial nesting birds or electrofishing surveys of wadable streams), others are moderately sensitive (e.g., benthic macroinvertebrates), and still others are rather insensitive (e.g., fish community surveys in large reservoirs and small mammal surveys). Sensitivity is not just a matter of precision. For example, surveys of territorial breeding birds can be quite precise, but they are insensitive to toxic effects because the number of breeding pairs on a site is often

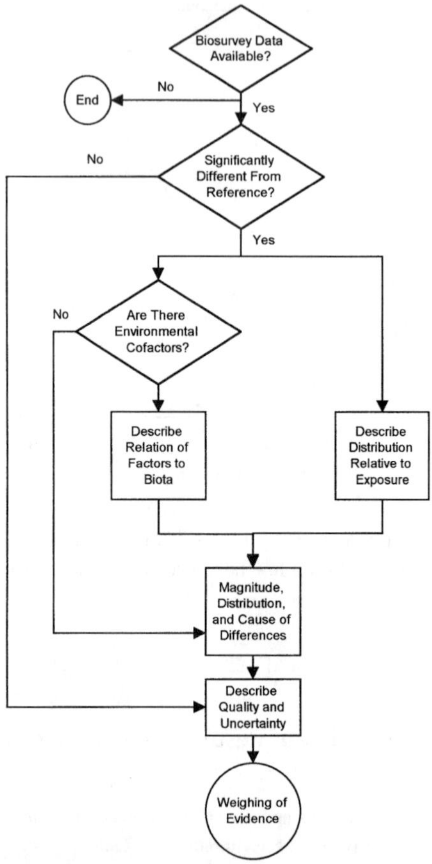

FIGURE 6.10 Risk characterization based on biological survey data.

low, and because any mortality or reduction in reproductive success is unlikely to be reflected in densities of breeding pairs. However, even relatively insensitive surveys may be useful in assessments. For example, if the concentrations of chemicals suggest that a medium should be highly toxic, but toxicity tests of the medium find no toxicity, even a relatively insensitive survey that found a community that was not highly modified could indicate that the chemical analyses were misleading and the toxicity test data were probably correct (e.g., the chemical was not in a bioavailable form or consisted of a less toxic species). Conversely, a highly modified community in the absence of high levels of analyzed chemicals would suggest that combined toxic effects, toxic levels of unanalyzed contaminants, episodic contamination, or some other disturbance had occurred. However, field surveys interpreted in isolation without supporting data can be misleading, particularly when the absence of statistically significant differences are inappropriately interpreted as an absence of effects.

If biological survey data are consistent with significant reductions in abundance, production, or diversity, associations of apparent effects with causal factors must be examined. First, the distribution of apparent effects in space and time must be compared with the distribution of sources, of contaminants, and of habitat variables. Second, the distribution of apparent effects must be compared with the distribution of habitat factors that are likely to affect the organisms in question, such as stream structure and flow, to determine whether they account for the apparent effects. For example, most of the variability in the benthic community of Poplar Creek embayment was found to be due to variance in sediment texture and organic matter (Jones et al., 1999). Only after that variance had been modeled by multiple regression could the residual variance be associated with contaminants. This process may be aided by the use of habitat models (available from the U.S. Fish and Wildlife Service) or habitat indexes (Rankin, 1995). For example, surveys of wildlife at the metal-contaminated Lehigh Gap, PA showed drastic reductions in abundance (Beyer and Storm, 1995). However, the effects of vegetation loss were removed using habitat suitability models to determine the reduction that could be attributed to direct toxic effects. Even when available habitat models or indexes do not characterize habitat effects at the site, they indicate what habitat parameters might be used to generate a site-specific model. Fourth, if there is an apparent effect of contaminants on the survey metrics, the relationship should be modeled or at least plotted. As with the ambient toxicity test results, the exposures may be expressed as concentrations of chemicals assessed individually, concentrations of individual chemicals used in a multiple regression, summed concentrations of related chemicals, ΣTU values, or HI values (Figure 6.11). Finally, if results are available for toxicity tests of ambient media, their relationships to the survey results should be determined. For example, soil from locations at the Naval Weapons Center in Concord, CA with reduced abundances of earthworm species or of total macroinvertebrates caused mortality and reduced growth in earthworm toxicity tests (Jenkins et al., 1995).

To facilitate the weight-of-evidence analysis and to make the bases clear to the reader, it may be useful to summarize the results of this integration for each reach or area using Table 6.3.

6.4 BIOMARKERS AND PATHOLOGIES

Biomarkers are physiological or biochemical measures, such as blood cholinesterase concentration, that may be indicative of exposure to or effects of contaminants. They are seldom useful for estimating risks by themselves, but they can be used to support other lines of inference. In particular, if the biota of a site are depauperate, biomarkers in the remaining resistant species may indicate what may have caused the loss of the missing species. The inference begins by asking if the levels of the biomarkers significantly differ from those at reference sites (Figure 6.12). If they do, then it is necessary to determine whether they are diagnostic or at least characteristic of any of the COPECs or of any of the habitat factors that are thought to affect the endpoint biota. If the biomarkers are characteristic of contaminant exposures, then the distribution and frequency of elevated levels must be compared with the distributions and concentrations of contaminants. Finally, to the extent that the biomarkers are known

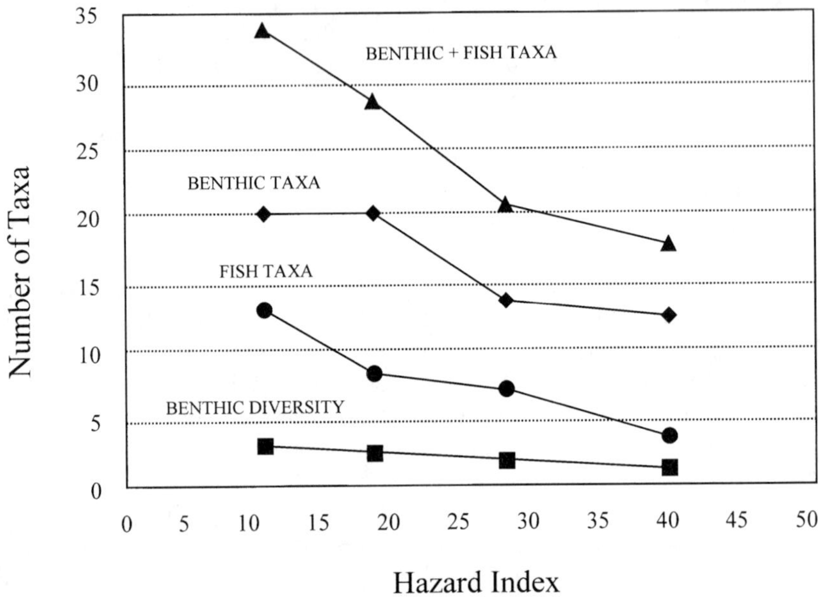

FIGURE 6.11 Results of biological surveys (number of fish taxa, number of benthic invertebrate taxa, and total taxa) as functions of the chronic hazard index from a risk assessment of the Trinity River. (Redrawn from Parkhurst, B. R. et al., Project 91-AER-1, Water Environment Research Foundation, Arlington, VA, 1996. With permission.)

to be related to overt effects such as reductions in growth, fecundity, or mortality, the implications of the observed biomarker levels for populations or communities should be estimated.

A novel but potentially powerful use of biomarkers is as bioassays to estimate external or internal exposure to mixtures of chemicals. If chemicals in a mixture have a common mechanism of action, then a biochemical measure of some step in that mechanism may be used as a measure of the total effective exposure. This approach has been applied to dioxin-like compounds. Generation of 7-ethoxyresorufin-*o*-deethylase (EROD) in rat hepatoma cell cultures has been used as a bioassay for 2,3,7,8-TCDD-equivalents (TCDD-EQs) in the food of endpoint organisms or in the organisms themselves (Tillitt et al., 1991). These can then be compared with TCDD levels that are known to cause effects in toxicity tests. The method has been used to determine the cause of deformities and reproductive failures in piscivorous birds of the Great Lakes (Ludwig et al., 1996). The technique is equivalent to the use of TCDD TEFs discussed above (Section 6.1), but eliminates the uncertainties associated with application of the factors to different species and conditions.

Pathologies include lesions, tumors, deformities, and other signs of disease. The occurrence of visible gross pathologies may be an endpoint itself because of public

TABLE 6.3
Summary Table for Integration of Biological Survey Results

Issue	Result
Taxa and properties surveyed	List species or communities and measures of effect
Nature and severity of effects	List types and magnitudes of apparent effects
Minimum detectable effects	For each measure of effect, define the smallest effect that could have been distinguished from the reference condition
Spatial extent of effects	Delineate meters of stream, square meters of land, etc., that are apparently affected
Number and nature of reference sites	List and describe reference sites including habitat differences from the contaminated site
Association with habitat characteristics	Describe any correlations or qualitative associations of apparent effects with habitat variables
Association with source	Describe any correlations or qualitative associations of apparent effects with sources
Association with exposure	Define relationships to ambient contaminant concentrations, body burdens, or other measures of exposure
Association with toxicity	Define relationships to toxicity of media
Most likely cause of apparent effects	Based on the associations described in previous items, present the most likely cause of the apparent effects
Estimated effects	Summarize the estimated nature and extent of effects and credible upper bounds
Confidence in results	Provide rating and supporting comments

concern. However, they are more often used like biomarkers to help diagnose the causes of effects on organisms. Manuals are available for this purpose including Friend (1987), Meyer and Barklay (1990), and Beyer et al. (1998). This type of evidence is particularly useful for identifying alternative potential causes of observed effects such as epizootics or anoxia. Greater diagnostic power can be obtained by combining pathologies with condition metrics and even population properties (Goede and Barton, 1990; Gibbons and Munkittrick, 1994; Beyer et al., 1998).

To facilitate the weight-of-evidence analysis and to make the bases clear to the reader, it may be useful to summarize the results of this integration for each reach or area using Table 6.4.

6.5 WEIGHT OF EVIDENCE

The weighing of evidence begins by summarizing the available lines of evidence for each endpoint (Figure 6.13). Given that one has estimated risks based on each line of evidence, the process of weighing the evidence amounts to determining what estimate of risks is most likely, given those results. If the assessment endpoint is defined in terms of some threshold for significance, then the process can be conducted in two steps. First, for each line of evidence determine whether it is consistent with exceedence of the threshold, inconsistent with exceedence, or ambiguous. Second, determine whether the results as a whole indicate that it is likely or unlikely

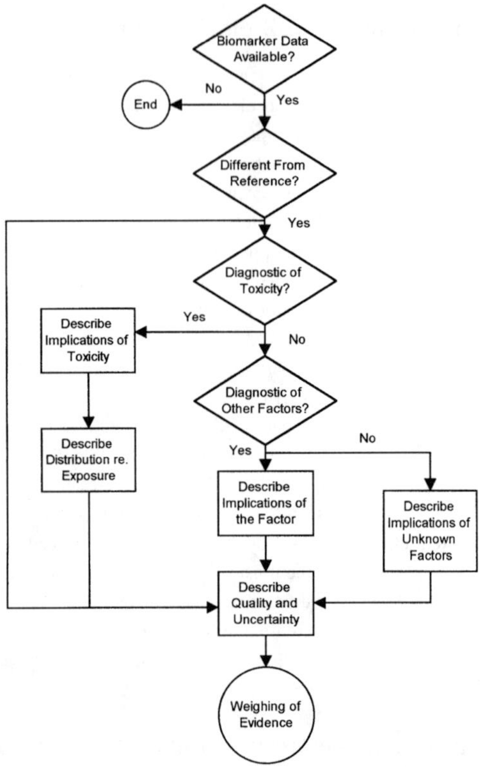

FIGURE 6.12 Risk characterization based on biomarker data.

that the threshold is exceeded. If the results for all lines of evidence are consistent, the result of the weighing of evidence is clear. If there is no bias in the assessment that affects all lines of evidence, agreement among multiple lines of evidence is strong evidence to support a conclusion. However, if there are inconsistencies, the true weighing of evidence must occur. The weights are determined based on the following considerations adapted from Menzie et al. (1996) and Suter (1998b).

Relevance — Evidence is given more weight if the measure of effect is more directly related to (i.e., relevant to) the assessment endpoint.

- Effects are relevant if the measure of effect is a direct estimate of the assessment endpoint or if validation studies have demonstrated that the measurement endpoint is predictive of the assessment endpoint. Note that a measure of effect based on statistical significance (e.g., a NOEC) is less likely to bear a consistent relationship to an assessment endpoint than one that is based on biological significance (e.g., EC*x*).
- The mode of exposure may not be relevant if the media used in a test are not similar to the site media. Normalization of media concentrations may

TABLE 6.4
Summary Table for Integration of Biomarker or Pathology Results

Issue	Result
Taxa and response	List the species and specific responses
Implications of responses for organisms and populations	Describe, as far as possible, the relationship between the biomarkers or pathologies and population or community endpoints
Causes of the observed response	List chemicals, chemical classes, pathogens, or conditions (e.g., anoxia) that are known to induce the biomarker or pathology
Number and nature of reference sites	List and describe reference sites including habitat differences from the contaminated site
Association with habitat or seasonal variables	List habitat or life cycle variables that may affect the level of the biological response at the site
Association with sources	Describe any correlations or qualitative associations of the responses with sources
Association with exposure	Define relationships to contaminant concentrations or other measures of exposure
Most likely cause of response	Based on the associations described in previous items, present the most likely cause of the apparent responses
Estimated effects	Summarize the estimated nature and extent of effects associated with the biomarker or pathology and credible upper bounds if they can be identified
Confidence in results	Provide rating and associated comments

increase the relevance of a test if the normalization method has been validated. Similarly, the relevance of tests in solution to sediment or soil exposures is low unless the models or extraction techniques used to estimate aqueous phase exposures have been validated.

- Measures of effect derived from the literature rather than site-specific studies may have used a form of the chemical that is not relevant to the chemical detected in the field. For example, is it the same ionization state and has the weathering or sequestration of the field contaminant changed its composition or form in ways that are not reflected in the test?

In some cases, available information may not be sufficient to evaluate the relevance of a line of evidence. In such cases, relevance may be evaluated by listing the ways in which the results could be fundamentally inappropriate or so inaccurate as to nullify the results, and evaluate the likelihood that they are occurring in this case. For single-chemical toxicity tests, such a list could include the possibility that the test was (1) performed with the wrong form of the chemical, (2) performed in media differing from the site media in ways that significantly affect toxicity, or (3) insensitive due to short duration, a resistant species, or the lack of measures of sublethal effects.

Exposure–Response — As in all toxicological studies, a line of evidence that demonstrates a relationship between the magnitude of exposure and the effects is more convincing that one that does not. For example, apparent effects in media toxicity tests may be attributed to the chemical with measured concentrations that

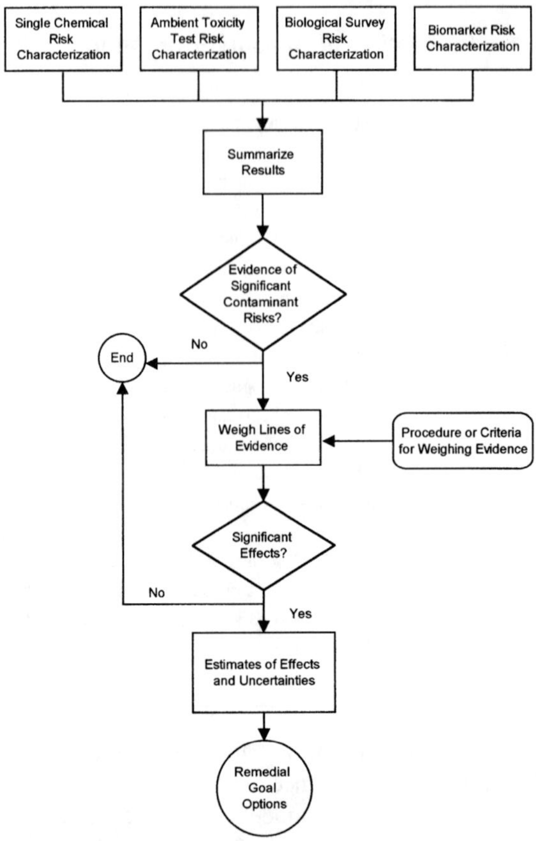

FIGURE 6.13 Risk characterization based on weighing of multiple lines of evidence.

exceed benchmarks by the greatest margin, but unless the tested medium is analyzed and an exposure–response relationship demonstrated, it may be suspected that effects are a result of other contaminants, nutrient levels, texture, or other properties. If an exposure–response relationship has not been demonstrated, then consideration should be given to the magnitude of the observed differences. For example, if medium test data include only comparisons of contaminated and uncontaminated soils, the observed differences are less likely to be due to extraneous factors if they are large (e.g., 100% mortality rather than 25% less growth).

Temporal Scope — A line of evidence should be given more weight if the data encompass the relevant range of temporal variance in conditions. For example, if contaminated and reference soils are surveyed during a period of drought, few earthworms will be found at any site, so toxic effects will not be apparent. Temporal scope may also be inadequate if aqueous toxic effects occur when storm events flush contaminants into streams, but water for chemical analysis or toxicity testing is not collected during such events. This phenomenon has occurred on the ORR and is probably a widespread problem. For example, studies of risks to fish from metals in the Clark Fork River, MT, focused on chronic exposures, but fish kills occurred due to episodes of low pH and high metal concentrations following thunderstorms (Pascoe et al., 1994).

Spatial Scope — A line of evidence should be given more weight if the data adequately represent the area to be assessed, including directly contaminated areas, indirectly contaminated areas, and indirectly affected areas. In some cases the most contaminated or most susceptible areas were not sampled because of access problems or because of the sampling design (e.g., random sampling with few samples).

Quality — The quality of the data should be evaluated in terms of the protocols for sampling, analysis, and testing; the expertise of the individuals involved in the data collection; the adequacy of the quality control during sampling, sample processing, analysis, and recording of results; and any other issues that are known to affect the quality of the data for purposes of risk assessment. Although the use of standard methods tends to increase the likelihood of high-quality results, they are no guarantee. Standard methods may be poorly implemented or may be inappropriate to a site. In contrast, a well-designed and well-performed site-specific measurement or testing protocol can give very high quality results.

Quantity — The adequacy of the data should be evaluated in terms of the number of observations taken. Results based on small sample sizes are given less weight than those based on large sample sizes. The adequacy of the number of observations must be evaluated relative to the variance as in any analysis of a sampling design, but it is also important in studies of this sort to consider their adequacy relative to potential biases in the sampling (see spatial and temporal scope, above).

Uncertainty — A line of evidence that estimates the assessment endpoint with low uncertainty should be given more weight. Uncertainty in a risk estimate is in part a function of the data quality and quantity, discussed above. In most cases, however, the major source of uncertainty is the extrapolations between the measures of effect and the assessment endpoint. In addition, the extrapolation from the measures of exposure to the exposure of the endpoint entities may be large due to considerations such as bioavailability and temporal dynamics.

These and other considerations can be used as points to consider in forming an expert judgment or consensus about which way the weight of evidence tips the balance. Table 6.5 presents an example of a simple summary of the results of weighing evidence based on this sort of process. The lines of evidence are listed, and a symbol is assigned for each: + if the evidence is consistent with significant effects on the endpoint, – if it is inconsistent with significant effects, and ± if it is too ambiguous to assign to either category. The last column presents a short summary of the results of the risk characterization for that line of evidence. If indirect effects are part of the conceptual model, they should be summarized in a separate line of the table. For example, effects on piscivorous wildlife could be due entirely or in part to inadequate food which may be due to toxicity to fish. The last line of the table presents the weight-of-evidence-based conclusion concerning whether significant effects are occurring and a brief statement concerning the basis for the conclusion. This conclusion is not based simply on the relative number of + or – signs. The "weight" component of weight of evidence is the relative credibility and reli-

TABLE 6.5
A Summary of a Risk Characterization by Weight of Evidence for a Soil Invertebrate Community in Contaminated Soil

Evidence	Result[a]	Explanation
Biological surveys	–	Soil microarthropod taxonomic richness is within the range of reference soils of the same type, and is not correlated with concentrations of petroleum components
Ambient toxicity tests	–	Soil did not reduce survivorship of the earthworm *Eisenia foetida;* sublethal effects were not determined
Organism analyses	±	Concentrations of PAHs in depurated earthworms were elevated relative to worms from reference sites, but toxic body burdens are unknown
Soil analyses/ single-chemical tests	+	If the total hydrocarbon content of the soil is assumed to be composed of benzene, then deaths of earthworms would be expected; relevant toxicity data for other detected contaminants are unavailable
Weight-of-evidence approach	–	Although the earthworm tests may not be sensitive, they and the biological surveys are both negative, and are both more reliable than the single-chemical toxicity data used with the analytical results for soil

[a] Results of the risk characterization for each line of evidence and for the weight of evidence: + indicates that the evidence is consistent with the occurrence of a 20% reduction in species richness or abundance of the invertebrate community; – indicates that the evidence is inconsistent with the occurrence of a 20% reduction in species richness or abundance of the invertebrate community; ± indicates that the evidence is too ambiguous to interpret.

ability of the conclusions of the various lines of evidence as discussed above. Additionally, those considerations can be used to grade the weight to be assigned to each line of evidence (e.g., high, moderate, or low weight) (Table 6.6). This still leaves the inference to a process of expert judgment or consensus but makes the bases clearer to readers and reviewers. Finally, a scoring system could be developed that would formalize the weighing of evidence. For example, a numerical weight could simply be assigned to each line of evidence based on quality, relevance, and other factors; a + or – sign assigned depending on whether the evidence is consistent or inconsistent with the hypothesized risk, and the weights summed across lines of evidence. A quantitative system of that sort has been developed by a group consisting of representatives of Massachusetts, the private sector, and U.S. government agencies (Menzie et al., 1996). Such systems have the advantage of being open, consistent, and less subject to hidden biases, but they may not give as reasonable a result in every case as a careful ad hoc weighing of the evidence would. However, the weighing of evidence is performed, it is incumbent on the assessment scientist to make the basis for the judgment as clear as possible to readers and reviewers. Where multiple units or reaches are assessed, it is helpful to provide a summary table for the weighing of evidence across the entire site as in Table 6.7 so that the consistency of judgment can be reviewed.

TABLE 6.6
Example of a Table Summarizing the Risk Characterization for the Species Richness and Abundance of a Fish Community in a Stream at a Waste Site

Evidence	Result[a]	Weight[b]	Explanation
Biological surveys	–	H	Fish community productivity and species richness are both high, relative to reference reaches; data are abundant and of high quality
Ambient toxicity tests	±	M	High lethality to fathead minnow larvae was observed in a single test, but variability is too high for statistical significance; no other aqueous toxicity was observed in ten tests
Water analyses/ single-chemical tests	+	M	Only zinc is believed to be potentially toxic in water and only to highly sensitive species
Weight-of-evidence approach	–		Reach 2 supports a clearly high quality fish community; other evidence which suggests toxic risks is much weaker (single-chemical toxicology) or inconsistent and weak (ambient toxicity tests)

[a] Results of the risk characterization for each line of evidence and for the weight of evidence: + indicates that the evidence is consistent with the occurrence of the endpoint effect; – indicates that the evidence is inconsistent with the occurrence of the endpoint effect; ± indicates that the evidence is too ambiguous to interpret.

[b]Weights assigned to individual lines of evidence: high (H), moderate (M), and low (L).

In some cases, the primary assessment problem is to determine the causation of observed effects such as fish kills or communities with clearly reduced species richness. This is in contrast to the more common situation described above in which the existence of effects has not been clearly established. Determination of causation requires the application of epidemiological inference to ecological risk assessment or ecoepidemiology (Bro-Rasmussen and Lokke, 1984; Suter, 1990; Fox, 1991). Considerable thought and argument has gone into the issue of establishing causation in epidemiology. Koch's postulates provide a standard of proof that may be applied to toxicants as well as pathogens (Adams, 1963; Woodman and Cowling, 1987; Suter, 1993a). More commonly, sets of criteria are used that provide greater confidence in causal relations without attempting to prove causation (Hill, 1965; Susser, 1986; Fox, 1991). A synthesis of these criteria is presented in Table 6.8. In situations where ecoepidemiological inferences must be made, the evidence could be evaluated against each of the criteria in Table 6.8, and the appropriate score assigned on the −−− to +++ scale. That result could be used like the results in Tables 6.6 and 6.7 to support and justify a weighing of the evidence. Examples of the application of causal inference in ecoepidemiology can be found in the papers from the three Cause

TABLE 6.7
Summary of Weight-of-Evidence Analyses for Reaches Exposed to Contaminants in the Clinch River/Poplar Creek Operable Unit

Reach	Biological Surveys	Bioindicators	Ambient Toxicity Tests	Fish Analyses	Water Analyses/ Single-Chemical Toxicity	Weight of Evidence
Upper Clinch River Arm	±	±		±	–	–
Poplar Creek Embayment	+	±	+	±	+	+
Lower Clinch River Arm	–	±	–	±	+	–
McCoy Branch Embayment			±		–	–

[a] Results of the risk characterization for each line of evidence and for the weight of evidence: + indicates that the evidence is consistent with the occurrence of the endpoint effect; – indicates that the evidence is inconsistent with the occurrence of the endpoint effect; ± indicates that the evidence is too ambiguous to interpret; blank cells indicate that data were not available for that line of evidence.

Effect Linkage Workshops organized in support of the Great Lakes Water Quality Initiative and published in the August 1991 issue of *Journal of Toxicology and Environmental Health* and in volume 19, number 4 and volume 22, number 2 of the *Journal of Great Lakes Research.*

The use of quantitative weighing of evidence or of an equivalent expert judgment about which lines of evidence are most reliable is based on an implicit assumption that the lines of evidence are logically independent. Another approach to weighing multiple lines of evidence is to determine whether there are logical relationships among the lines of evidence. Based on knowledge of site conditions and of environmental chemistry and toxicology, one may be able to explain why inconsistencies occur among the lines of evidence. For example, one may know that spiked soil tests tend to overestimate the availability and hence the toxicity of contaminants, and one may even be able to say whether the bias associated with this factor is sufficient to account for discrepancies with tests of site soils. This process of developing a logical explanation for differences among lines of evidence is potentially more convincing than simple weighing of the evidence because it is mechanistic. However, it is important to remember that such explanations can degenerate into just-so stories if the relevance of the proposed mechanisms is not well supported. Therefore, studies should be designed and carried out specifically to support inference concerning reality and causality of inferred effects. For example, one might analyze aqueous extracts of site soils and spiked soils to support the inference that differences are due to relative bioavailability.

In general, a logical analysis of the data should proceed from most realistic (i.e., site-specific) to most precise and controlled (e.g., single chemical and species

TABLE 6.8

Format for a Table to Summarize Results of an Inference Concerning Causation in Ecoepidemiology

Criterion	Results	Effect on Hypothesis
Strength of association	Strong	+++
	Moderate	++
	Weak	+
	None	0
Consistency of association	Invariate	+++
	Regular	++
	Most of the time	+
	Seldom	0
	(Preferably, present numeric results)	
Specificity		
of cause	High	++
	Moderate	+
	Low	0
of effect	High	++
	Moderate	+
	Low	0
Temporality	Compatible	++
	Incompatible	−−−
	Uncertain	0
Biological gradient	Clearly monotonic	+++
	Weak or other than monotonic	+
	None found	−
Plausibility	Plausible	+
	Implausible	−
Coherence	Evidence all consistent	+++
	Most consistent	+
	Many inconsistencies	−−−
Experiment	Experimental studies: Concordant	+++
	Ambiguous	+
	Inconcordant	−−−
	Absent	0
Analogy	Analogous cases: Many or few but clear	++
	Few or unclear	+
	None	0
Probability	Probability association occurred by chance: Very low	++
	Low	+
	High	−
	(Or present numeric results)	++, +, −

continued

TABLE 6.8 (continued)

Format for a Table to Summarize Results of an Inference Concerning Causation in Ecoepidemiology

Criterion	Results	Effect on Hypothesis
Predictive performance	Prediction: Confirmed	+++
	Failed	–
Internal exposure	Detected	++
	Undetected	– –
	Undetermined	0

Note: In an application, one result and a corresponding effect on hypothesis rating would be selected for each criterion (Suter, 1998b).

toxicity tests). Field surveys indicate the actual state of the receiving environment, so other lines of evidence that contradict the field surveys, after allowing for limitations of the field data, are clearly incorrect. For example, the presence of plants that are growing and not visibly injured indicates that lethal and gross pathological effects are not occurring but does not preclude reductions in reproduction or growth rates. Those other effects could be addressed by more detailed field studies of growth rates and seed production and viability. The presence of individuals of highly mobile species such as birds indicates almost nothing about risks because dispersal replaces losses of individuals or reduced reproduction.

Ambient media toxicity tests indicate whether toxicity could be responsible for differences in the state of the receiving environment, including differences that may not be detectable in the field. However, effects in the field are usually more credible than negative test results, because field exposures are longer and otherwise more realistic, and species and life stages from the site may be more sensitive than test species and life stages.

Single-chemical toxicity tests indicate which components of the contaminated ambient media could be responsible for effects. Because they are less realistic than other lines of evidence, single-chemical toxicity tests are usually less credible than the other lines of evidence. They do not include combined toxic effects, the test medium may not represent the site media, the exposure may be unrealistic, and the chemicals may be in a different form from that at the site. However, because these studies are more controlled than those from other lines of evidence, they are more likely to detect sublethal effects. In addition, single-chemical toxicity tests may include longer exposures, more sensitive responses, and more sensitive species than tests of contaminated ambient media. These sorts of logical arguments concerning the interpretation of single-chemical toxicity test results must be generated ad hoc, because they depend on the characteristics of the data and the site.

6.6 TRIAD ALTERNATIVES

The weight-of-evidence approach described above is general in the sense that it is applicable to any combination of lines of evidence. An alternative approach is the development of inferential rules based on a standard set of lines of evidence. The best developed example is the sediment quality triad (Long and Chapman, 1985; Chapman, 1990). The three components of the triad are sediment chemistry, sediment toxicity, and sediment invertebrate community structure. Assuming that all three components are determined with sufficient sensitivity and data quality, the rules in Table 6.9 can be used to reach a conclusion concerning the induction of effects by contaminants. The assumptions are critical. If any of the data are not sufficiently sensitive or are not of high quality, one must weigh the evidence as described above. The sediment quality triad was developed for estuarine sediments and has primarily been applied in those systems. However, it may be adapted to the soft sediments of streams and riverine systems (Canfield et al., 1994).

An alternative triad, the exposure–dose–response triad, has been proposed for assessment of risks from contaminated water or sediments by Salazar and Salazar (1998). Exposure is estimated by analysis of the ambient medium, dose by analysis of tissue chemistry, and response by surveys of community properties or toxicity tests of the contaminated media. Significant contaminant effects may be assumed to be occurring if effects are detected in the tests or surveys and if the effects are linked to the ambient contamination by body burdens that are sufficient to indicate a toxic dose. In the examples provided by Salazar and Salazar (1998), the effects

TABLE 6.9
Inference Based on the Sediment Quality Triad

Situation	Chemicals Present	Toxicity	Community Alteration	Possible Conclusions
1	+	+	+	Strong evidence for pollution-induced degradation
2	–	–	–	Strong evidence that there is no pollution-induced degradation
3	+	–	–	Contaminants are not bioavailable, or are present at non-toxic levels
4	–	+	–	Unmeasured chemicals or conditions exist with the potential to cause degradation
5	–	–	+	Alteration is not due to toxic chemicals
6	+	+	–	Toxic chemicals are stressing the system but are not sufficient to significantly modify the community
7	–	+	+	Unmeasured toxic chemicals are causing degradation
8	+	–	+	Chemicals are not bioavailable or alteration is not due to toxic chemicals

Note: Responses are shown as either positive (+) or negative (–) indicating whether or not measurable (e.g.. statistically significant) differences from control/reference conditions/measures are determined.

Source: Chapman, P. M., *Sci. Total Environ.*, 97/98, 815, 1990. With permission.

tests use growth responses of bivalves exposed in cages in the field. The tissue concentrations were from the same caged bivalves. Toxic tissue concentrations for the contaminants of potential concern were determined from controlled exposures. As the authors acknowledge, if the lines of evidence are not concordant, one must weigh the evidence, taking into consideration data quality and factors such as combined toxic effects, temperature, food availability, and variance in toxic tissue concentrations due to growth dilution.

6.7 RISK ESTIMATION

After the lines of evidence have been weighed to reach a conclusion about the significance of risks to an assessment endpoint, it is usually appropriate to proceed to estimate the nature, magnitude, and distribution of any effects that were judged to be significant. A significant risk is sufficient to prompt consideration of remedial actions, but the nature, magnitude, and distribution of effects determine whether remediation is justified, given remedial costs and countervailing risks. In general, it will be clear that one line of evidence provides the best estimate of effects. Some lines of evidence may be eliminated as inconsistent with the conclusion, and others may support the conclusion but not provide a basis for quantifying effects. If more than one line of evidence can provide apparently reliable estimates of effects, their results should be presented, and any discrepancies explained. If one best estimate is identified, other lines of evidence may contribute by setting bounds on the estimate.

If a representative species has been chosen for one or more assessment endpoints (see Box 2.5), it is important to estimate risks to the entire endpoint. That is, if night herons have been used to represent piscivorous birds, risks to all piscivorous birds on the site should be estimated. For example, one might estimate that complete reproductive failure is occurring in half of the nesting pairs in a night heron rookery, and therefore that reproductive failure is occurring in half the kingfisher territories that occur in the same area. If there is reason to believe that the kingfishers are less sensitive or less exposed, one might estimate that their reproduction is reduced by some lesser percentage. Such extrapolations may be performed using the extrapolation models discussed in Section 4.1.9. Alternatively, each species of the endpoint group may be independently assessed.

6.8 FUTURE RISKS

Baseline ERAs typically focus primarily on current risks as estimators of the risk that would occur in the near future in the absence of remediation. However, baseline risks in the far future should be characterized also when:

- Contaminant exposures are expected to increase in the future (e.g., a contaminated ground water plume will intersect a stream)
- Biological succession is expected to increase risks (e.g., a forest will replace a lawn)

- Significant recovery is expected to occur in the near term without remedial actions (i.e., the expense and ecological damage associated with remedial actions may not be justified)

Although these future baseline risks cannot be characterized by measuring effects or by testing future media, all lines of evidence that are useful for estimating current risks may be extended to them. As in human health risk assessments, risk models derived by epidemiological methods can be applied to future conditions and even applied to different sites. For example, if concentrations are expected to change in the future, the exposure–response relationship derived from biosurvey data (e.g., a relationship between contaminant concentration and fish abundance) may supply a better estimate of future effects than a concentration–response relationship derived from laboratory test data. Results of toxicity tests of currently contaminated media may also be used to estimate future effects. For example, contaminated groundwater may be tested at full strength and diluted in stream water to generate an exposure–response relationship that may be used to estimate the nature and extent of future effects. The utility of the various risk models depends on their reliability, as suggested by the weight-of-evidence analysis, and their relevance to the future conditions.

6.9 INTERPRETATION

The results of ecological risk characterization require interpretation to aid the risk manager in making a remedial decision and to promote understanding by stakeholders and the public. The risk characterization should have determined, for each assessment endpoint, which risks exceed the threshold for significance and estimated the magnitude and probability of effects associated with the significant risks. Those assessment endpoints should be valued properties of environmental entities which should have been defined as important by the risk manager during the problem formulation (Chapter 2). That determination may have been made after considering the expressed values of stakeholders or the public. Therefore, this step should begin by reviewing the bases for having declared each of the assessment endpoints to be important. The adversity of an effect depends in part on the nature of the value attached to it and the strength with which that value is held by the parties involved in the decision. For example, an increase in phytoplankton production may be considered adverse if the concern is with the aesthetics of swimming and boating or with blooms of noxious algae. However, at other sites, an equal percentage increase in phytoplankton production may be considered a benefit because it causes an increase in fish production. Therefore, in some cases, the endpoint property may be affected, but the nature of the effect may not be considered adverse.

Given that the assessment endpoint is estimated to change or have been changed in a way that is considered adverse, the significance of that change must be interpreted in terms of the intensity of the effects, its spatial and temporal scale, and the potential for recovery (EPA, 1998). These issues of significance should not be resolved by appeals to statistical significance which has no relation to ecological or anthropocentric significance (Suter, 1996a). Another often used criterion for determining whether the intensity of effects is significant is comparison to natural variation. When used, this criterion must be carefully defined. Over moderate time scales, sources of natural variation include drought, floods, fires, late freezes, and other events that cause larger variation than would normally be acceptable for contaminant effects.

Recovery is a more difficult issue than is generally recognized. The difficulty arises from the fact that ecosystems never recover to exactly the same state as existed prior to the contamination, and, even if they did, one could not be sure of it because of limits on what can be measured. Therefore, it is necessary to define explicitly what would constitute sufficient recovery. For example, recovery of a forest might be defined as restoration of the pre-contamination canopy height and 80% of the pre-contamination diversity of vascular plants. Given such a definition, one may then estimate the time to recovery based on successional studies in the literature (Cairns et al., 1977; Cairns, 1980; Yount and Niemi, 1990; Detenbeck et al., 1992; Wiens, 1995). For populations, it is possible to model the recovery process (Samuels and Ladino, 1983; Barnthouse, 1993). If the contaminants are persistent, the time required for degradation, removal, or dilution to nontoxic concentrations must be included in the recovery time. If recovery is an important component of the interpretation of risk for a site, that fact should be noted in the problem formulation. Because estimation of recovery is not simple, it may require site-specific studies or a significant modeling effort.

In general, the best interpretive strategies are comparative. That is, the intensity, spatial and temporal extent, and the time to recovery can be compared with defined sources of natural variation, with other instances of contamination, or with relevant disturbances. The most relevant comparison is of the baseline effects of the contamination to the effects of the remediation (Chapter 9). Such comparisons provide a context for decision making.

The interpretation of ecological risk characterizations must include presentation of uncertainties. The estimation and interpretation of uncertainty is discussed in the following chapter. Here, it is necessary to emphasize the importance of correctly interpreting the results of uncertainty analyses. It is not sufficient to say that the probability is *x*.

6.10 REPORTING ECOLOGICAL RISKS

The form in which ecological risks are reported is an often neglected aspect of the practice of ecological risk assessment. The EPA internal guidance for risk characterization states that a report of risk assessment results must be clear, transparent, reasonable, and consistent (memo cited in EPA, 1998). Considerations for achieving those goals are listed in Box 6.2. However, the goals of being brief and being transparent conflict. If sufficient detail is presented for the reader to fully understand how the results were derived, the resulting multi-volume report will be far thicker than anyone would care to read. As discussed in Chapter 7, simply justifying the assignment of distributions to parameters may result in a sizable report. The usual solution to this problem is the executive summary. Unfortunately, executive summaries attempt to summarize the entire assessment and are seldom sufficient to stand alone if the "executive" is the risk manager. A report of conclusions that neglected methods but presented results in adequate detail for decision making would probably be more useful in most cases. Ideally, the contents and level of detail would be worked out between the risk assessors and risk manager.

BOX 6.2

Clear, Transparent, Reasonable, and Consistent Risk Characterizations

For clarity:

- Be brief; avoid jargon.
- Make language and organization understandable to risk managers and the informed layperson.
- Fully discuss and explain any unusual issues specific to a particular risk assessment.

For transparency:

- Identify the scientific conclusions separately from policy judgments.
- Clearly articulate major differing viewpoints of scientific judgments.
- Define and explain the risk assessment purpose (e.g., regulatory purpose, policy analysis, priority setting).
- Fully explain assumptions and biases (scientific and policy).

For reasonableness:

- Integrate all components into an overall conclusion of risk that is complete, informative, and useful in decision making.
- Acknowledge uncertainties and assumptions in a forthright manner.
- Describe key data as experimental, state-of-the-art, or generally accepted scientific knowledge.
- Identify reasonable alternatives and conclusions that can be derived from the data.
- Define the level of effort (e.g., quick screen, extensive characterization) along with the reason(s) for selecting this level of effort.
- Explain the status of peer review.

For consistency with other risk characterizations:

- Describe how the risks posed by one set of stressors compare with the risks posed by a similar stressor(s) or similar environmental conditions.
- Indicate how the strengths and limitations of the assessment compare with past assessments.

Source: EPA (1998).

7 Uncertainty

> First, we must insist on risk calculations being expressed as distributions of estimates and not as magic numbers that can be manipulated without regard to what they really mean.
>
> —William Ruckelshaus (1984)

This chapter presents guidance for the analysis of uncertainties associated with the methods presented in the preceding chapters. The guidance is consistent with policy and guidance from the EPA (Hansen, 1997; Risk Assessment Forum, 1997), the DOE (Office of Environmental Policy and Assistance, 1996), and the National Research Council (National Research Council, 1994). However, it goes beyond those documents in attempting to provide guidance on how to address uncertainty in ecological risk assessment. Nominally, EPA policy on uncertainty analysis is more applicable to ecological than to human health assessments; uncertainty analysis may be applied to all components of ecological risk assessment but only to exposure estimation in human health risk assessment (Hansen, 1997). However, none of the sources of policy or guidance cited above addresses the particular problems of estimating uncertainty in ecological risk assessments. In addition to the discussion presented here, the reader is referred to Smith and Shugart (1994) and Warren-Hicks and Moore (1998).

7.1 CONCEPTS AND DEFINITIONS

Uncertainty is a seemingly simple concept that has caused confusion and conflict in the field of risk assessment. The concept of risk implies uncertainty. That is, if we had perfect knowledge, there would be no risks, because all current and future states would be perfectly specified. Conventionally, the undesired events that are the subjects of risk assessments have been treated as random draws from a specified probability distribution. For example, the risk that a ship will sink on a given voyage or that a given individual worker exposed to a chemical will die of cancer is treated as a random draw from an underlying distribution of ships and workers. There is no acknowledged uncertainty concerning the distribution, but the uncertainty concerning whether the individual ship will sink or the individual worker will die is of concern. This has been referred to as *identity uncertainty* because the only uncertainty in the assessment is the identity of the individuals affected (Suter, 1993a). The source of this uncertainty is random variance or variance that is treated as random. For example, the variance in susceptibility among individual humans is treated as random because the underlying genetic and phenotypic sources of variance are unknown or unspecified.

Not all assessment endpoints are expressed as probabilities. For example, in human health risk assessments, the EPA specifies that variance in susceptibility to noncarcinogens is not explicitly represented. Noncarcinogenic effects are specified as a toxic threshold value, the reference dose (RfD), that is applied to all exposed individuals. Most ecological risk assessments also use deterministic endpoints such as species richness rather than probabilistic ones such as probability of extinction. Parameters that have an actual value that is undetermined have been referred to as "true but unknown values" (Hammonds et al., 1994). For example, the number of species on a site relative to the number that was present immediately before contamination has a true value, but it is estimated with uncertainty.

Clearly, we may be uncertain about assessment endpoints, whether they are expressed as probabilities or fixed values. Dose–response relationships, exposure models, measurements, toxicological thresholds, interspecies extrapolations, and other components are subject to uncertainty. We may be interested in estimating the uncertainty concerning an endpoint that has a true but unknown value such as the number of cancers actually caused by Alar or the loss of loblolly pine production due to ozone. We may also be interested in the uncertainty concerning a true but unknown random variate such as the individual cancer risk (i.e., in estimating probability that the probability of contracting cancer is at least x). These options are presented in Table 7.1.

The definitions of health effects endpoints and risks in risk assessment for humans in the United States can be plotted on this grid. Health effects for noncarcinogenic chemicals have been treated as a deterministic constant (Cell 1); there is no variability because the threshold for effects (RfD) is the same for all humans. Health effects of carcinogens are treated as deterministic distributions (Cell 2); the dose required to cause cancer is randomly distributed among individuals, but that distribution is treated as certain. Because exposure is typically estimated for a single, realistic, maximally exposed individual, risk estimates for noncarcinogens and carcinogens also fall in Cells 1 and 2. Cells 3 and 4 include human noncarcinogenic and carcinogenic effects endpoints when risk assessors take the additional step of incorporating uncertainty concerning the RfDs or dose–response relationships. If exposures are estimated as probabilities of achieving a dose, then risks move to Cells 3 and 4 even if the effects are still deterministic. (Ecological examples are presented below, after more terminology is introduced.)

TABLE 7.1
Types of Endpoints for Risk Assessment Categorized in Terms of Their Acknowledged Uncertainties

State of knowledge	Endpoint	
	Single Value	**Probability**
Determined	1. Specified value	2. Probability from a specified distribution
Uncertain	3. Probability of an uncertain value	4. Probability of a probability from an uncertain distribution

Source: Suter (1997).

The vast majority of ecological risk assessments express risks like health risk assessments for noncarcinogens, as determined point values (Cell 1). However, because regulatory practice is less prescribed by guidance and tradition, a small but growing minority of ecological risk assessments use uncertainty analysis to estimate distributions on the risk estimate (Cell 3). Few ecological risk assessments have endpoints defined as probabilities (Cells 2 and 4). Population viability analyses, which estimate probabilities of extinction, are an exception.

Probabilistic endpoints are a result of some variability in the entities that comprise the assessment endpoint. The variability may be inherent to the entities. For example, probabilities in conventional cancer risk assessments are due to the assumed inherent variability in the susceptibility of individuals, and the probabilities of system failure in engineering risks are due to variability in time to failure of individual parts. Probabilistic endpoints may also be due to the inherent variability of exposure. For example, the probability of sinking of early merchant ships was largely due to the variability in the weather, leading to exposure to severe storms. Similarly, the probability of exposure to lethal levels of aqueous effluents may be largely determined by variance in receiving flows. This variability is assumed to be either inherently random (i.e., stochastic) or effectively random. It cannot be reduced.

Uncertainty concerning either probabilistic or deterministic endpoints is assumed to be due to lack of knowledge that is potentially available (i.e., ignorance). For example, the dose–response relationship for a test species is used to estimate the probability of response by an individual organism, but the application of this probability is uncertain due to the various extrapolations that are necessary to relate the test to endpoint organisms in the field. The ignorance may be reduced by testing the endpoint species or testing under more realistic conditions.

The utility of the distinction does not depend on whether any of these processes are truly stochastic. It also does not depend on there being an absolute distinction between stochasticity and ignorance. For example, it has been conventional and useful to treat variance in susceptibility to carcinogens as stochastic even though we know that there are potentially identifiable susceptibility factors. Some authors treat this distinction between variability and uncertainty as purely operational (Hammonds et al., 1994). That is, rather than distinguishing stochasticity from ignorance, they treat all distributions among individuals as due to variability and all distributions of those interindividual distributions as due to uncertainty. The distinction between variability and uncertainty has become conventional in risk assessments (IAEA, 1989; Morgan and Henrion, 1990; Hammonds et al., 1994; Hattis and Burmaster, 1994; MacIntosh et al., 1994; National Research Council, 1994; Risk Assessment Forum, 1997).

Note that both variability and uncertainty can be expressed as probability. Since this can lead to confusion, the terms *likelihood* and *likelihood distribution* are used to designate probabilities and their distributions that specifically result from variability (Suter, 1998c). Those that are due specifically to uncertainty are termed *credibility.* Note that this distinction is different from that between objective and subjective probabilities, discussed below. Both likelihoods and credibilities may be treated as either objective or subjective.*

---

*There is no standard terminology for different sorts of probabilities. For a thorough discussion see Chapter 6 in Good (1983).

By using this terminology, examples of the four classes of endpoints (Table 7.1) could be described as follows:

1. The reduction in population size;
2. The likelihood of extinction of the population;
3. The credibility of a 25% reduction in population size;
4. The credibility that the likelihood of extinction is greater than 50%.

Note that this terminology eliminates the confusing phrase "probability of a probability."

Another distinction that causes controversy in risk assessment is that between subjectivism and objectivism. Objectivists argue that distributions should not be assigned to the parameters of risk assessments unless they are well defined from relevant data. Objectivists often argue against the use of uncertainty analysis in particular cases because of the lack of sufficient data to develop objective distributions (Risk Assessment Forum, 1996). Subjectivists argue that all distributions express degrees of belief in the possible occurrence of values. Subjectivists tend to argue further that, since no distributions are objective distributions of the frequencies that are to be estimated, one may as well assign distributions in the best way that available information permits. Although this distinction is related to the distinction between frequentist and Bayesian statistics (Box 7.1), many practitioners treat distributions as subjective without employing the computational methods of Bayesian statistics (Hammonds et al., 1994).

This chapter resolutely straddles the fence between subjectivism and objectivism. That is, when objective distributions can be derived from relevant data, they should be, and such distributions should be preferred to subjectively derived distributions. However, this rule should be applied carefully. In particular, the objectively definable variance is often a small fraction of the relevant variance. Assessors should carefully consider the case in point. For example, the variance in flow estimated from long-term hydrological or weather records can be treated as an objective estimate of the future variance in flow, unless actions in the watershed will significantly change the hydrology. However, the variance in bioaccumulation among individual laboratory rats should not be treated as an estimate of the variance in that parameter among shrews in the field, much less as an estimate of the total uncertainty concerning shrew bioaccumulation. In this case, the apparently relevant objective distribution is inappropriate, and a subjective distribution developed by appropriate experts would be more defensible.

This chapter is limited to conventional techniques. The number of ways of analyzing and expressing uncertainty is immense. In addition to Bayesian updating, they include fuzzy arithmetic, fault-tree analysis, possibility theory, and others. The lack of discussion of those techniques should not be taken to imply rejection of them. It reflects a judgment that current conventional methods are appropriate and much more likely to be accepted than exotic methods.

7.2 WHY ANALYZE UNCERTAINTY?

The first step in an uncertainty analysis must be to determine motivation. The form and content of the analysis depend on the desired output of the analysis. However,

BOX 7.1
Bayesian Statistics and Ecological Risk Assessment

While the Reverend Bayes is best known for equating probability with degree of belief, there is more to Bayesian statistics than subjectivism. Bayesian statistics is based on the prescription of a prior distribution followed by updating based on new data. Although the iterative nature of most ecological risk assessments would seem to lend them to Bayesian updating techniques, it may not make sense in practice. In many cases, the choice of the prior is subjective, which may reduce the credibility of the result. However, even objective priors may have questionable effects on results. For example, one might perform a screening assessment using a distribution of plant uptake factors from the literature (Section 3.9), and for the definitive assessment derive distributions from actual measurements of contaminant concentrations in plants at the site. It would not make sense (at least to these authors) to condition that site-specific distribution on the prior, literature-derived distribution unless the prior was particularly relevant to the site and the site-specific data were few and of questionable quality.

A second aspect of Bayesian statistics that has appeal to ecological risk assessment is its assumption concerning the relationship between data and the model. Traditional frequentist statistics assume that sampling can be repeated and the samples vary, but the model is fully specified. Bayesian statistics, in contrast, assumes that the sample is fixed, but the model is uncertain. This is effectively the situation of risk assessors who are presented data from a sample and wish to define both the distribution of a variable population and the uncertainty concerning that variability. For simulation models, this problem of specifying variability and associated uncertainty is handled by nested Monte Carlo analysis (Section 7.3.3). However, Bayesian statistics provides an appropriate calculus for describing uncertain distributions of sets of measurements.

most guides to uncertainty analysis assume a particular motivation and desired output and proceed from that assumption. Reasons include the following.

7.2.1 Desire to Ensure Safety

Because of uncertainty, true effects may be larger than estimated effects. Therefore, if the goal of the assessment is to ensure that all credible hazards are eliminated or at least accounted for in the decision, uncertainty must be incorporated into the analysis. This may be done in at least four ways.

First, one may determine that the uncertainties are so large and poorly specified that no quantitative uncertainty analysis is possible. In such cases, a risk management decision may be made that all members of an allegedly hazardous class of chemicals or other hazardous entities should simply be banned. This is known as the precautionary principle. Once the risk management decision is framed in that way, the output of the risk analysis is a conclusion that a chemical or technology belongs or does not belong to a banned category.

Second, one may make conservative assumptions. For example, in human health risk assessments it is assumed that an individual drinks 2 liters of water a day from a contaminated source for a lifetime, consumes fish caught in contaminated waters, consumes vegetables grown on contaminated soil irrigated with contaminated water, etc. Following that example, ecological risk assessors commonly assume that an entire population of a wildlife species occupies the most contaminated portion of a site. By hypothesizing levels of exposure higher than are credible for any real human or wildlife population, these conservative assumptions ensure that exposure is not underestimated, even though the exposure is uncertain. The product of stringing conservative assumptions together is a "worst-case" or "reasonable-worst-case" estimate of risk.

Third, one may apply safety factors to the components or results of the assessment. These are factors (usually 10, 100, or 1000) that are applied to ensure an adequate margin of safety. Nearly always they are based on expert judgment and simple analyses of past cases. For example, to regulate chemicals under the Toxic Substances Control Act, the EPA applies a factor of 1000 to an LC50 to estimate a safe concentration for aquatic life (Environmental Effects Branch, 1984). The output of analysis using safety factors is a conservative risk estimate. However, because of the way factors are derived and the inappropriate way in which they combine multiple sources of uncertainty, the degree of conservatism that results from safety factors is unclear.

Finally, one may perform a formal quantitative uncertainty analysis and choose as an endpoint a likelihood of effects that is very low. For example, one might declare that the credibility must be less than 0.01 that the likelihood of extinction is as high as 0.0001 over the next 50 years.

7.2.2 Desire to Avoid Excessive Conservatism

As discussed above, the desire to ensure safety has led to the use of numerous conservative assumptions and safety factors in risk assessments. Some risk assessors, regulated parties, and stakeholders have objected that the resulting margins of safety are excessive. One response has been to argue for reduction of number and magnitude of factors and conservative assumptions or their elimination (i.e., use best estimates). An alternative is to develop anticonservative factors to correct, at least in part, for the compounding of conservatism (Cogliano, 1997). A third approach is to replace uncertainty factors and conservative assumptions with estimated distributions of parameters and to replace the compounding of factors with Monte Carlo simulation (Office of Environmental Policy and Assistance, 1996). Even if low percentiles of the distributions of risk estimates are used to ensure safety, it is assumed that the conservatism is less than with traditional regulatory approaches.

7.2.3 Desire to Acknowledge and Present Uncertainty

It is generally considered desirable to acknowledge and estimate the uncertainties associated with assessments. It is both safer to admit your uncertainties and more ethical than ignoring or hiding them. This is more the case with ecological risk assessments than human health risk assessments, because estimated ecological effects are often detectable, and therefore a conservative deterministic estimate may be refuted. A formal uncertainty analysis provides a clear and defensible method for estimating those uncertainties and justifying the estimates. However, many uncertainties are not estimated by conventional uncertainty analysis, such as the uncertainty associated with model selection or the uncertainty concerning assumptions about the future use of the site. Hence, presentations of uncertainty must include lists of issues, and qualitative judgments, as well as quantitative estimates.

7.2.4 Need to Estimate a Probabilistic Endpoint

Probably the least common reason for analysis of uncertainty in ecological risk assessment is the specification of a probabilistic endpoint by the risk manager. While probabilistic endpoints are conventional in other fields of risk assessment, ecological assessment endpoints typically have been too poorly specified for probabilistic considerations to arise. Probabilistic endpoints have been estimated by ecological risk assessors since the founding of the field, but the impetus has come from the assessors, not the risk managers (Barnthouse et al., 1982). A conspicuous exception is population viability analysis, which estimates probabilities of extinction of species or populations given prescribed management practices (Marcot and Holthausen, 1987). Such analyses should be done in the event that site contamination may pose a threat of extirpation. A more likely impetus for probabilistic endpoints is cases in which the ecological risks are driven by the probability of occurrence of an extreme event. Examples include the failure of the dam forming a waste lagoon or an extremely wet period that brings contaminated groundwater to the surface. Finally, any ecological assessment endpoint can be expressed as a probability, given either variability or uncertainty in exposure or effects.

7.2.5 Planning Remedial Investigations

Ideally, the field and laboratory investigations that provide the data for risk assessments should be prioritized and planned on the basis of an analysis of uncertainty. The goal of the quantitative DQO process is to gather enough data to reduce uncertainty in the risk estimate to a prescribed, acceptable level (Chapter 2). This formalism is not directly applicable to ecological risk assessment, but one can still allocate resources on the basis of expected reduction in uncertainty. This use of uncertainty analysis requires a separate analysis of the uncertainty that can be reduced by feasible sampling, analysis, or testing, rather than the total uncertainty. For example, a model of mink and heron exposure to PCBs and mercury was used in the Clinch River assessment to determine that the greatest source of reducible uncertainty in the exposure estimates was the distribution of PCB concentrations in water and sediment (MacIntosh et al., 1994).

7.2.6 Aiding Decision Making

Finally, the results of an uncertainty analysis may aid the risk manager in making a decision concerning the remedial action. Some decision-support tools require estimates of the likelihood of various outcomes, which must be derived by a probabilistic analysis of risks. More generally, the additional information provided by an uncertainty analysis may lead to a better-informed and more defensible decision, even without quantitative decision analysis.

7.2.7 Summary of Reasons

These reasons for evaluating uncertainty are not mutually exclusive, so an assessor may analyze uncertainty for multiple reasons. However, the chosen method must be able to satisfy the most restrictive reason. For example, if one wishes to ensure safety, any analysis will do; but, if one wishes to ensure safety and present a full disclosure of uncertainties in the assessment, only a quantitative uncertainty analysis will serve. And, if one is using uncertainty analysis to help plan a program of sampling, testing, and analysis, only a quantitative uncertainty analysis that distinguishes sources of uncertainty and variability will serve.

7.3 QUANTITATIVE TECHNIQUES FOR ANALYSIS OF UNCERTAINTY

Although recent discussions of uncertainty in risk assessment have focused on Monte Carlo analysis, it is only one of many potentially useful techniques. This section will discuss three classes of methods which are used in ecological risk assessment.

7.3.1 Uncertainty Factors

The most common technique for incorporation of uncertainty is uncertainty factors (also referred to as safety factors). These are numbers that are applied to either parameters of a risk model or the output of a model to ensure that risks are not underestimated. For example, the NOAEL values used to calculate wildlife toxicity benchmarks are divided by a factor of 10 if they are based on subchronic studies, because of uncertainties concerning subchronic endpoints as estimators of chronic toxicity (EPA, 1993e; Sample et al., 1996b). This factor is based on expert judgment that the threshold for chronic toxicity is unlikely to be more than a factor of 10 lower than a subchronic NOAEL. Most other uncertainty factors are also multiples of 10 and also based on judgment informed by experience or simple analyses of relevant data.

An exception is the factors used to calculate the Tier II water quality values (Host et al., 1991; EPA, 1993c). These are conservative estimates of National Ambient Water Quality Criteria (NAWQC) that are calculated when too few data are available to calculate an NAWQC. The calculation employs factors which were derived by repeatedly taking samples of different sizes from the data sets used to calculate the existing criteria. These samples were used to determine factors for each sample size that would ensure with 80% confidence that the Tier II value would not be higher than an NAWQC. Similar quantitative techniques could and should be used more often to derive factors.

Besides the informality of their derivation, the chief complaint against uncertainty factors is the way that they propagate uncertainty through a model. If a model contains four parameters which are multiplied, and each has an associated uncertainty factor of 10, then the total uncertainty is a factor of 10,000. However, this implies that in the case being modeled, all things are simultaneously, individually as bad as they can credibly be. The uptake factor is much higher than has been observed, the organisms have extremely small foraging ranges, the endpoint species is much more sensitive than the test species, etc. To avoid obtaining absurdly extreme estimates when using this method of uncertainty analysis, one should estimate a maximum credible uncertainty (i.e., an overall uncertainty factor) in addition to factors for the individual parameters.

Uncertainty factors are operationally equivalent to the extrapolation factors discussed in Chapter 4. The distinction is simply that extrapolation factors account for identified systematic differences between measures of effect and assessment endpoints, while uncertainty factors account for uncertainties when systematic differences are not identifiable. In the example of the subchronic tests discussed above, they are designed to be equivalent to chronic tests so we do not expect them to be different, but we are uncertain as to the truth of that assumption in any particular case. Therefore, an uncertainty factor is employed. If we knew that there was a predictable difference between subchronic and chronic test results, we might develop an extrapolation factor.

7.3.2 Data Distributions

Uncertainty often appears in the form of a distribution of realized values of a parameter. For example, concentrations of chemicals in repeated water samples or concentrations of chemicals at which individual organisms respond in a toxicity test have some distribution on the scalar, chemical concentration. These distributions can be used to estimate uncertainties concerning the mean water concentration during the period sampled, a future distribution of water concentrations, the concentration that will result in a fish kill, etc. As discussed below, it is important to carefully consider the relationship between any data distribution and the distributions to be estimated in the assessment. In general, distributions can serve two functions. First, they can be used to represent the uncertainty or variability of a parameter in a mathematical model of exposure or effects. Second, when exposure or effect metrics are directly measured, the distribution of the measurements may directly represent the uncertainty or variability of exposure or effects.

An important decision to be made is whether to fit a function to a data distribution and, if so, which. Conventionally, one describes a distribution by fitting a mathematical function to it, such as the normal, lognormal, uniform, or logistic. The result is referred to as a parametric distribution function or PDF. However, if the distribution is not well fit by a PDF or if the purpose is to display the actual distribution of the data, an empirical distribution function (EDF) may be used. In software for Monte Carlo analysis, EDFs are typically referred to as custom functions. One limitation of EDFs is that they do not describe the distribution beyond the data. This is a problem if the data set is not large and extreme values are of concern (e.g., estimating infrequent high exposure levels or responses of sensitive species). However, PDFs may also poorly represent extreme values due to their infinite tails and the fact that their fit is influenced primarily by the bulk of data near the centroid. Issues related to choosing and fitting functions are discussed in a workshop report (Risk Assessment Forum, 1999) and in two special issues of the journal *Risk Analysis* (vol. 14, no. 5 and vol. 19, no. 1). Published distribution functions are available for great blue heron exposure parameters, and such distributions may be expected in the future for other species (Henning et al., 1999).

The following strategies may be employed to develop distribution functions:

- If the data set is large, one can use statistical software packages to select the function that gives the best fit using goodness-of-fit criteria. However, if there are few data points in the distribution, fitting algorithms may fail to give appropriate results. In addition, one must beware of the fact that models with more parameters fit data better than those with fewer. Hence, one would not choose a three-parameter logistic over a two-parameter logistic model unless the fit was significantly better, or the additional parameter had some mechanistic significance (e.g., to fit a variable maximum value).
- One may choose a function to fit based on experience or on knowledge of the underlying distribution from which the data are drawn. For example, one may know from experience that flow rates are nearly always lognormally distributed when sufficient data are available to confidently define the distribution for a site. Therefore, one would use that function at a new site even though the data set is too small for its form to be clearly defined.
- One may choose a distribution based on underlying mechanisms. The addition of a large number of random variables results in a normal distribution due to the central limit theorem. The multiplication of a large number of random variables results in a lognormal distribution. Counts of independent random events result in Poisson distributions. Time to failure or to death of organisms results in Weibull distributions.
- One may use parsimonious strategies. If one feels that the shape of the distribution cannot be specified but the bounds can, a uniform distribution may be defined. If one feels that the bounds and centroid can be estimated, they can be used to define a triangular distribution.
- Finally, if the form of the distribution is unclear or clearly does not conform to any simple function (e.g., is polymodal), an empirical distribution may be used. Even if the form is clear and conforms reasonably well to a function, empirical distributions may be preferable because they reveal the actual form and variability of the data. The only technical difficulty is the proper choice of bins to avoid excessive smoothing (too few bins) or irregularity (too many bins with too few data per bin).

For mechanistic reasons (above) and because many environmental data sets are approximately lognormal in shape, the lognormal distribution is the most commonly used distribution in human health and ecological risk analysis (Koch, 1966; Hattis and Burmaster, 1994; Burmaster and Hull, 1997). The selection of a distribution should proceed by a logical process of determining what functions are likely based on knowledge of the type of data that the sample represents, the mechanism by which the variance in the data were generated, and the goodness-of-fit statistics. The function should not be selected by applying hypothesis testing inappropriately. It is common practice to assume a function and then test the null hypothesis that the data have the assumed functional form. However, this practice is inappropriate for two reasons. First, it is logically inappropriate to accept the null hypothesis because one has failed to reject it, although this is the most common conclusion drawn from such analyses. Second, it is inappropriate to apply a tool developed to prove the occurrence of treatment effects, where there is reason to favor the null, to a problem in estimation, where there is not. Rather the assessor should choose the best distribution based on prior knowledge and relative goodness of fit of those functions that are logically plausible.

7.3.3 Monte Carlo Analysis and Uncertainty Propagation

When mathematical models are employed with multiple uncertain or variable parameters, appropriate error propagation techniques are required. Many risk models are simple enough to perform the propagation analytically (IAEA, 1989; Morgan and Henrion, 1990; Hammonds et al., 1994). However, the availability of powerful personal computers and user-friendly software packages for Monte Carlo analysis has resulted in the displacement of analytical solutions by this numerical technique. Monte Carlo analysis is a resampling technique that samples from the distributions assigned to model parameters, solves the model, saves the solution, and repeats the process until a distribution of results is generated. Guidance for Monte Carlo analysis can be found in a recent EPA workshop, in EPA guidance (Risk Assessment Forum, 1996, 1997), in appropriate texts (Rubinstein, 1981), and in a special issue of the journal *Human and Ecological Risk Assessment* celebrating the 50th anniversary of the technique (Callahan, 1996). Several software packages are available for performing Monte Carlo simulations (Metzger et al., 1998).

7.3.4 Nested Monte Carlo Analysis

As discussed above, situations involving risks can be thought of as involving both variability and uncertainty. Both contribute to the estimation of the probability that a specified effect will occur on a particular site, but in conceptually different ways. This distinction matters when one is estimating a probabilistic endpoint and wishes to estimate the associated uncertainties or when one is using models to plan a sampling and analysis program. In such cases, the parameters of models should be divided into those that are well-specified constants, those that are uncertain constants, those that are variable but well specified, and those that are variable and uncertain. The nested Monte Carlo analysis (also called two-stage or two-dimensional Monte Carlo analysis) is begun by assigning distributions to the inherent variability of the variable parameters (e.g., dilution flow in a stream), uncertainty distributions to the uncertain parameters including the uncertain variable parameters (e.g., the uncertainty concerning the true distribution of flow), and constants to the well-specified parameters (e.g., molecular weight of the contaminant). Then, one performs the Monte Carlo analysis by first sampling from the variability distributions and then sampling from the uncertainty distributions for the uncertain variables and constants, and solving the model. By iterating the sampling, one generates a distribution of the model output based on variability and a distribution of the percentiles of that distribution based on uncertainty. An example of output from such an analysis is presented in Figure 7.1. Examples of the use of nested Monte Carlo analysis are presented in MacIntosh et al. (1994), McKone (1994), and Price et al. (1996).

Although this nested analysis is computationally complex, the greater difficulty is the conceptual problem of deciding how to classify the parameters. As discussed above, the assessor must determine how uncertainty relates to the goals of the assessment and use that knowledge to apply the analytical techniques consistently. A nested analysis increases the conceptual complexity, but it may increase the likelihood of performing the analysis appropriately by compelling a more thorough review of the problem.

7.3.5 Sensitivity Analysis

Sensitivity analysis estimates the relative contribution of parameters to the outcome of an assessment. It may be performed *a priori* or *a posteriori*. *A priori* sensitivity

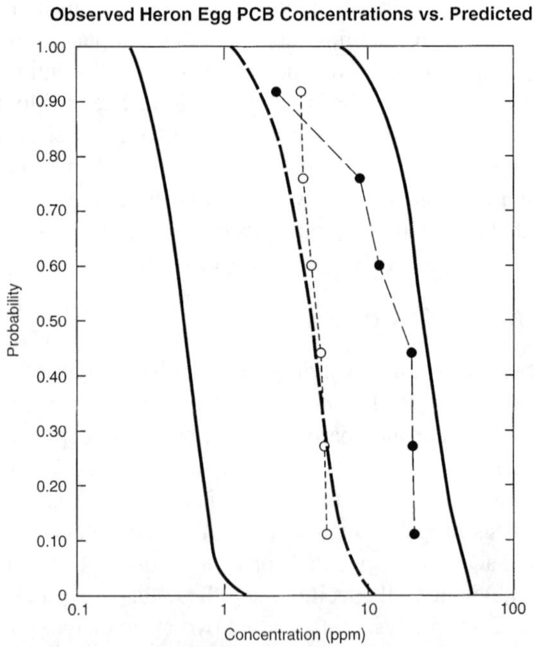

FIGURE 7.1 Complementary inverse cumulative distribution functions for concentrations of PCBs in great blue heron eggs. The dashed central curve represents the expected variance among eggs from different nests (i.e., due to variance among females). The solid outer curves represent the 5th and 95th percentiles on that distribution based on uncertainty in the parameters. The dots connected by dashed lines are measured concentrations in heron eggs. (Redrawn from MacIntosh et al., 1994.)

analyses determine the inherent sensitivity of the model structure to variation in the parameter values. That is, if one knows the model structure but not the parameter values or their variance structures, one can still determine the rate of change in the model output with respect to an input parameter at any nominal value of the parameter by calculating the partial derivative. More commonly, the individual parameter values are displaced by a prescribed small percentage from a nominal value, and the magnitude of change in the output is noted. The ratio of the change in the output to the change in the input variable is termed the *sensitivity ratio* or the *elasticity.* The model is more sensitive to parameters with higher sensitivity ratios (at least in the vicinity of the assigned value). This technique is applicable even when no quantitative uncertainty analysis is performed as a means of identifying influential parameters, and it has been recommended for determining which of the parameters should be treated probabilistically in the uncertainty analysis (Section 7.8). However, the sensitivity of the model to a parameter depends on the value assumed by the parameter (for models that are not fully linear) and its variance (for all models). Hence, the relative importance of parameters may differ from that predicted by the sensitivity analysis (Gardner et al., 1981).

A posteriori sensitivity analyses determine the relative contribution of parameters to the model estimates. These analyses are typically done by performing a Monte Carlo analysis, recording the pairs of input and output values for each parameter, and regressing the input parameter values against the model output values. Greater slopes indicate greater sensitivity. Various specific techniques have been employed including single, multiple, and stepwise multiple regression (Bartell et al., 1986; Brenkert et al., 1988; IAEA, 1989; Morgan and Henrion, 1990; MacIntosh et al., 1994). This sort of sensitivity analysis incorporates the expected values of the parameters and their assigned distributions due to variability or uncertainty. It can be used for communication of uncertainty in the risk estimates as well as to indicate which parameters of the model may be important to address in future iterations of the assessment. For example, despite the great complexity of the models estimating exposure of mink and great blue herons to PCBs in sediments, most parameter uncertainty was due to uncertainty concerning sediment PCB concentrations (MacIntosh et al., 1994). Good discussions of sensitivity analysis for risk models can be found in Iman and Helton (1988), Morgan and Henrion (1990), and Rose et al. (1991a).

7.3.6 Analysis of Model Uncertainty

Although most guidance for analysis of uncertainty is concerned with estimating uncertainty and variability of parameters, significant uncertainty can result from the choice of models used in risk assessments. However, it is usually acknowledged without being quantified, and there is little guidance for its estimation (Risk Assessment Forum, 1996). When there are multiple credible models, model uncertainty can be estimated from the variance among model results; otherwise it can be estimated from the variance due to changes in individual assumptions of a model (Gardner et al., 1980; Rose et al., 1991b).

7.4 MAKING SENSE OF UNCERTAINTY AND PROBABILITY

Unfortunately, uncertainty analysis techniques are often applied without sufficient thought to whether they make sense, given the goals of the assessment. In some cases, the results are misrepresented, or they are presented ambiguously. As in any other component of the risk assessment, it is appropriate to begin with a clearly defined assessment endpoint and determine the most appropriate way to estimate it (Chapter 2). For the analysis of uncertainty, one should begin by determining which of the cases in Table 7.1 is applicable. That is, must the assessment estimate a probability (e.g., probability of extinction) or a value (e.g., percent reduction in species richness), and are uncertainties of those probabilities or values to be quantitatively estimated? If probabilities derived from either variability or uncertainties (i.e., likelihood or credibility) are to be estimated, then distributions must be derived.

The next step is to define those distributions, which requires answering two questions.

- What is distributed?
- With respect to what is it distributed?

These questions need to be answered as clearly as possible. Distributed endpoints can be distributed because of variance or uncertainty in exposures, effects, or both.

7.4.1 Defining Exposure Distributions

In ecological risk assessments, exposure distributions are distributions of exposure metrics (e.g., concentration or dose) with respect to space, time, organisms, or credibility. The specific type of space, time, or credibility needs to be specified.

Space may be defined as arrays of points, as linear units, or as areas. Points are appropriate for immobile or nearly immobile organisms such as plants or benthic invertebrates if the endpoint is defined in terms of individuals. For example, an assessor may be asked to determine whether any plants are exposed to toxic soils or to estimate the proportion of plants exposed to toxic soils. In those cases, one would estimate the distribution of point concentrations from the distribution of sample concentrations (assuming an appropriate sampling design). Streams are typically defined as linear units called reaches, and wildlife associated with streams such as kingfishers have territories defined in linear units. For example, an assessment of belted kingfishers would consider the distribution of exposures experienced in 0.4 to 2.2 km territories. Most wildlife are exposed within areas defined as territories or foraging areas. Other areas of exposure include distinct plant communities and distinct areas with a particular land use.

Time may be defined as a succession of instants, as intervals, or as incidents. Most samples are instantaneous, and distributions of such instants in time may be appropriate when temporal variance is purely random. However, few relevant exposures are instantaneous, so such distributions are most often of use when one is interested in estimating an average exposure over some period and its uncertainty. Most relevant exposures occur over some interval. For example, one may be interested in determining whether a chemical will cause a chronic effect which is known to require an exposure to relevant concentrations during a time *x*. One would then be interested in the distribution of concentrations over time intervals with duration *x* (i.e., moving averages). Another relevant interval is the seasonal exposure experienced by migratory species or sensitive life stages. That is, one would estimate the distribution of doses received during the seasonal interval when a life stage or species occupies the site. Finally, one may be interested in incidents that result in exposure or elevated exposure, such as storm events that flush contaminants into a stream or suspend contaminated sediments. These might be expressed as the distribution of concentrations over incidents of some specified duration or the joint distribution of concentration and duration of incidents.

Exposure may be distributed across individual organisms as in human health risk assessments, either because the endpoint is risks to individuals of an endangered species or other highly valued species, or because the endpoint is risks to populations expressed as the proportion of individuals experiencing some effect. Exposures of individuals may be distributed due to variance in the areas they occupy, the food they consume, or inherent properties such as weight or food preferences.

Distributions of the credibility of exposure arise when the distributions are defined in terms of uncertainties or some mixture of uncertainties and variances. For example, a polyphagous and opportunistic species like a mink may feed primarily on muskrats at one site, fish at another, and a relatively even mixture of prey at a third, depending on prey availability. Hence, the uncertainty concerning the mink diet at a site may be much greater than the variance among individuals at the site, in which case the fractiles of the distribution of individual dietary exposures are credibilities rather than proportions of individuals.

7.4.2 Defining Effects Distributions

In ecological risk assessments, effects distributions are distributions of responses of organisms, populations, or communities (e.g., death or species richness) with respect to exposure. It is necessary to clearly specify the interpretation of the effects metric and the sort of exposure with respect to which it is it distributed (see above). Four general cases will be considered: effects thresholds, exposure–response relationships from toxicity tests, distributions of measures of effects, and outputs of effects simulation models.

Effects thresholds are typically defined by thresholds for statistical significance such as NOAELs or LOAELs. These values do not have associated variances or other uncertainty metrics and are conventionally treated as fixed values. However, while their inherent variance is unspecified, the uncertainty associated with extrapolating them between taxa, life stages, durations, etc. can be estimated (Chapter 4). The most common approach is uncertainty factors.

When conventional toxicity tests are performed, organisms are exposed to a series of chemical concentrations or doses, and the number of organisms responding at each concentration is recorded. Functions are fit to those data that permit the calculation of either the exposure causing a certain frequency of the effect or the frequency of effect at a given exposure. In a risk assessment that estimates risks to individual organisms, that frequency can be treated as a likelihood of effect in the endpoint organisms. Alternatively, the frequency can be treated deterministically as the proportion of an endpoint population experiencing the effect. If one is concerned about the uncertainties associated with these results, one might begin with the variance in the model estimate (e.g., confidence intervals) or the variance of the observations around the model (e.g., prediction intervals). However, these variances are generally smaller than the variance among tests, which is a more relevant measure of uncertainty. In general, variance among well-performed acute tests using the same protocol and species results in test endpoint values within a factor of ±2 (Canton and Adema, 1978; McKim, 1985; Gersich et al., 1986). However, ranges of results in less uniform test sets may be more than a factor of 10. In addition to this variance, which is inherent in the test, extrapolations between test species, life stages, response parameters, etc. should be represented by subjective uncertainty factors, empirically derived factors, or extrapolation models (Chapter 4).

If we assume that the sensitivity of species exposed to a chemical is a random variate, then distributions of test endpoints for multiple species can be treated as an estimate of either the distributions of exposure levels at which members of an endpoint community respond or as likelihoods of response of an untested species. The former, deterministic interpretation has been used by the EPA and others to derive water, soil, or sediment quality criteria in terms of concentrations that would affect a small fraction (usually 5%) of species in exposed communities (Erickson and Stephan, 1985; Stephan et al., 1985; Wagner and Lokke, 1991; Aldenberg, 1993; Aldenberg and Slob, 1993). As with functions fit to individual test results, most emphasis has been placed on the uncertainty associated with fitting the function to the data. However, the uncertainties associated with biases in the selection of test species or the extrapolations from the test endpoint to the assessment endpoint may be much larger (Smith and Cairns, 1993; Suter, 1993a). These uncertainties could be treated by applying factors, but the only factor currently used with these distributions is the factor of 2 applied to the 5th percentile of acute values when calculating the acute NAWQC (Stephan et al., 1985). That factor is a correction factor intended to generate a concentration that corresponds to a low percent mortality rather than 50% mortality, not to compensate for uncertainty in the distribution.

It should be noted here that there is an implicit assumption that has not been adequately recognized in discussions of the use of sensitivity distributions in ecological risk assessments. The distributions include not only biological variance but also physical variance and methodological variance. The conventional interpretation that these distributions are distributions of species sensitivity or community sensitivity is based on the assumption that biological variance is dominant. For aqueous toxicity that is probably true. The test methods and endpoints for aquatic toxic effects are reasonably consistent, so methodological variance should be relatively low. In addition, variance in test water chemistry is relatively low, particularly when hardness and pH normalization is used for metals and ionizable compounds, so physical variance should be relatively low. However, for both sediments and soils the testing and survey methods and the endpoints are highly variable, the media have highly variable textures and chemistries, and reliable normalization methods are not available. Therefore, the physical and methodological variances may be significant contributors to the effects distributions in sediments and soils. The methodological variance is extraneous. The physical variance is an actual property of soils and sediments. It could be thought of as extraneous as well. However, if one takes an ecosystem perspective, then the distributions resulting from the combination of biological and physical variance can be thought of as distributions of benthic ecosystem sensitivity, soil-plant system sensitivity, etc. That concept has been adopted in ecological risk assessments performed at Oak Ridge National Laboratory, and methodological variance has been assumed to be small relative to the combined biological and physical variance. However, it would be highly desirable to disaggregate those sources of variance by standardizing methods and developing methods to normalize soils and sediments (Section 3.4.1).

Finally, mathematical simulation models may be used to estimate effects, particularly effects on ecosystem properties or effects on populations mediated by ecosystem processes (O'Neill et al., 1982; Bartell et al., 1992). The uncertainties associated with these effects estimates are generated using Monte Carlo analysis. Frequencies are seldom explicitly considered because the endpoints are not expressed in those terms and because the uncertainties are large relative to variability.

7.4.3 Defining Risk Distributions

Risk is a function of exposure and effects. If only one of those components of risk is treated as a distributed variate, the estimation of risk as a distributed variate is relatively straightforward. For example, if the dose has been estimated to be distributed over organisms due to variance among individuals and the effects are assumed to have a determinate threshold, then the output of the risk characterization is the likelihood that an individual in the exposed population will receive an effective dose (Figure 7.2). However, if the exposure and effects are both estimated probabilistically, the concordance of the distributions must be ensured. For example, if one is estimating the probability that organisms of a particular species are affected, then both the exposure and effects metrics must be distributed with respect to organisms. Even though ecological risks have been defined as the joint probability of an exposure distribution and an effects distribution since the earliest published methods (Figure 7.3), little attention has been devoted to determining and explaining what those probabilistic risks represent. Methods for addressing this issue are discussed below (Section 7.7.2).

7.4.4 Data Distributions and Risk

Once the types of distributions that are needed to estimate the assessment endpoint have been specified, it is necessary to examine how the available data are distributed. For example, if a data set consists of a set of cadmium concentrations in water, they may be time-averaged or instantaneous concentrations, they may be at one or several points, they may be at one or more times which may extend over a season or a year, etc. Distributions derived from these data need to be created and interpreted in such a way as to correspond to the distributions needed to estimate the endpoint. This problem is addressed below for specific ecological risk assessment methods.

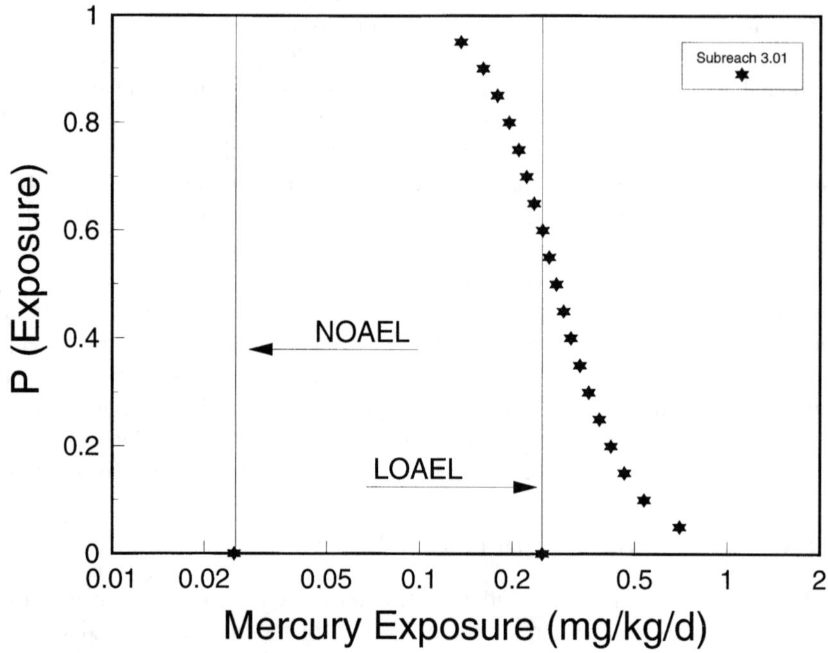

FIGURE 7.2 Inverse cumulative distribution function for exposure of rough-winged swallows to mercury in Poplar Creek, derived by Monte Carlo simulation of an oral exposure model. The vertical lines are the NOAEL and the LOAEL. The probability that an individual would receive a dose greater than the LOAEL is 0.6.

7.5 METHODS FOR EXPOSURE

7.5.1 Distributions of Measured Concentrations

The most common expression of exposure in ecological risk assessment of contaminated sites is measured chemical concentrations in ambient media. The most common probabilistic treatment of these data is to fit a distribution to the individual observations. However, as discussed above, these distributions often do not make sense as expressions of exposure. Assessors must determine the appropriate temporal and spatial units of exposure and how they are distributed given the definition of risk to the endpoint in question.

Concentrations in water may be treated as spatially constant (i.e., a fish community is assumed to occupy a pond or stream reach), and the critical variable is time. For acute effects of observed pulse exposures such as occur during storm events that flush pollutants into the system or repeated failures of treatment or containment systems, the distribution of aqueous concentrations with respect to duration of the event should be derived. Choosing an appropriate duration for chronic exposures is more difficult. A default value is 7 days, the duration of the standard subchronic toxicity test developed by the EPA and employed at the Oak Ridge

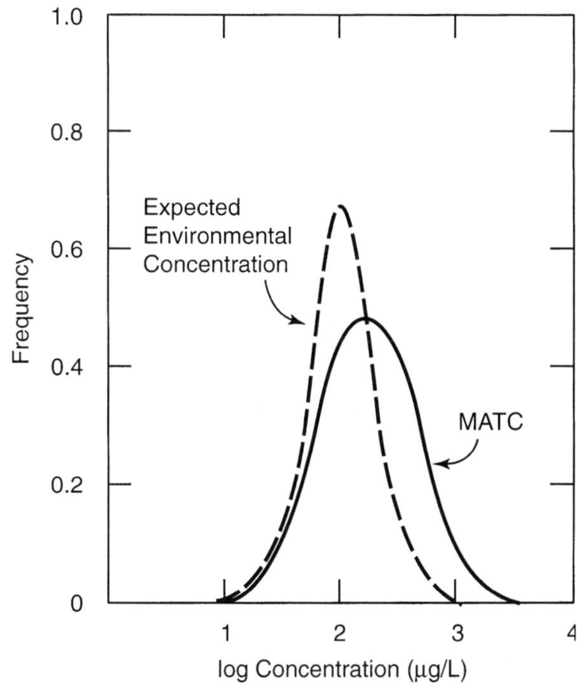

FIGURE 7.3 Probability density functions for a predicted *Salvelinus* maximum acceptable toxicant concentration (MATC) (solid line) and an expected environmental concentration (dashed line). (From Barnthouse, L. W. and Suter II, G. W., ORNL-6251, Oak Ridge National Laboratory, Oak Ridge, TN, 1986.)

Reservation and other contaminated sites. The time period is based on the time required for most chemicals to induce effects on survival and growth of larval fish and survival and reproduction of planktonic crustaceans. Some chemicals induce effects on those organisms much more quickly, and some organisms, particularly larger ones, respond much more slowly. Ideally, the durations should be set at the time required to induce the endpoint response. The selection of that interval requires a careful study and analysis of the original toxicological literature. Once the duration has been chosen, the appropriate measure of variability is the variance in the x day moving average concentration, where x is the time of the episodic exposure or the time required to induce chronic effects in routine exposures.

An additional concern is lack of correspondence between the measures of exposure and the appropriate exposure metric for the risk model. While it is generally recognized that measures of effects must be extrapolated, exposure extrapolations are less often acknowledged. Examples from ecological risk assessment include the following.

- Estimation of forage fish contamination from game fish,
- Estimation of whole fish concentrations from fillets,

- Estimation of dissolved concentrations from whole-water concentrations,
- Estimation of undetected concentrations from limits of detection,
- Estimation of pore water concentration from sediment concentration, and
- Estimation of annual average concentration from summer samples.

In some cases, data and techniques are available to estimate the uncertainties associated with these extrapolations. Examples include maximum likelihood estimators for undetected concentrations and variance estimates on the empirical models of the fillet to whole-fish extrapolation (Bevelhimer et al., 1996). In other cases, expert judgment must be employed.

7.5.2 Multimedia Exposure Models

The ecological analysis to which the Risk Assessment Forum (1997) guidance is most relevant is the estimation of multimedia exposure of wildlife. Like humans, wildlife drink potentially contaminated water, eat various proportions of foods, consume potentially contaminated soil, etc. These exposures are estimated using simple mathematical models which are subject to Monte Carlo analysis (Chapter 3). The reader is referred to the many guidance documents for probabilistic analysis in risk assessment which are concerned with this sort of analysis (Burmaster and Anderson, 1994; Hammonds et al., 1994; Hansen, 1997; Risk Assessment Forum, 1997).

The trick in performing probabilistic analyses of these models is representing the parameter distributions in such a way that the results are those required by the risk characterization. In general, the probabilities should be distributions of dose with respect to individuals, credibilities of doses with respect to individuals, or both (i.e., from nested Monte Carlo analysis). Consider, for example, mink exposed to contaminants in a stream. The model is:

$$E_j = \sum_{i=1}^{m} p_{ik}\left(\frac{\mathrm{IR}_i \times C_{ijk}}{\mathrm{BW}}\right)$$

where

E_j = total exposure to contaminant (j) (mg/kg/d)
m = total number of ingested media (e.g., food, soil, or water)
IR_i = ingestion rate for medium (i) (kg/d or l/d)
p_{ik} = proportion of type (k) of medium (i) consumed (unitless)
C_{ijk} = concentration of contaminant (j) in type (k) of medium (i) (mg/kg or mg/l)
BW = body weight of endpoint species (kg)

Body weight estimates may be obtained from the literature or summaries such as EPA (1993b), from which four distributions may be derived: (1) the distribution of weights of individual mink at an individual site (e.g., the mean and standard deviation of individual weights from a study); (2) the distribution of mean weights across sites (e.g., the mean and standard deviation of mean weights from multiple studies); (3) the distribution of individual mink weights across sites (e.g., the mean and standard deviation of individual weights from all studies); and (4) the distribution across sites of the moments of the distributions (e.g., means and standard deviations of means and standard deviations of individual weights across multiple studies). If one of the sites at which mink have been weighed is similar to the contaminated site, we may use the distribution of weights from that site to estimate the distribution of individual weights in the model (distribution 1). If none of the sites in the literature resembles the contaminated site, we may be tempted to use the distribution of means (distribution 2). However, assessments are concerned with the distribution of effects across the individual mink comprising a population at a site, not a hypothetical mean mink. Alternatively, one might use the distribution of individual weights across all studies to represent the distribution of individuals at the contaminated site, but that would inflate the variance by the amount of systematic variation in weight among populations (distribution 3). The most complete description of the situation would be to perform a nested Monte Carlo analysis with the distributions of the means and standard deviations (assuming normality or lognormality) from all studies that are believed to be potentially representative of the site (e.g., wild mink only) used to estimate the variance among individuals and the uncertainty concerning that variance (distribution 4).

Ingestion rates should be treated in the same manner as body weights. For example, if the mean and standard deviation of individual weights is taken from the study that is most similar to the site being assessed, then the same should be done for ingestion rates.

Contaminant concentrations also require careful interpretation. The input data for this parameter are likely to be concentrations in individual fish collected at various points on the site. The variance in concentration among individual fish is not the appropriate measure of variance in this parameter unless some individual fish are so contaminated as to cause acute toxic effects. Rather, because effects are due to chronic exposures in nearly all cases, the relevant exposure metric is the mean concentration in fish. The next issue is over what set of fish the mean should be calculated. The appropriate mean is the mean across fish within a foraging range of a mink. This assumption leads to alternative expressions of the parameter, depending on the size of the operable unit.

- If an operable unit is approximately as large as the foraging range of a mink, there is a single mean concentration. The variance is the variance on that mean, given sampling error, which can be estimated statistically if sampling is replicated. This can be taken as an estimate of the sampling variance among individual mink. The major uncertainty is the uncertainty concerning how representative a sample of fish taken by electrofishing or netting is of the sample taken by a foraging mink. That uncertainty may be estimated by expert judgment guided by the observed variation among fish of different species and sizes. Other uncertainties may be relevant depending on the sample preparation and analysis. For example, if the whole fish concentrations are estimated from fillet concentrations, the uncertainty concerning that conversion may be important (Bevelhimer et al., 1996).

- If the operable unit is smaller than the foraging range of a mink, the procedure is as above, except that the fish from inside the unit and outside the unit must be treated as different dietary items in the exposure model.
- If the operable unit is much larger than the foraging range of a mink, the unit should be divided into subunits that correspond approximately to a foraging range. In most cases, a river or stream will already have been divided into subunits, termed reaches, based on differences in contamination and physical features (Chapter 2). For the mink exposure model, one would simply aggregate or subdivide those reaches to approximate a foraging range. This approach requires the assumption that the bounds of a foraging range correspond to the bounds of the assessment's spatial units. This is not unreasonable for the purposes of the assessment, since the actual bounds are unknown and would vary from year to year, and the bounds of the units are relevant to the remedial decision. However, it is possible to eliminate the assumption about the bounds on foraging ranges. If there are sufficient sampling locations, one may use a moving average concentration where the average is of concentrations at sampling sites and the window within which they are averaged is the length of a foraging range. In either case, the most important variance is the variance among foraging ranges (i.e., among individual mink). The most important uncertainty is likely to be, as above, the uncertainty concerning the representativeness of the sample.

Alert readers will have noticed that we have assumed that spatial variance is important to mink, but temporal variance is important to fish, even though they are both exposed to aqueous contaminants. The difference results from their modes of exposure. Fish are exposed directly to water by gill ventilation and are susceptible to effects of short-term variance due to spills, storms, and other such events which make fish kills a common occurrence on industrialized streams. Mink, in contrast, are exposed primarily through diet, which results in considerable temporal averaging of contaminant concentrations. Spatial variance is assumed to be relatively unimportant for fish, because they are assumed to reside within a reach, which is designed to have uniform contamination (Chapter 2). In contrast, because mink have foraging ranges that may be large relative to reaches and divide space among the members of a population, space is a critical variable in estimating risks to mink. Hence, for fish, temporal variance is much more important than spatial variance, but for mink, space is more important. These assumptions hold for most sites with which we are familiar, but they are not inevitable. For example, at some sites aqueous concentrations may be effectively constant over time (i.e., no significant episodic exposures), but may be highly variable in space relative to the range of a fish or fish population.

7.5.3 Transport and Fate Models

Chemical transport and fate models are required to estimate future concentrations in baseline ecological risk assessments if the contaminant concentrations may increase in the future. Transport and fate models are also needed for estimation of risks following remediation, accidents, or other scenarios that may change conditions in the future. In those cases, conventional Monte Carlo analysis or nested Monte Carlo analysis should be employed. In addition, because transport and fate models may be relatively complex, it is important to consider the model uncertainty (Section 7.3.6).

7.6 METHODS FOR EFFECTS

7.6.1 Screening Benchmark Values

The simplest expression of effects is test endpoints and other benchmark values used in screening assessments (Chapter 4). Uncertainties in these benchmarks are handled in various ways depending on the type of toxicity data and the way it is used to derive benchmarks. For water and sediment, sets of benchmarks are used that vary in their conservatism. In screening assessments, the uncertainty concerning the most appropriate benchmark for a site may be treated by choosing the lowest benchmark for each chemical. In addition, some of the individual benchmarks incorporate variability among species by using species sensitivity distributions, and uncertainty by applying factors to compensate for small data sets (see below). For plants, soil invertebrates, and soil microbial processes, and for some of the sediment benchmarks, variability is incorporated by creating distributions across studies (see below). For wildlife, uncertainty and variability are incorporated by selecting the most sensitive test, by applying safety factors when tests are not reliably representative of chronic reproductive effects, and by using a no-effects level as a benchmark for screening.

The irregularity of these practices, while justifiable given the data sets and information available for their derivation, is disconcerting. We have found these benchmarks to be appropriately conservative for screening, in the sense that, based on experience using them in conjunction with site-specific toxicity tests and biological surveys, they have produced many false positives and few false negatives. This is in part because of the strategies for incorporating variability and uncertainty described above and in part because of the inherent conservatism of many of the toxicity tests on which they are based. However, the benchmarks have not been validated and at least one case of a false negative, aquatic toxicity of nickel in Bear Creek in Oak Ridge, has been detected (Kszos et al., 1992; Suter, 1996b).

Although the benchmarks are believed to be adequately conservative, additional conservatism may be achieved by using uncertainty factors if demanded by risk managers. An example of the use of uncertainty factors for this purpose is the ecological risk assessment for the Rocky Mountain Arsenal, in which factors were applied to account for intrataxon variability, intertaxon variability, uncertainty of critical effect, exposure duration, endpoint extrapolation, and residual uncertainty (Banton et al., 1996). For each of those six issues, a factor of 1, 2, or 3 was applied signifying low, medium, or high uncertainty, respectively. Clearly, the magnitudes of these factors are not related to estimates of actual variance or uncertainty associated with each issue, and the multiplication of factors bears no relationship to any estimate of the total uncertainty in the benchmarks. However, uncertainty factors provide an assurance of conservatism without appearing completely arbitrary.

An alternative is to derive uncertainty factors based on estimates of actual variance or uncertainty. An example is the prediction intervals on the intertaxon extrapolations and the uncertainty factors on the prediction intervals (PIs) for a given taxonomic level presented in Tables 4.5 and 4.7.

7.6.2 Species Sensitivity Distributions

Species sensitivity distributions may be interpreted in two ways (Suter, 1993a, 1998c). First, they may be interpreted as distributions of the probability that a particular species will be affected at a particular concentration. Based on this interpretation, at a concentration of 100 μg/L, the probability of effects on any exposed aquatic species is 0.28 (Figure 7.4). Second, they may be treated as an estimate of the distribution of sensitivities of species in the exposed community. Based on this interpretation at a concentration of 100 μg/l the proportion of the community affected by the exposure is 0.28 (Figure 7.4). Note that the results are a probability in the first case when the endpoint is effects on a population, but in the second case, when the endpoint is a community property, the result is deterministic. This distinction has been missed by those who interpret species sensitivity distributions as both a representation of community-level effects and as probabilistic (Solomon, 1996; Solomon et al., 1996).

The distinction may be clarified by analogy to exposure–response curves from conventional single-species toxicity tests. The percentiles of those curves can be

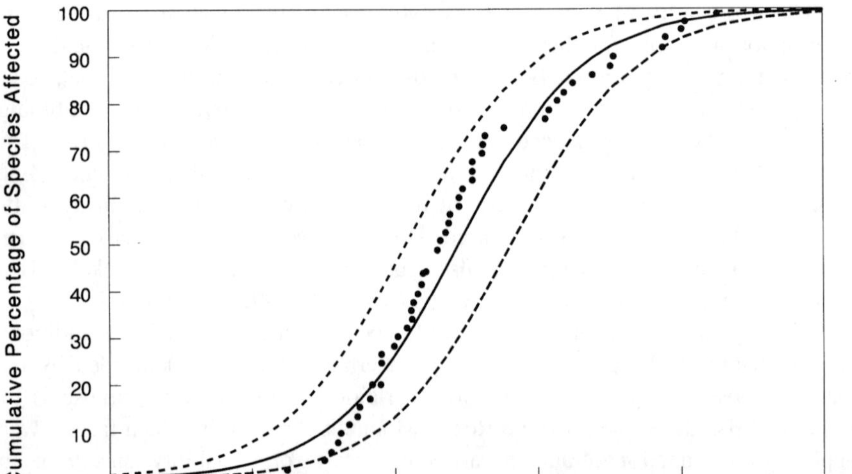

FIGURE 7.4 Cumulative species sensitivity distribution function for acute toxicity of copper. The curves are a logistic model fit to the data points and upper and lower 95% prediction limits, generated using the WERF software. (From Parkhurst, B. R. et al., Project 91-AER-1, Water Environment Research Foundation, Arlington, VA, 1996.)

interpreted as probabilities of effects on individuals or as proportions of exposed populations. The former interpretation, which is used in human health risk assessments, is probabilistic, like the population-level interpretation of species sensitivity distributions. The latter interpretation, which is more characteristic of ecological risk assessments, is deterministic, like the community-level interpretation of species sensitivity distributions.

The interpretation is further complicated if uncertainty concerning the distributions is considered. Uncertainties concerning the percentiles of species sensitivity distributions have been used to calculate conservative environmental criteria (Kooijman, 1987; Aldenberg, 1993; Aldenberg and Slob, 1993) and to estimate risks as probabilities of effects (Parkhurst et al., 1996). For a population endpoint, the percentiles of the distribution about the species sensitivity distribution are credibilities of a probability of effects on the endpoint population. That is, we partition variability among species from uncertainty concerning that variable parameter. Based on this interpretation, at a concentration of 100 μg/l the credibility is 2.5% that the probability of effects on any exposed species is as much as 0.46 (Figure 7.4). This is conceptually equivalent to the output of a nested Monte Carlo simulation of an exposure model where variability in exposure is partitioned from uncertainty. However, the uncertainty is simply the residual variance from the fitting of the distribution function to the toxicity data. This is a very incomplete estimate of the actual uncertainty (Suter, 1993a). For a community endpoint, the percentiles of the distribution could be probabilities based on the variability among communities or credibilities based on the uncertainty concerning the distribution as a representation of community response. However, as in the population-level interpretation, the variance from the fit is an incomplete estimate of the actual uncertainty. The total uncertainty is a result of biases in the selection of test species, differences between laboratory and field sensitivities, and differences between the laboratory responses and the endpoint properties.

In ecological risk assessments for the Oak Ridge Reservation, the aquatic endpoints have been defined at the community level, where the endpoint properties include reductions in species richness or abundance (Suter et al., 1994). Hence, the appropriate interpretation is that the percentiles of the distribution are estimates of the proportion of species affected. In assessments of risks to aquatic communities, actual measurements of fish species richness and abundance as well as toxicity tests of ambient waters may be available. In such cases, the species sensitivity distribution is used to determine which chemicals are likely causes of any observed toxicity or community degradation rather than to estimate risks. For attributing cause, empirical distributions have been used and uncertainty analysis has not been judged important (see Figure 4.1). However, if the risks are characterized on the basis of analyses of the species sensitivity distribution, uncertainty should be quantified. If sufficient data are available and if they are reasonably well fit by a sigmoid distribution, then a normal or logistic function may be fit and the uncertainty on the fit calculated. The aquatic ecological risk assessment software developed for the Water-Environment Research Foundation (WERF) provides distributions for common contaminants and provides an easy means of calculating others (Parkhurst et al., 1996), but any statistical software will serve. However, assessors should be aware that the estimate of uncertainty provided by these statistical fitting algorithms is minimal, particularly when the WERF software option of deriving chronic sensitivity distributions using the acute–chronic ratio is employed. Subjective uncertainty factors can be employed to estimate total uncertainty, but these should be based on a careful consideration of the data in the distribution and their relationship to the site-specific endpoint. A factor of 10 can be considered minimal. Research is needed to develop more objective estimates of the uncertainties concerning risk estimates derived from these distributions.

7.6.3 Sediment Effects Distributions

Benchmark concentrations for effects of contaminants in whole sediments are commonly derived using distributions of sediment concentrations that have been associated with various effects at various locations (MacDonald et al., 1994; Long et al., 1995). These distributions are difficult to interpret for ecological risk assessment because of the numerous variables that are aggregated into the chemical-specific distributions including location; laboratory and field conditions; lethal and sublethal responses; and individual, population; and community responses. Hence, it is difficult to say what the percentiles of these distributions represent. This problem can be reduced by subsetting the input data by the type of response. For assessments of sediment risks, the data from MacDonald et al. (1994) have been subset into laboratory mortality responses and field community responses (Jones, 1999). We may assume that the sites are randomly selected with respect to the properties that determine the toxicity of a chemical, and the species and communities are randomly drawn with respect to sensitivity. In that case, the distribution of mortality may be treated like the species sensitivity distributions discussed above, except that the medium as well as the species are significant sources of variance. Hence, the distributions may be treated as generating probabilities of effects on an individual species or as proportions of species affected given variance in species sensitivity and sediment properties. The community distributions can be interpreted analogously as generating probabilities of effects on an individual community or a proportion of communities affected, given variance in community sensitivity and sediment properties. The uncertainties associated with these distributions have the same problem of incomplete specification of the relevant uncertainties as those of species sensitivity.

7.6.4 Biological Surveys

Biological surveys potentially provide direct estimates of effects at sites receiving various levels of exposure. The primary uncertainties to be considered in such cases are sampling variance and biases in the survey results as estimators of the assessment endpoint. Sampling variance is estimated by conventional statistics. Biases must, in general, be estimated by expert judgment.

Additional uncertainties arise when biological survey results are used to estimate effects at sites other than those surveyed. Such estimates may require interpolation or extrapolation. An example of interpolation would be the use of fish surveys at certain locations in a stream to estimate effects at locations lying between sampled locations. This might be done by interpolation, by spatial statistics, or by process modeling. An example of extrapolation would be the use of survey data for one contaminated stream to estimate effects on another stream with the same contaminant. At minimum, the uncertainty in such extrapolations would be equal to the variance among fish communities at uncontaminated sites (i.e., upstream reference communities in the case of interpolation or regional reference communities in the case of extrapolation). Additional uncertainty results from variance in the effective contaminant exposure due to differences in chemical form, patterns of temporal variance, etc.

7.6.5 Media Toxicity Tests

As with the other methods of estimating effects, media toxicity tests have an uncertainty component that is easily estimated using statistical analysis (variance among replicates) and a less easily estimated component due to imperfect correspondence between the measure of effect and the assessment endpoint. Sources of the latter include the following.

- Species sensitivities in the receptor community may be poorly represented by test species.
- Not all life stages are included in most tests.
- Not all responses that are relevant to population or community response are measured.
- Durations are much shorter in laboratory than field exposures. This is a particular problem for contaminants with slow uptake and depuration such as PCBs.
- Chemical concentrations are typically determined for only one sample out of the three used in the standard EPA aquatic tests.

Good methods for estimating these uncertainties are not currently available.

7.7 METHODS FOR RISK CHARACTERIZATION

Risk characterization integrates the analysis of exposure and the analysis of effects to estimate the nature and magnitude of risks. If only one of these two components is treated as probabilistic, the estimate of variance or uncertainty is simply the variance or uncertainty in that component. However, if a probabilistic analysis is performed for both components, then care must be taken to ensure that they are combined in such a way as to provide an appropriate expression of risk. In particular, the dimensions of the functions must be concordant. Methods for cases in which both components are probabilistic are discussed below.

7.7.1 Simple Quotient Methods

Although some assessors have used Monte Carlo techniques to perform probabilistic analyses of hazard quotients, they are easily performed analytically (IAEA, 1989; Hammonds et al., 1994). The quotient model can be expressed as:

$$\ln Q = \ln D - \ln E \quad (7.2)$$

where Q is the hazard quotient, E is the exposure level causing a specified effect, and D is the dose or other exposure metric estimated to occur at the site. Q will be approximately lognormal even if the distributions assigned to E and D are not (IAEA, 1989; Hammonds et al., 1994). Hence, the geometric mean of Q is the antilog of the difference of the means of the logs of E and D, and the geometric variance is the antilog of the sum of the variances of the logs of E and D.

Assessors must ensure that E and D are concordant. For example, if E is the dose causing decreased reproductive success in individual mink, D must be the dose to individual mink, and, if the output is to be probability of effects on individual mink, then the variances associated with both D and E must be defined with respect to individuals. For example, the effects distribution might be the mean and variance in the dose causing a 10% reduction in weight of young per individual female relative to controls from a mink reproductive toxicity test. The concordant exposure distribution would be the distribution across foraging areas of estimated doses (Section 7.5.2). The result could be used to estimate the quotient and its confidence bounds or to estimate the proportion of mink on the site experiencing a reproductive decrement of as much as 10%, given variance among individuals (i.e., probability that the quotient exceeds 1). If an estimate of the uncertainty is provided, the credibility could be estimated as well. For example, one might subjectively judge that the uncertainty concerning the quotient is estimated by a uniform distribution with bounds of 0.1 and 10 times the mean, and apply that to the result of the estimate based on variance among mink. Alternatively, rather than variance among individuals, one might estimate the distributions of both D and E given both variance and uncertainty. The results of the analysis could be the credibility distribution of the quotient or the credibility that the quotient exceeded 1.

7.7.2 Joint Probabilities

If the exposure and effects are expressed as probability distributions, the distributions may be interpreted in various ways during the risk characterization. Consider, for example, a species sensitivity distribution for aquatic animals and a distribution of aqueous concentrations.

The simplest mode of interpretation is that provided by the Aquatic Risk Assessment and Mitigation Dialog Group (1994) (see also Solomon et al., 1996). They convert all distributions to a dichotomous criterion; there is a significant risk if the 90th percentile of the distribution of aqueous concentrations exceeds the 10th percentile of the species sensitivity distribution. Although this method has been recommended by a reputable group of scientists, it is not recommended here. The criterion is arbitrary, and it does not interpret the distributions in terms of either variability or uncertainty. It is simply easy and consistent.

Given two distributions with respect to a common variable, one may calculate from the joint probability the likelihood that a random draw from one distribution exceeds a random draw from the other. For example, if the distributions are an exposure distribution and an effects distribution with respect to concentration, one may calculate the probability that a realization of exposure exceeds a realization of effects (Suter et al., 1983). This concept has been applied to effects expressed as species sensitivity distributions in the WERF method (Parkhurst et al., 1996). The difficulty lies in interpreting the resulting probability. As discussed above, the species sensitivity distribution can be interpreted in terms of probability of effects on species or proportion of species in a community. Similarly, distributions of concentrations may be distributions over instants of time, distributions of temporal averages, distributions of estimates with respect to credibility, etc. (Section 7.4.1). For the probabilities to be meaningful, the distributions must be concordant. That is, they must not only be distributions with respect to concentration but the same sort of concentration: both concentrations averaged over sufficient time for chronic effects to occur, credibilities of concentration given uncertainties, etc.

When exposure and effects distributions are used as part of a logical weighing of evidence, it may be appropriate to interpret them logically rather than to calculate joint probabilities. For example, the following interpretation occurs in the risk assessment for fish community of the Poplar Creek embayment of the Clinch River (Suter et al., 1999).

> **Copper**. The distributions of ambient copper concentrations and aqueous test endpoints are shown in Figure 6.2. The ambient concentrations were dissolved phase concentrations in the subreaches with potentially hazardous levels of copper (3.04 and 4.01). The toxic concentrations were those from tests performed in waters with hardness approximately equal to the site water. The ambient concentrations fall into two phases. Concentrations below 0.01 mg/L display a fairly smooth increase suggestive of a lognormal distribution. The upper end of this phase of the distribution (above the 75th percentile of 4.01 and the 80th percentile of 3.04) exceeds the lowest CV (a bluntnose minnow CV for reproductive effects). However, the distributions above the 90th percentile are not continuous with the other points. The break in the curve suggests that some episodic phenomenon causes exceptionally high concentrations. The two points in 4.01 and one in 3.04 that lie above this break exceed approximately 90% of the CVs, approximately 30% of the acute values, and both the acute and chronic NAWQCs. These results are suggestive of a small risk of chronic toxicity from routine exposures, but a high risk of short-term toxic effects of copper during episodic exposures in lower Poplar Creek embayment and the Clinch River.

This sort of interpretation is a mixture of quantitative and qualitative analysis that can, as in this case, provide more information than a purely quantitative analysis. Had the Aquatic Risk Assessment and Mitigation Dialog Group method or WERF method been applied, the results would have been less ad hoc, but would have provided less basis for inference concerning the cause of observed effects and toxicity.

7.8 PARAMETERS TO TREAT AS UNCERTAIN

If the number of parameters in a risk model is large, or if research is required to define distributions, it is necessary to decide which parameters to treat probabilistically. The number of parameters is likely to be large in the early stages of a risk assessment when few data are available or in assessments that never perform much sampling and analysis. For example, once contaminant concentrations in plants have been measured, the plant uptake submodel of the wildlife exposure model can be eliminated. If a probabilistic analysis is performed that does not include all parameters, the following criteria should be used in the selection.

- If a probabilistic analysis is replacing an analysis that included uncertainty factors or conservative assumptions, the parameters to which those factors or assumptions were applied should be treated as uncertain.
- If the regulators, responsible parties, or other stakeholders have expressed concern that misspecification of a variable or uncertain parameter may affect the outcome of the assessment, that parameter should be treated as variable or uncertain.
- If the probabilistic analysis is performed in support of a planning decision, the parameters relevant to the decision must be treated as uncertain. For example, if the analysis is performed to aid development of a sampling and analysis plan, those parameters that may be measured must be treated as uncertain.
- The EPA and others have specified that parameters determined to be influential by the sensitivity analysis should be treated as uncertain (Hansen, 1997; Risk Assessment Forum, 1997). This requirement should be applied when other, more relevant criteria are not applicable. This requirement could cause one to select parameters, such as the molecular weight of the chemical, that are not significantly uncertain or variable and are not directly relevant to the decision.

7.9 QUALITY ASSURANCE

The quality of quantitative uncertainty analyses has been a major concern of risk assessors and regulators. The 14 principles of good practice for Monte Carlo analyses proposed by Burmaster and Anderson (1994) are often cited and reproduced (Risk Assessment Forum, 1996). They are all reasonable and sound, but they are more often cited than applied. This is in large part because the requirements are labor intensive, in part because they require a real understanding of probability and the Monte Carlo technique, and, in part, because they inflate the volume of assessments (which are already criticized as excessively voluminous) with information that is incomprehensible to most readers. Other sources of guidance on quality assurance include Ferson (1996) and Risk Assessment Forum (1996). Recently, the EPA has issued a less technically demanding set of requirements for acceptance of Monte Carlo or equivalent analyses (Box 7.2).

The requirements of both the EPA and Burmaster are clearly intended for use with human exposure analyses, but the EPA requirements are said to be applicable to ecological risk assessments as well. The following points provide an adaptation and expansion for ecological risk assessment of the eight EPA requirements, with numbers corresponding to those in Box 7.2.

1. The call for clearly defined assessment endpoints is particularly important for ecological risk assessments. In particular, it is important to specify whether each endpoint is defined as a probability. If so, it is important to know what source of variance or uncertainty is of concern to the risk managers.

BOX 7.2
Conditions for Acceptance by the EPA

When risk assessments using probabilistic analysis techniques (including Monte Carlo analysis) are submitted to the Agency for review and evaluation, the following conditions are to be satisfied to ensure high-quality science. These conditions, related to the good scientific practices of transparency, reproducibility, and the use of sound methods, are summarized here and explained more fully in the Attachment, "Guiding Principles for Monte Carlo Analysis."

1. The purpose and scope of the assessment should be clearly articulated in a "problem formulation" section that includes a full discussion of any highly exposed or highly susceptible subpopulations evaluated (e.g., children, the elderly). The questions the assessment attempts to answer are to be discussed and the assessment endpoints are to be well defined.
2. The methods used for the analysis (including all models used, all data upon which the assessment is based, and all assumptions that have a significant impact upon the results) are to be documented and easily located in the report. This documentation is to include a discussion of the degree to which the data are representative of the population under study. Also, this documentation is to include the names of the models and software used to generate the analysis. Sufficient information is to be provided to allow the results of the analysis to be independently reproduced.
3. The results of sensitivity analyses are to be presented and discussed in the report. Probabilistic techniques should be applied to the compounds, pathways, and factors of importance to the assessment, as determined by sensitivity analyses or other basic requirements of the assessment.
4. The presence or absence of moderate to strong correlations or dependencies between the input variables is to be discussed and accounted for in the analysis, along with the effects these have on the output distribution.
5. Information for each input and output distribution is to be provided in the report. This includes tabular and graphical representations of the distributions (e.g., probability density function and cumulative distribution function plots) that indicate the location of any point estimates of interest (e.g., mean, median, 95th percentile). The selection of distributions is to be explained and justified. For both the input and output distributions, variability and uncertainty are to be differentiated where possible.
6. The numerical stability of the central tendency and the higher end (i.e., tail) of the output distributions are to be presented and discussed.
7. Calculations of exposures and risks using deterministic (e.g., point estimate) methods are to be reported if possible. Providing these values will allow comparisons between the probabilistic analysis and past or screening level risk assessments. Further, deterministic estimates may be used to answer scenario specific questions and to facilitate risk communication. When comparisons are made, it is important to explain the similarities and differences in the underlying data, assumptions, and models.

BOX 7.2 (continued)

8. Since fixed exposure assumptions (e.g., exposure duration, body weight) are sometimes embedded in the toxicity metrics (e.g., Reference Doses, Reference Concentrations, unit cancer risk factors), the exposure estimates from the probabilistic output distribution are to be aligned with the toxicity metric.

Source: Hansen (1997).

2. The disclosure of methods called for in this condition is good practice whether or not the methods are probabilistic.
3. The EPA does not specify the type of sensitivity analysis to be performed. See Section 7.8 for a discussion of "basic requirements" to be considered when deciding which parameters to treat as distributed.
4. Moderate to strong correlations among parameters are common in risk models, and, if ignored, they inflate the output distribution. For example, body weights, feeding rates, and water consumption rates are all highly correlated. If correlations are believed to occur but cannot be estimated from available data, assessors should perform Monte Carlo simulations with correlations set to zero and set to high but plausible values to determine their importance and present the results (Burmaster and Anderson, 1994).
5. These requirements for disclosure of the nature of the distributions (Box 7.2) are appropriate. The demand for plots of both probability density and cumulative density functions for each input and output parameter seems excessive, but they do provide different views of the functions that give a more complete understanding than either alone. The tabular presentation should include the following:

    - Name of the parameter;
    - Units of the parameter;
    - If variable, with respect to what it varies;
    - Formula for the distribution of variability;
    - Basis for the distribution of variability;
    - If uncertain, the sources of uncertainty that are considered;
    - Formula for the uncertainty distribution; and
    - Basis for the distribution of uncertainty.

Distributions that are developed ad hoc may require considerable explanation of their bases. These may include the data from which they are derived or the elicitation techniques for expert judgments plus an explanation of how the data or judgments relate to the assumed sources of the variability or uncertainty. If the expert judgments of individuals are used to derive distributions, any information or logic that went into the judgment should be described as far as is possible. Burmaster and Anderson (1994) indicate that a typical justification for a distribution would require five to ten pages.

6. The sixth condition refers to the changes in the variability of the moments of the output distribution as the number of iterations of a Monte Carlo analysis increases. Most software packages provide criteria for termination of the analysis based on the stability of the output distribution. This condition specifies that the stability of both central tendency and an extreme should be noted and recorded. The specific mention of "high end" is based on the EPA assumption that only exposure will be probabilistically analyzed. For effects distributions, the lower extreme is of greater interest.
7. In general, deterministic analyses using conservative assumptions will have been performed in the screening assessment and therefore will be available for comparison with the probabilistic analysis. In addition, a deterministic analysis using realistic or best estimate values for the parameters is called for in condition 7. Finally, a deterministic analysis may be performed using assumptions and parameters favored by a regulatory agency. In some cases, discrepancies among conservative point estimates, best point estimates, regulatory estimates, and medians of probabilistic results will be quite large. The causes of these differences should be explained.
8. Reference doses, slope factors, or equivalent effects parameters have not been provided by the EPA for ecological risk assessment. Rather, ecological risk assessors must assure themselves and the reader that the expressions of exposure and effects are concordant as well as individually making sense, given the site conditions and assessment endpoint (Section 7.4). Note that, for distributions, this requirement goes beyond simple checking of units. The assessor must consider not only what is distributed, but with respect to what is it distributed.

The following additional considerations are not among the EPA conditions, but they are also important to ensuring the quality of probabilistic ecological risk assessments.

9. As far as possible, use empirical information to derive distributions (Burmaster and Anderson, 1994).
10. The correlation matrix must have a feasible structure. For example, if parameters *a* and *b* are both strongly positively correlated with *c*, they cannot be strongly negatively correlated with each other (Ferson, 1996).
11. Multiple instances of the same variable in a model must be assigned the same value in an iteration of a Monte Carlo analysis (Ferson, 1996). This is a particular problem in stepwise or nested analyses in which different components of risk are estimated by separate simulations.
12. Care must be taken to avoid nonsensical values of input and output distributions. For example, negative values should not be generated for parameters such as concentrations or body weights, herbivore consumption rates should not exceed plant production rates, and contaminant concentrations should not exceed a million parts per million. This can be

prevented by truncation, by the appropriate selection of the distribution, or by constraints on the relationships among variables.

13. In general, for parameter uncertainty, it is most important to treat the parameters correctly (e.g., do not treat variables as constant), next most important to get the magnitude of variability or uncertainty right, and least important to get the form of the distribution right (e.g., triangular vs. normal).
14. For species sensitivity distributions, distributions of measures of exposure, and similar parametric distribution functions, goodness-of-fit statistics and prediction intervals should be reported as estimates of model uncertainty.
15. In general, model uncertainty cannot be well or reliably estimated because the range of models cannot be well defined. At least, model uncertainty should be acknowledged. The acknowledgment should list specific issues in model selection or design that are potentially important sources of error. That list should include any issues about which there was initial disagreement among the parties or issues about which there is no consensus in ecological risk assessment practice. When there are clear differences of opinion concerning assumptions, models should be run with each of the alternatives to determine their influence on the results.
16. As far as possible, specify whether model assumptions introduce an identifiable bias. Examples include:

    - Assuming 100% bioavailability introduces a conservative bias.
    - Assuming independent toxic effects (i.e., response addition) introduces an anticonservative bias.
    - Assuming additive toxic effects (i.e., concentration addition) introduces a conservative bias.
    - Assuming that the chemical occurs entirely in its most toxic form introduces a conservative bias.
    - Assuming that the most sensitive species of a small number of test species is representative of highly sensitive species in the field introduces an anticonservative bias.

Note that a bias does not mean that there is a consistent direction of error in every case. For example, strong antagonistic or synergistic effects could negate the bias associated with assuming additive toxicity. However, the bias is real because such effects are relatively uncommon.

17. Quantification of uncertainties due to model assumptions is difficult, but should be done as far as is reasonably possible. For example, the uncertainty due to assuming that the chemical occurs entirely in its most toxic form can be bounded by presenting results for the least toxic form.

8 Remedial Goals

Diseases desperate grown by desperate appliance are relieved, or not at all.

—William Shakespeare, *Hamlet*

The first step of a feasibility study in the RI/FS process is to identify remedial goals (also called remedial action objectives) for protecting human health and the environment. The remedial goals specify contaminants and media of concern, potential exposure pathways, and cleanup criteria (EPA, 1990a). The criteria are typically concentrations of chemicals in the specified media that are expected to protect human health and the environment adequately, based on risk assessments of the specified routes of exposure. Chemical concentrations are the usual criteria because they are the single line of evidence used in a human health risk assessment. However, ecological risk assessment offers the possibility that remedial goals can be defined more broadly than chemical criteria. These remedial goals are developed iteratively, beginning with the DQO process in the problem formulation and ending with site-specific goals that are set by regulatory agencies, and agreed to by site managers if the site is a U.S. federal facility. The remedial goals are the basis for the selection of candidate remedial alternatives by engineers and site managers. Remedial goals must specify a receptor and exposure route, because the EPA acknowledges that protection can be attained by actions that decrease exposure as well as by decreasing concentrations of chemicals in environmental media (EPA, 1990a).

The risk assessor's primary input to risk management is proposed cleanup criteria, alternatively termed *preliminary remediation goals* (EPA, 1991d), *treatment endpoints* (Alexander, 1995), *corrective action goals* (ASTM, 1999), or *remediation objectives* (CCME, 1996a). The term *remedial goal options* (RGOs), used by some EPA Regions, is preferable to the other four, because it emphasizes (1) that reducing risks from contamination to minimal levels is only one of the risk manager's options when making a remedial decision and (2) that risk assessors may present multiple options for remedial goals, based on different levels of risk, different endpoints, or different definitions of the remedial action objective. Thus, we use the term RGO throughout this chapter. In this chapter, the term *preliminary remedial goals* (PRGs) is restricted to toxic concentrations of individual chemicals that are generically derived and serve as default RGOs (Section 8.1). Thus, the following definitions apply: preliminary remedial goals are concentrations in media that are starting points for developing cleanup targets; remedial goal options are the assessors' recommendations concerning ways that remediation might achieve protection of the assessment endpoints; and remedial goals are the ultimate cleanup targets set by risk managers that engineers attempt to achieve. The final rule for the *National Oil and Hazardous Substances Pollution Contingency Plan* states that the EPA sets remedial goals (EPA, 1990a). The EPA and states set these goals at many sites; however, for U.S. federal facilities, these goals are negotiated between site managers and the EPA, with a high degree of involvement by risk assessors.

8.1 PRELIMINARY REMEDIAL GOALS

PRGs are upper concentration limits for specific chemicals in generic soils, waters, or sediments that are anticipated to protect human health or the environment. The following discussion is focused on the development and use of PRGs for ecological endpoints. PRGs are more generic than RGOs and can be used as a starting point for the development of RGOs for remedial investigations at multiple facilities, sites, or units. There are two important questions to the risk assessor: (1) Which chemical concentrations in environmental media should be used as PRGs? (2) How may PRGs be modified to generate site-specific RGOs?

The EPA has published guidance entitled "Risk Assessment Guidance for Superfund: Volume I—Human Health Evaluation Manual, Part B" (RAGS), which is a useful aid in developing PRGs intended to protect human health (EPA, 1991d). However, no guidance is available in the United States on how to develop PRGs based on ecological risk or even what level of protection is analogous to the 10^{-6} risk for human carcinogens. (See Section 9.2.3 for a discussion of balancing health and ecological risks.) PRGs should not be higher than any numerical applicable or relevant and appropriate requirements (ARARs) for the chemical of concern. For ecological endpoints, the only ARARs are National Ambient Water Quality Criteria (NAWQC) that are available for many contaminants in surface waters. Other government entities have published guidelines that are recommended for use as PRGs. The Canadian Council of Ministers of the Environment (CCME), for example, has published a protocol for the derivation of soil quality guidelines (CCME, 1996b) and guidelines for 20 chemicals derived using the protocol (CCME, 1997). These guidelines are intended to be used or modified by site managers as remediation criteria (CCME, 1996a, b, 1997); thus, they are PRGs.

Risk assessors may use existing PRGs (often national or state guidelines) or derive PRGs that are less generic. In the latter case a risk assessor may, for example, derive soil PRGs using generic toxic doses and site-specific wildlife food uptake factors (LMES, 1997). Many ecological risk assessors do not distinguish between screening benchmarks (Section 4.1.8) and PRGs; however, PRGs are useful if (1) more than one assessment endpoint that is exposed to a single medium (e.g., piscivorous birds and fish in water) requires protection; (2) screening benchmarks are no observed adverse effects levels (NOAELs) and therefore cannot serve as PRGs (PRGs should be minimal effects levels unless the endpoint is a threatened or endangered species); or (3) multiple screening benchmarks for an endpoint exist. In the latter case, the selection of general PRGs serves to assure that particular benchmarks (e.g., among aquatic or sediment toxicity benchmarks) are consistently the starting points for developing final remedial goals. It is acknowledged that screening ecotoxicity benchmarks are biased to avoid eliminating contaminants that are possible contributors to risk; thus, excessive cleanups could result if they were used as final remedial goals (TNRCC, 1996). Sheppard et al. (1992) provide a short history of the development of "generic guidelines" that are applicable to remediation of a broad range of sites. Most of these focus on health effects, but environmental effects are incorporated in others.

When considering the use of existing PRGs, the risk assessor must be aware of (1) the intended use and (2) how they were derived. For example, ecotoxicity benchmarks are often not appropriate PRGs, particularly if they represent no-effects levels. Also, if PRGs for an arid state in the United States are derived using arid soil data, they are not appropriate PRGs for humid locations. Similarly, PRGs derived from toxicity data for organisms that are not related to endpoint species should not be used. For example, PRGs based on toxicity to osprey should not be used for remediation of a small stream. Toxicity tests on which the PRGs are based vary, depending on needs of the institution or facility that developed them. For Canadian soils, test endpoints include mortality, reproduction, growth, development, behavior, activity, lesions, physiological changes, respiration, nutrient cycling, decomposition, genetic adaptation, and physiological acclimatization (CCME, 1997). Since the PRGs may be used as default remedial goals, the assessor should be aware of the level of conservatism associated with the goals.

In general, existing PRGs correspond to small effects on individual organisms, and these chemical concentrations would be expected to cause minimal effects on populations and communities. Far more studies that concern effects on individual organisms are available than those that demonstrate effects at higher levels of organization. PRGs developed in this manner may not be sufficiently protective of individual organisms among threatened and endangered populations if they are also sensitive species. Because these species are protected at the individual level, remedial goals for such species should be developed ad hoc and should be based on NOAELs.

PRGs may apply to one of three environmental media: surface water, sediment (including pore water), and soil. At hazardous waste sites where cleanup is contemplated, it is unlikely that an air source is the major contributor to the ecological risk and even less likely that air will be remediated. Similarly, ecological PRGs for groundwater exposure have not been developed. Groundwater contamination has greater consequences for human health than for nonhuman organisms, data on microscopic and other small biota of groundwater are scarce, and regulatory agencies do not typically advocate their protection. Although contaminants of potential concern at a site can be identified based on concentrations in wildlife food or in the assessment endpoint organism's tissues, ultimately one of the three abiotic media is remediated. Therefore, the media for which PRGs have been developed do not include "foods" and are limited to surface water, sediments, and soil. In addition, indirect effects of contamination (such as avoidance of contaminated food) are not typically considered in the derivation of PRGs (CCME, 1996b).

8.1.1 PRGs for Surface Water

In the United States, PRGs for surface waters should be at least as conservative as NAWQC, unless the particular NAWQC are based on effects on humans. Numerous other benchmarks for aquatic toxicity may also serve as PRGs. At Oak Ridge National Laboratory, PRGs for surface waters are chosen by comparing the ORNL benchmarks for screening toxicity of contaminants to aquatic life (chronic NAWQC or secondary chronic values; Suter and Tsao, 1996) with those for toxicity to piscivorous wildlife (LOAEL; Sample et al., 1996b). The lower of the two benchmarks is the PRG (LMES, 1997). It should be noted that this PRG may be too conservative in the case of small streams, where avian or mammalian piscivores may not forage. The PRG for a particular chemical cannot be assumed to protect piscivorous wildlife if information on the toxicity is not available. As in the risk assessment for aquatic organisms, the filtered concentration of a chemical in water should be assumed to be more representative of exposure than the total concentration. The only exception would be in the drinking water concentrations used in the calculation of PRGs for piscivorous wildlife.

8.1.2 Sediment

Both the concentrations of chemicals in the solid phase of sediments and concentrations in the pore water are relevant to the exposure of benthic (sediment-inhabiting) organisms, and PRGs may be developed for both media (LMES, 1997). If PRGs are available for both sediment and pore water, the PRG that is determined by the remedial investigation to be the best estimate of risk to sediment biota should take precedence. At ORNL, PRGs for sediments have been defined as the lowest concentration of six types of sediment toxicity benchmarks (LMES, 1997). Most PRGs for nonionic organic chemicals are based on equilibrium partitioning. In the United States, PRGs for chemicals for which sediment quality criteria have been proposed should be at least as low as those values.

Few peer-reviewed publications exist in which PRGs are derived. However, one is worth mentioning. Comber et al. (1996) provide sediment guideline values (which may be used as PRGs) for dioxins and dibenzofurans. They derived values using two independent sets of parameters: (1) an aquatic guideline value corresponding to a toxic level for rainbow trout fry and the K_{oc} and (2) a toxic residue level for benthic organisms and the sediment–organism uptake factor. The values for 2,3,7,8-TCDD were within a factor of 6 of each other.

8.1.3 Soil

Standard benchmarks for soil do not exist in the United States, although they are currently being developed by the EPA. At ORNL, PRGs for soil are protective of avian and mammalian wildlife, plant communities, and soil invertebrate communities (LMES, 1997). Microbial processes are not included. The EPA and state regulatory agencies in the United States rarely make remedial decisions based on the protection of soil invertebrates, and to the authors' knowledge have never based a decision on protecting microbial processes. Thus, regulatory priorities and precedents should be considered when the assessor is selecting or developing PRGs for use at a particular site. The CCME adds nutrient cycling processes to the ecological receptors of concern for the derivation of its effects-based soil quality guidelines or PRGs (CCME, 1996b). Soil PRGs should be derived under the assumption that plants and soil invertebrates are exposed through direct contact with the chemical in soil and that wildlife are exposed primarily through ingestion. Interestingly, the CCME (1996b) assumes that PRGs developed for soil-dependent organisms should be protective of wildlife exposed via ingestion and dermally, except in the case of a few specific chemicals (e.g., molybdenum and selenium).

The CCME (1996b) provides a methodology for estimating the threshold effects concentration (TEC) using single-chemical toxicity tests, where the first method is preferred to the second, and the second is preferred to the third: (1) weight-of-evidence analysis (percentile of the combined effects and no-effects distributions of concentrations), (2) extrapolation from the LOEC, or (3) extrapolation from the EC50 or LC50. A microbial effects concentration (nitrogen fixation, nitrification, nitrogen mineralization, respiration, and decomposition) is compared with the TEC. If the microbial concentration is lower than the TEC, the geometric mean between the two is used as the guideline for soil contact. As stated above, guidelines for soil contact are assumed to protect wildlife exposed through ingestion (CCME, 1996b). An exception is ingestion by herbivores in an agricultural scenario (CCME, 1997).

A methodology for calculating soil PRGs for terrestrial wildlife has been developed for the Oak Ridge Reservation (LMES, 1997). PRGs were calculated as a concentration in soil that would result in an estimated dose by all oral routes equal to the contaminant-specific and species-specific LOAEL (Figure 8.1). Exposure estimates were iteratively calculated using varying soil concentrations and soil-to-biota uptake models. The soil concentrations were manipulated to produce an exposure estimate equivalent to the wildlife endpoint-specific and contaminant-specific LOAEL. Because different diets may dramatically influence exposures, and sensitivity to contaminants varies among species, PRGs were developed for six species present on the Oak Ridge Reservation: short-tailed shrew, white-footed mouse, red fox, white-tailed deer, American woodcock, and red-tailed hawk. For each chemical, the PRG for each of the wildlife species was compared, and the lowest soil concentration was selected as the final wildlife PRG. Estimates of oral exposure to contaminants were generated using a generalized exposure model (Section 3.10). Among the 18 chemicals and six wildlife species for which PRGs were derived, the final PRG for protection of wildlife was always based on either the short-tailed shrew or the American woodcock (LMES, 1997).

Reports in which PRGs are compiled or derived do not typically recommend a soil depth to which the quantities should apply (e.g., CCME, 1996b). The relevant depth of exposure is left to the assessor to determine. The considerations related to the depth of sampling for ecological risk assessments (e.g., Section 3.4.3.1) apply here.

8.1.4 Modification of PRGs

The specific, numerical PRGs that an assessor uses are less important than how they are modified (below) or how ultimate RGOs are chosen (Section 8.2). PRGs that are not ARARs or based on ARARs may be modified during the remedial investigation and feasibility study using site-specific data (EPA, 1991d). These modified PRGs may be recommended by assessors as RGOs for the site. The use of the same remedial goal at sites with varying soils or exposure pathways would result in variable risks (see Labieniec et al., 1996, for an analysis of this risk variability with respect to human health). The Canadian Council of Ministers of the Environment

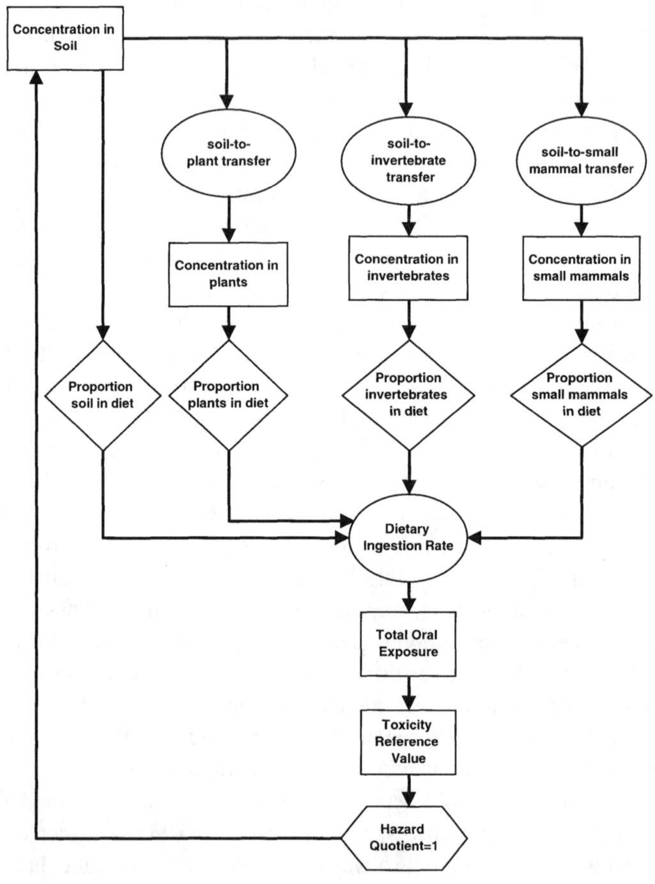

FIGURE 8.1 Procedure for calculation of PRGs for soil based on toxicity to wildlife (LMES, 1997)

(CCME, 1996a) suggests that generic guidelines (PRGs) should be modified if (1) a particular site has high background concentrations of a chemical; (2) contaminants may move from soil to groundwater or air; (3) toxicological data used to derive the guidelines are not relevant to the site (different receptors, complex mixtures of chemicals); or (4) land uses necessitate modification of PRGs. The EPA emphasizes that multiple chemicals and multiple exposure pathways may justify the modification of PRGs (EPA, 1990a). California adds that if multiple media contribute to exposure, the media should be assumed to be in equilibrium for the development of remedial goals (Cal EPA, 1996).

In summary, modifications of PRGs may be based on:

- Land-use assumptions;
- Exposure assumptions and habitat considerations (e.g., fraction of land that is suitable habitat);

- Environmental assumptions used to derive PRGs (e.g., water hardness, soil pH, and organic content);
- Absence of type of organism (e.g., wide-ranging predator) which the PRG is intended to protect (or exclusion from list of assessment endpoints);
- Synergistic, antagonistic, or additive effects of multiple pollutants;
- Form of chemical different from assumption in derivation of PRG (e.g., nonaqueous-phase liquids);
- Exposure from multiple media;
- Impacts of contamination of one medium on another (EPA, 1991d);
- Impacts of remediation of one medium (such as sediments) on contamination of another medium (such as surface water);
- Effects of remediation on organisms and their habitat (Chapter 9);
- Desirable level of protection;
- Background concentration of element higher than PRG; or
- Indirect effects of contamination.

8.1.5 Land Use

Land-use scenarios play a different role in ecological risk assessments than they do in human health risk assessments. For human health risk assessments, remediation depends on the land-use scenario because land use determines human exposure. Exposure pathways for humans can change, for example, depending on whether the land is industrial, agricultural, or residential. Soil ingestion by children may occur in residential areas and not in industrial areas; plants for human consumption are not grown on industrial sites; and inhalation of soil particles would be more significant in agricultural than in residential areas. Therefore, because humans engage in different activities in different locations, exposure depends on land use, and risk-based remediation should depend on land use. In ecological risk assessments, three issues must be considered with respect to post-remedial land use.

1. **What habitat will occur on the site given the proposed land uses?** Plants and most animals are more likely than humans to engage in all activities on a particular site. Therefore, land-use scenarios are important primarily because they determine which receptors will find habitat on the site. Land use determines habitat, which determines the presence of endpoint populations and communities, which determines exposure and risk. Even for migratory species, the question arises: Will the site provide habitat for the species during the time it might spend on the site? Therefore, for an industrial site, one might develop a soil PRG for a continued industrial use scenario based on leaching and runoff to an off-site stream, because there would be no on-site ecological endpoint receptors. However, if the site is to be converted to a park with a plant community, birds, and small mammals, the soil PRG should be protective of on-site as well as off-site endpoint receptors. In some cases, as in human health risk assessment, land use may eliminate habitat for some activities but not all. For example, waterfowl may rest on and drink from an industrial pond during

migration but would not breed there. Therefore, a PRG might be developed for that limited exposure.

2. **Will land use affect the value of ecological endpoint receptors?** In some land uses, endpoint populations or communities are not as highly valued as they would be with other uses. For example, earthworms in soil adjacent to a factory are less valued than in agricultural or natural land uses where maintenance of soil texture and fertility is important. These differences in value could result in different degrees of protection. An example of this concept is the identification of "designated uses" for receiving waters under the Clean Water Act. The various designated uses, which vary among states, require different water qualities, which in turn require different levels of pollutant regulation.
3. **Will land use affect the sensitivity of endpoint receptors?** Land use may modify the species composition of an endpoint community or the life stages of an endpoint population on the site. For example, streams in industrial, agricultural, or residential areas are physically disturbed and tend to have little riparian vegetation and highly variable flows, resulting in low species richness. Similarly, stream reaches with no spawning habitat (e.g., suburban channelized reaches) may contain adult fish but very few or no fish eggs or larvae. Such stream reaches may require less protection, because (a) sensitive species or life stages are absent, (b) no amount of waste remediation will result in recovery, and (c) the land use effectively precludes habitat restoration.

For any of those reasons, risk managers may decide that different levels of protection should be associated with different land uses, thus modifying the PRG for a medium. However, it should be noted that no particular land-use scenario is assumed for the application of ARARs (e.g., NAWQC) to a site.

Protection levels for different land uses may be determined generically for a nation or other political entity. Soil quality guidelines developed by the CCME (1997) and used as PRGs are specific to different land uses: agricultural, residential/parkland, commercial, and industrial. Agricultural lands include agricultural edge habitats. The residential/parkland use assumes that the land serves as a buffer zone between residences but is not broad wilderness. Commercial land includes managed systems, such as cultivated lawns and flowerbeds (CCME, 1996b). However, it is assumed that the normal range of activities on commercial lands does not rely as much on ecological services as agricultural or residential/parkland. Although the same receptors are generally examined in the derivation of the PRGs, the level of protection for commercial and industrial land uses is lower than for the other two (CCME, 1997).

8.2 REMEDIAL GOAL OPTIONS

Differing quantitative or qualitative definitions of RGOs are possible because of the multiple lines of evidence that are available in ecological risk assessment. Conventionally, an RGO is defined as a concentration of a particular chemical that constitutes a threshold for unacceptable risk. The risk assessor modifies the PRG to derive the RGO. Media with chemical concentrations below the RGO are assumed to be acceptable, but those with concentrations above the RGO may be remediated. Indeed, the EPA definition of a remedial goal is a concentration of a chemical in an environmental medium (EPA, 1990a).

Alternatively, RGOs may be defined in terms of media toxicity test results (Office of Emergency and Remedial Response, 1994a). That is, one may specify that areas where a particular test endpoint (e.g., >20% mortality of earthworms) or any one of a set of test endpoints is exceeded are candidates for remediation. For example, an RGO may be defined in terms of toxicity if the risk assessment identifies a medium as toxic without clearly isolating the cause. Potential reasons for the use of a toxicity RGO would be inadequate data on concentrations of all chemicals, poor relationship between bioavailability at the site and in laboratory media, failure of any of the chemicals measured in the medium to be toxic by themselves, or high variance in the relative contributions of individual chemicals to toxicity at different sample locations. In such cases an appropriate RGO could be a direction to remediate all toxic areas. This was the case for a sediment depositional area in Bear Creek in Oak Ridge, where the sediments were unambiguously toxic (i.e., *Hyalella azteca* survival was reduced by 37% relative to reference and control), but a causative agent could not be determined from the available chemical data. Alternatively, an RGO may be to perform a toxicity identification and evaluation (TIE) procedure and remediate the chemicals that are identified by the TIE to be causing the toxicity (Section 4.2). Finally, one may specify that areas where biological surveys indicate levels of effects in exceedence of some measure of effect (e.g., dead plants or fewer than x earthworms per square meter) are candidates for remediation.

Derivation of chemical-concentration-based RGOs should incorporate site-specific exposure and effects data, to the extent practical. One way to do this is to derive a site-specific no apparent effects level (SSNAEL) for an environmental medium (Jones et al., 1999). The SSNAEL is the highest measured concentration of each chemical at which toxicity was never observed in standard tests of the ambient medium from the contaminated site. If the tests are sufficiently sensitive, the RGO for a particular chemical should not be lower than the concentrations which were shown to be nontoxic in the site medium. This approach can also be used for survey data (e.g., species richness or abundance of benthic invertebrates), provided that the surveys are sufficiently sensitive and the measured chemical concentrations are representative of the exposures associated with the observed effects. An example of the application of this approach to the assessment of risks to benthic invertebrates in Poplar Creek, TN is presented by Jones et al. (1999). The recommended RGO was the higher of the generic probable effects level from the literature (e.g., the effects range–median; see Section 4.1.8) and the SSNAEL.

As the word *options* in the phrase suggests, multiple RGOs may be provided to the risk manager as alternative levels of protection. If the best basis for remediation is unclear, the assessors may provide a set of concentrations or other criteria from which the risk managers could select or derive the final remedial goals. For example, RGOs for water might include (1) the chronic NAWQC for the chemical that is believed to cause significant toxic effects on an endpoint community, (2) a threshold for toxicity of that chemical in a toxicity test such as the EPA subchronic *Ceriodaphnia* test performed with site water as the diluent, (3) chemical concentrations derived by performing a TIE on the contaminated water (Section 4.2), (4) a concentration of a chemical bioaccumulated to toxic levels in fish tissue but not detected in water, and (5) a requirement that toxicity be eliminated, as determined by one or more specified test endpoints. Combinations of these types of RGOs may be used. For example, to confirm that apparent effects are due to contamination, risk managers may require that areas to be remediated show some level of toxicity and have some minimum level of a chemical that is the primary contaminant of concern.

In the Netherlands, site-specific risk assessment is not used to determine whether or not remediation is required at a particular site. Under the Dutch Soil Protection Act, remediation is required if "serious soil contamination" is present, i.e., if risk-based intervention values are exceeded (Swartjes, 1997). The intervention value incorporates both human health and ecotoxicological screening criteria; the ecotoxicological criterion is the hazardous concentration 50 (HC50), the concentration at which 50% of species is assumed to be protected.

Risk assessment is used to determine the priority for remediation of sites where concentrations exceed the intervention values. The prioritization of sites may be based on the factors above, such as the diversity of ecological receptors, soil characteristics, and results of ambient media toxicity tests.

The CCME (1996a) provides five alternative procedures for developing RGOs (termed *site-specific objectives*). They are:

- **Adopt a PRG (termed generic guideline) directly as site-specific objective**. This alternative provides a conservative level of protection for ecological and human receptors under known land uses.
- **Modify a PRG.** PRGs may be modified if receptors or other site conditions are somewhat different from the assumptions used in deriving the PRG (see Section 8.1.4). For example, if specific toxicity data used to derive the generic guidelines are not relevant to the site, PRGs may be recalculated without them. Data used in the calculation of PRGs may be eliminated for ecological receptors not present at the site, but the adjusted data set must retain values for plants, vertebrates, and invertebrates from families that are or could be represented at the site. Also, properties of the medium such as organic carbon content may be used to modify the PRGs.
- **Develop RGO using risk assessment.** If the site is considerably different from the assumptions used in the derivation of PRGs, risk assessment is recommended. More specifically, risk assessment is recommended: (a) if critical habitats are on or near the site; (b) if a large degree of uncertainty is associated with the fate and transport of contaminants (e.g., periodic flooding); (c) if sensitive populations or endangered species are present; (d) if a large degree of uncertainty is associated with the fate or toxicity of contaminant mixtures or metabolites; or (e) if multiple sources of contamination or exposure pathways exist and were not considered in the derivation of PRGs.

- **Derive site-specific objective with the CCME (1996b) protocol.** If minimal acceptable data requirements are met, RGOs may be developed for contaminants for which no effects-based guidelines exist.
- **Consult other jurisdictional options.** Government entities may have particular requirements or may recommend the use of background levels.

Final remedial goals for different environments may vary over orders of magnitude. Although some variance is due to political or policy differences, others are based on real differences in bioavailability and toxicity among sites. Zhang (1992) shows that the final remedial goals for arsenic and pentachlorophenol in soils in Records of Decisions for U.S. Superfund sites have ranged from 1.1 to 300 mg/kg for arsenic and 0.0012 to 83,000 mg/kg for pentachlorophenol. These goals have typically been based on human health exposure scenarios. The wide ranges in values reflect the large number of approaches to the derivation of remedial goals (Sheppard et al., 1992).

8.3 SPATIAL CONSIDERATIONS

Soil concentrations that constitute remedial goals are usually applied on a point-by-point basis, rather than averaged over the area of exposure (Bowers et al., 1996). Although this criticism of risk assessment by Bowers and colleagues is based on human health risk, the same statistical argument applies to wildlife assessment endpoints. Remedial action objectives for wide-ranging wildlife can be met through the iterative use of geographic information systems (GIS, Clifford et al., 1995). Wildlife are exposed to multiple contaminant concentrations at multiple locations, and some contamination may be irrelevant if it occurs in a nonhabitat area. Optimal remedial alternatives for different areas may be different. Clifford et al. (1995) suggest that future contaminant concentrations at different locations in the particular medium be estimated, based on what each remedial alternative can reasonably achieve. Then the associated residual risk can be determined through the use of GIS. Thus, if only chemical concentrations in soil were available as the line of evidence, the RGO could be defined in terms of risk reduction.

An approach for determining remedial goals in a spatial context is presented by Sample (1996). At large contaminated sites such as the Oak Ridge Reservation and component watersheds, population-level risks to wildlife are determined by performing Monte Carlo simulations of the average contaminant exposure over the entire contaminated site. It may be assumed that wildlife are equally likely to forage at any location at the site and that each site contributes equally to risk. This approach is biased (wildlife are not likely to use all areas equally), but in the absence of comparative habitat-use data, the procedure is adequate. To determine which sites should be remediated to reduce the estimated population-level risk to an acceptable level (e.g., proportion of population experiencing exposures greater than the LOAEL is <20%), exposures at the most contaminated sites (potential remedial units) are set to the estimated exposure at background, to represent the results of remediation. Sites are added to the remediated set until an acceptable level of risk has been achieved. Exposures at sites recommended for remediation are set to background because inorganic contaminants that present risks are frequently present at low concentrations in uncontaminated background soils which are likely to be used to remediate the contaminated sites. (For organic contaminants, such as PCBs, exposure goals should be set to zero. Clean soils should have unmeasurable concentrations of most organic toxicants.) This method assumes that the primary means of remediation is the removal or capping of soil, rather than the incomplete reduction of contamination by biological or chemical means.

The approach described above focuses on the most contaminated sites, with the goal of reducing population-level risk to an acceptable level. As a consequence, some sites, where point estimates of exposure indicate risks may be present, will not be recommended for remediation. While these sites may present a hazard to individuals that use these sites extensively, their remediation is not required to prevent risks at the population level. However, if threatened and endangered (T&E) species are present in the area for which these RGOs are being developed, remediation should be considered for these areas. An example of the application of this method to the Bear Creek watershed in Oak Ridge may be found in DOE (1996a).

8.4 HUMAN HEALTH

As stated above, the EPA provides guidance for developing PRGs related to human health risks associated with different chemicals (EPA, 1991d). The text above has addressed the development and use of ecological PRGs, specifically. It is advisable to develop remedial goals for ecological endpoints separately from those for human health, and to compare the two when all site-specific considerations (such as those discussed in Section 8.1.3) have been factored into the final ecological remedial goals (e.g., CCME, 1996b). The ecological risk assessor's role is to provide the risk manager with remedial goal options based on ecological risks. However, the human health and ecological risk assessors should have a common understanding of the differences in exposure and sensitivity of the endpoints that are responsible for discrepancies in the two sets of PRGs.

9 Remedial Decisions

Good policy analysis recognizes that physical truth may be poorly or incompletely known. Its objective is to evaluate, order, and structure incomplete knowledge so as to allow decisions to be made with as complete an understanding as possible of the current state of knowledge, its limitations, and implications.

—Granger Morgan (1978)

The remedy is worse than the disease.

—Francis Bacon, *On Seditions*

Following the CERCLA remedial investigation, the feasibility study (FS) is conducted to analyze the benefits (i.e., risk reduction), costs, and risks associated with remedial alternatives. The use of ecological risk assessment should not end with the baseline risk assessment for the site or even with the recommendation of remedial goal options in the remedial investigation (Chapter 8). Risk assessment is integral to (1) the analysis of individual remedial alternatives; (2) the ultimate, balanced remedial decision for the site; (3) prioritization of the remediation sequence for multiple sites; and (4) the assessment of the efficacy of remediation. The baseline assessment in the RI addresses only the no-action alternative. This chapter addresses the use of ecological risk assessment in the FS and in the subsequent decision-making process.

In remedial alternative analysis, the following questions should be asked: (1) How will present and future risks associated with contaminants be mitigated by each alternative? (2) What new risks may be associated with each alternative? The first question can be answered based on the baseline risk characterization, remedial goals, and the proposed remedial alternatives. No new risk assessment is necessary. The analysis required to answer the second question fully should often be a complete risk assessment: problem formulation, exposure assessment, effects assessment, risk characterization, and description of uncertainties. Stressors (most often physical) are new, and some assessment endpoints and exposure pathways are likely to be different from those in the original assessment. Some of the hazards associated with remedial actions are listed in Table 9.1. Recovery is an important part of the risk characterization for effects of stressors associated with remedial actions. Unfortunately, remedial risks are rarely given due attention in the feasibility study because (1) the FS is often under a severe time constraint; (2) the FS is often performed by the engineers who design the remedial alternatives, not risk assessors; and (3) in the United States, regulators do not require or expect rigorous assessments of remedial actions. The EPA guidance for assessment of human health risks of remedial actions is much less demanding than that for baseline risk assessments, and it makes quantitative assessment optional (EPA, 1991d). The guidance for assessment of ecological risks from remedial actions is less than a page in length (Sprenger and Charters, 1997).

TABLE 9.1
Examples of Hazards Posed by Remedial Actions

From Chemicals:
- Mobilization by dredging of contaminants buried in sediments
- Increased availability of contaminants due to use of chelating agents
- Exposure of consumers to high contaminant levels in hyperaccumulator plants
- Release of contaminants during incineration or thermal desorption
- Use of biocides to eliminate contaminated communities

From Physical Disturbance:
- Destruction of benthic communities by dredging
- Destruction of terrestrial ecosystems by:
  - Removal of contaminated soil
  - Creation of roads, parking areas, laydown areas, and other support facilities
  - Creation or expansion of waste burial grounds
  - Creation or expansion of borrow pits for caps or fill
- Mowing to maintain lawns
- Paving to eliminate hydrological and biotic exposure to soil
- Rerouting, channelization, and lining of streams
- Destruction of soil structure by cleaning
- Soil erosion and compaction

From Biotic Introductions:
- Revegetation with exotic species
- Invasion of ecosystems by microbes or plants introduced for bioremediation

Indirect:
- Opening the site to human use and development
- Encouraging development in the surrounding area by removing the stigma of contamination

Following the remedial alternative analysis, risk managers must finally decide which remedial option is best. Risks from remediation must be balanced against the baseline risks that would be mitigated. It is also advisable for the final remedy to balance human health and ecological risk—that is, for the remedial action to be as protective of ecological receptors as it is of human health. Finally, the decision, of necessity, includes the costs of each alternative.

At large facilities such as the DOE Oak Ridge Reservation, multiple contaminated sites require remediation. The process of prioritizing these sites should incorporate principles of ecological risk assessment. For example, if all sites cannot be remediated immediately, it may be appropriate to evaluate the risk associated with a delay in remediation of each site.

9.1 REMEDIAL ALTERNATIVE ANALYSIS

The best remedial option is chosen by balancing costs and benefits of the various alternatives, the latter including reduction of the ecological risks described in the remedial investigation. According to the National Contingency Plan, the detailed analysis of alternatives consists of using nine criteria to evaluate each one, and then identifying relative advantages and disadvantages of each (EPA, 1990a). The criteria are

1. Protection of health and the environment
2. Compliance with federal applicable or relevant and appropriate requirements (ARARs)
3. Long-term effectiveness and permanence
4. Reduction of toxicity, mobility, or volume through treatment
5. Short-term effectiveness
6. Implementability
7. Cost
8. State acceptance
9. Community acceptance

Additional criteria that risk managers consider prior to making a decision about remediation are existing background levels of chemicals, present and future land uses, present and future resource uses, and ecological significance of the site, both locally and nationally (Sprenger and Charters, 1997).

The first two criteria are threshold criteria and should be weighed more heavily than others (Sprenger and Charters, 1997). Clearly, the first criterion should include effects of the remedial action. Few references to risks from remediation are made in the National Contingency Plan, although potential negative impacts of remediation are alluded to in a discussion regarding the EPA expectation that the preferred alternative will often be treatment of contaminated media. Treatment will be limited when "implementation of a treatment-based remedy would result in greater overall risk to human health and the environment due to risks posed to workers and the surrounding community during implementation" and "severe effects across environmental media resulting from implementation would occur" (EPA, 1990a). The third criterion addresses "any residual risk remaining at the site after completion of the remedial action" (discussed in Chapter 10). Short-term effectiveness, the fifth criterion, refers to adverse and beneficial effects of the action during implementation and construction.

State acceptance and community acceptance criteria relate to ecological risk assessment to the extent that the state regulatory agency and community value ecological receptors. Indeed, for three units on the Oak Ridge Reservation, Lower East Fork Poplar Creek and two ponds at the K-25 site, members of the public insisted that the DOE balance risks from the proposed remedial action against risks identified in the remedial investigation to choose the appropriate alternative. These representatives of the community were in favor of maximum ecological protection at reasonable cost. It is also notable that state and local communities may accept the risks associated with minimal clean-up if the site is designated a "brownfield" (i.e., an industrial land-use site), particularly if the alternative is to construct a facility on cleaner land.

Finally, the risk manager chooses the most appropriate remedy for the site. The balancing of factors involved in this decision is discussed in Section 9.2. Although final remedial actions always require the consideration of the factors above, interim actions do not. At any time during the remedial investigation and feasibility study process, the risk manager may decide that a chemical release or the threat of a release of pollutants necessitates an interim action. This time-critical response, termed a *removal action*, is not a comprehensive or final remedy for the site. Thus the nine criteria do not apply (EPA, 1990a). For example, a decision was made to treat a TCE plume beneath the K-25 site on the Oak Ridge Reservation, based on human health concerns and the potential for the plume to migrate off-site. Ecological concerns, such as the impacts of the reduced flow to a stream, were not required to be considered prior to the removal action decision.

9.1.1 Risks Associated with Remedial Alternatives

The conventions of ecological risk assessment are rarely followed to identify and characterize ecological risks that may be associated with remedial alternatives. Thus, risk assessment associated with remediation is discussed at length here. During negotiations among stakeholders concerning remediation, it is often expected that health and safety issues will arise, including risks to construction workers, risks to the public from incinerator emissions, and risks to the public from dump truck traffic. However, the prospect of new ecological risks is rarely a concern. It is probably assumed that any ecological risks from remediation are short-lived. EPA and state regulatory agencies do not typically require well-structured, prospective ecological risk assessments as part of Superfund remedial feasibility studies. Although the EPA definition of stressor in its "Guidelines for Ecological Risk Assessment" is broad and includes physical stressors, risks associated with CERCLA remediation are not a focus of the document (EPA, 1998). Nonetheless, the "Ecological Risk Assessment Guidance for Superfund" (Sprenger and Charters, 1997) states that the ecological impacts of remedial options are an important aspect of protecting the environment. Given the often haphazard ecological analyses in feasibility studies, decision makers are at risk of unknowingly substituting ecological risks from remedial alternatives for human health and ecological risks that have been identified in the remedial investigation.

Ecological risks from remediation may be classified into two categories: (1) the exacerbation of existing contaminant risks or (2) the physical destruction or transformation of ecological habitats and associated ecological communities. In the first instance, a removal action may cause further contamination of groundwater and surface water, or remedial technologies may increase the bioavailability of contaminants. An addendum to the baseline exposure assessment may be necessary to characterize these risks from chemicals. In contrast, it is primarily the problem formulation phase of the risk assessment that must be improved if ecological risk assessment is to contribute to the assessment of nonchemical remedial alternatives, such as excavation and dredging. Components of the problem formulation that merit discussion are the identification of stressors and assessment endpoints and the development of conceptual models. Exposure and exposure–effects relationships may be obvious if ecosystems or portions of them are eliminated by physical disturbance. In addition, the risk characterization for physical stressors associated with remediation should evaluate the recovery of the affected ecological receptors.

If a large number of activities are associated with a single remedial action, or if multiple remedial actions are undertaken concurrently and in close proximity, it may be necessary to use an ecological risk assessment framework that has been developed for multiple activities. The standard ecological risk assessment framework was developed for assessments of individual chemicals and other individual agents and does not incorporate a logical structure for assessing multiple agents and integrating their risks (EPA, 1998). Similarly, indirect or secondary effects are not addressed. Suter (1999b) developed a framework for the assessment of military testing and training programs that would be applicable to complex remedial actions. Suter recommends that impacts of each activity be assessed separately and integrated in the risk characterization, and provides a conceptual approach for addressing combined effects.

The sections below are organized according to the EPA ecological risk assessment framework. Although environmental impacts of remedial actions are required to be assessed in the feasibility study, the EPA framework is not required to be used. Nonetheless, the authors believe that the framework is helpful in organizing the process of analyzing and characterizing risks from remediation.

9.1.2 Problem Formulation

9.1.2.1 The Nature of Stressors

Physical, chemical, and biological stressors may be introduced as a result of particular remedial actions. Technologies that may introduce new chemical stressors include microbial bioremediation, phytoremediation, solvent extraction, chemical oxidation, and poisoning of contaminated fish prior to removing them. Chemical stressors associated with bioremediation could include toxic metabolites of the process (e.g., vinyl chloride from TCE), nutrients added to enhance the process, surfactants added to enhance the process, and peroxide added to provide a source of oxygen to bacteria. The microorganisms themselves could be biological stressors, if their multiplication and dispersal would constitute a hazard. Similarly, some plants introduced for phytoremediation or revegetation could become weeds. A summary of chemical emissions from conventional remedial technologies is presented in EPA (1991d).

Physical stressors associated with remediation might be the most harmful, at least in the short term. Examples include removal of vegetation and topsoil and soil compaction by heavy equipment and human activity. Similarly, the maintenance of lawn would be a stressor to the plant community and wildlife populations. In aquatic systems, changes in water flow, erosion of stream banks, dredging of sediments, and decreased riparian vegetation would be potential stressors. In all environments, the removal of habitat is a stressor that would be expected to result from a physical removal action.

9.1.2.2 Conceptual Models for Alternatives Assessment

The presentation of conceptual models could potentially increase the clarity and rigor of the alternatives assessments. For no-action alternatives or alternatives that are intended as human health rather than as ecological remedies (e.g., fences, fishing advisories, land-use controls), conceptual models for the baseline risk assessments are applicable. In addition, these conceptual models may be used if the remedial action may mobilize chemical contaminants. However, the remedial alternatives that involve removal, isolation, or treatment of soil or sediment require disturbance not only of the contaminated areas but also of uncontaminated areas used for roads, structures, laydown areas, borrow pits, landfills, or treatment facilities. Hamby (1996) reviews common remedial technologies for soils, surface water, and groundwater. The Federal Remedial Technologies Roundtable provides a Web site listing *in situ* and *ex situ* technologies that have been used in over 100 case-study remediations (http://www.frr.gov).

Generic conceptual models for potential impacts of these activities on components of terrestrial ecosystems, aquatic ecosystems, and wetlands are presented in Figures 9.1, 9.2, and 9.3, respectively. These conceptual models for physical disturbance differ from typical models for chemicals in that the arrows represent chains of causal processes rather than flows of chemicals. Additionally, the receptors are defined broadly because the consequences of physical disturbances tend to be less discriminatory than those of chemicals. Because of the great diversity of physical disturbances that could occur during remediation, these generic models require substantial adaptation to specific cases. The generic models should be modified as remediated sites are monitored and unexpected links emerge. For example, on the Oak Ridge Reservation a TCE-contaminated aquifer is being remediated through a pump-and-treat technology. To prevent Mitchell Branch, a neighboring stream, from being drained by the remedial measure, a length of the stream has been altered to a culvert. Damage to the stream, soil compaction, and the trampling of the riparian community along the stream should be included in the conceptual model or models for the remedial action.

Large-scale physical or chemical remedial measures may impact neighboring sites. Thus, remedial decisions should be considered in the context of the management of neighboring sites. For example, on the Oak Ridge Reservation land managers have proposed draining a contaminated pond to mitigate risks to trespassing fishermen and avian piscivores. Hydrologically connected to this pond is a waste burial ground, contaminated groundwater, and an associated spring. All of these elements should be components of the conceptual model for the remedial action. For example, rotenone added to the pond to kill PCB-contaminated fish may escape from the pond to hydrologically connected water bodies. Similarly, Garten (1999) has found that forest vegetation mitigates leaching of strontium-90 from soils at locations where transport is controlled mainly by subsurface flow. Thus, a conceptual model for the removal of trees from a similar strontium-90-contaminated site should include a pathway to groundwater and possibly to surface water and aquatic organisms.

More indirect effects of remediation may interfere with goals for protection of ecological receptors. For example, chelation agents added to soil to aid in phytoremediation may strip the soil of particular nutrients (Entry et al., 1996), thus causing adverse effects on the plant community. These agents could also increase the uptake of contaminants by soil invertebrates and increase food web transfer. Thermal cleaning of soil destroys its structure and organic matter, raises its pH, and has a sterilizing effect (Tamis and Udo de Haes, 1995). Not only are the native soil flora and fauna affected, but plants seeded on the cleaned soil are likely to be adversely impacted.

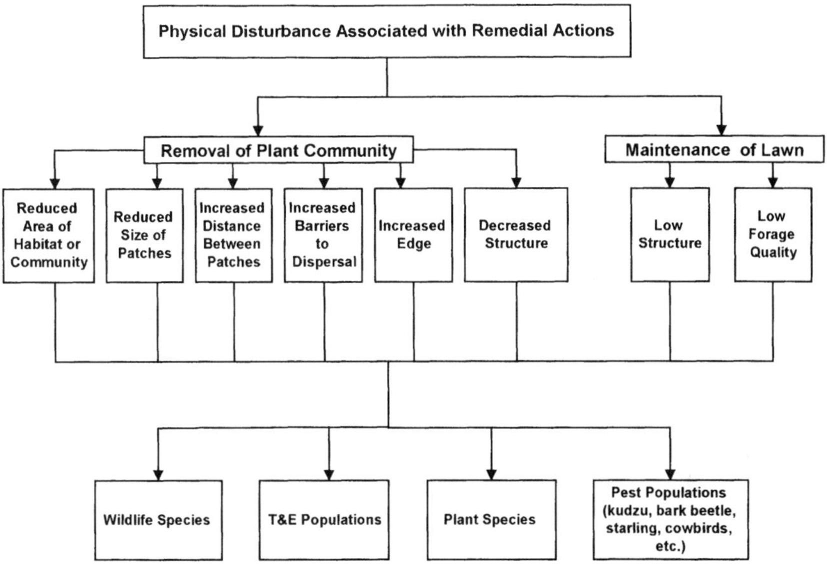

FIGURE 9.1 Generic conceptual model of the effects of physical disturbance on terrestrial ecosystems.

A more complex example involves the remediation of the Rocky Mountain Arsenal, including the demolition of chemical factories at the site. This reduction in the stigma of contamination and the improvement in aesthetics are leading to increased development on nearby lands. The development threatens the wildlife habitat that exists in the vicinity of the site and provides a habitat corridor between the site and the Front Range (Baron, 1997). Risk assessors and managers must decide which stressors have been created indirectly by the remedial action and pose likely or potentially high-magnitude risks.

If conceptual models are consistent with remedial alternatives, the models may include the ultimate environmental fate of the contaminated medium. Where will dredged sediments be deposited? Is treated soil proposed to be returned to its site of origin? Tamis and Udo de Haes (1995) note that in the Netherlands cleaned soil has a stigma associated with it. Soil that has been cleaned through thermal processes is generally used as fill in construction because of the loss of structure and organic matter (Tamis and Udo de Haes, 1995). Soil cleaned through the use of chemical extraction is used in the concrete and asphalt industries. Biologically remediated soil is often used to cover waste dumps (Tamis and Udo de Haes, 1995).

Suter (1999a) notes that conceptual models that represent multiple activities, multiple agents, nonchemical agents, and indirect effects, all of which may be associated with remedial actions, can be difficult to develop. Because these conceptual models do not simply represent flows of contaminants, it is advisable to define the processes that link the physical components of the models. Because these models may become quite complex, it is often desirable to structure them hierarchically in

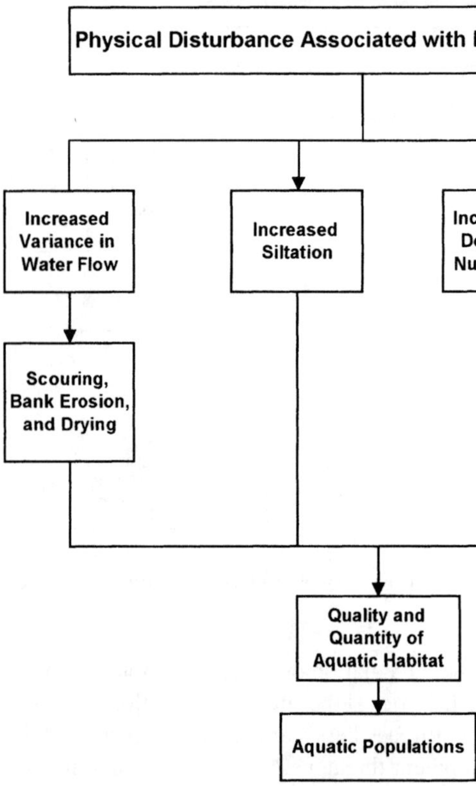

FIGURE 9.2 Generic conceptual model of the effects of physical disturbance on aquatic ecosystems.

both detailed and aggregated form. Suter (1999a) also recommends that risk assessors create modular component models that can be reused in different combinations for different assessments.

9.1.2.3 Assessment Endpoints

As stated in Section 2.5, ecological assessment endpoints are statements of environmental values, i.e., entities and associated properties that are to be protected. Candidate assessment endpoints in the remedial feasibility study should include both those that were selected as endpoints for the baseline ecological risk assessment and those that were excluded because they were not deemed to be exposed to contaminants at the site. Some of these receptors could be at risk from the physical disturbances associated with remediation. In addition, assessment endpoints should include receptors at neighboring sites, such as terrestrial or aquatic communities

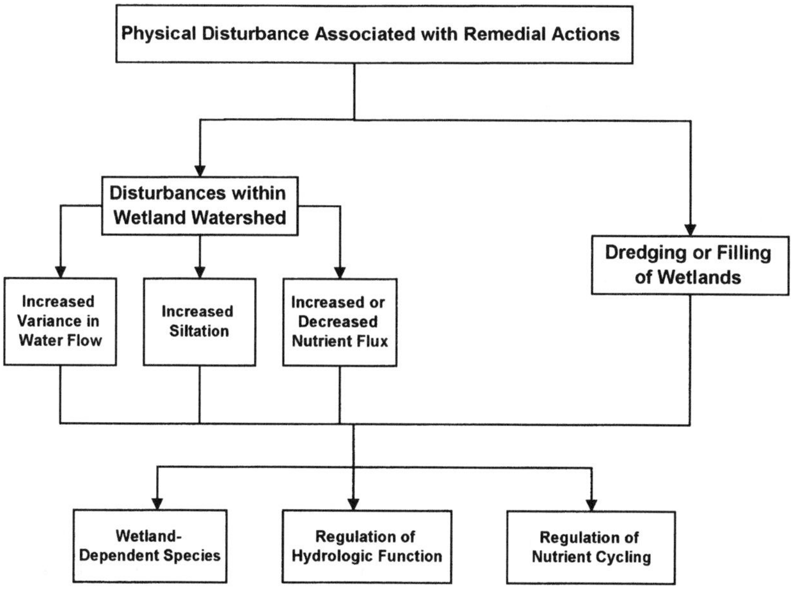

FIGURE 9.3 Generic conceptual model of the effects of physical disturbance on wetlands.

affected by off-site activities such as road construction and creation of borrow pits and waste disposal sites. As in the baseline ecological risk assessment, ecosystems and organisms with special regulatory status, such as wetlands and threatened and endangered species, should be assessment endpoints if an exposure pathway exists.

Some physical remedial actions are likely to severely disrupt habitat for endpoint populations or communities by virtue of their severity or large spatial scales. Although hazardous waste sites that are entirely denuded of vegetation because of contamination are rare, the removal of soil and the associated plant community as a remedial measure is common. Thus, the physical disruption of habitat, which could put associated populations of organisms at risk, should dictate that these populations be selected as assessment endpoints. Appropriate assessment endpoints in the context of physical disturbance might include diversity of the plant community and populations of wildlife that might be affected by the new arrangement of patches of habitat and forage vegetation (see Figure 9.1). The boundary of a proposed action may determine whether a rare or highly valued plant community would be entirely removed. It is notable that some of the more subtle properties of assessment endpoints for contaminated sites, such as reproductive potential, would not be appropriate for locations where the removal of an entire community is planned.

9.1.2.4 Reference

The assessment of risks from remediation must be performed with respect to a reference condition, which should be chosen in the problem formulation (or equivalent) for the feasibility study. Two alternatives are possible: (1) an uncontaminated and relatively undisturbed reference site or (2) the contaminated site prior to remediation. If baseline risks from the contaminants are balanced against remedial risks, as is suggested in Section 9.2 below, then the conditions resulting from both types of risks should be compared with the reference conditions (e.g., background soils) that are used in the baseline risk assessment (Section 2.7.3).

9.1.3 Exposure Assessment

The feasibility study should include an estimate of exposure of all assessment endpoints to all significant stressors for each remedial alternative. If any new exposure pathways are identified in the conceptual model for the remedial action, a new exposure assessment may be required. An example is the introduction of new chemicals into soil or water, either as reactants in the remedial technology or as degradation products of the initial toxicant. The volatilization of organic contaminants during water evaporation from dredged sediments is another example (Chiarenzelli et al., 1998). If contaminated media are moved (e.g., dredge spoil transported upland for disposal), new assessment endpoints may be appropriate. In the case of dredge spoil, the models used to estimate uptake of contaminants by wildlife foods in soil (Section 3.5.2) may not be appropriate for estimating accumulation of chemicals from disposed sediment (Edwards et al., 1998).

If the remedial technology has the potential to increase the bioavailability of the remaining contaminants, an amendment to the baseline exposure assessment is required. For example, the hyperaccumulation of contaminants by plants in phytoremediation could increase the availability of the contaminants to herbivores. Solvent extraction, if performed *in situ*, could increase the bioavailability of aged organic chemicals or reduce the ability of the soil to support a community of microorganisms that could otherwise degrade the chemical. For example, Inoue and Horikoshi (1991) found that in liquid culture, none of 61 bacteria tested could grow in the presence of organic solvents with values for the log octanol–water partition coefficient (log K_{ow}) less than 3.1, and most could not grow in the presence of a solvent of log K_{ow} less than 4.0. Also, dredging suspends contaminated and anoxic sediments in the water column, increasing exposure of aquatic organisms in the water column to contamination.

Information concerning the release of contaminants or changes in their form due to remediation may be obtained from the results of treatability studies. These are bench-scale or small field trials of proposed technologies using the actual contaminated medium from the site. Another source of information is monitoring conducted at sites where the remedial technology has been previously applied.

The estimated exposure of assessment endpoint receptors to physical stressors (such as soil removal or trampling) may consist of only a description of spatial extent, intensity, frequency, and duration, if known. Exposures to any indirect stressors, such as impacts on habitat, should be considered, even if only qualitative assessment is possible. Only quantities that can be related to effects are used to estimate risks; other exposures are noted as having uncertain impacts in the risk characterization. Because an exposure assessment for a remedial action would be a prospective assessment, actual data reflecting exposure would not be available, and some estimates of exposure could be highly uncertain.

9.1.4 Effects Assessment

If the relationship between exposure and effects is not obvious for the various remedial technologies, a formal effects assessment may be performed. This may include toxicity tests of soil from bench-scale tests of treatment technologies. Of course, toxicity tests are more useful than concentrations of contaminants in soil alone as evidence of risk to most population- or community-level assessment endpoints. Thus, toxicity tests may be thought of as measures of the effectiveness of a remedial technology as well as evidence for new risks. (The assessment of efficacy following the implementation of remedial actions is described in Section 10.4.) Surveys of biota cannot be performed at the site of concern in a prospective risk assessment for a remedial action. However, the spatial extent of the remedial action is known, so a significant level of effects can be defined spatially (Box 9.1). In addition, records of the monitoring of sites following historical uses of the particular technologies or actions may contribute to the effects assessment. Also, habitat suitability models may be used to compare the suitability of wildlife habitat before and after a proposed remedial action or to compare alternative remedial actions (Rand and Newman, 1998). In the United States, habitat evaluation procedure models are available for more than 100 terrestrial and aquatic species which can be used to estimate changes in habitat suitability or abundance of the endpoint species (U.S. Fish and Wildlife Service, 1988). Alternatively, habitat models may be developed *ad hoc*. For chemical stressors, the use of exposure-response models to describe the relationship between contaminant concentrations in environmental media and effects is advisable (Chapter 4).

Typical exposure–response relationships are based on severity rather than duration. The duration of effects has not normally been considered in the regulation of ecological risks (Suter, 1993a). However, the acceptability of the no-action alternative, also known as "natural attenuation" (i.e., natural dilution and degradation of chemicals), depends on the acceptability of the duration of effects. Similarly, the acceptability of engineered remedial alternatives may depend on the time to recovery of the remediated ecosystem (Section 9.1.5). Therefore, the duration of effects is a more important dimension of ecological risk in remedial assessments than in assessments of most purely toxicological risks.

9.1.5 Risk Characterization

As in other risk characterizations (Chapter 6), the available evidence concerning exposure to and effects of each remedial action on associated assessment endpoints should be weighed in the feasibility study. The weight of evidence process has been described in Section 6.5, and the discussion does not need to be repeated here. However, two concepts that merit attention are the spatial context of risk and time required for recovery of assessment endpoints affected by physical disturbance. The concept of recovery is discussed at length because the recovery times of affected

BOX 9.1
The *de Minimus* Effects Level with Respect to Spatial area

In Chapter 2, a 20% criterion was discussed as the upper bound for a *de minimis* level of ecological effects. The criterion was used to refer to a level of effects on a population or other assessment endpoint that was deemed significant by regulatory agencies. There are no known regulatory precedents for establishing a 20% criterion for identifying a significant area affected by physical disturbance, such as that associated with a remedial action. Nonetheless, the 20% rule for severity of effects may be applicable to areal scales of ecological effects by analogy. In particular, the loss of all individuals from 20% of the range of a population can be considered equivalent to loss of 20% of individuals from the entire range of a population (if the issue of relative rates of recovery is temporarily ignored). Therefore, for actions such as dredging, which lead to loss of all members of a population or community within a prescribed area, the 20% criterion may be assumed to apply to spatial area. Capping, dredging, or paving of >20% of the range a population or community would be a potentially significant loss. The effects assessment should reflect this assumption if federal and local regulatory agencies and other risk managers approve.

The use of the 20% severity criterion to signify a *de minimus* level of disturbance to particular areas may be questioned for two reasons. First, the regulatory precedents for the criterion are based on toxic effects that are not readily observed or measured. In contrast, when the 20% criterion is applied to areas physically disturbed rather than areas experiencing toxic effects, the results are likely to be readily observed. Hence, because physical disturbances are more apparent than toxic effects, the 20% criterion may be less acceptable for the former. Second, studies of risk perception indicate that familiar risks are more acceptable than unfamiliar risks (Slovic, 1987). Indeed, McDaniels et al. (1997) used a principal components analysis to identify four factors that characterize perceived ecological risk: ecological and human impact, human benefit, controllability, and knowledge. Therefore, the familiar clearing of 20% of a forest may be more acceptable than unfamiliar toxic effects in 20% of that same forest. Thus, we have simply assumed that in the case of physical disturbance, the observability and risk perception factors described above are negligible or cancel, permitting the use of a 20% criterion for physical disturbance.

populations, communities, or ecosystem processes affect the comparison of risks associated with contaminants and those induced by remedial actions.

9.1.5.1 Spatial Considerations

Several spatial considerations should arise during the characterization of remedial risks in the feasibility study. This spatial context of risk is particularly critical for proposed remedial actions that include a physical disturbance component, such as excavation of soil or dredging of sediments. If the plant community is an assessment endpoint, a relevant question is whether the plants that will be excavated are surrounded by similar vegetation. The severity of the risk is increased if the answer is no. Not only the total area, but the spatial pattern of physical disturbance is an important determinant of risk. If similar habitat is available outside of the area of disturbance, mobile organisms may find refuge at neighboring sites, but habitat is often limiting. The literature of landscape ecology provides substantial information on the sizes and connections of habitat patches necessary to support particular wildlife populations (Forman and Godron, 1986). Reviewers of the EPA "Guidelines for Ecological Risk Assessment" have suggested that the principles of this discipline be more fully incorporated into the guidelines (EPA, 1998), but research and guidance development are needed.

9.1.5.2 Recovery

Although adverse ecological effects may be associated with the physical or chemical disturbance regimes of remediation, many of the effects may correct themselves with time. One measure of the severity of remedial hazards is the time to recovery. Fisher and Woodmansee (1994) list factors that influence the recovery of ecosystems from disturbance: current state, disturbance severity and frequency, successional history, history of disturbance, preferred state, management of the disturbance, and chance. Recolonization time is dependent on the size of the site and the proximity to a recolonization source. Few studies of ecological recovery have been performed in the context of remediation of contaminated sites. Examples of times to recovery from disturbance are presented in Table 9.2, although this is by no means an exhaustive list. For a more thorough review and analysis of recovery of aquatic systems, see Niemi et al. (1990). Niemi and colleagues reviewed 139 publications and extracted recovery times for various aquatic ecological receptors. In 14 of the studies, dredging was a stressor.

One example of ecological recovery from effects of remedial actions may be found in Tamis and de Haes (1995). They present results of research on the recovery of earthworms at 11 sites in the Netherlands where soils were cleaned by either thermal cleaning (heating the soil and burning evaporated contamination, nine sites) or bioremediation (two sites). The ages of the cleaned soil ranged from 1 month to almost 7 years. Grassland was chosen as the reference ecosystem because either grass developed naturally or the soil was seeded with grass at all sites. Reference information from three sources was deemed to provide evidence regarding recovery: (1) biological and physicochemical characteristics of the adjacent soil, (2) physical and chemical properties of the cleaned soil compared with properties of grasslands in the literature, and (3) the literature on colonization and succession of earthworms. Recovery of the climax community in thermally cleaned soils was estimated to occur after 100 years, the number of years for the accumulation and decomposition of organic matter to reach equilibrium. The recovery of population size and species composition took or was estimated to take less than 10 years for each of two biologically cleaned soils and for a thermally cleaned soil to which organic matter was added (Tamis and Udo de Haes, 1995).

TABLE 9.2
Select Studies of Ecological Recovery

Endpoint	Disturbance	Time to Recovery (yr)	Notes	Ref.
Species composition, distribution, and abundance of Collembola	Reclamation of coal overburden in Luscar, Alberta	Unknown — no reference site	Comparable densities of Collembola and Acari in 2-, 4-, and 8-year-old field sites	Parsons and Parkinson, 1986
Earthworm populations (*Allolobophora chlorotica* and *Lumbricus rubellus*)	Open-cast coal mining and reclamation (ploughing and reseeding)	3–15	Community differed from that at control sites; *L. terrestris* especially suppressed following reseeding	Rushton, 1986
Marine benthic fauna community structure	Suction dredging to harvest cockles	0.15		Hall and Harding, 1997
Population of *Hexagenia* (mayfly) equivalent to carrying capacity	Eutrophication and pollution in Lake Erie	48–81	Time to recovery estimated with model	Kolar et al., 1997
Rodent species diversity	Fire in Mediterranean pine forest, northern Israel	<4	Diversity greater than at unburned site; earlier successional stage	Haim et al., 1996
Return of dominant forest rodent species	Fire in Mediterranean pine forest, northern Israel	20	Recovery dependent on recovery of vegetation	Haim et al., 1996
Density of microorganisms	Various in lentic systems	0.01–0.07		Niemi et al., 1990
Density of zooplankton	Various pulse (limited and definable) disturbances in aquatic systems	0.03 to >3.00		Niemi et al., 1990
Presence of climax community of earthworms	Thermal cleaning of soil	100	Estimated time for accumulation and decomposition of organic matter to reach equilibrium (reduced to less than 10 with the addition of organic matter to soil)	Tamis and Udo de Haes (1995)
Presence of climax community of earthworms	Bioremediation	<10	Estimated time of succession	Tamis and Udo de Haes (1995)

Little can be generalized from the examples in Table 9.2, except that certain groups of aquatic organisms are quick to recover in abundance. Recovery of aquatic systems generally proceeds faster than recovery of terrestrial systems. Few studies of the recovery of wildlife from any disturbance exist, probably because changes in wildlife population densities are difficult to detect. The rodent study in Table 9.2 is an exception.

The term *recovery* requires definition in the context of remedial actions. Recovery may be defined as return to an explicit set of conditions prior to the physical disturbances associated with remediation, conditions prior to the contamination event, or conditions at a neighboring, uncontaminated site. Perhaps the remedial goal is for an ecosystem or its components to return to a specified point on the "predisturbance trajectory" (Fisher and Woodmansee, 1994). If colonization of a certain population is the objective, a reference site is not needed. Niemi et al. (1990) provide a list of recovery endpoints for aquatic systems.

1. Recovery to mean size for an individual
2. Recovery to predisturbance density
3. Recovery of species or genera richness to a level of 80% of the original quantity of taxa
4. Recovery to predisturbance biomass
5. First reappearance of species
6. Return to a prestressor population level that is considered to be within normal seasonal fluctuations

The property chosen should be consistent with the property of the ecological receptor that has been selected as the assessment endpoint.

It is notable that the success of some remedial alternatives relies on the inhibition of recovery. In particular, the maintenance of caps requires that deeply rooted vegetation and burrowing mammals be kept off a site (Suter et al., 1993). This maintenance may be accomplished by paving, mowing, or rip-rapping. These site requirements should be considered when remedial alternatives are balanced.

Methods to hasten or stimulate ecological recovery are more prevalent in literature related to environmental restoration of physically disturbed areas than in the toxicity literature. Examples include riffle reconstruction (Kondolf, 1996) or the addition of fertilizer to cleaned soils (Tamis and de Haes, 1995). If recovery is defined as revegetation rather than restoration of native plant communities, the time to recovery can be reduced. For example, a 135-ha area in northeastern Belgium that had been denuded of vegetation by contaminants from a zinc smelter was revegetated (Vangronsveld et al., 1995). Metals were immobilized with beringite and compost, and metal-tolerant commercial cultivars of the grasses *Agrostis capillaris* and *Festuda rubra* were used. After 5 years the species richness was higher on the treated plot, with several perennial forbs having colonized the revegetated land because of the growth of mycorrhizae (Vangronsveld et al., 1996). Whether this result constitutes recovery would depend on the assessment endpoint for plants. Is diversity important for the site, or does the land use necessitate that the site be maintained as a mowed lawn?

9.2 RISK BALANCING

In the United States, the remedial decisions for hazardous waste sites tend to emphasize reduction of certain types of risks to some prescribed level. In such cases, the only decision-support analysis that is required is a cost-effectiveness analysis to determine which option achieves that risk level at the lowest cost. However, it is argued here that the goal of risk managers should be, and often is, broadened to selection of the option that results in least net injury. Indeed, the goal of the remedy selection process in CERCLA is "to select remedies that will be protective of human health and the environment, that will maintain protection over time and that will minimize untreated waste" (EPA, 1990a). Competing goals must be balanced. This balancing is done at multiple levels.

First, the risks associated with the remedial actions must be balanced against the baseline risks from the existing contaminants without remediation. The emphasis in this volume is on ecological risks, in both cases. Conceptually, this balancing of risks to a single assessment endpoint can be considered in terms of the time integral of effects. For example, both a petroleum spill and removal of contaminated surface soil may kill all plants on the site, so remediation is preferable if succession on the exposed subsoil is more rapid than the time required for degradation of the petroleum to nontoxic levels plus succession on the surface soil. The balance is less clear if only a portion of the plants are killed by the petroleum spill. Clearly, the balance of the equation also depends on the type of ecosystem. For the analysis of remedial actions for Lower East Fork Poplar Creek on the Oak Ridge Reservation, relative durations of effects were not considered. The time to recovery of the floodplain ecosystems from remediation (i.e., succession on denuded soil) and from the contamination (i.e., burial of the contaminated soil by alluvial deposition to below the root zone) were judged to be approximately equal. If risks associated with remediation are mitigated (e.g., recovery is facilitated), the mitigated remedial risks should be compared with the baseline risks from contaminants (Sprenger and Charters, 1997).

In addition to time, the spatial extent of the various risks should be considered. If the area disturbed by remedial activities is different from the contaminated area (e.g., if roads or a treatment facility are constructed on the site), the ecological loss of the additional disturbed areas must be considered. If the areas affected by remediation are assumed to correspond to the areas that are contaminated to toxic levels, the area parameters cancel out of the balancing equation.

It is notable that the public may not accept an equal weighting of the risks from contaminants and risks from remediation. In typical communities, risks from contaminants are unfamiliar, involuntary risks, and risks from remediation are familiar and voluntary. (In Oak Ridge, a community with many retired scientists, sometimes the risks from contaminants may be the more familiar.) Starr (1969) concluded that the acceptability of risk from an activity is roughly proportional to the third power of the benefits for that activity. Although his data were not from remedial actions, his equation suggests that the public may often not accept an equal weighing of benefits (reduction of baseline risks) and remedial risks.

Second, risks to different ecological assessment endpoints may need to be balanced. One endpoint, such as earthworm abundance, may be significantly affected by the contaminants in soil, but others, such as plant or soil microarthropod diversity, may not be. Since remedial actions often damage all components of the ecosystem in the affected area, benefits to some endpoints must be balanced against injury to others. For example, aquatic macrophytes may grow quite well in sediments that contain sufficient PCB concentrations to lead to death or reproductive deficits in some fish. Protection of a fraction of the fish community would require removal of the sediments, macrophytes, and habitat for other fish that were not sensitive to or did not accumulate the contaminants. The decision to kill PCB-contaminated fish in a pond on the Oak Ridge Reservation implied that the certain mortality of several tons of fish was not as great an adverse effect as the probable reproductive decrement to individual herons, ospreys, and kingfishers feeding at the pond currently and in the future. Needless to say, this was a value-laden decision, and one that received much public comment. The difficulty of balancing risks to different endpoints increases as the nature of the endpoints and/or stressors diverge.

Third, ecological risks may need to be balanced against human health risks. Remedial actions such as the removal of contaminated soil may reduce risks to humans but increase ecological risks. In such cases, remedial alternatives such as the control of land use until adequate degradation has occurred may be preferable to more ecologically injurious actions.

Historically, the assessment of ecological risks has been given less attention in the remedial process than the assessment of human health risks. As a result, remedial actions for waste sites have not necessarily protected the environment. For the reasons discussed previously, nonhuman organisms are often more highly exposed and more sensitive than humans to environmental contaminants. It is the policy of the EPA and is stated in the National Contingency Plan that both human health and the nonhuman environment are to be protected. There is, however, no guidance from the agency on how to achieve that goal, given that there may be conflicts between the two component goals. One could calculate remedial goal options (contaminant concentrations in soil, sediment, or water — Chapter 8) for both human health and ecological receptors, and remediate to whichever goal is lower. However, factors such as cost and potential risks from remediation would recommend against such a procedure. If the policy of protecting both health and ecological receptors is to be implemented, a mechanism for combining these factors into a common decision structure must be developed. For the reasons below, such balancing is not simple.

9.2.1 Different Risk Metrics

If health and ecological risks could be placed on a common scale, it would be possible to compare their relative magnitudes to determine which should drive the remedial decision making. That is, remediation could mitigate or prevent the largest risk. However, differences in health and ecological assessment endpoints are a barrier to such a common scale. The endpoint for health risk assessments is the health of individual humans. The probability of cancer is the endpoint for a carcinogen, and the threshold for effects is the endpoint for a noncarcinogen. In contrast, ecological risk assessments are concerned with the protection of many populations (e.g., woodcock, tall delphinium), communities (e.g., soil heterotrophs, stream fish), and ecosystems (e.g., forests, streams), as well as various properties of those entities (e.g., production, diversity). Multiple ecological assessment endpoints are usually selected for a single, contaminated site, and the receptors are likely to have different metrics (e.g., density, number of species). As a result, the apples of health risk must be balanced against the oranges, pineapples, and kumquats of ecological risk.

In addition, assessment endpoints related to human health and various ecological receptors have qualitatively different values. That is, no weighting factor or other function may be used to determine a percent reduction in the fecundity of particular small mammal populations that is equivalent to the loss of 5 ha of bottomland hardwood forest, much less to determine the level of either ecological risk that is equivalent to a 10^{-6} human cancer risk. Moreover, the greater conservatism in health risk assessments compared with typical ecological risk assessments makes the comparison difficult.

9.2.2 Land-Use Conflicts

Issues of current and future land use are also relevant to the potential conflict between the protection of human health and the environment. A common assumption in human health risk assessments is that the upper-bound human exposure, which may be associated with an improbable land use, will occur at some time in the future on the contaminated site. However, this conservative assumption may not always be appropriate to identify the best remedial alternative. For example, if the contaminated land is a riparian wetland forest, the risk manager would be ill-advised to select a remedial alternative that results in the destruction of that valuable ecosystem so that a hypothetical future homesteader could safely dike, drain, and occupy it. Similarly, ecological risk assessors are often asked to assess risks to ecosystems that would result if natural succession were allowed to occur on the site. However, if the contaminated site is intended to be mowed or paved for industrial or commercial use, the risk manager should not select a remedial option to protect a hypothetical future forest ecosystem.

The balance of health and ecological risks is not the same at all contaminated sites, primarily because of the different current and potential future land uses at various sites. Further, the balance may be somewhat altered across different areas of a site if it includes various land uses or ecosystem types. As shown in Table 9.3, the possible land uses and habitats can be presented on a scale which extends from areas that are dominated by human uses and in which human health risks are the most significant concern to areas in which human use is not important, ecological value is paramount, and, consequently, ecological risks should be the major concern. The balancing of health and ecological risks is most difficult in areas where both types of risks are potentially significant. These areas would include pastures, hay fields, recreational lakes, and estuaries.

9.2.3 An Approach to Balancing Risks

In the following sections, an approach for creating a common scale of health and ecological risks and a method for applying the approach to make remedial decisions

TABLE 9.3
Relationship of Human to Ecological Dominance of Land Uses and Ecological Habitats

Relationship of Human to Ecological Dominance	Types of Space	
	Ecological Habitats	**Human Land Use**
Human-dominated land use	Pavement and buildings	Industrial
	Bare earth	Commercial
	Closely mowed grasslands	Residential
		Recreational (e.g., ball fields)
Intermediate dominance	Tilled fields	Agricultural
	Meadows, pastures	
	Ecologically important habitats (e.g., forests, streams)	Recreational (e.g., fishing)
Ecologically dominated land use	Habitats of T&E species	Minimal or open space
	Wetlands	(low impact recreation)
	Old-growth forest	

Source: Modified from Suter II, G. W. et al., *Risk Anal.*, 15, 221, 1995. With permission .

are presented. The approach was developed for the mercury-contaminated floodplain of East Fork Poplar Creek in Oak Ridge, TN (Suter et al., 1995). The objective of this approach was to increase the likelihood that acceptable health and ecological risks would result from remedial decisions. It provided a formal methodology for organizing the results of the risk assessment so that the risks that must be balanced are clearly presented together in similar terms.

9.2.3.1 A Common Scale

Although no common metric for health and ecological risks is in use, it is possible to assign the risks to a common classification based on their consequences. Some health and ecological risks are so severe that a remedial action should always be taken (*de manifestis* risks), some are so insignificant that remediation would never be required (*de minimis* risks), and some possibly require remediation, depending on considerations such as costs and competing risks (intermediate risks). This categorical scale has been applied to health risks (Travis et al., 1987; Whipple, 1987; Kocher and Hoffman, 1991), and recently to ecological risks (Suter et al., 1995). The categorization has been used to create a common scale for balancing health and ecological risks.

9.2.3.2 Human Health Categories

The following classification of human health risks is based on accepted U.S. regulatory practice (Suter et al., 1995).

De minimis human health risk
- Excess cancer risk less than or equal to 10^{-6}
- Hazard quotient below 1 for any exposure to a single contaminant or combination of contaminants associated with similar toxicological effects

Intermediate human health risk
- Excess cancer risk between 10^{-4} and 10^{-6}

De manifestis human health risk
- Excess cancer risk greater than or equal to 10^{-4}
- Hazard quotient greater than or equal to 1 for any exposure to a single contaminant or combination of contaminants associated with similar toxicological effects

9.2.3.3 Ecological Categories

The following classification of ecological risks is based on the authors' experience, analysis of regulatory practice, and discussions with relevant regulatory authorities. However, it is for illustrative purposes only, and may not be acceptable at other sites.

De manifestis ecological risk
- Risk to an ecological entity (e.g., threatened and endangered species or wetlands) that has a high level of specific legal protection
- Risk to an ecological component of a site that has extraordinary local value

Intermediate ecological risk
- Ecological risk of magnitude between *de minimis* and *de manifestis* risk

De minimis ecological risk
- Mild, transient, or localized effect(s) on one or more ecological entities that are not highly protected (e.g., loss of a gravel bar plant community, which is naturally destroyed and reformed by each major storm, or increased mutation rate in a local population of a high fecundity, short-life-span species, such as mosquito fish, for which the effects on the population of selective removal of mutation-bearing individuals is lost in the high natural mortality)

De manifestis ecological risks would require remediation unless such an action would conflict with protection of humans from *de manifestis* health risks or some other compelling goal. Intermediate ecological risks are nontrivial but are not so compelling as to require remediation without balancing against costs, health risks, and other considerations. *De minimus* ecological risks are those that would not normally require remediation because they are considered to be trivial.

If a significant level of effects and the spatial concepts discussed in Box 9.1 are added, the definition of *de minimis* ecological risk is revised to a more quantitative form.

De manifestis ecological risk
- Risk to an ecological receptor (individual, population, or community) that has high local value or legal protection. The level of effects (severity, area,

or duration) that would be considered significant would depend on the specifics of the legal protection or the local values

Intermediate ecological risk

- Greater than 20% reduction in the abundance or production of an endpoint population within suitable habitat within a unit area, if the endpoint population does not have special legal or local protection
- Loss of greater than 20% of the species in an endpoint community in a unit area, if the community does not have special legal or local protection
- Loss of greater than 20% of the area of an endpoint community in a unit area, if the community does not have special legal or local protection

De mimimus ecological risk

- Less than 20% decrement in the abundance or production of an endpoint population on suitable habitat within a unit area
- Loss of less than 20% of the species in an endpoint community within a unit area
- Loss of less than 20% of the area of an endpoint community within a unit area
- Loss of more than 20% of a community in a unit area if the community has negligible ecological value (e.g., a baseball field, see Table 9.3) or if the loss is brief because the community is adapted to physical disturbances (e.g., the plant communities of stream gravel bars)
- Threshold concentrations for significant ecotoxic effects (i.e., those that are likely to affect population productivity or abundance) are found in less than 20% of the habitat in a unit area (This last option assumes that the level of effect cannot be determined through biological surveys or estimated from effects models.)

Clearly, other definitions are possible which would be equally defensible. For example, based on professional judgment and experience in the development of the EPA biological criteria for streams, Hughes (1995) has proposed the following categories. Differences of exposed communities from reference communities >50% are "clearly unacceptable," those between 25 and 50% are "marginal," and those <25% are "acceptable."

The definition of *de minimis* ecological risk requires identification of a unit area for which proportional losses of ecological communities are estimated. As discussed in Chapter 2, these units should be defined on the basis of physical, ecological, or land-use discontinuities.

9.2.3.4 Risk Balancing Based on a Common Scale

Three categories of risks—ecological risks from exposure to contaminants, ecological damage from remediation, and human health risks—must be factors in the final remedial decision. The combinations of these sets of risk categories resulted in nine possible ecological risk categories in the case of East Fork Poplar Creek (Table 9.4). Only two health risk categories were possible; risks to human health from remediation were deemed to be negligible, and no intermediate health risk category was

TABLE 9.4
Human Health and Ecological Risk Integration

Ecological Risk		Human Health Risk	
		de minimis	*de manifestis*
Baseline Contaminant Risk	**Remedial Risk**	(HI ≤ 1)	(HI > 1)
de minimis	*de minimis*	NR	R
	Intermediate	NR	RLU
	de manifestis	NR	RLU
Intermediate	*de minimis*	R	R
	Intermediate	B	B
	de manifestis	NR	RLU
de manifestis	*de minimis*	R	R
	Intermediate	B	B
	de manifestis	B	B

Note: For each combination of health and ecological risk categories, a risk-based remedial option is presented
HI = hazard index; R = remediation required; NR = no remediation required; RLU = remediation by land-use controls; B = balancing of risks and nonrisk factors by the risk managers is required

Source: Suter II, G. W. et al., *Risk Anal.*, 15, 221, 1995. With permission.

possible, since the primary human health concern was the noncancer effects of mercury. Therefore, 18 combinations of health and ecological risks were possible (Table 9.3). If the health effects of remediation were considered with the baseline health, baseline ecological, and remedial ecological risks, the full matrix would be $9 \times 9 = 81$ combinations.

Once the matrix of risk categories is defined, the decision about whether or not to choose the remedial action is obvious for most combinations of categories (Table 9.4). For example, if both human health and ecological risks are *de minimis,* remediation is not necessary if the methods and results of the risk assessment are acceptable to the risk managers and the public. For other examples of combinations, it is advisable for decision makers to balance health and ecological risks with cost and other considerations. For example, if the health risk is *de minimis,* but the baseline ecological risks due to contaminants are *de manifestis* and the remedial ecological risks are *de manifestis*, then the risk managers must balance the two conflicting ecological risks, the cost of remediation, public satisfaction, and any other considerations. If the ecological risk from remediation is greater than the ecological risk from contamination, but the health risks are potentially *de manifestis*, then a land-use control remedy is recommended. In the East Fork Poplar Creek case study, health risks are significant only when crops that accumulate mercury are assumed to be grown on the site. Thus, the risk can be eliminated by prohibiting crop production and excluding residential housing from the most contaminated portions of the floodplain.

9.2.3.5 Remedial Units

The recommendations regarding the balancing of risks assume that the land use and associated habitat type are the same across the site, or that they are similar enough that a single risk category (e.g., *de minimus* baseline ecological risk) can be applied to the entire site. The level of aqueous or soil contamination must be sufficiently uniform for a single risk category to be applied to the entire site. Similarly, the proposed remedial alternatives (and zones of impact) should be appropriate for the entire site. Such units should have been defined in the problem formulation (Chapter 2). In the case of the East Fork Poplar Creek operable unit, however, these requirements were violated. Therefore, the operable unit was subdivided into remedial units that met those assumptions.

If ecological and/or health risks are different in current and future land-use scenarios, the decision logic above may need to be carried out independently for each scenario.

9.2.3.6 Summary of Risk Balancing

The selection of remedial alternatives requires a complex balancing of risks: ecological vs. human health risks, baseline contaminant risks vs. remedial risks, and current risks vs. future, land-use-dependent risks (with potentially different levels of conservatism and different levels of confidence). The framework described above may be used by risk assessors to aid risk managers in the balancing process. The framework includes the following components (Suter et al., 1995).

1. Categorization of all risks into *de minimis, de manifestis,* or intermediate
2. Implementation of a clear methodology for recommending no action, remediation, or land-use controls, given particular combinations of ecological, human health, and remedial risk categories for which the relative risks suggest one alternative
3. Specification of combinations of risk categories that do not suggest a particular action and that require the risk manager to balance risks with costs and other considerations
4. Partitioning of heterogeneous sites into remedial units that are essentially uniform with respect to land use, habitat quality, and level of contamination

In addition to a clear presentation of the decision logic, the risk assessor should present enough information for the risk manager to reach an independent judgment. Although the entire risk assessment documentation provides that basis, concise summaries of assessment results are more helpful than the hundreds of pages of data and detailed analysis. A tabular summary of the risk characterization (Section 9.1.1.4) is useful. The table could include the current baseline state, a hypothesized future baseline in which no institutional controls minimize human exposure, conditions during implementation of each proposed remedial alternative, and the conditions of the units following completion of each proposed remedial alternative. The descriptions of toxic effects or physical disturbances should include the estimated time to recovery. Differences in the uncertainties associated with the risks should be noted.

This method was developed for a particular waste site in the absence of national guidance for risk balancing. Its applicability to other sites depends on its acceptability to local risk managers. Any such decision-support system is inevitably too simple to characterize the actual decision-making process. In the case of East Fork Poplar Creek, considerations that were not formally incorporated include public concerns about the loss of the greenbelt and recreation area provided by the floodplain, public concerns about human health risks due to the remediation, public belief that the EPA was not sufficiently willing to consider the bioavailability of the mercury, and the concern of the EPA about setting a precedent for accepting the choice of the no-action alternative where mercury concentrations are high. Although real decisions are more complicated than any decision-support system, such systems can help to clarify inputs to the decision.

9.3 LIFE-CYCLE ASSESSMENT

Life-cycle assessment is a method for determining the relative environmental impacts of alternative products and technologies based on the consequences of the life cycle from extraction of raw materials to disposal of the product following use. It has recently been proposed that life-cycle analysis be used as a basis for choosing among remedial alternatives for contaminated sites (Diamond et al., 1999; Page et al., 1999). Like the method discussed above, it ultimately leads to a qualitative balancing of diverse risks. However, it is more complete in that it systematically identifies and quantifies considerations such as the energy costs of excavating and hauling contaminated soils that would not normally be included in a site assessment.

9.4 COST–BENEFIT ANALYSIS

Cost–benefit analysis adds an additional level of complexity to risk management. It is based on the assumption that the best decision is one that, rather than simply choosing an alternative that reduces risks to an acceptable level or choosing the alternative with the least total risk, ensures that the economic benefits of remediation exceed their cost. Although quantitative cost–benefit analysis is seldom applied to remedial decisions, risk managers make qualitative judgments concerning the cost-effectiveness of proposed remediation. Therefore, it is incumbent upon risk assessors to include in the risk characterization a description of the importance and implications of changes in the condition of the endpoint properties.

9.5 PRIORITIZATION OF SITES FOR REMEDIATION

Once the decision has been made to remediate a site, an immediate question arises: When? If multiple waste sites are in line for remediation, as is the case for multiple DOE facilities, the chronology should reflect technical factors, even if politics is acknowledged as important. One decision-support system that was developed for high-level prioritization of environmental restoration projects by DOE determined that the weight attached to the ecological risk component of risk reduction benefits associated with remediation should be 13% of the total, where other categories included health risk (36%), socioeconomic impact (9.5%), regulatory responsiveness (9.5%), and uncertainty reduction (32%) (Jenni et al., 1995). The weights were elicited from senior managers within the DOE Office of Environmental Restoration.

In the Netherlands, risk assessment is used to determine the priority of remediation of sites with "serious soil contamination" (Swartjes, 1997). "Serious soil contamination" is defined as an unsaturated soil volume of at least 25 m^3 that exceeds the intervention value for soil, a screening criterion based on risks to human health and ecological receptors and processes (Swartjes, 1997). The ecotoxicological component of the intervention value is the hazardous concentration 50 (HC50), the concentration at which 50% of species are assumed to be protected. Thus, the decision to remediate soils is made after the screening assessment and before a more definitive site assessment. However, because only "urgent" cases require cleanup within 20 years (Swartjes, 1997), it is conceivable that some "nonurgent" sites may never reach a high-priority position.

Ecological risk factors that should determine the prioritization of remediation of individual sites include the size of the affected site, the number of assessment endpoints affected, whether the endpoints are legally protected entities, the time until the effect is expected to occur (if no current risk exists), and the time until the effect is estimated to become insignificant, as in natural attenuation. A useful calculation is the integral of effects during the period of time prior to remediation. It is advisable for the prioritization to reflect the logic of the conceptual model. That is, a waste burial ground should be remediated and storm drain inputs should be discontinued prior to the covering of contaminated sediments in a hydrologically connected pond. Otherwise, the clean sediment cap will become contaminated.

A prioritization of projects at a facility, including remedial actions and remedial investigations, may be required for budgeting purposes. In this case, priorities must be selected (and risks for uncharacterized sites must be grossly estimated) without much information.

10 Post-Remedial Assessments

The man who is certain he is right is almost sure to be wrong; and he has the additional misfortune of inevitably remaining so.

—Michael Faraday, quoted in Johnson (1991)

The assessment process does not necessarily end with the completion of a remedial action. If the contaminants are not all removed or destroyed or if the environment is not restored to an acceptable state by the remedial action, it may be necessary to perform additional sampling, analysis, and assessment. These monitoring and assessment activities have two purposes: (1) to ascertain whether additional remedial actions are needed and (2) to estimate the residual contaminant-induced injuries to the environment to determine what restoration activities are needed. Although these purposes are closely related, they are procedurally distinct under U.S. regulations and therefore are discussed separately below. A potential collateral purpose of these post-remedial assessments is to provide a basis for testing the predictions made in the prior assessments. Like any other activity based on scientific principles, the predictions of risk assessments must be tested to prevent the complacency of which Faraday warned.

10.1 REMEDIAL EFFICACY

Remedial actions seldom remove or destroy all contaminants, and some simply control uses of the site. It is advisable to monitor incompletely cleaned environments to establish that the remedial actions were adequate and that unacceptable risks do not persist. The EPA notes that the efficacy of ecological risk reduction is sometimes easier to observe than that of human health risk reduction. The agency recommends that the effectiveness of remedial actions be assessed, following their implementation (Sprenger and Charters, 1997). In many cases, biological monitoring will provide a better indication of risk reduction than simple contaminant monitoring (Fox, 1993). In addition, the extent and rate of ecological recovery following the selection of the no-action alternative should be determined (Sprenger and Charters, 1997). In the United States under CERCLA, such monitoring must be carried out for 5 years, at which time an assessment must be performed.

Requirements for post-remedial monitoring and assessment should be included in the record of decision along with the requirements for remediation. The extent and intensity of this activity should be inversely related to the extent and intensity of the sampling, analysis, and assessment performed prior to remediation. For example, if it was clearly established that the only significant risks prior to remediation were due to consumption of contaminated fish by piscivorous wildlife, monitoring could be limited to the analysis of the contaminants of concern in fish from the areas where risks had been significant. The assessment would document the trends in contaminant levels and associated risks over time and would identify when risks dropped below prescribed threshold levels. In contrast, if an action is taken without an ecological risk assessment or with an inadequate assessment, and if all of the contaminant has not been removed or destroyed, the site should be treated like a newly discovered contaminated site. The process of problem formulation, analysis of exposure and effects, risk characterization, and risk management should be carried out as described in prior sections. Kondolf and Micheli (1995) provide recommendations for the post-project evaluation of stream restoration projects. Some of this guidance, including the relationship of initial restoration objectives and specific evaluation criteria, may be useful for remedial efficacy assessment.

The importance of considering new toxic exposures and effects on new assessment endpoints from remedial measures is illustrated in the tests of the efficacy of remediation in Table 10.1. For many endpoints and actions, soil was more toxic following remediation than preceding it. The adverse effects resulted from (1) increased bioavailability following washing with solvents or (2) alteration of the wetting properties of soil. In most of these studies, the technologies are still in the development stage. The toxicity of environmental media following the use of remedial technologies has been studied only for the past few years. Nonetheless, these cases illustrate the importance of monitoring population parameters and ecological processes following remediation. In addition to the plant and earthworm endpoints described in Table 10.1, toxicity of remedial actions to aquatic and sediment organisms and terrestrial wildlife should be tested in the laboratory, observed in field surveys, or estimated using chemical concentration data.

10.2 RESIDUAL RISKS

Even when remedial actions are rapid and thorough, some injury to the environment is likely to occur when sites have been contaminated by spills or improper waste disposal. Under CERCLA, the responsible party must pay "damages for injury to, destruction of, or loss of natural resources" (CERCLA, §107(a)(4)(c)). The liability is to natural resource trustees, which include agencies of the United States, individual states, and native American tribes. The trustees are required to recover monetary damages for residual injuries, which are injuries that are not eliminated by the remediation. The assessment process for determining those damages is termed Natural Resource Damage Assessment (NRDA). The primary federal trustees are the U.S. Fish and Wildlife Service (FWS) and the National Oceanic and Atmospheric Administration (NOAA), but any landholding agency is also a trustee for the resources on their lands. The compensation may be for either the costs of restoring the injured natural resource services or the diminution of value of the natural resource services, plus the costs of performing the assessment. NRDAs follow a distinct set of regulations, procedures, and terminology (Box 10.1). The following discussion describes the procedure defined in Department of Interior's (DOI) regulations (40 CFR 11). This procedure need not be followed, but NRDA results produced by trustees who follow the procedure have a

TABLE 10.1
Studies of the Efficacy of Remedial Technologies in Protecting Ecological Endpoints

Problem	Remedial Technology	Test for Efficacy	Principal Conclusion	Ref
Soil contaminated with PCBs, other organics, and heavy metals from electrical transformers	Washing with isopropanol	Toxicity to earthworms (*Lumbricus terrestris* and *Eisenia fetida andrei*) and common onion (*Allium cepa*)	Extraction removed 99% of PCBs, with 2 mg/kg remaining; reproductive toxicity to earthworms was not altered; phytotoxicity increased following washing, but when isopropanol was washed away with water, phytotoxicity was reduced	Meier et al., 1997
Soil contaminated with lead to 700–800 mg/kg (also cadmium, zinc, chromium, nickel, silver)	Soil washing/soil leaching process	Toxicity to earthworms (*L. terrestris* and *E. fetida*), lettuce (*Latuca sativa*), and local grasses blue grama and sideoat grama	Following washing, earthworm mortality increased in a mixture of contaminated soil and artificial soil; root elongation of the grasses was reduced; germination and root elongation were reduced in lettuce; the adverse effect on plants was eliminated after soils were rinsed	Chang et al., 1997
Soil contaminated with PAHs	Bioremediation using white-rot fungus, *Phanerochaete chrysosporium*	Toxicity to lettuce, millet, oat, and *Tradescantia* — micronucleus test	Plant tests showed that soil was significantly detoxified following treatment; the micronucleus test showed that soil was less genotoxic than untreated soil	Baud-Grasset et al., 1993
Explosives-contaminated sediment from a lagoon at an army depot in Oregon	Sediment composted with a 70% organic amendment comprising cow and chicken manure, sawdust, alfalfa, and potato waste	Toxicity to lettuce (*L. sativa*), radish (*Raphanus sativus*), and soybean (*Glycine max*), earthworms (*E. foetida*), and isopods (*Armedillidium vulgare*) in mesocosms; toxicity to cabbage (*Brassica oleracea*), clover (*Trifolium repens*), and *Arabidopsis thaliana*.	Germination of lettuce, clover, and *Arabidopsis thaliana* was reduced in the contaminated compost, relative to reference compost; soybean leaves were chlorotic, and plant biomass was reduced compared with the reference compost; few adverse effects to soil invertebrates were observed	Gunderson et al., 1997

continued

TABLE 10.1 (continued)
Studies of the Efficacy of Remedial Technologies in Protecting Ecological Endpoints

Problem	Remedial Technology	Test for Efficacy	Principal Conclusion	Ref
Agricultural topsoil contaminated with oil	Bioremediation of soil to total extractable hydrocarbon level of 2% (from 4%)	Toxicity to barley (*Hordeum vulgare*)	The barley yield was not improved following bioremediation; shoot dry mass decreased slightly following bioremediation; the lack of improvement was attributed to poor soil water sorption because of the remaining hydrocarbons	Li et al., 1997
Soil denuded of vegetation by contaminants from zinc smelter in Belgium	Metals immobilized with beringite and compost	Toxicity to *Phaseolus vulgaris*	Growth inhibition to seedlings in treated soil was not observed	Vangronsveld et al., 1995
Soil contaminated with uranium	Treatment with carbonate and bicarbonate or citric acid	Toxicity to oats, radish	Treatment with sodium carbonate or ammonium carbonate reduced the seed germination fraction for all plants (unless soil was washed with $CaCl_2$); germination of radish and oats in soils treated with citric acid and $CaCl_2$ were reduced; treated soils were resistant to wetting and had a concrete-like surface	Edwards, 1994

"rebuttable presumption." That is, the results are legally assumed to be correct unless the responsible party can prove otherwise. For a guide to current NRDA regulatory documents and guidance, visit http://www.doi.gov/oepc/frlist.html. The currently ongoing NRDA for the Fox River and Green Bay, WI can be followed at http://www.fws.gov/r3pao/nrda/. Because of the high profile of this site, including effects of PCBs on birds, it should be an influential case.

BOX 10.1

Key Terms in Natural Resource Damage Assessment

Natural resources: "Land, fish, wildlife, biota, air, water, groundwater, drinking water supplies and other such resources. . . ." The resources must be public property.
Injury: "A measurable adverse change, either long- or short-term, in the chemical or physical quality or the viability of a natural resource resulting either directly or indirectly from exposure to a . . . release of a hazardous substance. . . ."
Services: "Physical or biological functions performed by natural resources, including human uses of those functions."
Damages: Money paid to a trustee in compensation for the injuries to natural resources.
Restoration: Return of the natural resource services to a baseline condition.
Replacement/Acquisition of the Equivalent: Substitution of another resource that provides equivalent levels of services for the injured resource.
Source: DOI (1986).

The NRDA procedures have been developed to encompass assessments performed under the Clean Water Act (1977 Amendments), the Outer Continental Shelf Act (1978 Amendments), and the Oil Pollution Act of 1990 as well as CERCLA. The provisions of NRDA differ under the other acts in that the injured site would not have been subject to a remedial investigation and feasibility study but may have been subject to prespill planning to predict and minimize injuries (Deis and French, 1998). Ideally, the needs of the NRDA would be anticipated in the planning of a remedial investigation so that the costs of sampling, analysis, and assessment could be minimized. In addition, it would be advisable to consider natural resource damages when planning the remedial action so that the total cost of remediating the contaminated site and the natural resource damages could be minimized. However, this integration seldom occurs in practice, because different agencies are involved. The EPA is required to notify natural resource trustees of CERCLA sites that potentially involve natural resource injuries, and collaboration between the EPA and trustees could and should continue (Office of Emergency and Remedial Response, 1992d). The responsible parties should want to integrate the two assessments, because they bear the cost for both remediation and restoration. However, in the authors' experience, responsible parties do not want to think about NRDA during the remedial investigation.

There are two distinct NRDA procedures. Type A assessments use simulation models to characterize injuries and estimate damages for spills of oil and other hazardous materials in coastal marine waters and in the Great Lakes (French et al., 1994). The trustee's assessors must paramaterize the appropriate model with characteristics of the receiving environment, conditions at the time of the spill, and characteristics of the spilled material. The Type A model then generates estimates of injuries and associated damages such as loss of commercial and recreational fisheries and loss of recreational use of beaches. Regulations for performing a Type A NRDA were published in DOI (1996). Type B assessments estimate injuries on a site-specific basis using methods similar to a remedial investigation and calculate damages using local values. The Type B method, which is far more generally useful, is discussed below.

10.2.1 Preassessment Screen Phase

The preassessment screen phase is equivalent to the scoping and screening steps in a remedial investigation. Its purpose is to determine whether to proceed with a full NRDA. The criteria for proceeding to full NRDA are as follows:

- Release of a hazardous substance has occurred
- Natural resources for which a federal or state agency or native American tribe may assert trusteeship under CERCLA may have been adversely affected by the release
- The quantity or concentration of the released substance is sufficient for the potential injury of those natural resources
- Data to pursue an assessment are readily available or are likely to be obtained at a reasonable cost
- Response actions, if any, do not or will not sufficiently remedy the injury to natural resources without further action (i.e., residual injury)

It is intended that the preassessment screen be carried out rapidly with existing information, and that any sampling follow the screening phase. However, sampling may be performed before completion of the preassessment screen if necessary data or materials would otherwise be lost. Such sampling is likely to be needed for spills or CERCLA emergency actions, but is unlikely to be needed for typical CERCLA NRDAs. Where an RI/FS has been performed, the ecological risk assessments in those documents should provide a more-than-adequate basis for a preassessment screen.

Another part of the preassessment process is notification of trustees. In many cases, there are multiple natural resource trustees (e.g., a federal land manager, the U.S. Fish and Wildlife Service for migratory birds and threatened or endangered species, a state agency managing nonmigratory species, and a tribe with treaty rights to fisheries). In such cases, a lead trustee should be designated to conduct the NRDA, while cotrustees are consulted in planning the assessment and other important decisions. The lead trustee appoints an "authorized official" with authority to make decisions on behalf of the agency, including the decision whether to proceed with the NRDA. In a few major NRDAs (e.g., the *Exxon Valdez* case), agreement on a lead trustee is difficult, so a trustee council manages the assessment.

10.2.2 Assessment Plan Phase

The assessment plan for an NRDA is equivalent to the analysis plan for a remedial investigation. The discussion of analysis plans in Section 2.7 is applicable here. The DOI procedure specifies that the plan should ensure that the NRDA is performed in a well-organized and systematic manner, and that the methods selected can be conducted at reasonable cost. The emphasis on cost of assessment is based on a concern that trustees not use the NRDA process as a means of supporting staff or acquiring equipment at the expense of the responsible party. The cost of the assessment should not be a large fraction of the expected damages. However, in cases with large natural resource damages, such as the *Exxon Valdez* oil spill, the assessment may be quite extensive, intensive, and expensive.

10.2.3 Assessment Phase

The assessment phase is equivalent to the analyses of exposure and effects and the risk characterization in a conventional ecological risk assessment. It consists of three steps: injury determination, quantification of effects on services, and damage determination.

10.2.3.1 Injury Determination Phase

The injury determination phase must demonstrate the occurrence of adverse effects on a property of a natural resource that constitutes a service and the causal relationship of those effects to the release. The identification of effects that constitute a loss of services is equivalent to the selection of assessment endpoints in a remedial investigation (Section 2.5), but is restricted by the need to assess damages. That is, NRDAs do not simply estimate effects on societally or ecologically significant properties. They must estimate effects on properties that constitute services to some component of society that can be associated with monetary values. In some cases, the estimation is straightforward. For example, if the contaminant levels exceed Safe Drinking Water Act standards, the inability to use that water for drinking is an injury that has clearly associated costs. However, the term *services* has been broadly interpreted to include nonuse values along with the more conventional market and nonmarket use values. Definitions of injuries contained in the NRDA regulations are presented in Box 10.2. Note that this is not a list of all compensable natural resource injuries. However, trustees who select injuries from this list maintain their rebuttable presumption. The process is complicated by the concept of services of one resource to another. For example, water provides a service to the fish community by providing habitat, and the fish provide recreation, food, and nonuse values.

To prove that a resource has been injured, the NRDA must confirm the potential exposures and effects identified in the preassessment screen and must demonstrate that the injury is associated with the subject release. Exposure may be confirmed by characterizing the release and modeling the transport and fate processes. More commonly, exposure is determined by measuring contaminant concentrations in biota or abiotic media. Confirmation of effects is most commonly performed by directly measuring reductions in the quality of the resource, but less direct measures, including toxicity tests, may also be used.

BOX 10.2

Natural Resource Injuries Specified in 43 CFR 10.62

Surface water resources

- Contaminants in excess of Safe Drinking Water Act criteria or standards
- Contaminants in excess of Clean Water Act criteria or standards
- Contaminant concentrations sufficient to cause bed, bank, or shoreline sediments to exhibit characteristics identified under the Resource Conservation and Recovery Act
- Contaminant concentrations sufficient to injure other resources

Groundwater resources

- Contaminants in excess of Safe Drinking Water Act criteria or standards
- Contaminants in excess of Clean Water Act criteria or standards
- Contaminant concentrations sufficient to injure other resources

Air resources

- Emissions in excess of National Environmental Standards for Hazardous Air Pollutants
- Emissions sufficient to injure other resources

Geologic resources

- Contaminant concentrations sufficient to cause bed, bank, or shoreline sediments to exhibit characteristics identified under the Resource Conservation and Recovery Act
- Soil pH below 4.0 or above 8.5
- Sodium absorption ratio above 0.176
- Decreased water-holding capacity
- Impedance of microbial respiration
- Inhibition of carbon mineralization
- Restricted mineral access, development, or use
- Physical/chemical changes in unsaturated zone groundwater
- Toxicity to soil invertebrates
- Phytotoxicity
- Contaminant concentrations sufficient to injure other resources

Biological resources

- Adverse changes in viability: death, disease, behavioral abnormalities, cancer, mutations, physiological (including reproductive) malfunctions, and physical deformations
- Exceedence of Food, Drug, and Cosmetics Act levels in edible portions
- Exceedence of state health agency directives in edible portions

Selection of measures of effects for many injuries is straightforward. If the measured contaminant concentrations exceed criteria or standards, or if they change the properties of a sediment or soil so that it constitutes a hazardous waste under RCRA, then an injury has occurred (Box 10.2). However, appropriate measures of effects on biological resources are less obvious. The regulations provide the following acceptance criteria for those measures.

- The measured response often results from exposure to the contaminants (i.e., is well documented in the literature)
- The response has been demonstrated in free-ranging organisms
- The response can be produced by exposure to the contaminant in controlled experiments
- The response measurement is practical and produces scientifically valid results

The demonstration that the injury is associated with the release requires (1) the identification of a pathway from the source to the receptor and (2) a qualitative or quantitative analysis of the transport and fate processes sufficient to indicate that the exposure that is causing the injury could result from the release. A conceptual model is sufficient for the first step and either a simple transport and exposure model or measurements of contaminant concentrations in intermediate media are sufficient for the second. For example, if the confirmation of exposure is based on measurement of the contaminant in fish tissues, a pathway from the release to the fish must be identified (e.g., leaching of contaminants in soil to groundwater, input to a gaining reach of a stream, and uptake by fish), and its magnitude must be shown to be sufficient to result in the fish tissue levels (e.g., by measuring concentrations in groundwater and stream water and showing that they are sufficient to cause the concentrations in fish).

10.2.3.2 Quantification Phase

In effect, the DOI procedure requires that, having completed a qualitative assessment of injuries in the injury determination phase, the assessors then perform a quantitative assessment. The procedure for quantification calls for identification of the services provided by the injured resource, estimation of the baseline service level (i.e., services provided in the absence of the release and contamination), and quantification of the reduction in services due to the release.

Factors to be considered in the selection of resource services and measurement methods according to the DOI include

- The degree to which the resource is affected
- The degree to which the resource may represent a range of related resources and their services
- The consistency of the method of measuring services with the requirements of the economic methodology
- Technical feasibility of the methods
- Preliminary estimates of the services

As in the injury determination, the DOI procedure provides considerations for selection of biological measures to be used in the quantification step.

- Emphasize population and ecosystem levels
- Choose representative species or habitats
- Choose sensitive species, habitats, or ecosystems that provide especially significant services

- Focus on injured resources
- Provide data that can be interpreted in terms of services

Quantification of reduction in services requires quantification of a baseline state. The baseline data may come from existing sources, including prior environmental assessments, scientific literature, databases, and other scientific studies. In the absence of adequate existing data, measurements should be taken from reference areas which are comparable to the affected area but are not exposed to the release. Shaw and Bader (1996) argue that the greatest problem with NRDA is its requirement for baseline data. For example, in Prince William Sound, baseline data prior to the *Exxon Valdez* oil spill were sparse, and some data reflected conditions resulting from a major earthquake in 1964 (Shaw and Bader, 1996). Disturbances that could have influenced apparent effects include winter storms, changes in salinity, changes in suspended sediment, changes in otter populations due to the sediment loads or due to the Marine Mammal Protection Act, and releases of salmon from hatcheries. Shaw and Bader (1996) suggest that the lack of baseline data and high natural variability may have contributed to the trustees' willingness to settle the damage claims. In the CERCLA context, managers of large sites, especially those managed by government entities, may be expected to have more baseline data than managers of small, privately owned sites.

Factors to be considered in the quantification of reduction in services include

- The total area, volume, or numbers of the resource affected
- The degree to which the resource is affected
- The ability of the resource to recover
- The proportion of the available resource affected
- The services normally provided by the resource that have been reduced as a result of the release

The choice of baseline can have a large effect on the estimate of injury. For example, the NRDA for the Blackbird Mine focused on the loss of spawning salmon in the creek due to metal contamination beginning in the early 1900s (Renner, 1998). The population prior to the mine of 2000 to 3000 salmon might have been used as a baseline, but that would not reflect the fact that salmon runs have been diminished by dam construction and commercial fishing since that time. Instead, a baseline of 200 salmon was established, based on current runs in similar streams in the region.

10.2.3.3 Damage Determination

The natural resource damages include the cost of restoring the resources, the diminution of the value of resource services prior to restoration of those services, and the cost of performing the assessment (Figure 10.1). Restoration costs and assessment costs may be controversial, if the responsible party believes that the trustee is performing a gold-plated assessment or restoration, but the quantification of these components of the damages is not controversial. The difficulties arise in the quantification of the diminution of the value of services. Those values may

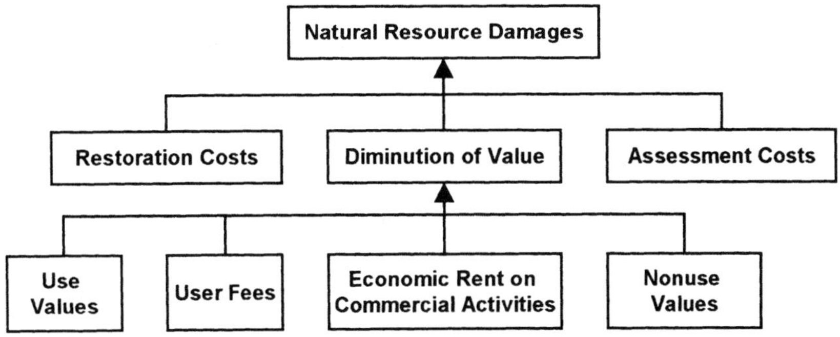

FIGURE 10.1 Components of natural resource damages.

include use values, nonuse values, user fees, and economic rent on commercial activities (Figure 10.1).

Use values are consumer surpluses associated with uses of natural resources, including consumptive uses such as wood harvesting, fishing, and hunting, and nonconsumptive uses such as boating, swimming, and hiking. Use values include not only market values but also travel costs (e.g., how much more does it cost to travel to an uncontaminated beach), hedonic pricing (e.g., how much more do people pay for a house that does not adjoin a contaminated site), and contingent valuation (a survey technique that asks how much one would be willing to pay to prevent loss of a resource). Nonuse values are consumer surpluses that people may have for resources that they are not using. They include willingness to pay for the option to use the resource in the future and willingness to pay to preserve the resource, even though they have no intention of using it.

Estimates of lost-use values may be based on behavior and therefore are reasonably reliable. Even when contingent valuation is used to estimate use values, the reasonableness of the results can be checked against the actual amounts spent on vacations, etc. Because there is no market for nonuse values and no behavior is involved in not using a resource, only survey techniques can be used for measuring nonuse values. The only accepted survey method is contingent valuation, and contingent valuation has been legally accepted for determining natural resource damages (NOAA Panel on Contingent Valuation, 1993). However, the method is highly controversial. Studies have shown that it is unreliable as an estimator of how much people will actually pay for preservation of a resource, and results are sensitive to the context and wording of the questions.

10.2.4 Post-Assessment Phase

The post-assessment phase consists of documenting the results of the assessment, creating an account for the damages awarded, developing a restoration plan, and performing the restoration. Restoration may include additional removal or destruction of contaminants beyond the remediation performed under the CERCLA record of decision. Alternatively, restoration may include a variety of other activities, such as rehabilitation of the resources, replacement of the resources, or acquisition of equivalent resources. Examples include stocking fish, replanting vegetation, restoring or improving habitat structure, creating new habitat, or purchasing land to be set aside as parks or refuges.

10.2.5 Alternative Approaches

Conventional NRDA requires that the assessors quantify the magnitude and extent of the loss of natural resource services and convert that loss of services to an amount of monetary damages. Because the damages can amount to many millions of dollars, the methods used by the trustees are often challenged by the responsible parties, leading to protracted legal battles and delays in restoration. In addition, the NRDA process has focused on monetary damages rather than the ultimate goal of restoring natural resource services. In some cases such as the New Bedford Harbor, MA, NRDA, trustees have been awarded large damages without knowing how they would restore the lost services of the natural resources.

An alternative approach is to skip the conversion of injuries to damages and go directly from quantifying injuries to developing a restoration plan. This approach has the obvious advantage of eliminating economics as a field of contention. In addition, it is easier to reach agreement about what constitutes a feasible restoration than about the abstract concepts of injured services and consumer surpluses for nonuse values. In particular, restoration often involves the relatively straightforward problem of finding a way to physically restore, create, or purchase an ecosystem equivalent to the baseline state of the injured ecosystem. If the restoration plan is developed jointly by the trustee and the responsible party, it should be possible to reach agreement relatively rapidly. If, as in the case of the Blackbird Mine, the restoration is performed by the responsible party, even the payment of damages can be eliminated (Renner, 1998). NOAA (1996a) has included provisions for collaboratively moving toward restoration in its regulations for NRDA under the Oil Pollution Act, and has developed guidance for identifying equivalent habitats (NOAA, 1996b). Reportedly, the DOI is considering similar provisions in revised regulations for NRDA under CERCLA (Renner, 1998).

10.2.6 Conclusions

The relationship of the CERCLA RI/FS risk assessments to the CERCLA NRDAs has been slight to nonexistent. Logically, there should be a single process of determining the risks to the natural environment posed by contaminants and selection of a remedial action that not only removes or destroys contaminants that pose significant risks, but also restores the quality of the environment, including the services that the environment performs for human society. As discussed in the preceding chapter, the remediation performed following the RI/FS may actually result in a net degradation of the environment. However, the NRDA regulations do not allow for restoration of the loss of natural resource services due to the prior remediation. Damages may be collected only for injuries due to contamination. Because of regulatory obstacles, the integration of the risk assessment and remediation with restoration is seldom possible in the United States. However, it would be in the interest of responsible parties to reduce redundancy and inefficiency by introducing the idea of restoring the environment to an acceptable state from the beginning rather than simply decontaminating and then trying to restore the environment. The trend to skipping the collection of damages and going directly from estimation of injury to compensatory restoration (Section 10.2.5) suggests that the NRDA system may be evolving toward greater rationality and efficiency.

The relationship of NRDA to risk assessment in general is unclear. The NRDA procedure was published in 1986, the same year as the first ecological risk assessment framework (Barnthouse and Suter, 1986; DOI, 1986). As a result, the structure of the NRDA procedure is quite different from the ecological risk framework, as discussed above. However, the two are not incompatible, and three of the phases (preassessment screen, injury determination, and quantification) could be thought of as tiered assessments with increasingly rigorous analysis. NRDAs might benefit from the logic of the ecological risk framework and from the ability to adopt risk assessments performed for the remedial investigation, but the legal advantage of the rebuttable presumption makes deviation from the DOI procedure undesirable. In addition, ecological risk assessment has not focused on the monetary value of the endpoint entities and processes as has NRDA. However, as cost–benefit analysis becomes an increasingly important adjunct to risk assessment, NRDA and ecological risk assessment practices may converge.

The NRDA process is particularly problematical for federal agencies that have contaminated sites for which they are trustees. For example, the DOE activities have contaminated the Oak Ridge Reservation, making it a responsible party under CERCLA, but it is also a trustee for the natural resources of that reservation. In theory, the DOE could seek damages from itself to restore lost natural resource services. The situation is further complicated by the facts that (1) federal agencies cannot sue each other for damages; and (2) the federal facility agreements that bind the EPA, the states, and the responsible federal agencies together in CERCLA remedial actions limit the abilities of the states to sue; but (3) if trustees do not collect damages for natural resource injuries, they can be sued by third parties such as environmental advocacy groups. The DOE has developed guidance for integrating the RI/FS and NRDA processes (Office of Environmental Guidance, 1991, 1993), but such integration is not occurring at DOE sites.

The obvious lesson of the NRDA process for countries other than the United States is to avoid the development of overly complex and legalistic processes. Rather, environmental remediation should focus from the beginning on carrying out an efficient process of restoring, as far as possible, the quality of the contaminated environment.

Appendix

Biota Sampling and Survey Methods

The biota of contaminated and reference sites may be sampled to obtain tissues for residue or biomarker analysis (Section 3.1.2) and may be sampled or surveyed for estimation of effects (Section 4.3). Methods for sampling or surveying biota that are potentially applicable for residue analyses at contaminated sites are briefly discussed below (Sections A.1 to A.11). Methods for performing habitat surveys as a means of determining what wildlife species should be present are presented in Section A.12. Sampling considerations, including selection of tissues to collect, appropriate methods of killing collected organisms, and health concerns for sampling personnel, are presented in Section A.13.

A.1 FISHES

Sampling techniques for fish include electrofishing, nets, and traps. Selection of the appropriate method depends on the species of interest and the type of aquatic system being sampled.

In electrofishing, an electric current, supplied by a gasoline- or battery-powered generator to a set of probes placed in the water, is employed to stun fish which are then captured with a net. The advantages of electrofishing include that it is effective for both juveniles and adults of most species and for sampling structurally complex habitats, and it efficiently samples large areas in a relatively limited time while capturing a large percentage of individuals within an area. Numerous studies indicate that under proper conditions, electrofishing can be the most effective sampling technique (Jacobs and Swink, 1982; Wiley and Tsai, 1983; Layher and Maughan, 1984). Disadvantages include potential mortality (which may be a significant issue if sampling is repeated or if highly valued species are present); low efficacy for benthic or deep-water species, for very low- or high-conductivity water, and for turbid water; and potential hazards to users. Additional information on electrofishing can be found in Hartley (1980) and Reynolds (1983).

A wide variety of nets and traps are used to sample fish populations or communities. Two basic types exist, nets that snag or entangle fish and traps or net arrangements that provide a holding area into which fish are enticed. The most common entanglement nets are gill nets and trammel nets that use an open mesh through which fish attempt to swim. As the fish attempts to pass through, gill covers or fins become snagged on the fine filament netting. Gill nets are generally more effective in turbid water and areas without snags (Hubert, 1983) and are effective for sampling deep areas not accessible by other techniques. Gill nets are also highly effective for a variety of larger fish sizes (depending on mesh size used), and for fast-swimming or schooling species. Disadvantages include potential injury or mortality of snagged fish; the limited range of fish sizes sampled by any one gill net mesh size; the high rate of capture of nontarget species, resulting in an increase in sampling time and total mortality; low success for fish species with low mobility (e.g., sunfish); and highly variable results. Further details are given in Hartley (1980), Hamley (1980), and Hubert (1983).

Stationary fish traps include fyke nets, hoop nets, trap nets, and pot gear (e.g., slat baskets and minnow traps). All of these devices work by allowing the movement of the fish to take them through a small opening into a larger holding area. Stationary traps are available in small (minnow traps) to large (fyke nets) sizes, allowing multiple species and life stages to be sampled. Because fish remain alive while in the trap, the traps do not need to be checked as frequently as entanglement nets. Stationary traps are effective for cover-seeking species (e.g., sunfish) or benthic species (e.g., catfish). Disadvantages of these traps include high variance in efficiency across species and susceptibility of catch rates to changes in temperature and turbidity. The larger fyke, trap, and hoop nets are most effective in reservoirs, ponds, lakes, and river backwaters. Pot gear and smaller hoop nets can be more effective in smaller streams or faster water. In both cases, traps can be combined with weirs or directional structures that channel fish into areas where the traps are deployed. Additional discussions can be found in Craig (1980) and Hubert (1983).

A.2 PERIPHYTON

Sampling techniques for periphyton entail scraping, coring, or suctioning the periphyton from the substrate. The substrate may be natural or artificial. Method details are presented in APHA (1999).

Periphyton samples are generally collected from hard substrates, for which most of the available techniques are appropriate and relatively straightforward. Periphyton are less commonly collected from soft sediments, because this is more difficult and time-consuming than sampling hard substrates (Warren-Hicks et al., 1989). Soft substrates are collected via suction and then the individual algae must be sorted from the sediment material for identification and quantification.

Artificial substrates are placed in the water and the periphyton are allowed to colonize the substrate. Typically, the samples are collected after 2 to 4 weeks, although a longer colonization period may be needed in nutrient-limited systems (Rosen, 1995). A common and widely accepted artificial substrate is frosted glass slides. The slides are held in a frame that can be suspended in the water at a given depth (APHA, 1999). Other commonly used materials include ceramic tiles, plastic strips, and granite slabs (Warren-Hicks et al., 1989). The periphyton are removed from a measured area of the substrate by scraping, coring, or suctioning and are preserved for analysis.

Natural substrates are generally less uniform in surface texture than are artificial substrates, and the periphyton are generally more patchily distributed (Rosen, 1995). If a sample of a specified area is to be removed from a rough substrate, the sampler should be designed to fit snugly against the surface. This is generally accomplished by using samplers that have neoprene rubber seals around the edge of the sampler. Alternatively, the periphyton can be scraped or brushed from the entire piece of natural substrate, provided the sample units are suitably small. In this case the surface area of each sampling unit of natural substrate must be measured, which can be accomplished using aluminum foil following the procedures outlined by Coler et al. (1989).

A.3 PLANKTON

Sampling equipment for phytoplankton and zooplankton include closing samplers, traps, pumps, and nets (APHA, 1999; Office of Emergency and Remedial Response, 1994b; Warren-Hicks et al., 1989). Samples may be collected from discrete depths or be integrated over a range of depths or horizontal distances. Method selection depends on the target organisms, target depths, and desired sample quality.

Discrete-depth samplers include closing tubes or bottles, traps, and pumps. Closing tubes (e.g., Van Dorn and Kemmerer models) are quantitative for all sizes of plankton, including nanoplankton and ultraplankton (Warren-Hicks et al., 1989). The tube (or bottle) is lowered to the desired depth and closed via a weighted messenger. Multiple discrete depths can be sampled simultaneously by hanging multiple samplers in series.

Trap samplers operate on the same principle as the closing samplers, but are generally much larger (10 to 30 liters). The larger volume helps ensure that less common species and agile zooplankton are collected (Warren-Hicks et al., 1989; APHA, 1999). However, their large size also makes them ungainly to operate.

Pump samplers consist of a weighted sampling hose that is lowered to a selected depth and a submerged or boat-mounted pump. Pumps can be motorized or manual, and common types include diaphragm pumps, peristaltic pumps, and centrifugal pumps. Sample volume can be determined by using a receptacle of known volume or a flow meter, thus allowing the operator to increase or decrease the sample size easily depending on the apparent organism density at the time of sampling (Warren-Hicks et al., 1989). Volume can be measured accurately, resulting in quantitative samples of most plankton. The exception is agile zooplankton, which may be able to avoid the pump head (APHA, 1999). Disadvantages of pumps include the large size of the typical pump, high cost, and damage to the organisms.

Integrated-depth samplers include pumps, depth-integrating column samplers, and nets. Pumps are operated as for discrete-depth samples, except that the pump head is moved through the water column at a specified rate. Depth-integrated column samplers are long closing samplers used in shallow water. They collect a quantitative sample, but their ungainly size (to several meters in length) makes them difficult operate (Warren-Hicks et al., 1989). Towed nets provide quantitative samples of zooplankton. Net samples are only qualitative for phytoplankton, because the mesh size (e.g., 60 to 80 μm) is too large for nanoplankton and ultraplankton (Warren-Hicks et al., 1989). Net samplers can be towed vertically or horizontally, and specific depths or distances can be sampled by using closing net samplers (e.g., Birge closing net).

A.4 BENTHIC INVERTEBRATES

Many techniques are suitable for the collection of benthic macroinvertebrates for exposure evaluation. These methods include grab-and-core samplers for standing waters and kick sampling, Surber samplers for running water, and artificial substrates (Murkin et al., 1994). Exposure of benthic invertebrates may also be evaluated using *in situ* exposure of organisms, particularly bivalve mollusks, maintained in a holding device.

Grab samplers such as the Ekman, Petersen, Ponar, and Smith-McIntyre samplers may be used to collect organisms from deep-water habitats. These devices engulf a portion of sediment (and its associated organisms), which is then hauled to the surface for processing. Organisms are separated from the sample material by washing the sediment in a box screen. Grab samplers are generally easy to use and are suitable for a variety of water depths. Depth of sediment penetration may vary with sediment type, and rocks or other obstructions may prevent complete closure, resulting in partial sample loss. Because grab samplers tend to produce large samples, processing effort may be considerable (Murkin et al., 1994). Isom (1978) reviewed several types of grab samplers, their specifications, the type of substrate each was designed for, and advantages and disadvantages associated with each type. Standard methods for the collection of benthic invertebrates using various types of grab samplers are also presented in ASTM (1999).

Core samplers may be employed in both shallow and deep water and consist of a metal or plastic tube which is inserted into the substrate. When the tube is removed, samples of both the sediment and organisms are obtained (Murkin et al., 1994). The samples are then washed in a sieve and the organisms are removed from the remaining sample debris. Core samplers are inappropriate for loose or unconsolidated sediment, sand, or gravel (Murkin et al., 1994). Additional information on core sampling can be found in Smock et al. (1992) and Williams and Hynes (1973).

Kick sampling is a simple method used in running waters. A net is placed against the streambed, and the substrate upstream of the mouth of the net is agitated for a defined time period to suspend the organisms, which are then washed into the net by the current (Murkin et al., 1994). While this method is easy, the exact area sampled is undefined, and therefore it is unsuitable in instances when quantitative samples are needed.

When quantitative samples from running water are needed, Surber samplers should be used. Surber samplers consist of a frame with an attached net. The frame is placed on the streambed, the substrate within the frame is disturbed, and rocks and other debris are rubbed to dislodge invertebrates. Water current carries invertebrates into the sampling net (Murkin et al., 1994). Standard methods for the collection of benthic invertebrates using Surber and related samplers are presented in ASTM (1999).

Artificial substrates do not provide estimates of actual benthic community properties but can provide quantitative estimates of artificial community metrics relative to artificial substrates in reference streams. The most common artificial substrate is the Hester–Dendy multiple-plate sampler (APHA, 1999), or modified versions thereof. These samplers have a known surface area (generally about a square foot) consisting of tempered hardboard plates and spacers mounted on an eyebolt creating multiple parallel surfaces separated by spaces of one, two, and three spacer thicknesses. Replicate samplers are deployed at each sampling location. Care is taken to ensure that the samplers are completely submerged and oriented with the plates perpendicular to the current. Artificial substrates are selective of certain species and do not represent rare species or the actual taxa richness of a system. However, they are relatively quick and easy to use, provide standardized and repeatable results, and are often recommended or accepted by regulatory agencies (DeShon, 1995).

The basket sampler is a variant of the artificial substrate sampler (APHA, 1999). It consists of a wire basket filled with rocks or rocklike material. It is deployed in the same manner as multiplate samplers. However, rocks similar to those found in-stream can be used as the substrate, possibly reducing the bias associated with the artificial materials. Standard materials (e.g., limestone rocks) eliminate much of this advantage over multiplate samplers. The surface area of all nonuniform substrates must be measured, which adds effort and some uncertainty to the sampling process.

D-framed or rectangular nets can be used for kick, sweep, or jab sampling (Barbour et al., 1997). The major advantage of these nets is that all habitats can be sampled relatively easily. But the results are, at best, semiquantitative. Qualitative samples require little or no consideration of effort or distance sampled. Semiquantitative samples are generated when a standard distance or duration of effort is used for all sites (Barbour et al., 1997).

Contaminant exposure of benthic invertebrates may also be evaluated through *in situ* exposure of individuals of a surrogate species (Peterson and Southworth, 1994; Salazar and Salazar, 1998; ASTM, 1999). The selected organisms are held in polypropylene mesh cages, which are placed in the area of potential contamination and each reference site. After the prescribed period of exposure (generally 4 weeks), the organisms are analyzed for contaminants and levels are compared with those from organisms caged at the reference sites. Indigenous organisms should be used to prevent the unintentional introduction of exotic species where they do not exist.

A.5 TERRESTRIAL PLANTS

Collection of plant material for residue analyses is a simple procedure. After plants of the appropriate species are identified, they may be sampled either as whole organisms (roots plus aboveground parts), as aboveground parts, or as discrete parts (roots, foliage, seeds, fruit, etc.). Samples may be collected by stripping or breaking parts from the plant, by cutting plant parts with shears, or by digging up plants with a spade. Additional information on vegetation sampling for residue analysis may be found in Sprenger and Charters (1997), Environmental Response Team (1996), DOE (1987), and Temple and Wills (1979).

A.6 TERRESTRIAL MOLLUSKS

Methods for the collection of terrestrial mollusks (snails and slugs) are not as well defined as those for other terrestrial invertebrates. Collection methods include the use of bran- or metaldehyde-baited traps or refuge traps (boards placed at a site to attract slugs; Newell, 1970). Snails or slugs may also be extracted from litter or soil collected from the site. Snails will generally float and slugs sink when the samples are immersed in water. Although population estimates of snails may be made by counting their abundance within randomly placed quadrants, this method is likely to be biased toward adults and against immatures (Newell, 1970). Additional discussion of sampling and extraction of terrestrial mollusks may be found in Newell (1970) and Southwood (1978).

A.7 EARTHWORMS

The primary methods for collection of earthworm samples are hand sorting of soil, wet sieving, flotation, and application of expellants. Hand sorting is regarded as the most accurate sampling method, and is frequently used to evaluate the efficacy of other methods (Satchell, 1970; Springett, 1981). While accurate, hand sorting is very laborious and may underestimate the abundance of small individuals. Efficiency is dependent on the density of the root mat, clay content of the soil, and weather conditions if sorting is done in the field. Wet sieving uses a water jet and a sieve to separate earthworms from the soil (Satchell, 1970). The efficiency of this method is not documented, and worms may be damaged during washing. Flotation consists of placing soil samples in water and collecting earthworms as they float to the surface (Satchell, 1970). This method may be used to extract egg capsules and adults of species too small to recover efficiently by hand sorting.

In contrast to methods that require excavation and processing of soil, expellants are applied *in situ* to collect earthworms. In practice, an expellant solution is applied to the soil surface within a sampling frame and allowed to percolate. Earthworms are then collected as they emerge from the soil. To enhance absorption of the expellant by the soil and to facilitate collection of earthworms as they emerge, vegetation at each sampling location should be clipped down to the soil surface. Expellants have traditionally consisted of formaldehyde or potassium permanganate solutions (Raw, 1959; Satchell, 1970). Drawbacks to these expellants include carcinogenicity, phytotoxicity, and toxicity to earthworms. In addition, these expellants may introduce additional contamination and interfere with residue analysis. As an alternative, Gunn (1992) suggested the use of a mustard solution as an expellant. A commercially available prepared mustard emulsion was mixed with water at a rate of 15 ml/l and applied to soil within a 1-m^2 frame (to confine the expellant). Efficacy of mustard was found to be superior to formaldehyde and equivalent to potassium permanganate (Gunn, 1992). Recent work at Oak Ridge National Laboratory indicates that a suspension of dry mustard (1 tsp/l) is also an effective expellant (B. Sample, personal observation). If worm samples are being collected for residue analysis, analyses should be performed on samples of the mustard expellant. These data will indicate if any contamination can be attributed to the extraction method.

A.8 TERRESTRIAL ARTHROPODS

Many methods are available to sample terrestrial arthropods. Because of the great diversity of life-history traits and habitats exploited by arthropods, no single method is efficient for capturing all taxa (Julliet, 1963). Every sampling method has some associated biases and provides reliable population estimates for only a limited number of taxa (Kunz, 1988a; Cooper and Whitmore, 1990). Reviews of sampling methods for insects and other arthropods were given by Southwood (1978), Kunz (1988a), Cooper and Whitmore (1990), and Murkin et al. (1994). Descriptions of

12 commonly employed methods, arthropod groups for which they are appropriate, and advantages and disadvantages of each are summarized in Table A.1.

A.9 BIRDS

A.9.1 Sampling Birds

Methods to collect birds include firearms, baited traps, cannon nets, mist nets, drive and drift traps, decoy and enticement lures, and nest traps (Schemnitz, 1994). Methods employed depend upon the species to be sampled. Additional information concerning methods for capturing birds may be found in Schemnitz (1994), the "North American Bird Banding Manual" (U.S. Fish and Wildlife Service and Canadian Wildlife Service, 1977), *Guide to Waterfowl Banding* (Addy, 1956), and *Bird Trapping and Bird Banding* (Bub, 1990).

Firearms used to collect birds may include rifles, shotguns, or pellet guns. This method, while highly dependent on the skill of field personnel, may be used for all groups of birds. However, because samples may be extensively damaged during collection, projectiles or shot may interfere with residue analyses, and because of safety considerations, the use of firearms is not a recommended sampling method. In addition, the use of firearms for sample collection precludes repeated sampling of the same individual.

Baited traps are most useful for gregarious, seed-eating birds. In their simplest form, a wire-mesh box is placed over bait (generally seeds or grain), and one side is supported by a stick. Once birds enter the box to feed, the operator pulls a string attached to the support stick, the side falls, and the birds are entrapped. Other types of baited traps include funnel or ladder traps. These traps are designed with entrances through which birds can easily enter but not easily exit.

Cannon nets may be used for birds that are too wary to enter traps. This type of trap is frequently used for wild turkey and waterfowl and has been successfully used for sandhill cranes and bald eagles (Schemnitz, 1994). Cannon nets consist of a large, light net that is carried over baited birds by mortars or rockets. In use, nets are laid out and baited for 1 to 2 weeks to allow the birds to become accustomed to the net and bait. Once birds make regular use of the bait, the trap may be deployed.

Mist netting is a method useful for some species that are not attracted to baits. A detailed review of the use and application of mist nets is provided by Keyes and Grue (1982). This method may be used for birds as large as ducks, hawks, or pheasant but is most applicable to passerines and other birds under approximately 200 g. Mist nets are constructed from fine, black silk or nylon fibers; the nets are usually 0.9 to 2.1 m wide by 9.0 to 11.6 m long, attached to a cord frame with horizontal cross braces called shelfstrings (Schemnitz, 1994). The net is attached to poles at either end such that the shelfstrings are tight but the net is loose. The loose net hangs down below the shelf strings, forming pockets. When properly deployed, birds (or bats) strike the net and become entangled in the net pocket. Mist nets may be employed passively or actively. In a passive deployment, nets are set across flight corridors, and birds are caught as they fly by. For an active deployment, a group of nets is set and birds are driven toward the nets. Another effective approach is to use recorded calls of conspecifics or distress calls to attract birds to the net.

The following must be considered when using mist nets.

- Avoid windy conditions; wind increases the visibility of the net.
- Check nets frequently; unintended mortality may result from stress if birds are left in the net for >1 h.
- Do not operate nets during rain; birds may become soaked, and mortality may result from hypothermia.

Drive and drift traps consist of nets or low wire-mesh fencing erected at ground level. Birds are driven or herded into the fence, which then guides them into an enclosure. This method is most frequently used to capture waterfowl while they are molting and flightless. Drift traps have also been used successfully with upland game birds, rails, and shorebirds (Schemnitz, 1994). Because many birds are reluctant to flush and fly when birds of prey are present, trapping success may be enhanced by playing recorded hawk calls.

Decoy and enticement lures are used most frequently for birds of prey. The most common trap of this type is the bal-chatri trap. This trap consists of a wire-mesh cage, on top of which are attached numerous monofilament nooses. A small bird or rodent is placed in the trap as bait. When a hawk or owl attempts to attack the bait, its feet become entangled in the nooses.

Nest traps are useful to capture birds at the nest for reproductive studies. For ground-nesting birds, drop nets erected over the nest are sometimes effective. For cavity-nesting birds, trip doors may be devised that can be closed once the adult enters the nest. Other types of nest traps are summarized by Schemnitz (1994).

A.9.2 Avian Population Survey Methods

Many methods are available to determine the abundance, density, and spatial distribution of birds. These methods may be used to census populations of a single species or to census the entire avian community in a given area. The commonly used methods are territory mapping, transects, point counts, mark-recapture, song tapes, aerial counts, and habitat-focused surveys.

A.9.2.1 Territory Mapping

Territory mapping is among the most accurate and reliable methods for determining bird population density (Wakely, 1987a). This method consists of plotting (by individual species) the locations of birds seen or heard during repeated visits (generally eight to ten). A gridded sampling plot is used for this purpose (Verner, 1985; Ryder, 1986; Wakely, 1987a). Clusters of observations are assumed to represent the center of activity for individual territories. The total number of birds on a plot is then estimated by summing the number of clusters (i.e., territories) and multiplying by 2 (assuming an even sex ratio) (Verner, 1985). This method works best for species that sing conspicuously from within their territories (e.g., most passerines). It is not well suited for birds that frequently sing within the boundaries of a conspecific's territory, or quiet or secretive species, or nonterritorial birds (called floaters), or species with territories larger than the study plot (Verner, 1985). Also, because the efficacy of this method depends on territorial behavior, it is useful only during the breeding season (except for birds that maintain year-round territories). This method also requires considerable time to lay out and mark the sampling plot and for repeated visits. Additional limitations of territory mapping are summarized by Oelke (1981). Falls (1981) reports that detection of individuals may be enhanced by using playback of recorded songs. Birds defend their territories in response to the song tape and their singing locations provide an indication of the boundary of a territory.

The consecutive-flush technique (Whitmore, 1982; Verner, 1985) may be used to reduce the number of plot visits needed to complete a territory map. An observer simply approaches a singing bird until it flushes. Its initial position, line of flight, and landing position are then recorded on the plot map. The observer again approaches and flushes the bird and records its movement. The process is repeated until at least 20 consecutive flushes have been mapped. This technique is most applicable in open habitats such as grasslands or marshes, where an observer may keep an individual bird under constant observation. Flushing may also help delineate territory boundaries in forested habitats (Verner, 1985).

A.9.2.2 Transects

Transect census methods consist of counting birds either seen or heard along one or both sides of a line through one or more habitats (Ryder, 1986). Transects are more flexible than are mapping methods. Because they do not depend on territoriality, their use is not restricted to the breeding season. In addition, they may detect both floaters and juveniles. Verner (1985) defines three general types of transects.

1. Line transects without distance estimates. The observer simply walks a preset line and records all birds seen or heard, without measuring or estimating distances to the birds. This is an efficient method for generating lists of species. However, the results cannot be used to estimate density because the area sampled is unknown. Data may be used for intraspecies or interspecies comparisons (either temporal or spatial), if it is assumed that all individuals or species are equally detectable in all samples and factors that affect detectability are similar among all samples.
2. Variable-width line transects. This is the most commonly used transect method. Perpendicular distances from the transect line to birds detected are measured or estimated. These observations are then used to estimate the area sampled and, thus, bird density.
3. Belt transects. This method is essentially a line transect with fixed boundaries (usually 25 to 50 m on either side of the line), within which all birds seen or heard are counted. This is a simpler method than the variable-width transect method because the observer need only estimate one distance, the belt width. Density estimates are obtained by dividing the total number of birds observed by the area of the belt.

Burnham et al. (1980) provide a very detailed discussion of line-transect techniques, applications, and data analysis methods. Additional discussion is provided by Wakely (1987b). Analytical methods for line-transect data are discussed by Krebs (1989).

A.9.2.3 Point Counts

Point counts consist of counting the number of birds seen or heard for a fixed time in all directions from a single point. Similar to transects, distances around the sampling point may be undefined, fixed, or variable (Verner, 1985). With the variable circular plot method (Reynolds et al., 1980), the distance from the sampling point to the bird is estimated. This distance is then used to estimate the population density. Because point counts do not depend on territorial behavior, they may be performed year-round. Best results, however, are obtained during the breeding season. Although point counts may be performed in any habitat where transect sampling would be applicable, point counts are best suited for steep, rugged, or thickly vegetated habitats where observer movement along the transect may disturb birds and interfere with their detection (Reynolds et al., 1980; Ryder, 1986; Wakely, 1987c). Use of point counts to survey birds in bottomland hardwood forests is discussed by Smith et al. (1993).

A.9.2.4 Mark–Recapture

The ratio of marked individuals to unmarked individuals may be used to estimate population size. Population size and area sampled can then be used to estimate density. Karr (1981) suggests using mist nets (see Section A.9.1) to capture and color-band birds for population studies. Although mark-recapture is not considered an efficient population census method for birds (Verner, 1985; Ryder, 1986), it may provide very useful information, particularly in studies of threatened and endangered (T&E) species. For example, mark-recapture data may be used to identify the number of pairs of a species that are present, or to distinguish migrants from residents and breeders from nonbreeders, or to identify ranges or territorial boundaries for individual birds (Ryder, 1986). Additional discussion of the use of mark-recapture to estimate avian populations is presented by Nichols et al. (1981) and Jolly (1981). Analytical methods for mark-recapture data are discussed by Krebs (1989).

A.9.2.5 Song Tapes

Censusing inconspicuous or secretive birds (i.e., nocturnal, marsh, or some forest birds) may be very difficult. Johnson et al. (1981) and Marion et al. (1981) suggest that song tapes may be employed to perform relative or absolute censuses for these species. By playing recordings in different areas and recording occurrence and number of responses, presence, abundance, and density may be estimated.

A.9.2.6 Aerial Counts

Large flocks of waterfowl and shorebirds may be photographed from the air and later counted (Verner, 1985). Aerial counts are also suggested for breeding osprey (Swenson, 1982). Because osprey nests are large and conspicuous and generally placed in trees or atop artificial structures, they may be clearly observed from the air. Census flights should be made during the incubation period (generally April through June) using a high-winged aircraft or a helicopter. It should be noted, however, that aerial counts are suitable only for very large contaminated sites. Analytical methods for aerial survey data are discussed by Krebs (1989).

A.9.2.7 Habitat-Focused Surveys

Habitat-focused surveys are particularly suited for T&E species. First, areas with critical habitat are identified, and then the presence, abundance, and distribution of the target species is determined. By focusing on a particular, critical habitat, usually nesting habitat, the likelihood of finding the T&E species and collecting data relevant to ecological risk assessment is increased. As an example, Thompson (1982) describes a habitat-focused survey method for the red-cockaded woodpecker. Red-cockaded woodpeckers are a colonial-nesting T&E species that require mature, open, fire-maintained pine forests (Thompson, 1982). Survey methods for this species rely on identification of appropriate habitat (old-growth pine forest) and nest trees within the habitat (large-diameter trees with clear boles and flattened crowns). Habitat and trees within habitat may be identified using a combination of remote sensing and ground truthing. Presence of red-cockaded woodpeckers in an area is indicated by

- Excavated 2-in.-diameter cavities in living sapwood
- Chipping of small wounds (resin wells) in the pine bark
- Flow of pine resin from cavity and resin wells, giving the tree a glazed appearance
- Flaking of loose bark from the tree cavity

Once the presence of red-cockaded woodpeckers in an area has been verified, the population size may be determined by observing the activity at the cavities and counting the number of individuals observed (Thompson, 1982).

A.9.2.8 Additional Information

Much has been written on avian censusing techniques. Detailed discussions and comparisons of census methods, methods for analysis of census data, sampling designs for avian censuses, and factors that affect census results are presented in Ralph and Scott (1981). Chapters concerning census.methods for songbirds, raptors, shorebirds, waterfowl, colonial waterbirds, and upland game birds may be found in Cooperrider et al. (1986). Davis (1982a) presents census methods specifically for 43 species of birds and 14 groups of birds or birds in specific habitats.

A.9.3 Avian Nest Study Methods

The nesting stage is a critical life stage for all birds. Any environmental factors that affect birds during this stage and reduce recruitment may have adverse population effects. One way to evaluate whether recruitment is being affected is to calculate nest success.

The most common method of calculating nest success is the Mayfield method (Mayfield, 1975). This approach considers the survival of a nest over the period of time that the nest is observed. In practice, the daily survival rate is estimated by dividing the total number of young or eggs lost by the total number of days the nest has been observed and subtracting this quotient from 1. This value represents the probability of survival for the nest during that time period. By analyzing the time-frame of the different nesting stages (i.e., laying, incubating, nestling, etc.), investigators can identify the stage at which mortality is occurring. Applications and mathematical validity of the Mayfield method are discussed by Miller and Johnson (1978), Johnson (1979), Hensler and Nichols (1981), and Winterstein (1992).

Nest attentiveness is another factor that may affect nest success and, thus, recruitment. Grue et al. (1982) observed that European starlings exposed to a sublethal organophosphate insecticide dose fed their nestlings less frequently and were away from the nest longer. Nestlings in nests of exposed birds lost weight. Because fledging weight is correlated with survival (Perrins, 1965), altered nest attentiveness may cause negative impacts to avian populations.

Methods to monitor nest attentiveness or activity include visual observations (e.g., Heagy and Best, 1983), time-lapse cameras (e.g., Grundel and Dahlsten, 1991), telemetric eggs (e.g., Varney and Ellis, 1974), and radio-equipped birds (e.g., Licht et al., 1989). Baron et al. (1997) present a method for monitoring nests of burrowing birds such as kingfishers. Additional methods for cavity-nesting birds are discussed by Mallory and Weatherhead (1992).

A.9.4 Avian Food Habit Study Methods

Food habit studies have two primary applications in risk assessment. First, they may be used to identify and quantify contaminant exposure pathways through the food web. Samples of food consumed, excreta, or rejecta may be collected and analyzed for residues and to determine diet composition. Second, use-availability studies or foraging behavior studies may be performed to evaluate if indirect effects are occurring that may affect the energetic status of the species in question.

Methods for performing avian diet analysis have been reviewed and summarized by Rosenberg and Cooper (1990). Data may be presented as percentage occurrence (number of samples in which a food item appears), frequency (number of times a food item appears in a sample), or percentage volume or weight (proportion of total sample volume or weight accounted for by a food item). To prevent confusion and to minimize bias, both frequency and volume data should be reported. For example, an important food type may be consumed in high volume but low frequency. Conversely, a food of minimal importance that is highly abundant may be observed in high frequency but low volume.

For additional discussion of methods and approaches to investigating avian food habits, consult Morrison et al. (1990). This volume includes papers that discuss approaches to quantifying diets, design and analysis of foraging behavior studies, use-availability analysis, energetics, and foraging theory. Additional methods for analysis of use-availability data, niche overlap, and dietary data are described in Krebs (1989).

A.10 MAMMALS

A.10.1 Sampling Mammals

Numerous methods are available for the collection of mammals. Suitable methods vary by species and habitat, with multiple methods often being suitable for the same species (Jones et al., 1996). For risk assessment purposes, small mammals, primarily within the orders Rodentia and Insectivora, are most commonly collected. This is because they are frequently assessment endpoints themselves, are important food items for endpoint predators, and are more likely to be present in sufficient numbers than larger mammals. Methods discussed will therefore focus on these taxa. Discussion of methods for the collection of other mammalian taxa are presented in Wilson et al. (1996), Schemnitz (1994), Kunz (1988b) and Nagorsen and Peterson (1980).

Small mammals are generally collected by one of three methods: snap traps, box traps, or pitfall traps. Snap traps are the familiar "mousetrap" and consist of a spring-powered metal bale that is released when the animal contacts the baited trigger pan (Jones et al., 1996). These traps are lethal, with animals being killed by cervical dislocation. Nagorsen and Peterson (1980) report snap traps to be the most successful trapping method for small rodents and insectivores. However, because they are nonselective, snap traps may collect any animal that is attracted to the bait. This may be a serious concern if T&E species are believed to reside in the study area.

Box traps represent the most effective method to capture small mammals unharmed (Jones et al., 1996). The use of box traps allows the selection of only those species of interest and the release of nontarget species. Box traps are typically rectangular metal or wooden boxes with openings at one or both ends and a baited trip pan. Animals are captured when they contact the trip pan, causing spring-loaded doors to close. Captured animals may be maintained in box traps for several hours if food and bedding are provided and temperatures are not extreme. The size of the trap, trap type, ambient conditions at the trapping site, and body size of animals to be trapped all influence trapping success (Jones et al., 1996). Because some animals are reluctant to enter box traps (shrews in particular), box traps are not as effective as snap traps (Nagorsen and Peterson, 1980).

Pitfall traps consist of a container buried into the ground so that its rim is flush with the surface. Animals are captured when they fall into the container. Pitfall traps are among the most effective traps for collecting shrews (Jones et al., 1996). Success rates for pitfall traps may be dramatically increased by employing drift fences. Drift fences consist of barriers of metal, plastic, fiberglass or wood that direct small mammals into the pitfall trap. Pitfall traps may be employed as either live or killing traps. Killing pitfall traps are partially filled with water to drown animals. Live pitfall traps must be at least 40 cm deep to prevent small mammals from jumping out (Jones et al., 1996).

Both snap traps and box traps must be baited. Baits that are employed depend on the species sought. Generally, peanut butter and oats or other seeds are effective for most granivorous or omnivorous small mammals (Jones et al., 1996). Because small mammals simply fall into pitfall traps, these traps do not need to be baited (Nagorsen and Peterson, 1980). Trapping success is generally enhanced if traps are set but locked open within the sampling area for several days prior to trapping. This allows the animals to become accustomed to the presence of the traps. Once traps are baited and set, both snap and box traps should be checked daily. Pitfall traps should be checked more frequently (twice daily) to prevent shrews from starving or consuming each other (Jones et al., 1996).

Trap placement for the purposes of collecting animals for residue analysis differs from that for a population survey. Sampling for residue analyses does not require a trapping array suitable to determine density. Sampling along transects is adequate. Jones et al. (1996) recommend that traps be placed along transects that are at least 150 m long with traps placed every 10 to 15 m. Regardless of spacing, traps should be placed at habitat features favored by or indicative of small mammals (e.g., logs, trees, runways, burrow entrances, dropping piles, etc.; Nagorsen and Peterson, 1980; Jones et al., 1996). In addition, sampling must be appropriately distributed with respect to concomitant distributions and near locations where media are sampled.

A.10.2 Surveying Mammals

In contrast to birds, it is difficult to observe most mammals in their natural environment. This is because many mammals are nocturnal (e.g., most carnivores), crepuscular (active at dusk or dawn), or otherwise cryptic or secretive in behavior (e.g., shrews). Three general approaches to biological surveys of mammals have been widely used: direct observation, indirect observation, and habitat evaluation methods. Habitat evaluation methods focus primarily on measurements of plant distribution and structure of the vegetation community and are discussed in Section A.12. Representative methods for the remaining categories are described below.

A.10.2.1 Direct Observation

Direct observation methods consist of those methods in which the animals of interest are actually seen, heard, or captured. Because animals are directly observed, these methods produce the highest-quality population estimates with the lowest amount of uncertainty. However, direct methods are generally more time, personnel, and cost intensive than indirect observation or habitat-survey methods. Examples of direct observation methods include drives, silent detection, thermal scanners, vocalization surveys, mark-recapture surveys, and trapping. Trapping methods for mammals are discussed in Section A.10.1.

Drives. In this method, animals within an area are surrounded and then counted as they are forced to leave the area (Rudman et al., 1996). Drives require a large number of people and target species must be clearly visible within the area. This method is most appropriate for diurnal, medium- or large-sized, terrestrial mammals with a conspicuous flight reaction (most ungulates and lagomorphs) and is most effective when survey areas are small. This method is inappropriate for species that hide, for large predators, or for fossorial or arboreal species. Other disadvantages of this method are that it is stressful to the animals being counted, and animals that move ahead of drivers may not be counted. Lancia et al. (1994) and Rudman et al. (1996) provide additional information on performing drives and on data analysis methods.

Silent detection. Silent detection consists of observers quietly approaching animals and counting them (Rudman et al., 1996). This method is suitable for a wide range of mammals, including marine mammals, diurnal, nocturnal, arboreal, and fossorial species. Counting of nocturnal animals is generally performed using a spotlight. Silent detection sampling may be performed by observers on foot or from mobile platforms (vehicles or boats) and may consist of either total counts or sample counts. Total counts are performed for small areas or for social animals, and may consist of counts of individuals in groups, or at dens or burrows. Sample counts are employed for larger areas and may consist of surveys along transects or within quadrants. These methods have been employed for deer (Teer, 1982) and squirrels (Bouffard, 1982) and are similar to those employed for birds (see Section A.9.1). For very large areas, transect surveys from aircraft may be employed. Aerial surveys are best suited for marine mammals (Braham, 1982; Odell, 1982), wolves (Fuller, 1982), many ungulates (Kufeld et al., 1982; Rolley, 1982) and species that live in open or patchy habitat. The advantage of silent detection methods are that they are not stressful to the animals being surveyed. The primary disadvantages are that the techniques are time-consuming and some animals may be missed, particularly in dense habitats or rough terrain. Additional detail on silent detection methods are presented in Rudman et al. (1996).

Vocalization surveys. Vocalization surveys consist of counts of animals that respond to broadcast recordings of animal calls or sirens. This method has been used for canids (wolves, coyotes, and dogs; Lancia et al., 1994) and ground squirrels (Lishak, 1982). While this method can be highly species specific, special equipment and significant time and effort are required (Lancia et al., 1994). A variation of vocalization surveys is the use of ultrasonic detecting devices to identify the presence and activity of bats (Cross, 1986; Thomas and LaVal, 1988; Kunz et al., 1996). This method may be used to identify some bats by species-specific call patterns, it does not cause any stress to the animals, and it is useful in virtually any habitat. Some bats, however, are not easily detected by ultrasonic devices, and the method cannot be used to estimate population densities.

Thermal scanners. Some large mammals may be counted based on detection of their heat output using infrared thermal scanners (Lancia et al., 1994; Naugle et al., 1996). This method is best applied during winter when contrast between animals and the ambient environment is best. As this is still a very new method, information on the successful application of this method is limited.

Mark-recapture. Mark-recapture data are useful to estimate population size, structure, and survivorship. In principle, mark-recapture methods for mammals are similar to those for birds (see Section A.9.2); the ratio of marked to unmarked individuals is used to estimate population size, which coupled with the area sampled is used to estimate population density. Mark-recapture methods are among the most widely used and reliable methods for estimating mammalian population sizes (Lancia et al., 1994). Methods for capturing small mammals have been discussed here. Discussions of methods for capturing other mammalian taxa are presented in Jones et al. (1996), Schemnitz (1994), Kunz (1988b), and Nagorsen and Peterson (1980). Davis (1982a) presents specific mark-recapture methods for a variety of species. Nietfield et al. (1994) discuss methods that are available for marking mammalian wildlife. Methods for analyzing mark-recapture data are discussed by Nichols and Dickman (1996), Lancia et al. (1994), Krebs (1989), and Davis (1982a).

Radiotelemetry. Radiotelemetry is a method in which small radio transmitters are attached to individual animals so that their movements and activities can be monitored. As animals are relocated over time, their foraging area and home range is delimited. Estimates of exposure can be generated by overlaying home range data on maps of contaminant distribution in a geographic information system (GIS). Although quality data of use in risk assessments may be generated by radiotelemetry, because this method is expensive and time-consuming, its use can generally be justified only for assessments of large and complex sites or for endangered or otherwise important species such as the San Joaquin kit fox at the Elk Hills Naval Reserve in California. Extensive literature is available concerning the use of radiotelemetry in wildlife studies. Hegdal and Colvin (1986) and Samuel and Fuller (1994) provide reviews of design of telemetry studies, selection of equipment, field procedures, and analysis of telemetry data. Brewer and Fagerstone (1998) discuss radiotelemetry for studies of toxic effects on wildlife.

A.10.2.2 Indirect Observation

Indirect observation methods are those in which signs of the animals' presence or activity are employed as an index of their abundance and distribution. Because animals are not directly observed, these methods often produce more uncertain estimates than direct methods. Indirect methods are generally less time, labor, and cost intensive than direct methods, however. Examples of selected indirect observation methods are outlined below.

Structure or habitat features. Because many mammals construct structures for protection or to raise their young or create trails or other features, the abundance of these structures or features can be used as an index of population abundance (Wemmer et al., 1996). For example, hay piles provide an index of pika abundance (Smith, 1982) and muskrat populations may be estimated by counting houses within a given area (Danell, 1982). If a more detailed population estimate is needed, muskrats occupying several houses can be trapped, with the mean number of muskrats per house used to estimate the muskrat populations. McCaffery (1982) report that relative abundance of deer can be estimated by counting the number of deer trails that cross a 0.4-km transect.

Scent stations. Scent stations were designed to census foxes and have proved useful for other carnivores including coyotes, bear, raccoons, bobcats, otter, and mink (Phillips, 1982; Spowart and Sampson, 1986; Wemmer et al., 1996). Scent stations consist of a scent capsule containing either a synthetic attractant or a natural scent (e.g., fermented egg or bobcat urine) attractant located in the center of a 1-m circle of sifted dirt. Animals visiting the station are identified by the tracks they leave in the sifted dirt. The stations are inspected for 5 consecutive days for animal visits. Surveys are generally conducted in the fall, using a series of scent stations set up along unpaved or secondary roads or transect lines at 0.3- to 0.5-km intervals (Wemmer et al., 1996). A relative population index can be calculated: Index = Total number of visits/total station nights × 1000. These indexes can be compared among sites to determine how predator populations vary. While population indexes from scent stations were well correlated with population estimates from other methods for bobcats, raccoon, and foxes, they were not for opossums (Wemmer et al., 1996).

Scat surveys/pellet counts. The presence of scat, feces, or pellets provides an indication of the use of an area by the animal of interest. For example, the use by elk of vegetation damaged by metals at Anaconda, MT was estimated using fecal pellet counts (Galbraith et al., 1995). Collected fecal materials may also be analyzed for contaminants as an indication of exposure and dissected to determine site-specific food habits. In addition, if the rate of scat production is known, a population index may be generated from scat or pellet counts (Wemmer et al., 1996). This method is most generally applied for ungulates. For example, Longhurst and Connolly (1982) and Lautenschlager (1982) report that pellet counts provide a quick, fairly accurate, and relatively inexpensive population estimation method for deer. In contrast, Wolff (1982) recommends that while pellet counts can be used to estimate habitat use and population trends for snowshoe hare, they are not suitable to estimate population numbers. While relative abundance of coyotes has been estimated by scat counts, reliability of the estimates is limited by the unquantified effect of diet on scat production (Spowart and Sampson, 1986).

Browse surveys. The relative size of herbivore populations may be evaluated by surveying the amount of browse material consumed. For example, browse intensity (percent of available twigs that have been browsed) has been used as an index of numbers of snowshoe hares in relation to habitat carrying capacity (Wolff, 1982).

Remote cameras. Remote cameras consist of camouflaged still, movie, or video cameras located along trails or baited stations and attached to a trip plate, active infrared sensor, or other trigger to take pictures of animals within its field of view (Wemmer et al., 1996). This method is well suited for secretive species that use established trails, dens, or feeding sites. Advantages of this method are that it is nonintrusive, large areas can be monitored with few people, there is minimal human disturbance, animals do not have to be captured, and observers do not need to be in attendance. Disadvantages include cost of equipment and film, risk of theft, and size and species biases for many triggering devices (Wemmer et al., 1996). In addition, statistical methods for analyzing image data are poorly developed. A discussion of the problems and biases associated with analyses of these data is presented by Wemmer et al. (1996).

A.10.2.3 Additional Information

Much has been written on censusing techniques for mammals. Detailed discussions and comparisons of methods for measuring mammalian populations, sampling designs, and methods for analysis of these data are presented in Wilson et al. (1996) and Bookhout (1994). Chapters concerning census methods for rodents and insectivores, lagomorphs, carnivores, bats, and ungulates may be found in Cooperrider et al. (1986). Davis (1982a) presents census methods specifically for at least 65 species of mammals and 13 groups of mammals in specific habitats.

A.11 REPTILES AND AMPHIBIANS

Because reptiles and amphibians are not endpoint entities in most ecological risk assessments, a detailed discussion of field survey methods for these animals is not presented here. However, in at least one case, field surveys indicated that forest salamanders were much more sensitive to metal-contaminated soils than birds or mammals (Beyer and Storm, 1995). If the endpoints include reptiles or amphibians, field sampling methods are available. Additional discussion of sampling methods for reptiles and amphibians are presented in Heyer et al. (1994), Jones (1986), and Davis (1982a). In addition, Degraaf and Yamasaki (1992) present a nondestructive technique to monitor the relative abundance of terrestrial salamanders. If mark-recapture studies are to be performed, Nietfield et al. (1994) present marking techniques for reptiles and amphibians.

Opportunistic collection consists of searching suitable habitats for species of interest. Once found, individuals are collected by hand, net, or other devices that may facilitate immobilizing individuals.

Numerous types of nets and traps are available for sampling herpetofauna. Traps are generally effective for alligators, turtles, and snakes. Stebbins (1966), Conant (1975), and Shine (1986) discuss various aquatic trapping methods. Some traps may be set by one person. To prevent inadvertent mortality from trapping, traps should be checked at least daily (trap mortality is generally low if checked often). Aquatic traps should be set partially above the water line to permit the captured organisms to breathe.

Although developed for sampling fish, electrofishing may also be very effective for aquatic salamanders and aquatic snakes (Jones, 1986). This method occasionally yields turtles, sirens, and hellbenders. Electrofishing requires two or more people (a shocker and a netter) and is most effective in shallow water (streams, ponds, and shallow rivers). Deep-water habitats (lakes, reservoirs, and embayments) may be shocked from boats, but this approach is probably less effective for most herpetofauna than for fish. One disadvantage to electrofishing is that it may cause some mortality, especially in hot weather.

The use of small-mesh seines (7-mm or less) is moderately effective for sampling of aquatic salamanders, frogs, snakes, and turtles (Jones, 1986). This method generally requires at least two people to operate the seine. Other personnel are beneficial for disturbing the substrate, blocking potential escape routes, and handling the catch.

A.12 VEGETATION ANALYSIS TO IDENTIFY HABITAT SUITABLE FOR WILDLIFE SPECIES

Plants provide the most important component of habitat requirements for wildlife species. Identifying the presence of suitable habitat (over a large enough area) is the first step in determining whether a given species is likely to be present. Methods for sampling and analysis of vegetation communities are discussed in Hays et al. (1981), Anderson and Ohmart (1986), Higgins et al. (1994), and Environmental Response Team (1994d). Habitat evaluation methods are discussed by Anderson and Gutzwiller (1994). The assessment of the Anaconda, MT site provides an example of the use of habitat evaluation to assess effects of vegetation injury on wildlife (Galbraith et al., 1995).

Three basic habitat variables can be directly measured and used to predict habitat suitability: foliage density, species composition, and fruit production (Anderson and Ohmart, 1986). Of these, species composition is perhaps most useful for many rodent and bird species (Anderson and Ohmart, 1986). A number of other variables can be derived from these basic measurements, but these indirect variables may be less helpful in locating habitat for specific wildlife species.

Foliage density is the amount of foliage per unit area or to the extent of canopy cover. Plant density (the number of plants per unit area) is not the same as foliage density. Foliage density generally is measured at different vertical levels within the vegetation. Canopy-cover requirements for species may be related to types of vegetation such as herbs, shrubs (defined either by height or diameter limits), or overstory. Habitat requirements for some species (e.g., cavity nesters) may include a minimum number of snags (dead trees) or downed logs per area. Other species (e.g., small mammals, amphibians, and invertebrates) may require a degree of forest floor litter cover.

Foliage density can most easily be measured using a transect system. Transects are established either randomly or in representative areas. At predetermined points along each transect (e.g., every 5 m), the canopy cover or foliage density is measured at each desired vertical level. Quadrant methods may be used alone or in conjunction with transects. Quadrants are predetermined areas (frequently 1 × 1 m squares or 1-m-diameter circles) that are sampled to estimate the foliage density or canopy cover. Transect and quadrant methods are best suited for use with low-growing species or large areas. Further details on these methods may be found in Hays et al. (1981), Anderson and Ohmart (1986), and Higgins et al. (1994).

Fruit production refers to the quantity of fruit produced by plants. Mast surveys conducted by many state wildlife agencies are an example of this parameter. For many species of plants, fruit production does not correlate well with number of individual plants present (Anderson and Ohmart, 1986). In these cases, it may be necessary to measure fruit production for a representative number of individual plants. Higgins et al. (1994) discuss methods for sampling of fruit.

For certain species (e.g., some birds) the degree of patchiness or the amount of edge habitat per unit area may be important. These parameters are best measured from large-scale vegetation cover maps derived from aerial or satellite data. The use of computerized GIS procedures can greatly enhance habitat analysis.

A.13 SAMPLING CONSIDERATIONS

A.13.1 Spatial Components of Biota Sampling

To aid in interpretation of results and in identification of areas for potential remediation, samples should be situated with respect to the spatial distribution of contamination. In addition, to ensure that samples are relevant to the endpoint species of concern, samples should be collected from within areas that represent habitat for endpoint species. This will aid in the exposure estimation. Alternatives for incorporation of spatial components into the sampling design include sampling along transects through gradients of contamination, sampling areas with distinct levels of contamination, or the use of a combination of transects and reference locations.

Spatial considerations differ among species. For example, body burdens in highly mobile wildlife represent the contaminant concentration averaged over their foraging range and not necessarily the location from where they were collected. In contrast, body burdens within species with limited mobility (e.g., plants and earthworms) are likely to be highly representative of their sampling location.

When identifying areas for sampling of biota, it is important to consider where samples from the abiotic media were collected. Ideally, both biotic and abiotic samples should be collected from the same locations. By co-locating samples, contaminant concentrations in soil, sediment, and water may be compared with biota samples to provide an indication of bioavailability. In addition, co-located data may be used to develop site-specific contaminant uptake models that may be applied to sampling locations where only samples of abiotic media were collected. Development of these site-specific uptake models may help reduce project costs by reducing the number of biota samples that are collected and analyzed. It should be noted, however, that because wildlife may travel widely, body burdens may not be well correlated with the chemical concentration in media from individual locations.

A.13.2 Sample Handling

The manner in which biological samples are handled and prepared will profoundly influence the utility of the resulting data for risk assessment. Sample-handling issues include how samples are pooled (i.e., compositing), sample washing, and depuration.

If the amount of sample material is too small for accurate chemical analysis (e.g., individual earthworms or other invertebrates or organs from vertebrates), samples from multiple individuals may be composited to produce a sample of sufficient size. Samples may also be composited over a unit area in an effort to reduce analytical costs. While the resulting composited sample represents the mean chemical concentration from all included samples, it does not provide any information concerning the distribution among locations or individuals. Consequently, minimum and maximum values within the composite are unknown, a single high or low concentration may dominate the resulting composite value, and the composite value may over- or underestimate the concentrations present in the majority of samples. Compositing of samples must be appropriate for the intended use of the data. Compositing is generally suitable for biota samples to be used for dietary exposure modeling. This is because consumers are exposed to the average concentration in their diet. In contrast, if the samples are to represent internal exposures for endpoint species, compositing of samples will result in underestimation of the exposure of highly exposed individuals. Because compositing of samples results in loss of information and may result in biased estimates, it must be performed with caution.

In addition to containing contaminants within their tissue matrix, biota samples may have external contamination in the form of soil or dust adhering to their surface. Depending on the purpose of the analyses and the intended use of the analytical results, these external residues may or may not be washed off prior to analysis. If the contaminant of interest has a significant aerial deposition pathway, or soil ingestion is not being considered in the exposure model, then samples should not be washed. It should be recognized that these unwashed samples will be biased and will represent both bioaccumulation and external adhesion of contaminants. If, however, the soil ingestion is explicitly included in the exposure model, it is preferable that samples be washed prior to analysis.

Depuration refers to the voiding of the gastrointestinal tract of sampled animals and is a consideration primarily for earthworms. Undepurated earthworms may have higher or lower chemical concentrations than depurated earthworms from the same location. This depends on whether the chemical is bioaccumulated by worms to concentrations that are higher or lower than soil concentrations. Chemical concentrations in undepurated worms are lower in the former case and higher in the latter case than those in depurated worms. Chemicals in the soil in the gastrointestinal tract bias the body burden estimates for the worms and the dose estimates for vermivores. If the model being employed to estimate exposure of animals that consume earthworms does not include a term for soil ingestion, this bias is not critical (as long as food consumption is adjusted for the mass of soil). However, if there is a soil ingestion term in the model, the use of undepurated worms will result in some double-counting of the soil consumed.

A.13.3 Tissues to Analyze

Determination of which tissues to analyze is dependent on the intended use of the data in the assessment and on the pharmacokinetics of the chemical of interest. If the analyzed species is an endpoint entity, tissues should be analyzed that are measures of internal exposure that can be related to effects. These tissues include target organs for the chemical such as the liver, kidney, or brain. Among birds, if reproductive effects are a consideration, eggs may be analyzed. Hair and feathers may also be sampled and analyzed as a nondestructive measure of exposure. The organs to analyze depend on the pharmacokinetics of the chemical. As the loci for metabolism and excretion, the liver and kidney are target organs for many contaminants. If the chemical is known to be neurotoxic or lipophilic, brain and fatty tissues should be analyzed, respectively. Bone should be analyzed for those chemicals, such as lead and strontium, that have high affinities for bone. An important issue when deciding to perform target organ analyses is whether an appropriate exposure–response model (e.g., one in which exposure is expressed as target organ concentration) exists to allow the interpretation of the data. In the absence of a suitable model, target organ concentrations only provide information concerning exposure, not effects.

If the species is a food item consumed by an endpoint species, the tissues to be analyzed should be those that are consumed by that species. For example, because most predators consume their prey whole, whole-body analyses of fish or small mammals should be performed if piscivores or small mammal predators are the endpoint species. Seeds, fruit, or foliage should be analyzed for granivorous, frugivorous, and folivorous endpoints, respectively. Analysis of the food type and tissue that is most appropriate for the endpoint minimizes the uncertainty associated with the exposure estimates based on these data.

When fish are collected for analysis as part of a combined human and ecological risk assessment, fillet (muscle) tissues are frequently the only tissues analyzed. This is because these tissues represent the portion of the fish that is consumed by people. These data, however, are not appropriate for piscivorous wildlife which typically consume fish whole. There are two potential solutions to this problem. Sufficient samples of fish could be collected such that analyses could be performed on both fillets and on whole fish. The alternative is to analyze a sample of the carcasses that remain after filleting and calculate fillet-to-whole-fish-concentration ratios which can then be used to estimate whole-fish concentrations from fillet data. Models for this purpose have been developed by Bevelhimer et al. (1997).

For many persistent, high-molecular-weight chemicals such as PCBs, fat is the primary repository. For those chemicals, it is common to analyze the contaminant content of fat or to lipid-normalize the concentrations in whole organisms or organs. In theory, effects of these chemicals are less variable with respect to concentrations in fat or fat-normalized concentrations than whole-organism concentrations, because the concentration at the site of action is a result of equilibrium partitioning with fat. As a result, fatter animals require a larger dose to induce an effect (Lassiter and Hallam, 1990). Methods for determining the lipid content of an organism or component may be found in Bligh and Dyer (1959) and Herbes and Allen (1983).

In some cases, the sampled species is both an endpoint and a prey species. In such cases, both target organs and the carcass (remaining tissue after removal of target organs) should be weighed and analyzed. Target organ data can be used to estimate exposure for the species as an endpoint. Whole-body concentrations can be calculated by averaging concentrations in target organ and carcass with mass of tissue as a weighting factor.

A.13.4 Permits

In most states, collection of biota is regulated by fish and game laws. National and international statutes may also apply, depending upon the species of interest. As a consequence, before any biota collection program is initiated, all appropriate permits must be obtained. Failure to obtain the needed permits may result in the rejection of the data or civil or criminal actions against the parties involved. For example, taking of migratory waterfowl requires a U.S. Fish and Wildlife Service (FWS) permit or a state hunting license (in season) and a federal waterfowl stamp. Any activity involving T&E species requires a permit from the FWS, National Oceanic and Atmospheric Administration (NOAA), or the responsible state conservation agency. Permits for the collection of migratory birds must also be obtained from the FWS. All states regulate the collection of fur-bearing species, such as muskrats, and game mammals, such as deer. In many states, collection of large numbers of small mammals and lagomorphs requires special collection permits. Local FWS offices and state fish and wildlife agencies should provide assistance on regulations and permits that are required.

A.13.5 Euthanasia

Although most capture techniques described are designed to capture animals alive, animals generally must be sacrificed prior to preparation for contaminant residue analysis. (Exceptions include blood, fur, or feather residue analyses, which may be performed on live animals.) It is essential that humane methods be employed to sacrifice animals for analysis.

Gullet (1987) provides a detailed discussion of euthanasia methods for birds; these methods are also adaptable for mammals. Euthanasia may be achieved using either physical or chemical methods. Physical methods include cervical dislocation, decapitation, stunning, bleeding (exsanguination), and shooting. Chemical methods include lethal injection or inhalation of anesthetic or toxic gas. Questions to consider when choosing a technique (Gullet, 1987) include

- Will it interfere with residue analyses (chemical methods may confound results)?
- Is it appropriate for the size and type of animal?
- Does it present a risk to human health and safety?
- Are specialized equipment or training required?
- Is it time- and cost-effective?
- Will the technique offend the casual observer?

A.13.6 Health Concerns

Many wild animals either have parasites and pathogens that are communicable to humans or serve as vectors for them. These include ticks, mites, rabies, hantavirus, and histoplasmosis. Depending on the taxa being collected, anyone involved in collection or preparation may be exposed. To ensure the health and safety of personnel, it is imperative that disease be considered as part of the sampling protocol and that all appropriate protective measures be taken. Kunz et al. (1996) present an extensive discussion of human health concerns associated with mammalian sampling.

TABLE A.1
Comparison of Common Arthropod Sampling Techniques

Method	Method Description	Arthropods Sampled	Advantages	Disadvantages
Sticky trap	Adhesive material applied to a surface, usually cylindrical; arthropods adhere to surface upon contact	Flying or otherwise active arthropods	Simple, inexpensive, versatile, and portable	Messy; temperature affects adhesive; adhesive likely to interfere with residue analysis; removal of samples from adhesive difficult, requiring use of hazardous chemicals; quantification of area sampled difficult
Malaise trap	Fine mesh netting "tent" with baffles that guide arthropods into a collection jar that may or may not contain a killing agent/preservative	Primarily flying arthropods; crawling arthropods to a lesser degree	Versatile and simple to use; samples suitable for residue analysis (depends on use of preservative)	Expensive and bulky; catch strongly affected by trap placement; biased against Coleoptera; fewer catches per unit time; quantification of area sampled difficult
Shake-cloth	Cloth or catch basin placed beneath plant; when plant is beaten or shaken, arthropods drop onto sheet and are collected	Foliage-dwelling arthropods	Simple, fast, and easy to perform; requires minimal equipment; samples suitable for residue analysis	Biased against active arthropods and individuals that adhere tightly to vegetation; quantification of area sampled difficult
Sweep net	Among most widely used methods; insect net is swept through vegetation in a predetermined manner	Foliage-dwelling arthropods	Simple, fast, and easy to perform; requires minimal equipment; samples suitable for residue analysis	Sample efficacy highly dependent upon vegetation structure and sampling personnel; biased against arthropods that adhere tightly to vegetation; quantification of area sampled difficult

Pitfall trap	Cup or bucket (covered or uncovered) buried in ground up to rim; may or may not contain killing agent/preservative; may be employed with drift fences	Ground/litter arthropods	Simple and inexpensive; may estimate population density using mark-recapture; samples suitable for residue analysis (depends on use of preservative)	Biased against inactive arthropods; very active individuals may escape; captures affected by density and type of ground cover
Light trap	Light source (generally ultraviolet) attached to vanes and a collecting bucket; may or may not employ killing agent/preservative	Nocturnal, phototactic, predominantly flying arthropods	Portable; simple to use; collects many taxa, but Lepidoptera predominate; samples suitable for residue analysis (depends on use of preservative)	Catch affected by environmental conditions and trap placement; species-specific responses to light unknown; area sampled cannot be quantified
Pesticide knockdown	Pyrethroid insecticide applied to vegetation by a fogger; arthropods killed are collected on drop sheets	Foliage-dwelling arthropods	Simple, fast, and easy to perform; samples many arthropods with approximately equal probability	Foggers, pesticides expensive; affected by wind; may miss extremely active or sessile arthropods; pesticide may interfere with residue analysis; quantification of area sampled difficult
Emergence trap	Conical or box-shaped traps erected over water or soil to collect emerging adult arthropods	Arthropods emerging from soil or water	Inexpensive; simple to use; can estimate density of emerging arthropods; samples suitable for residue analysis	Large number may be needed to accurately estimate population
Pole pruning	Foliage samples clipped; arthropods on foliage manually removed and counted	Foliage arthropods (especially Lepidoptera larvae)	Inexpensive and easy to perform; good for inactive and tightly attached arthropods; population density can be calculated; samples suitable for residue analysis	Biased against active arthropods; few arthropods per sample; sample processing is labor intensive

continued

TABLE A.1 (continued)
Comparison of Common Arthropod Sampling Techniques

Method	Method Description	Arthropods Sampled	Advantages	Disadvantages
Portable vacuum samplers	Uses portable, generally backpack-mounted, vacuums to sample insects (Dietrick et al., 1959); widely used to sample agricultural pests	Foliage arthropods	Easy to use; population density can be calculated; samples suitable for residue analysis	Expensive (>$1000 each); best suited for low vegetation; application in forest is questionable; may not accurately sample all taxa
Stationary suction	Consists of fan that pushes air through a metallic gauze filter to remove insects (Johnson and Taylor, 1955)	Flying arthropods	Easy to use; population density can be calculated; samples suitable for residue analysis	Expensive; not very portable; use limited to areas with electrical power; difficult to sample large areas
Tree bands	Burlap bands are attached to trees; takes advantage of tendency of some arthropods to move vertically on tree trunks	Vertically mobile arthropods	Simple and inexpensive; population density may be calculated; samples suitable for residue analysis	Installation is time-consuming; biased against most flying species.

Source: Murkin et al. (1994), Cooper and Whitmore (1990), Kunz (1988a), and Southwood (1978) unless otherwise stated.

Glossary

aged chemical: A chemical that has resided in contaminated soil or sediment for a long period (e.g., years) or that was experimentally added to soil or sediment and permitted to diffuse and sorb for a period of time. Generally it is less bioavailable than a chemical freshly added to soil. Also termed a *weathered chemical.*

agent: Any physical, chemical, or biological entity that can induce an adverse or beneficial response (synonymous with stressor, but more general).

ambient media toxicity test: A toxicity test conducted with environmental media (soil, sediment, water) from a contaminated site.

analysis of effects: A phase in an ecological risk assessment in which the relationship between exposure to contaminants and effects on properties of endpoint entities are estimated along with associated uncertainties.

analysis of exposure: A phase in an ecological risk assessment in which the spatial and temporal distributions of the intensity of the contact of endpoint entities with contaminants are estimated along with associated uncertainties.

applicable or relevant and appropriate requirements (ARARs): Criteria, standards, or other regulatory requirements that might be applied to any of the media or receptors on a contaminated site. A remedial action under CERCLA must meet all ARARs independent of the associated risks.

assessment endpoint: An explicit expression of the environmental value to be protected. An assessment endpoint must include an entity and specific property of that entity.

assessor: An individual involved in the performance of a risk assessment.

background concentration: The concentration of a substance in environmental media that are not contaminated by the sources being assessed or any other local sources. Background concentrations are due to natural occurrence or regional contamination.

Baes factor: One of a commonly-used set of soil-to-plant uptake factors derived by C. F. Baes (Baes et al., 1984). We do not recommend their use.

bioaccumulation: The net accumulation of a substance by an organism due to uptake from all environmental media, including food.

bioavailability: The extent to which the form of a chemical occurring in a medium is susceptible to being taken up by an organism. A chemical is said to be bioavailable if it is in a form that is readily taken up (e.g., dissolved) rather than a less available form (e.g., sorbed to solids or to dissolved organic matter).

bioconcentration: The net accumulation of a substance by an organism due to uptake from aqueous solution.

Note: Many of these definitions are taken directly from or modified from the glossary in EPA (1998).

biosurvey: A process of counting or measuring some property of biological populations or communities in the field. An abbreviation of biological survey.

brownfield: A contaminated site that will remain in industrial or commercial use or, for some other reason, will not support natural or agricultural biotic communities.

canopy cover: A measure of the degree to which the surface is covered by above-ground vegetation. It is related to the interception of solar radiation.

carbon mineralization: The process of conversion of the carbon in organic compounds to the inorganic state (e.g., carbon dioxide).

cation exchange capacity: A measure of the capacity of clay and organic colloids to remove positive ions from soil solution.

chemicals of potential ecological concern: Chemicals that are believed to be site-related contaminants and to pose potentially significant risks to ecological endpoint receptors.

chlorosis: An abnormally yellow color of plant tissues resulting from loss of or partial failure to develop chlorophyll.

cleanup criterion: A concentration of a chemical in an environmental medium or other goal that is determined to be sufficiently protective of human health and ecological assessment endpoints.

community: The biotic community consists of all plants, animals, and microbes occupying the same area at the same time. However, the term is commonly used to refer to a subset of the community such as the fish community or the benthic macroinvertebrate community.

conceptual model: A representation of the hypothesized causal relationship between the source of contamination and the responses of the endpoint entities.

contaminant: A substance that is present in the environment due to release from an anthropogenic source and is believed to be potentially harmful.

corrective action goal: A concentration of a chemical in an environmental medium or other goal that is determined to be protective of human health and ecological assessment endpoints (cleanup criterion).

cost–benefit analysis: Methods for comparing estimates of the costs of an action or technology with its benefits.

***de minimus*:** Sufficiently small to be ignored, e.g., risks low enough to never require remediation.

***de manifestis*:** Sufficiently large to be obviously significant, e.g., risks so severe that actions are always taken to remediate them.

definitive assessment: An assessment that is intended to support a remedial decision by estimating the likelihood of endpoint effects. See **scoping assessment** and **screening assessment**.

direct effect: An effect resulting from an agent acting on the assessment endpoint or other ecological component of interest itself, not through effects on other components of the ecosystem (see **indirect effect**).

dredge spoil: Sediments dredged from a water body and deposited as waste to land or another aquatic location.

ecological risk assessment: A process that evaluates the likelihood that adverse ecological effects may occur or are occurring as a result of exposure to one or more agents.

ecosystem: The functional system consisting of the biotic community and abiotic environment occupying a specified location in space and time.

effects range–low (ER-L): The lower 10th percentile of effects concentrations in coastal marine and estuarine sediments (NOAA).

effects range–median (ER-M): The median effects concentration in coastal marine and estuarine sediments (NOAA).

efficacy assessment: Analysis of the effectiveness of remedial actions in reducing effects on human or nonhuman endpoint properties.

endpoint entity: An organism, population, species, community, or ecosystem that has been chosen for protection. The endpoint entity is, along with the endpoint property, a component of the definition of an assessment endpoint.

endpoint property: One of the set of attributes of an endpoint entity that have been chosen for protection. For example, if the endpoint entity is a fish community, endpoint properties could include the number of species, the frequency of deformities, the trophic structure, etc.

equilibrium partitioning: The transfer of a chemical among environmental media so that the relative concentrations in any two media are constant.

exposure: The contact or co-occurrence of a contaminant or other agent with a receptor.

exposure pathway: The physical route by which a contaminant moves from a source to a biological receptor. A pathway may involve exchange among multiple media and may include transformation of the contaminant.

exposure profile: The product of the characterization of exposure in the analysis phase of ecological risk assessment. The exposure profile summarizes the magnitude and spatial and temporal patterns of exposure for the scenarios described in the conceptual model.

exposure–response profile: The product of the characterization of ecological effects in the analysis phase of ecological risk assessment. The exposure-response profile summarizes the data on the effects of a contaminant, the relationship of the measures of effect to the assessment endpoint, and the relationship of the estimates of effects on the assessment endpoint to the measures of exposure.

exposure route: The means by which a contaminant enters an organism (e.g., inhalation, stomatal uptake, ingestion).

exposure scenario: A set of assumptions concerning how an exposure may take place, including assumptions about the exposure setting, stressor characteristics, and activities that may lead to exposure.

feasibility study: The component of the CERCLA assessment process that is conducted to analyze the practicality, benefits, costs, and risks associated with remedial alternatives

geographic information systems: Software that uses spatial data to generate maps or to model processes in space; commonly abbreviated as GIS.

geophagous: Consuming soil.

hyperaccumulator: An organism (usually plant) that accumulates unusually high concentrations of an element or compound, relative to concentrations in soil or another medium

indirect effect: An effect resulting from the action of an agent on components of the ecosystem, which in turn affect the assessment endpoint or other ecological component of interest (see **direct effect**). Indirect effects of chemical contaminants include reduced abundance due to toxic effects on food species or on plants that provide habitat structure.

intervention value: A screening criterion (Netherlands) based on risks to human health and ecological receptors and processes. The ecotoxicological component of the Intervention Value is the Hazardous Concentration 50 (HC50), the concentration at which 50% of species are assumed to be protected.

land farm: An area where organic wastes are tilled into the soil for disposal.

life-cycle assessment: Method for determining the relative environmental impacts of alternative products and technologies based on the consequences of their life cycle, from extraction of raw materials to disposal of the product following use.

line of evidence: A set of data and associated analysis that can be used, alone or in combination with other lines of evidence, to estimate risks. Each line of evidence is qualitatively different from any others used in the risk characterization. In ecological risk assessments of contaminated sites, the most commonly used lines of evidence are (1) biological surveys, (2) toxicity tests of contaminated media, and (3) toxicity tests of individual chemicals.

lowest observed adverse effect level (LOAEL): The lowest level of exposure to a chemical in a test that causes statistically significant differences from the controls in any measured response.

measure of effect: A measurable ecological characteristic that is related to the valued characteristic chosen as the assessment endpoint (equivalent to the earlier term **measurement endpoint**).

measure of exposure: A measurable characteristic of a contaminant or other agent that is used to quantify exposure.

mechanism of action: The process by which an effect is induced. It is often used interchangeably with **mode of action** but is usually more specific. For example, the mode of action of an agent on a population may be lethality and its mechanism of action may be crushing, acute narcosis, cholinesterase inhibition, or burning.

mechanistic model: A mathematical model that simulates the component processes of a system rather than using purely empirical relationships.

media toxicity test: A toxicity test of water, soil, sediment, or biotic medium that is intended to determine the toxic effects of exposure to that medium.

median effective concentration (EC50): A statistically or graphically estimated concentration that is expected to cause a prescribed effect in 50% of a group of organisms under specified conditions.

median lethal concentration (LC50): A statistically or graphically estimated concentration that is expected to be lethal to 50% of a group of organisms under specified conditions.

median lethal dose (LD50): A statistically or graphically estimated dose that is expected to be lethal to 50% of a group of organisms under specified conditions.

mode of action: A phenomenological description of how an effect is induced (see **mechanism of action**).

model uncertainty: The component of the uncertainty concerning an estimated value that is due to possible misspecification of a model used for the estimation. It may be due to the choice of the form of the model, its parameters, or its bounds.

Monte Carlo simulation: A resampling technique frequently used in uncertainty analysis in risk assessments to estimate the distribution of the output parameter of a model.

mycorrhiza: A symbiotic association of specialized mycorrhizal fungi with the roots of higher plants. The association often facilitates the uptake of inorganic nutrients by plants.

National Contingency Plan: National Oil and Hazardous Substances Pollution Contingency Plan. It is the regulatory framework for national response to hazardous substance releases and oil spills, including emergency removal actions.

natural attenuation: Degradation or dilution of chemical contaminants by unenhanced biological and physicochemical processes.

nitrification: The oxidation of ammonium to nitrate.

nitrogen fixation: The transformation of N_2 to ammonia by symbiotic or non-symbiotic biological processes.

no observed adverse effect level (NOAEL): The highest level of exposure to a chemical in a test that does not cause statistically significant differences from the controls in any measured response.

nonaqueous-phase liquid (NAPL): A chemical or material present in the form of an oil phase.

normalization: Alteration of a chemical concentration or other property (usually by dividing by a factor) to reduce variance due to some characteristic of an organism or its environment (e.g., division of the body burden of a chemical by the organism's lipid content to generate a lipid-normalized concentration).

octanol-water partition coefficient (K_{ow}): The quotient of the concentration of an organic chemical dissolved in octanol divided by the concentration dissolved in water if the chemical is in equilibrium between the two solvents.

parties: The organizations that participate in the decision making for a site. The representatives of all of the parties are risk managers.

phytoremediation: Remediation of contaminated soil via the accumulation of the chemicals by plants or the promotion of degradation by plants.

phytotoxicity: Toxicity to plants.

pitfall trap: A container buried in soil so that its top is flush with the surface and into which a vertebrate or invertebrate animal falls.

population: An aggregate of interbreeding individuals of a species occupying a specific location in space and time.

preliminary remedial goal (PRG): A concentration of a contaminant in a medium that serves as a default estimate of a remedial goal for receptors exposed to the contaminated medium.

probable effects level (PEL): The geometric mean of the 50th percentile of effects concentrations and the 85th percentile of no effects concentrations in coastal and estuarine sediments (Florida Department of Environmental Protection).

problem formulation: The phase in an ecological risk assessment in which the goals of the assessment are defined and the methods for achieving those goals are specified.

receptor: An organism, population or community that is exposed to contaminants. Receptors may or may not be assessment endpoint entities.

record of decision: The document presenting the final decision regarding selection of a remedial action, and justifying the decision on the basis of the results of the remedial investigation and feasibility study.

recovery: The return of a population, community, or ecosystem process to a previous, valued state. Due to the complex and dynamic nature of ecological systems, the attributes of a "recovered" system must be carefully defined.

reference, negative: An effectively uncontaminated site or the information obtained from that site used to estimate the state of the site being assessed in the absence of contamination.

reference, positive: A site (other than the site that is being assessed) or the information obtained from that site used for comparison of effects of prescribed contamination to the apparent effects of contaminants at the site being assessed.

reference value: A chemical concentration or dose that is a threshold for toxicity or significant contamination.

release: The movement of a contaminant from a source to an environmental medium.

remedial unit: An area of land or water to which a single remedial alternative applies.

remedial alternative: An action which is considered for remediation of a contaminated site. In addition to the usual engineered actions such as capping or thermal desorption, remedial alternatives may include controls on land use and the no action alternative (**natural attenuation**).

remedial goal: A contaminant concentration, toxic response, or other criterion that is selected by the risk manager to define the condition to be achieved by remedial actions.

remedial action objective: A specification of contaminants and media of concern, potential exposure pathways, and cleanup criteria (**remedial goal**).

remedial goal option: A contaminant concentration, toxic response, or other criterion that is recommended by the risk assessors as likely to achieve conditions that are sufficiently protective of the assessment endpoints.

remediation: Actions taken to reduce risks from contaminants including removal or treatment of contaminants and restrictions on land use. Remediation is the goal of the CERCLA RI/FS process. Note that, in contrast to restoration, remediation focuses on reducing risks from contaminants and may actually reduce environmental quality.

removal action: An interim remedy for an immediate threat posed by a release of hazardous substances.

restoration: Actions taken to make the environment whole, including restoring the capability of natural resources to provide services to humans. Restoration (or replacement) is the goal of the CERCLA NRDA process. Restoration goes beyond remediation to include restocking, habitat rehabilitation, reduction in harvesting during a recovery period, etc.

rhizosphere: The portion of a soil which is in the vicinity of and influenced by plant roots; includes enhanced microbial activity, nutrient mobilization, and other processes.

riparian: Occurring in or by the edge of a stream or in its floodplain.

risk assessor: An individual engaged in the performance of the technical components of risk assessments. Risk assessors may have expertise in the analysis of risk or specific expertise in an area of science or engineering relevant to the assessment.

risk characterization: A phase of ecological risk assessment that integrates the exposure and stressor response profiles to evaluate the likelihood of adverse ecological effects associated with exposure to the contaminants.

risk management: The process of deciding what remedial or restoration actions to take, justifying the decision, and implementing the decision.

risk manager: An individual with the authority to decide what actions will be taken in response to a risk. Usually risk managers are representatives of regulatory agencies, land managers, or other organizations.

rooting profile: The vertical spatial distribution of plant roots.

scoping assessment: A qualitative assessment that determines whether a hazard exists that is appropriate for a risk assessment. It determines whether contaminants are present and whether there are potential exposure pathways and receptors.

screening assessment: An assessment performed to determine the scope of a definitive assessment by eliminating from further consideration chemicals and receptors that are clearly not associated with a potential risk.

screening benchmark: A concentration or dose which is considered a threshold for concern for the elimination of contaminants from consideration in a risk assessment.

screening level concentration (SLC): An estimate of the highest concentration of a particular contaminant in sediment that can be tolerated by about 95% of benthic infauna (Neff et al., 1988).

single-chemical toxicity test: A toxicity test of an individual chemical administered to an organism or added to soil, sediment, or water to which an organism is exposed.

site: An area which has been identified as contaminated and potentially in need of remediation.

source: An entity or action that releases contaminants or other agents into the environment (primary source) or a contaminated medium that releases the contaminants into other media (secondary source). Examples of primary sources for contaminated sites include spills, leaking tanks, dumps, and waste lagoons. An example of a secondary source is contaminated sediments that release contaminants by diffusion, bioaccumulation, and exchange.

stakeholders: Individuals or organizations that have an interest in the outcome of a remedial action but are not officially parties to the decision making. Examples include natural resource agencies and citizens groups. A somewhat clearer synonym is "interested parties."

threshold effects concentration: A concentration derived from various toxicity test endpoints, on which Canadian guidelines for soil contact are based (CCME).

threshold effects level (TEL): The geometric mean of the 15th percentile of effects concentrations and the 50th percentile of no effects concentrations in coastal and estuarine sediments (Florida Department of Environmental Protection).

toxicity identification and evaluation (TIE): A process whereby the toxic components of a mixture (usually effluents) are identified by removing components of a mixture and testing the residue, fractionating the mixture and testing the fractions, or adding components of the mixture to background medium and testing the artificially contaminated medium.

treatment endpoint: A concentration of a chemical in an environmental medium or other goal for a treatment technology that is determined to be protective of human health and ecological assessment endpoints (a cleanup criterion).

unit: An area which is the object of a risk assessment. A site may be assessed as a single unit, or there may be multiple units in a site. Common variants are "operable unit," "remedial unit," and "spatial unit."

uptake: The process by which a chemical is incorporated into an organism.

uptake factor: The quotient of the concentration of a chemical assimilated in an organism divided by the concentration in an environmental medium.

vermiculite: Any of a group of hydrous silicates of aluminum, magnesium, and iron which are commonly used as a soil-substitute in horticulture.

water effect ratio: A factor by which a water quality criterion is multiplied to adjust for site-specific water chemistry.

weight of evidence: (1) A type of analysis that considers all available evidence and reaches a conclusion based on the amount and quality of evidence supporting each alternative conclusion; (2) The result of a weight-of-evidence analysis.

References

Adams, D. F. 1963. Recognition of the effects of fluorides on vegetation. *J. Air Pollut. Control Assoc.* 13:360–362.

Adams, S. M. and M. G. Ryon. 1994. A comparison of health assessment approaches for evaluating the effects of contaminant-related stress on fish populations. *J. Aquatic Ecosyst. Health* 3:15–25.

Adams, W. J. 1986. Toxicity and bioconcentration of 2,3,7,8–TCDD to fathead minnows (*Pimephales promelas*). *Chemosphere* 15:1503–1511.

Adams, W. J. 1987. Bioavailability of neutral lipophilic organic chemicals contained on sediments: a review. In K. L. Dickson, A. W. Maki, and W. A. Brungs (Eds.), *Fate and Effects of Sediment-Bound Chemicals in Aquatic Systems*. Pergamon Press, New York. 219–244.

Addy, C. E. 1956. *Guide to Waterfowl Banding*. U.S. Fish and Wildlife Service, Laurel, MD.

Akin, E. W. 1991. Supplemental Region IV Risk Assessment Guidance, U.S. EPA, Region IV, Atlanta, GA.

Alabaster, J. S. and R. Lloyd. 1982. *Water Quality Criteria for Freshwater Fish*, 2nd ed. Butterworth Scientific, London.

Aldenberg, T. 1993. E_TX 1.3a, a program to calculate confidence limits for hazardous concentrations based on small samples of toxicity data. RIVM, Bilthoven, the Netherlands.

Aldenberg, T. and W. Slob. 1993. Confidence limits for hazardous concentrations based on logistically distributed NOEC toxicity data. *Ecotoxicol. Environ. Saf.* 25:48–63.

Alexander, M. 1995. How toxic are toxic chemicals in soil? *Environ. Sci. Technol.* 29:2713–2717.

Aloi, J. E. 1990. A critical review of recent fresh-water periphyton field methods. *Can. J. Fish. Aquatic Sci.* 47(3):656–670.

Alsop, W. R., E. T. Hawkins, M. E. Stelljes, and W. Collins. 1996. Comparison of modeled and measured tissue concentrations for ecological receptors. *Hum. Ecol. Risk Assess.* 2:539–557.

Andersen, C. 1979. Cadmium, lead and calcium content, number and biomass, in earthworms (*Lumbricidae*) from sewage sludge treated soil. *Pedobiologia* 19:309–319.

Anderson, B. W. and R. D. Ohmart. 1986. Vegetation. In A. Y. Cooperrider, R. J. Boyd, and H. R. Stuart (Eds.), Inventory and Monitoring of Wildlife Habitat. U.S. Department of the Interior, Bureau of Land Management, Service Center, Denver, CO. 639–660.

Anderson, S. H. and K. J. Gutzwiller. 1994. Habitat evaluation methods. In T. A. Bookhout (Ed.), *Research and Management Techniques for Wildlife and Habitats*. 5th ed. The Wildlife Society, Bethesda, MD. 592–606.

Anderson, S. L. and T. J. Norberg-King. 1991. Precision of short-term chronic toxicity tests in the real world. *Environ. Toxicol. Chem.* 10:143–145.

Andrews, S. M., M. S. Johnson, and J. A. Cooke. 1989a. Distribution of trace element pollutants in a contaminated grassland ecosystem established on metalliferous fluorspar tailings, 1: Lead. *Environ. Pollut.* 58:73–85.

Andrews, S. M., M. S. Johnson, and J. A. Cooke. 1989b. Distribution of trace element pollutants in a contaminated grassland ecosystem established on metalliferous fluorspar tailings, 2: Zinc. *Environ. Pollut.* 59:241–252.

APHA. 1999. Standard Methods for the Examination of Water and Waste Water, 20th ed. American Public Health Association: Washington, D.C.

Aquatic Risk Assessment and Mitigation Dialog Group. 1994. Final Report. Society for Environmental Toxicology and Chemistry, Pensacola, FL.

Arenal, C. A. and R. S. Halbrook. 1997. PCB and heavy metal contamination and effects in European starlings (*Sturnus vulgaris*) at a Superfund site. *Bull. Environ. Contam. Toxicol.* 58:254–262.

Arthur, W. J., III and A. W. Alldredge. 1979. Soil ingestion by mule deer in North Central Colorado. *J. Range Manage.* 32:67–70.

Arthur, W. J., III and R. J. Gates. 1988. Trace elements intake via soil ingestion in pronghorns and in black-tailed jackrabbits. *J. Range Manage.* 41:162–166.

ASTM. 1994. Emergency standard guide for risk-based corrective action applied to petroleum release sites. ES 38-94. American Society for Testing and Materials, Philadelphia.

ASTM. 1999. *Annual Book of ASTM Standards, Sec. 11 Water and Environmental Technology.* American Society for Testing and Materials, Philadelphia, PN.

Aulerich, R. J., R. K. Ringer, and S. Iwamoto. 1973. Reproductive failure and mortality in mink fed on Great Lakes fish. *J. Reprod. Fertil.*, Suppl. 19:365–376.

Aulerich, R. J., R. K. Ringer, H. L. Seagran, and W. G. Youatt. 1971. Effects of feeding coho salmon and other Great Lakes fish on mink reproduction. *Can. J. Zool.* 49:611–616.

Baath, E., K. Arnebrant, and A. Nordgren. 1991. Microbial biomass and ATP in smelter-polluted forest humus. *Bull. Environ. Contam. Toxicol.* 47:278–282.

Baes, C. F., R. D. Sharp, A. L. Sjoreen, and R. W. Shor. 1984. A Review and Analysis of Parameters for Assessing Transport of Environmentally Released Radionuclides through Agriculture. ORNL-5786. Oak Ridge National Laboratory, Oak Ridge, TN.

Bailer, A. J. and J. T. Oris. 1997. Estimating inhibition concentrations for different response scales using generalized linear models. *Environ. Toxicol. Chem.* 16:1554–1559.

Baker, D. E. and M. C. Amacher. 1982. Nickel, copper, zinc, and cadmium. In A. L. Page, R. H. Miller, and D. R. Keeney (Eds.), *Methods of Soil Analysis, Vol. 2*, 2nd ed. American Society of Agronomy, Soil Science Society of America, Madison, WI. 323–336.

Banton, M. I., J. S. Klingensmith, D. E. Barchers, P. A. Clifford, D. F. Ludwig, A. M. Macrander, R. L. Sielken Jr., and C. Valdez-Flores. 1996. An approach for estimating ecological risks from organochlorine pesticides to terrestrial organisms at Rocky Mountain Arsenal. *Hum. Ecol. Risk Assess.* 2:499–526.

Banuelos, G. S., R. Mead, L. Wu, P. Beuselinck, and S. Akohoue. 1992. Differential selenium accumulation among forage plant species grown in soils amended with selenium-enriched plant tissue. *J. Soil Water Cons.* 47:338–342.

Barbour, M. T., J. Gerritsen, G. O. Griffith, R. Freydenborg, E. McCarron, J. S. White, and M. L. Bastian. 1996. A framework for biological criteria for Florida streams using benthic macroinvertebrates. *J. North Amer. Benthol. Soc.* 15:185–211.

Barbour, M. T., J. Gerritsen, B. D. Snyder, and J. B. Stribling. 1997. Revision to Rapid Bioassessment Protocols for Use in Streams and Rivers: Periphyton, Benthic Macroinvertebrates, and Fish. EPA 841–D-97–002. U.S. Environmental Protection Agency, Washington, D.C.

Barnthouse, L. W. 1993. Population-level effects. In G. W. Suter II (Ed.), *Ecological Risk Assessment.* Lewis Publishers, Boca Raton, FL. 247–274.

Barnthouse, L. W. 1996. Guide for Developing Data Quality Objectives for Ecological Risk Assessment at DOE Oak Ridge Operations Facilities. ES/ER/TM-815/R1. Environmental Restoration Risk Assessment Program, Lockheed Martin Energy Systems, Oak Ridge, TN.

Barnthouse, L. W. and J. Brown. 1994. Conceptual model development. Ch. 3 in Ecological Risk Assessment Issue Papers. EPA/630/R-94/009. U.S. Environmental Protection Agency, Washington, D.C.

Barnthouse, L. W. and G. W. Suter II, Eds. 1986. User's Manual for Ecological Risk Assessment. ORNL-6251. Oak Ridge National Laboratory, Oak Ridge, TN.

Barnthouse, L. W., G. W. Suter II, and A. E. Rosen. 1990. Risks of Toxic Contaminants to Exploited Fish Populations: Influence of Life History, Data Uncertainty, and Exploitation Intensity. *Environ. Toxicol. Chem.* 9:297–311.

Barnthouse, L. W., D. L. DeAngelis, R. H. Gardner, R. V. O'Neill, G. W. Suter II, and D. S. Vaughan. 1982. Methodology for environmental risk analysis. ORNL/TM-8167. Oak Ridge National Laboratory, Oak Ridge, TN.

Baron, D. 1997. All things considered. National Public Radio, March 18. Available at http://www.realaudio.com/rafiles/npr/password/nc7m1801–5.ram.

Baron, L. A., B. E. Sample, and G. W. Suter II. 1999. Ecological risk assessment of a large river-reservoir. 5. Aerial insectivorous wildlife. *Environ. Toxicol. Chem.* 18:621–627.

Baron, L. A., T. L. Ashwood, B. E. Sample, and C. Welsh. 1997. Monitoring bioaccumulation of contaminants in the belted kingfisher (*Ceryle alcyon*). *Environ. Monitor. Assess.* 47:153–185.

Barrett, G. W. 1968. The effects of an acute insecticide stress on a semi-enclosed grassland ecosystem. *Ecology* 49:1019–1035.

Bartell, S. M., J. E. Breck, R. H. Gardner, and A. L. Brenkert. 1986. Individual parameter perturbation and error analysis of fish bioenergetics models. *Can. J. Fish. Aquatic. Sci.* 43:180–188.

Bartell, S. M., R. H. Gardner, and R. V. O'Neill. 1992. *Ecological Risk Estimation.* Lewis Publishers, Ann Arbor, MI.

Barton, A., C. Berish, B. Daniel, S. Ells, T. Marshall, J. Messer, M. Powell, M. Rice, A. Sergeant, V. Serveiss, I. Sunzenauer, and M. Whitworth. 1997. Priorities for Ecological Protection: An Initial List and Discussion Document for EPA. EPA/600/S-97/002. U.S. Environmental Protection Agency, Washington, D.C.

Baud-Grasset, F., S. Baud-Grasset, and S. Safferman. 1993. Evaluation of the bioremediation of a contaminated soil with toxicity tests. *Chemosphere* 26:1365–1374.

Beall, M. L. and R. G. Nash. 1971. Organochlorine insecticide residues in soybean plant tops: root vs. vapor sorption. *Agron. J.* 63:460–464.

Bechtel Jacobs Company. 1998a. Biota-Sediment Accumulation Factors for Invertebrates: Review and Recommendations for the Oak Ridge Reservation. BJC/OR-112. Oak Ridge National Laboratory, Oak Ridge, TN.

Bechtel Jacobs Company. 1998b. Empirical Models for the Uptake of Inorganic Chemicals from Soil by Plants. BJC/OR-133. U.S. Department of Energy, Oak Ridge, TN.

Belfroid, A., J. Meiling, D. Sijm, J. Hermens, W. Seinen, and K. van Gestel. 1994. Uptake of hydrophobic halogenated aromatic hydrocarbons for food by earthworms. *Arch. Environ. Contam. Toxicol.* 27:260–265.

Belfroid, A., M. van den Berg, W. Seinen, J. Hermens, and K. van Gestel. 1995. Uptake, bioavailability, and elimination of hydrophobic compounds in earthworms (*Eisenia andrei*) in field-contaminated soil. *Environ. Toxicol. Chem.* 14:605–612.

Bell, G. P. 1990. Birds and mammals on an insect diet: A primer on diet composition analysis in relation to ecological energetics. *Stud. Avian Biol.* 13:418–422.

Bence, A. E. and W. A. Burns. 1995. Fingerprinting hydrocarbons in the biological resources of the Exxon Valdez spill area. In P. G. Wells, J. N. Butler, and J. S. Hughes (Eds.), *Exxon Valdez Oil Spill: Fate and Effects in Alaskan Waters.* American Society for Testing and Materials, Philadelphia. 84–140.

Bengtsson, G. and S. Rundgren. 1984. Ground-living invertebrates in metal-polluted forest soils. *Ambio* 13(1):29–33.

Bergman, H. L. and E. J. Dorward-King, Eds. 1997. *Reassessment of Metals Criteria for Aquatic Life Assessment*. SETAC Press, Pensacola, FL.

Bernstein, P. L. 1996. *Against the Gods: The Remarkable Story of Risk*. John Wiley & Sons, New York.

Bervoets, L., M. Baillieul, R. Blust, and R. Verheyen. 1996. Evaluations of effluent toxicity and ambient toxicity in a polluted lowland river. *Environ. Pollut.* 91:333–341.

Bevelhimer M. S., B. E. Sample, G. R. Southworth, J. J. Beauchamp, and M. J. Peterson. 1996. Estimation of Whole-Fish Contaminant Concentrations from Fish Fillet Data. ES/ER/TM-202. Oak Ridge National Laboratory, Oak Ridge, TN.

Beyer, W. N. and G. Storm. 1995. Ecotoxicological damage from zinc smelting at Palmerton, Pennsylvania. In D. J. Hoffman, B. Rattner, G. A. Burton, and J. Cairns (Eds.), *Handbook of Ecotoxicology*. Lewis Publishers, Boca Raton, FL. 596–608.

Beyer, W. N., R. L. Chaney, and B. M. Mulhern. 1982. Heavy metal concentration in earthworms from soil amended with sewage sludge. *J. Environ. Qual.* 11:381–385.

Beyer, W. N., E. Conner, and S. Gerould. 1994. Estimates of soil ingestion by wildlife. *J. Wildl. Manage.* 58:375–382.

Beyer, W. N., G. Hensler, and J. Moore. 1987. Relation of pH and other soil variables to concentrations of Pb, Cu, Zn, and Se in earthworms. *Pedobiologia* 30:172–187.

Beyer, W. N., L. J. Blus, C. J. Henny, and D. Audet. 1997. The role of sediment ingestion in exposing wood ducks to lead. *Ecotoxicology* 6:181–186.

Beyer, W. N., J. C. Franson, L. N. Locke, R. K. Stroud, and L. Sileo. 1998. Retrospective study of the diagnostic criteria in a lead-poisoning survey of waterfowl. *Environ. Contam. Toxicol.* 35:506–512.

Bilyard, G. R., H. Beckert, J. J. Bascietto, C. W. Abrams, S. A. Dyer, and L. A. Haselow. 1997. Using the data quality objectives process during the design and conduct of ecological risk assessment. DOE/EH-0544. Pacific Northwest National Laboratory, Richland, WA.

Bisessar, S. 1982. Effect of heavy metals on microorganisms in soils near a secondary lead smelter. *Water Air Soil Pollut.* 17:305–308.

Blacker, S. and D. Goodman. 1994a. Risk-based decision making: an integrated approach for efficient site cleanup. *Environ. Sci. Technol.* 28:466A-470A.

Blacker, S. and D. Goodman. 1994b. Risk-based decision making: case study: application at a Superfund cleanup. *Environ. Sci. Technol.* 28:471A-477A.

Bligh, E. G. and W. J. Dyer. 1959. Lipid extraction and purification. *Can. J. Biochem. Physiol.* 37:912–917.

Boersma, L., C. McFarlane, and F. T. Lindstrom. 1991. Mathematical model of plant uptake and translocations of organic chemicals: application to experiments. *J. Environ. Qual.* 20:137–146.

Bogomolov, D. M., S.-K. Chen, R. W. Parmelee, S. Subler, and C. A. Edwards. 1996. An ecosystem approach to soil toxicity testing: a study of copper contamination in laboratory soil microcosms. *Appl. Soil Ecol.* 4:95–105.

Bookhout, T. A. 1994. *Research and Management Techniques for Wildlife and Habitats*, 5th ed. The Wildlife Society, Bethesda, MD.

Bossert, I. and R. Bartha. 1984. The fate of petroleum in soil ecosystems. In R. Atlas (Ed.), *Petroleum Microbiology*. Macmillan, New York. 435–473.

Bouffard, S. H. 1982. Tree squirrels. In D. E. Davis (Ed.), *Handbook of Census Methods for Terrestrial Vertebrates*. CRC Press, Boca Raton, FL. 160–161.

Bowers, T. S., N. S. Shifrin, and B. L. Murphy. 1996. Statistical approach to meeting soil cleanup goals. *Environ. Sci. Technol.* 30:1437–1444.

Braham, H. W. 1982. Sea lions. In D. E. Davis (Ed.), *Handbook of Census Methods for Terrestrial Vertebrates*. CRC Press, Boca Raton, FL. 237–238.

Braunschweiler, H. 1995. Seasonal variation in the content of metals in the earthworm *Dendrobaena octaedra* (Sav.) in Finnish forest soils. *Acta Zool. Fenn.* 196:314–317.

Breda, N., A. Granier, F. Barataud, and C. Moyne. 1995. Soil water dynamics in an oak stand. *Plant and Soil* 172:17–27.

Brenkert, A. L., R. H. Gradner, S. M. Bartell, and F. O. Hoffman. 1988. Uncertainties associated with estimates of radium accumulation in lake sediments and biota. In G. Desmet (Ed.), *Reliability of Radioactive Transfer Models*. Elsevier Applied Science, London. 185–192.

Brewer, L. and K. Fagerstone, Eds. 1998. *Radiotelemetry Applications for Wildlife Toxicology Field Studies*. SETAC Press, Pensacola, FL.

Briggs, G. G., R. H. Bromilow, and A. A. Evans. 1982. Relationship between lipophilicity and root uptake and translocation of non-ionized chemicals by barley. *Pestic. Sci.* 13:495–504.

Bromilow, R. H. and K. Chamberlain. 1995. Principles governing uptake and transport of chemicals. In F. Trapp and J. C. McFarlane (Eds.), *Plant Contamination: Modeling and Simulation of Organic Chemical Processes*. Lewis Publishers, Boca Raton, FL. 37–68.

Bro-Rasmussen, F. and H. Lokke. 1984. Ecoepidemiology — A casuistic discipline describing ecological disturbances and damages in relation to their specific causes; exemplified by chlorinated phenols and chlorophenoxy acids. *Regul. Toxicol. Pharmacol.* 4:391–399.

Brueske, C. C. and G. W. Barrett. 1991. Dietary heavy metal uptake by the least shrew, *Cryptotis parva*. *Bull. Environ. Contam. Toxicol.* 47:845–849.

Bub, H. 1990. *Bird Trapping and Bird Banding*. Cornell University Press, Ithaca, NY.

Burger, J. 1993. Metals in avian feathers: bioindicators of environmental pollution. *Rev. Environ. Toxicol.* 5:203–311.

Burger, J. and M. Gochfeld. 1997. Risk, mercury and birds: relating adverse laboratory effects to field monitoring. *Environ. Res.* 75:160–172.

Burmaster, D. E. and P. D. Anderson. 1994. Principles of good practice for the use of Monte Carlo techniques in human health and ecological risk assessment. *Risk Anal.* 14:477–481.

Burmaster, D. E. and D. A. Hull. 1997. Using lognormal distributions and lognormal probability plots in probabilistic risk assessments. *Hum. Ecol. Risk Assess.* 3:235–255.

Burnham, K. P., D. R. Anderson, and J. L. Laake. 1980. Estimation of density from line transect sampling of biological populations. *Wildl. Monog.* 72:1–202.

Burns, L. A., D. M. Cline, and R. R. Lassiter. 1982. Exposure Analysis Modeling System (EXAMS): User Manual and System Documentation. EPA/600/3–82/023. U.S. Environmental Protection Agency, Environmental Research Laboratory, Athens, GA.

Burt, W. H. and R. P. Grossenheider. 1976. *A Field Guide to the Mammals of America North of Mexico*, 3rd ed. Houghton Mifflin, Boston.

Burton, G. A., Jr., Ed. 1992. *Sediment Toxicity Assessment*. Lewis Publishers, Boca Raton, FL.

Burton, K. W., E. Morgan, and A. Roig. 1984. The influence of heavy metals upon the growth of sitka-spruce in South Wales forests. II. Greenhouse experiments. *Plant Soil* 78:271–282.

Bysshe, S. E. 1988. Uptake by biota. In I. Bodek, W. J. Luman, W. F. Reehl, and D. H. Rosenblatt (Eds.), *Environmental Inorganic Chemistry: Properties, Processes, and Estimation Methods*. Pergamon Press, New York. 4-1–4.7-1.

Cadmus Group. 1996. Aquatic Ecological Risk Assessment Software and User's Manual, Version 1.1. Proj. 91-AER-1. Water Environment Research Foundation, Alexandria, VA.

Cairns, J. J., Ed. 1980. *The Recovery Process in Damaged Ecosystems*. Ann Arbor Science, Ann Arbor, MI.

Cairns, J., Jr., K. L. Dickson and E. E. Herricks, Eds. 1977. *Recovery and Restoration of Damaged Ecosystems*. University Press of Virginia, Charlottesville.

Cairns, J., Jr., K. L. Dickson, and A. W. Maki. 1979. Estimating the hazard of chemical substances to aquatic life. *Hydrobiologia* 64:157–166.

Calabrese, E. J. 1991. *Multiple Chemical Interactions*. Lewis Publishers, Chelsea, MI.

Calabrese, E. J. and L. A. Baldwin. 1993. *Performing Ecological Risk Assessments*. Lewis Publishers, Boca Raton, FL.

Calabrese, E. J. and L. A. Baldwin. 1994. A toxicological basis to derive a generic interspecies uncertainty factor. *Environ. Health Perspect*. 102:14–17.

Calabrese, E. J. and E. J. Stanek III. 1995. A dog's tale: Soil ingestion by a canine. *Ecotoxicol. Environ. Saf*. 32:93–95.

Calamari, D., M. Vighi, and E. Bacci. 1987. The use of terrestrial plant biomass as a parameter in the fugacity model. *Chemosphere* 16:2359–2364.

Calder, W. A. and E. J. Braun. 1983. Scaling of osmotic regulation in mammals and birds. *Am. J. Physiol*. 224:R601–R606.

Cal EPA (California Environmental Protection Agency). 1996. Guidance for Ecological Risk Assessment at Hazardous Waste Sites and Permitted Facilities. Department of Toxic Substances Control, Sacramento.

Call, D. J., L. T. Brook, M. L. Knuth, S. H. Poirier, and M. D. Hoglund. 1985. Fish subchronic toxicity prediction model for industrial organic chemicals that produce narcosis. *Environ. Toxicol. Chem*. 4:335–342.

Callahan, B. G., Ed. 1996. Special issue: Commemoration of the 50th anniversary of Monte Carlo. *Hum. Ecol. Risk Assess*. 2:627–1037.

Callahan, C. A. and B. D. Steele. 1998. Ecological risk assessment guidance for Superfund sites. In A. de Peyster and K. E. Day (Eds.), *Ecological Risk Assessment: A Meeting of Policy and Science*. SETAC, Pensacola, FL. 9–22.

Canfield, T. J., N. E. Kemble, W. G. Brumbaugh, F. J. Dwyer, C. G. Ingersoll, and J. F. Fairchild. 1994. Use of benthic community structure and the sediment quality triad to evaluate metal-contaminated sediment in the upper Clark Fork River, Montana. *Environ. Toxicol. Chem*. 13:1999–2012.

Canton, J. H. and D. M. M. Adema. 1978. Reproducibility of short-term and reproduction toxicity experiments with *Daphnia magna* and comparison of the sensitivity of *Daphnia magna* with *Daphnia cucullata* and *Daphnia pulex* in short-term experiments. *Hydrobiologia* 5:135–140.

Carlisle, D. M. and W. H. Clements. 1999. Sensitivity and variability of metrics used in biological assessments of running waters. *Environ. Toxicol. Chem*. 18:285–291.

Carlson, R. W. and F. A. Bazzaz. 1977. Growth reduction in American sycamore (*Plantanus occidentalis* L.) caused by Pb-Cd interaction. *Environ. Pollut*. 12:243–253.

Carlson, R. W. and G. L. Rolfe. 1979. Growth of rye grass and fescue as affected by lead-cadmium-fertilizer interaction. *J. Environ. Qual*. 8:348–352.

Cataldo, D. A. and R. E. Wildung. 1978. Soil and plant factors influencing the accumulation of heavy metals by plants. *Environ. Health Perspect*. 27:149–159.

CCME (Canadian Council of Ministers of the Environment). 1996a. Guidance Manual for Developing Site-Specific Soil Quality Remediation Objectives for Contaminated Sites in Canada. The National Contaminated Sites Remediation Program, Ottawa, Ontario.

CCME (Canadian Council of Ministers of the Environment). 1996b. A Protocol for the Derivation of Environmental and Human Health Soil Quality Guidelines. The National Contaminated Sites Remediation Program, Ottawa, Ontario.

CCME (Canadian Council of Ministers of the Environment). 1997. Recommended Canadian Soil Quality Guidelines, Ottawa, Ontario.

Chang, A. C., T. C. Granato, and A. L. Page. 1992. A methodology for establishing phytotoxicity criteria for chromium, copper, nickel, and zinc in agricultural land application of municipal sewage sludges. *J. Environ. Qual.* 21:521–536.

Chang, L. W., J. R. Meier, and M. K. Smith. 1997. Application of plant and earthworm bioassays to evaluate remediation of a lead-contaminated soil. *Arch. Environ. Contam. Toxicol.* 32:166–171.

Chapman, P. M. 1990. The sediment quality triad approach to determining pollution-induced degradation. *Sci. Total Environ.* 97/98:815–825.

Chapman, P. M., A. Fairbrother, and D. Brown. 1998. A critical evaluation of safety (uncertainty) factors for ecological risk assessment. *Environ. Toxicol. Chem.* 17:99–108.

Chapman, P. M., M. Cano, A. T. Fritz, C. Gaudet, C. A. Menzie, M. Sprenger, and W. A. Stubblefield. 1997. Workgroup summary report on contaminated site cleanup decisions. In C. G. Ingersoll, T. Dillon, and G. R. Biddinger (Eds.), *Ecological Risk Assessment of Contaminated Sediments*. SETAC Press, Pensacola, Florida. 83–114.

Charbonneau, P. and L. Hare. 1998. Burrowing behavior and biogenic structures of mud-dwelling insects. *J. North Am. Benthol. Soc.* 17:239–249.

Chatt, A. and S. A. Katz. 1988. *Hair Analysis: Applications in the Biomedical and Environmental Sciences*. VCH Publishers, New York.

Chiarenzelli, J., R. Scrudato, B. Bush, D. Carpenter, and S. Bushart. 1998. Do large-scale remedial and dredging events have the potential to release significant amounts of semivolatile compounds to the atmosphere? *Environ. Health Perspect.* 106(2):47–49.

Chung, N. and M. Alexander. 1998. Differences in sequestration and bioavailability of organic compounds aged in dissimilar soils. *Environ. Sci. Technol.* 32:855–860.

Clements, W. H. 1997. Ecological significance of endpoints used to assess sediment quality. In C. G. Ingersoll, T. Dillon, and G. R. Biddinger (Eds.), *Ecological Risk Assessment of Contaminated Sediments*. SETAC Press, Pensacola, FL. 123–134.

Clifford, P. A., D. E. Barchers, D. F. Ludwig, R. L. Sielken, J. S. Klingensmith, R. V. Graham, and M. I. Banton. 1995. An approach to quantifying spatial components of exposure for ecological risk assessment. *Environ. Toxicol. Chem.* 14:895–906.

Cogliano, V. J. 1997. Plausible upper bounds: are their sums plausible? *Risk Analysis* 17:77–84.

Coler, M., Y., C. Yi, and R. A. Coler. 1989. The determination of limestone surface-area by linear regression. *Hydrobiologia* 184:165–168.

Comber, S. D. W., A. J. Dobbs, and S. Lewis. 1996. Guideline values for sediments contaminated with dioxins and furans. *Ecotoxicol. Environ. Saf.* 35:102–108.

Committee on Pesticides and Groundwater. 1996. Risks of pesiticides to groundwater ecosystems. 1996/11E. Health Council of the Netherlands, Rijswijk.

Conant, R. 1975. *A Field Guide to Reptiles and Amphibians of Eastern and Central North America*. Houghton Mifflin, Boston.

Connell, D. W. 1990. *Bioaccumulation of Xenobiotic Compounds*. CRC Press, Boca Raton,FL.

Connell, D. W. and R. D. Markwell. 1990. Bioaccumulation in the soil to earthworm system. *Chemosphere* 20:91–100.

Cook, R. B., G. W. Suter II, and E. R. Sain. 1999. Ecological risk assessment of a large river-reservoir. 1. Introduction and background. *Environ. Toxicol. Chem.* 18(4):581–588.

Cooper, R. J. and R. C. Whitmore. 1990. Arthropod sampling methods in ornithology. *Stud. in Avian Biol.* 13:29–37.

Cooperrider, A. Y., R. J. Boyd, and H. R. Stuart. 1986. Inventory and Monitoring of Wildlife Habitat. U.S. Department of the Interior, Bureau of Land Management, Service Center, Denver, CO.

Corp, N. and A. J. Morgan. 1991. Accumulation of heavy metals from polluted soils by the earthworm, *Lumbricus rubellus*: can laboratory exposure of "control" worms reduce biomonitoring problems? *Environ. Pollut.* 74:39–52.

Cowan, C. E., D. Mackay, T. C. J. Feijtel, D. van de Meent, A. DiGuardo, J. Davies, and N. Mackay. 1995. *The Multi-Media Fate Model: A Vital Tool for Predicting the Fate of Chemicals.* SETAC Press, Pensacola, FL.

Craft, R. A. and K. P. Craft. 1996. Use of free ranging American kestrels and nest boxes for contaminant risk assessment sampling: a field application. *J. Raptor Res.* 30:207–212.

Craig, J. F. 1980. Sampling with traps. In T. Backiel and R. L. Welcomme (Eds.), *Guidelines for Sampling Fish in Inland Waters.* Food and Agricultural Organization of the United Nations, Rome, Italy. 55–70.

Cross, S. P. 1986. Bats. In A. Y. Cooperrider, R. J. Boyd, and H. R. Stuart (Eds.), *Inventory and Monitoring of Wildlife Habitat.* U.S. Department of the Interior, Bureau of Land Management, Service Center, Denver, CO. 497–518.

Crump, K. S. 1984. A new method for determining allowable daily intakes. *Fund. Appl. Toxicol.* 4:854–871.

Currie, R. S., W. L. Fairchild, and D. C. G. Muir. 1997. Remobilization and export of cadmium from lake sediments by emerging insects. *Environ. Toxicol. Chem.* 16:2333–2338.

Czarnowska, K. and K. Jopkiewicz. 1978. Heavy metals in earthworms as an index of soil contamination. *Pol. J. Soil Sci.* 11:57–62.

Danell, K. 1982. Muskrat. In D. E. Davis (Ed.), *Handbook of Census Methods for Terrestrial Vertebrates.* CRC Press, Boca Raton, FL. 202–203.

Davidson, I. W. F., J. C. Parker, and R. P. Beliles. 1986. Biological basis for extrapolation across mammalian species. *Regul. Toxicol. Pharmacol.* 6:211–237.

Davis, B. M. K. and N. C. French. 1969. The accumulation of organochlorine insecticide residues by beetles, worms, and slugs in sprayed fields. *Soil Biol. Biochem.* 1:45–55.

Davis, D. E. 1982a. *Handbook of Census Methods for Terrestrial Vertebrates.* CRC Press, Boca Raton, FL.

Davis, D. E. 1982b. Calculations used in census methods. In D. E. Davis (Ed.), *Handbook of Census Methods for Terrestrial Vertebrates.* CRC Press, Boca Raton, FL. 344–372.

Davis, W. S. and T. P. Simon, Eds. 1995. *Biological Assessment and Criteria: Tools for Water Resource Planning and Decision Making.* Lewis Publishers, Boca Raton, FL.

Dedrick, R. L. and K. B. Bischoff. 1980. Species similarities in pharmacokinetics. *Federation Proc.* 39:54–59.

DeGraaf, R. M. and M. Yamasaki. 1992. A nondestructive technique to monitor the relative abundance of terrestrial salamanders. *Wildl. Soc. Bull.* 20:260–264.

Deichmann, W. B., D. Henschler, B. Holmstedt, and G. Keil. 1986. What is there that is not a poison? A study of the *Third Defense* by Paracelsus. *Arch. Toxicol.* 58:207–213.

Deis, D. R. and D. P. French. 1998. The use of methods for injury determination and quantification from natural resource damage assessment in ecological risk assessment. *Hum. Ecol. Risk Assess.* 4:887–903.

Dennett, D. C. 1991. *Consciousness Explained.* Little, Brown, Boston.

De Pieri, L. A., W. T. Buckley, and C. G. Kowalenko. 1996. Micronutrient concentrations of commercially grown vegetables and of soil in the Lower Fraser Valley of British Columbia. *Can. J. Soil. Sci.* 76:173–182.

Depledge, M. H. and M. C. Fossi. 1994. The role of biomarkers in environmental assessment: 2. Invertebrates. *Ecotoxicology* 3:173–179.

DeShon, J. E. 1995. Development and application of the invertebrate community index (ICI). In W. S. Davis and T. P. Simon (Eds.), *Biological Assessment and Criteria: Tools for Water Resource Planning and Decision Making*. Lewis Publishers, Boca Raton, FL. 217–243.

Detenbeck, N. E., P. W. DeVore, G. J. Niemi, and A. Lima. 1992. Recovery of temperate-stream fish communities from disturbance: a review of case studies and synthesis of theory. *Environ. Manage.* 16:33–53.

Detenbeck, N. E., R. Hermanutz, K. Allen, and M. C. Swift. 1996. Fate and effects of the herbicide atrazine in flow-through wetland microcosms. *Environ. Toxicol. Chem.* 15:937–946.

Deutsch, C. V. and A. G. Journel. 1992. *GSLIB: Geostatistical Software Library and Users Guide*. Oxford University Press, Oxford, U.K.

Diamond, M. L., C. A. Page, M. A. Campbell, S. McKenna, and R. Lall. 1999. Life-cycle framework for assessment of site remediation options: method and generic survey. *Environ. Toxicol. Chem.* 18:788–800.

Dickson, K. L., W. T. Waller, J. H. Kennedy, and L. P. Ammann. 1992. Assessing the relationship between ambient toxicity and instream biological response. *Environ. Toxicol. Chem.* 11:1307–1322.

Dieter, C. D., L. D. Flake, and W. G. Duffy. 1995. Effects of phorate on ducklings in northern prairie wetlands. *J. Wildl. Manage.* 59:498–505.

Dietrick, E. J., E. I. Schlinger, and R. Van den Bosch. 1959. A new method for sampling arthropods using a suction collection machine and a Berlese funnel separator. *J. Econ. Entomol.* 52:1085–1091.

DiToro, D. M., C. S. Zarba, D. H. Hansen, W. J. Berry, R. C. Swartz, C. E. Cowan, S. P. Pavlou, H. E. Allen, N. A. Thomas, and A. P. R. Paquin. 1991. Technical basis for establishing sediment quality criteria for nonionic organic chemicals using equilibrium partitioning. *Environ. Toxicol. Chem.* 10:1541–1583.

DiToro, D. M., J. D. Mahony, D. J. Hansen, K. J. Scott, A. R. Carlson, and G. T. Ankley. 1992. Acid volatile sulfide predicts the acute toxicity of cadmium and nickel in sediments. *Environ. Sci. Technol.* 26:96–101.

DOE (U.S. Department of Energy). 1987. The Environmental Survey Manual. DOE/EH-0053. U.S. Department of Energy, Office of Environmental Audit, Washington, D.C.

DOE (U.S. Department of Energy). 1995. Remedial Investigation Report on the Chestnut Ridge Operable Unit 2 (Filled Coal Ash Pond/Upper McCoy Branch) at the Oak Ridge Y-12 Plant, Oak Ridge, Tennessee, Volume ICMain Text, DOE/OR/01-1268/V1&D2, Lockheed Martin Energy Systems, Oak Ridge, TN.

DOE (U.S. Department of Energy). 1996a. Report on the remedial investigation of Bear Creek Valley at the Oak Ridge Y-12 plant. Oak Ridge, Tennessee. Volume 6, Appendix G. Baseline Ecological Risk Assessment Report. DOE/OR/01-1455/V6&D1, Oak Ridge National Laboratory, Oak Ridge, TN.

DOE (U.S. Department of Energy). 1996b. Remedial Investigation/Feasibility Study for the Clinch River/Poplar Creek Operable Unit. DOE/OR/01-1393V1&D1. U.S. DOE, Oak Ridge, TN.

DOE (U. S. Department of Energy). 1998. Report on the Remedial Investigation of the Upper East Fork Poplar Creek Characterization Area at the Oak Ridge Y-12 Plant, Oak Ridge, Tennessee. DOE/OR/01-1641/V1&D1. U.S. DOE, Office of Environmental Management, Oak Ridge, TN.

DOI (U.S. Department of the Interior). 1986. Natural resource damage assessments; final rule. *Fed. Regist.* 51:27674–27753.

DOI (U.S. Department of the Interior). 1996. Natural resource damage assessments — Type A rule. *Fed. Regist.* 61:20560–20570.

Donker, M. H., H. Eijsackers, and F. Heimbach, Eds. 1994. *Ecotoxicology of Soil Organisms*. Lewis Publishers, Boca Raton, FL.

Donkin, S. G. and D. B. Dusenbery. 1993. A soil toxicity test using the nematode *Caenorhabditis elegans* and an effective method of recovery. *Environ. Contam. Toxicol.* 25:145–151.

Dourson, M. L. and J. F. Stara. 1983. Regulatory history and experimental support of uncertainty (safety) factors. *Reg. Toxicol. Pharmacol.* 3:224–238.

Dowdy, D. L. and T. E. McKone. 1997. Predicting plant uptake of organic chemicals from soil or air using octanol/water and octanol/air partition ratios and a molecular connectivity index. *Environ. Toxicol. Chem.* 16:2448–2456.

Dowdy, S. and S. Wearden. 1983. *Statistics for Research.* John Wiley & Sons, New York.

Dreicer, M., T. E. Hakonson, G. C. White, and F. W. Whicker. 1984. Rainsplash as a mechanism for soil contamination of plant surfaces. *Health Phys.* 46:177–187.

Dunning, J. B. 1984. Body weights of 686 species of North American birds. Western Bird Banding Association Monograph No. 1. Eldon Publ. Co., Cave Creek, AZ.

Dunning, J. B. 1993. *CRC Handbook of Avian Body Masses*. CRC Press, Boca Raton, FL.

Dymond, P., S. Scheu, and D. Parkinson. 1997. Density and distribution of *Dendrobaena octaedra* (Lumbricidae) in aspen and pine forests in the Canadian Rocky Mountains (Alberta). *Soil. Biol. Biochem.* 29:265–273.

Edwards, N. T. 1994. Effects of various uranium leaching procedures on soil: short-term vegetation growth and physiology. *Progress Report*. April 1994. Oak Ridge National Laboratory, Environmental Sciences Division. Oak Ridge, TN.

Edwards, S. C., C. L. Macleod, and J. N. Lester. 1998. The bioavailability of copper and mercury to the common nettle (*Urtica dioica*) and the earthworm *Eisenia fetida* from contaminated dredge spoil. *Water Air Soil Pollut.* 102:75–90.

Efroymson, R. A. and G. W. Suter II. 1999. Finding a niche for soil microbial toxicity tests in ecological risk assessment. *Hum. Ecol. Risk Assess.* 5:715–727.

Efroymson, R. A., M. E. Will, and G. W. Suter II. 1997b. Toxicological Benchmarks for Screening Contaminants of Potential Doncern for Effects on Soil and Litter Invertebrates and Heterotrophic Process: 1997 Revision, ES/ER/TM-126/R2. Oak Ridge National Laboratory, Oak Ridge, TN.

Efroymson, R. A., M. E. Will, G. W. Suter II, and A. C. Wooten. 1997a. Toxicological Benchmarks for Screening Contaminants of Potential Concern for Effects on Terrestrial Plants: 1997 Revision, ES/ER/TM-85/R3. Oak Ridge National Laboratory, Oak Ridge, TN.

Eganhouse, R. P. and J. A. Calder. 1976. The solubility of medium molecular weight aromatic hydrocarbons and the effects of hydrocarbon cosolutes and salinity. *Geochem. Cosmochim. Acta* 40:555–561.

Eisler, R. 1995. Electroplating wastes in marine environments: a case history at Quonset Point, Rhode Island. In D. J. Hoffman, B. Rattner, G. A. Burton, and J. Cairns (Eds.), *Handbook of Ecotoxicology.* Lewis Publishers, Boca Raton, FL. 539–548.

Emlen, J. M. 1989. Terrestrial population models for ecological risk assessment: A state-of-the-art review. *Environ. Toxicol. Chem.* 8:831–842.

Emmerling, C., K. Krause, and D. Schroder. 1997. The use of earthworms in monitoring soil pollution by heavy metals. *Z. Pflanzenernahr. Bodenk.* 160:33–39.

Entry, J. A., N. C. Vance, M. A. Hamilton, D. Zabowski, L. S. Watrud, and D. C. Adriano. 1996. Phytoremediation of soil contaminated with low concentrations of radionuclides. *Water Air Soil Pollut.* 88:167–176.

Environmental Effects Branch. 1984. Estimating "Concern Levels" for Concentrations of Chemical Substances in the Environment. U.S. Environmental Protection Agency, Washington, D.C.

Environmental Response Team. 1994a. Chlorophyll Determination. SOP#: 2030. U.S. Environmental Protection Agency, Edison, NJ.

Environmental Response Team. 1994b. Plant Biomass Determination. SOP#: 2034. U.S. Environmental Protection Agency, Edison, NJ.

Environmental Response Team. 1994c. Plant Peroxidase Activity Determination. SOP#: 2035. U.S. Environmental Protection Agency, Edison, NJ.

Environmental Response Team. 1994d. Terrestrial Plant Community Sampling. SOP#: 2037. U.S. Environmental Protection Agency, Edison, NJ.

Environmental Response Team. 1994e. Tree Coring and Interpretation. SOP#: 2036. U.S. Environmental Protection Agency, Edison, NJ.

Environmental Response Team. 1995a. Superfund Program Representative Sampling Guidance, Vol. 1: Soil, Interim Final. EPA 540/R-95/141. U.S. Environmental Protection Agency, Edison, NJ.

Environmental Response Team. 1995b. Superfund Program Representative Sampling Guidance, Vol. 2: Air (Short Term Monitoring), Interim Final. EPA 540/R-95/140. U.S. Environmental Protection Agency, Edison, NJ.

Environmental Response Team. 1995c. Superfund Program Representative Sampling Guidance, Vol. 5: Surface Water and Sediment, Interim Final. OSWER Directive 9360.4-16. U.S. Environmental Protection Agency, Edison, NJ.

Environmental Response Team. 1996. Vegetation Assessment Field Protocol. SOP#: 2038. U.S. Environmental Protection Agency, Edison, NJ.

Environmental Response Team. 1997. Superfund Program Representative Sampling Guidance, Vol. 3: Biological, Interim Final. EPA 540/R-94/XXX. U.S. Environmental Protection Agency, Edison, NJ.

Environmental Risk Characterization Work Group. 1996. *Method 3 — Environmental Risk Characterization, Guidance for Disposal Site Risk Characterization*. WSC/ORS-95-141. Massachusetts Department of Environmental Protection.

EPA (U.S. Environmental Protection Agency). 1985a. Water Quality Criteria; Availability of Documents. *Fed. Regist.* 50:30784–30796.

EPA (U.S. Environmental Protection Agency). 1985b. Statement of Work for Inorganics Analysis — Multimedia Multi-Concentration. Doc. No. OLM03.0. Contract Laboratory Program, Office of Emergency and Remedial Response, Washington, D.C.

EPA (U.S. Environmental Protection Agency). 1986a. The Risk Assessment Guidelines of 1986. EPA/600/8-87/045. Washington, D.C.

EPA (U.S. Environmental Protection Agency). 1986b. Superfund Public Health Evaluation Manual. EPA 540/1-86/060. Office of Emergency and Remedial Response, Washington, D.C.

EPA (U.S. Environmental Protection Agency). 1989a. Risk Assesment Guidance for Superfund, Vol. II: Environmental Evaluation Manual. EPA/540-1-89/001. U.S. Environmental Protection Agency, Washington, D.C.

EPA (U.S. Environmental Protection Agency). 1989b. Use of Starling Nest Boxes for Field Reproductive Studies: Provisional Guidance Document and Technical Support Document. EPA 600/8-89/056. U.S. Environmental Protection Agency, Office of Research and Development, Corvallis, OR.

EPA (U.S. Environmental Protection Agency). 1989c. Risk Assessment Guidance for Superfund, Vol. I: Human health evaluation manual. OSWER Directive 9285.701. EPA 540/1-89/002. Office of Emergency and Remedial Response, Washington, D.C.

EPA (U. S. Environmental Protection Agency). 1990a. 40 CFR Part 300. National Oil and Hazardous Substances Pollution Contingency Plan; Final Rule. *Fed. Regist.* 55:8665–8865.

EPA (U.S. Environmental Protection Agency). 1990b. Quality Assurance/Quality Control (QA/QC) Guidance for Removal Actions. EPA/540/G-90/004. Office of Solid Waste and Emergency Response, Washington, D.C.

EPA (U.S. Environmental Protection Agency). 1991a. Methods for Aquatic Toxicity Identification Evaluations: Phase I Toxicity Characterization Procedures, 2nd ed. EPA-600/6-91-003. U.S. Environmental Protection Agency, Duluth, MN.

EPA (U.S. Environmental Protection Agency). 1991b. Overview of Methods for Evaluating Effects of Pesticides on Reproduction in Birds. EPA 600/3-91/048. U.S. Environmental Protection Agency, Office of Research and Development, Corvallis, OR.

EPA (U.S. Environmental Protection Agency). 1991c. Statement of Work for Organics Analysis — Multimedia Multi-Concentration. Doc. No. OLM01.8. Contract Laboratory Program, Office of Emergency and Remedial Response, Washington, D.C.

EPA (U.S. Environmental Protection Agency). 1991d. Risk Assessment Guidance for Superfund: Volume 1 — Human Health Evaluation Manual (Part C, Risk Evaluation of Remedial Alternatives). Publication 9285.7–01C. U.S. Environmental Protection Agency, Washington, D.C.

EPA (U.S. Environmental Protection Agency). 1992a. Framework for Ecological Risk Assessment. EPA/630/R-92/001. Risk Assessment Forum, Washington, D.C.

EPA (U.S. Environmental Protection Agency). 1992b. Dermal Exposure Assessment: Principles and Applications. EPA/600/8–91/011B. Office of Health and Environmental Assessment, Washington, D.C.

EPA (U.S. Environmental Protection Agency). 1992c. Draft report: A cross-species scaling factor for carcinogen risk assessment based on equivalence of mg/kg^{3}/4/day; notice. *Fed. Regist.* 57(109):24152–24173.

EPA (U.S. Environmental Protection Agency). 1993a. Technical Basis for Deriving Sediment Quality Criteria for Nonionic Organic Contaminants for the Protection of Benthic Organisms by Using Equilibrium Partitioning. EPA-822-R-93-001. Office of Water, U.S. Environmental Protection Agency, Washington, D.C.

EPA (U.S. Environmental Protection Agency). 1993b. Wildlife Exposure Factors Handbook. Vol. I. EPA/600/R-93/187a. Office of Research and Development, Washington, D.C.

EPA (U.S. Environmental Protection Agency). 1993c. Water quality guidance for the Great Lakes system and correction; proposed rules. *Fed. Regist.* 58(72):20802–21047.

EPA (U.S. Environmental Protection Agency). 1993d. Great Lakes Water Quality Initiative Technical Support Document for the Procedure to Determine Bioaccumulation Factors. PB93–154664. National Technical Information Service, Springfield, VA.

EPA (U.S. Environmental Protection Agency). 1993e. Wildlife Criteria Portions of the Proposed Water Quality Criteria for the Great Lakes System. EPA/822/R-93/006. Office of Science and Technology, U.S. Environmental Protection Agency, Washington, D.C.

EPA (U.S. Environmental Protection Agency). 1993f. Methods for Aquatic Toxicity Identification Evaluations: Phase II Toxicity Identification Procedures for Samples Exhibiting Acute and Chronic Toxicity. EPA-600/6–92–080. U.S. Environmental Protection Agency, Duluth, MN.

EPA (U.S. Environmental Protection Agency). 1993g. Sediment Quality Criteria for Protection of Benthic Organisms — Acenaphthene. EPA-822-R-93-013. U.S. Environmental Protection Agency, Washington, D.C.

EPA (U.S. Environmental Protection Agency). 1993h. Sediment Quality Criteria for Protection of Benthic Organisms — Dieldrin. EPA-822-R-93-015. U.S. Environmental Protection Agency, Washington, D.C.

EPA (U.S. Environmental Protection Agency). 1993i. Sediment Quality Criteria for Protection of Benthic Organisms — Endrin. EPA-822-R-93-016. U.S. Environmental Protection Agency, Washington, D.C.

EPA (U.S. Environmental Protection Agency). 1993j. Sediment Quality Criteria for Protection of Benthic Organisms — Fluoranthene. EPA-822-R-93-0012. U.S. Environmental Protection Agency, Washington, D.C.

EPA (U.S. Environmental Protection Agency). 1993k. Sediment Quality Criteria for Protection of Benthic Organisms —Phenanthrene. EPA-822-R-93-014. U.S. Environmental Protection Agency, Washington, D.C.

EPA (U.S. Environmental Protection Agency). 1994a. CLP National Functional Guidelines for Inorganic Data Review. Pub. 9240.1-05. Office of Solid Waste and Emergency Response, Washington, D.C.

EPA (U.S. Environmental Protection Agency). 1994b. Methods for Measuring the Toxicity and Bioaccumulation of Sediment-Associated Contaminants with Freshwater Invertebrates. EPA-600/R-94/024. Office of Research and Development, Duluth, MN.

EPA (U.S. Environmental Protection Agency). 1994c. Water Quality Standards Handbook: 2nd ed., EPA 823/B94/005. Office of Water, Washington, D.C.

EPA (U.S. Environmental Protection Agency). 1996a. Proposed Testing Guidelines. *Fed. Regist.* 61:16486–16488.

EPA (U.S. Environmental Protection Agency). 1996b. Calculation and Evaluation of Sediment Effect Concentrations for the Amphipod *Hyalella azteca* and the midge *Chironomus riparius*. EPA 905-R96-008. Great Lakes National Program Office, Chicago, IL.

EPA (U.S. Environmental Protection Agency). 1997. Test Methods for Evaluating Solid Waste. SW-846. Office of Solid Waste, Washington, D.C.

EPA (U.S. Environmental Protection Agency). 1998. Guidelines for Ecological Risk Assessment. EPA/630/R-95/002F. Risk Assessment Forum, Washington, D.C.

EPA (U.S. Environmental Protection Agency). 1999. A Guide to Preparing Superfund Proposed Plans, Records of Decision, and Other Remedy Selection Decision Documents. EPA 540-R-98-031. Office of Solid Waste and Emergency Response, Washington, D.C.

EPA Region IV (U.S. Environmental Protection Agency). 1994. Draft Region IV Waste Management Division Sediment Screening Values for Hazardous Waste Sites. 2/16/94 version. Waste Management Division, U.S. EPA Region IV, Atlanta, GA.

EPA Region IV (U.S. Environmental Protection Agency). 1995. Ecological screening values. *Ecological Risk Assessment Bulletin No. 2*. Waste Management Division, U.S. EPA Region IV, Atlanta, GA.

Erickson, R. J. and C. E. Stephan. 1985. Calculation of the Final Acute Value for Water Quality Criteria for Aquatic Organisms. National Technical Information Service, Springfield, VA.

Ernst, W. H. O. 1998. Effects of heavy metals in plants at the cellular and organismic level. In G. Schuurmann and B. Markert (Eds.), *Ecotoxicology.* John Wiley & Sons, New York. 587–620.

Fairbrother, A. and L. A. Kapustka. 1996. Toxicity Extrapolation in Terrestrial Systems. California Environmental Protection Agency, Sacramento, CA.

Falls, J. B. 1981. Mapping of territories with playback: an accurate census method for songbirds. *Stud. in Avian Biol.* 6:86–91.

Feng, L., L. Wang, Y. Zhao, and B. Song. 1996. Effects of substituted anilines and phenols on root elongation of cabbage seed. *Chemosphere* 32:1575–1583.

Ferson, S. 1996. Automated quality assurance checks on model structure in ecological risk assessment. *Hum. Ecol. Risk Assess.* 2:558–569.

Finney, D. J. 1971. *Probit Analysis.* Cambridge University Press, Cambridge, U.K.

Fisher, S. G. and R. Woodmansee. 1994. Issue paper on ecological recovery. In Ecological Risk Assessment Issue Papers. EPA/630/R-94/009. U. S. Environmental Protection Agency, Washington, D.C. 7-1–7-54.

Fletcher, J. S., F. L. Johnson, and J. C. McFarlane. 1990. Influence of greenhouse versus field testing and taxonomic differences on plant sensitivity to chemical treatment. *Environ. Toxicol. Chem.* 9:769–776.

Ford, C. J., J. T. Byrd, J. M. Grebmeier, R. A. Harris, R. C. Moore, S. E. Madix, K. A. Newman, and C. D. Rash. 1995. Final Project Report on Arsenic Biogeochemistry in the Clinch River and Watts Bar Reservoir, Vol. 1: Main Text. ORNL/ER-206/V1. Oak Ridge National Laboratory, Oak Ridge, TN.

Forman, R. T. T. and Godron, M. 1986. *Landscape Ecology.* John Wiley & Sons, New York.

Fox, G. A. 1991. Practical causal inference for ecoepidemiologists. *J. Toxicol. Environ. Health* 33:359–373.

Fox, G. A. 1993. What biomarkers told us about the effects of contaminants on the health of fish-eating birds in the Great Lakes? The theory and a literature review. *J. Great Lakes Res.* 19:722–736.

Foxx, T. S., G. D. Tierney, and J. M. Williams. November 1984. Rooting Depths of Plants Relative to Biological and Environmental Factors. Los Alamos National Laboratory, Los Alamos, NM.

Foy, C. D., R. L. Chaney, and M. C. White. 1978. The physiology of metal toxicity in plants. *Annu. Rev. Plant Physiol.* 29:511–566.

Freireich, E. J., E. A. Gehan, D. P. Ral, L. H. Schmidt, and H. E. Skipper. 1966. Quantitative comparison of toxicity of anticancer agents in mouse, rat, hamster, dog, monkey, and man. *Cancer Chemother. Rep.* 50:219–244.

French, D. P. et al. 1994. The CERCLA Type A Natural Resource Damage Assessment Model for Coastal and Marine Environments (NRDAM/CME). Technical Documentation. U.S. Department of the Interior, Washington, D.C.

Freshman, J. S. and C. A. Menzie. 1996. Two wildlife exposure models to assess impacts at the individual and population levels and the efficacy of remediation. *Hum. Ecol. Risk Assess.* 2:481–498.

Friend, M. 1987. Field Guide to Wildlife Diseases. Resource Pub. 167. U.S. Fish and Wildlife Service, Washington, D.C.

Fries, G. F., and G. S. Marrow. 1981. Chlorobiphenyl movement from soil to soybean plants. *J. Agric. Food Chem.* 29:757–759.

Froese, K. L., D. A. Verbrugge, G. T. Ankley, G. J. Niemi, C. P. Larsen, and J. P. Giesy. 1998. Bioaccumulation of polychlorinated biphenyls from sediments to aquatic insects and tree swallow eggs and nestlings in Saginaw Bay, Michigan, USA. *Environ. Toxicol. Chem.* 17:484–492.

Fuller, T. K. 1982. Wolves. In D. E. Davis (Ed.), *Handbook of Census Methods for Terrestrial Vertebrates*. CRC Press, Boca Raton, FL. 225–226.

Galbraith, H., K. LeJeune, and J. Lipton. 1995. Metal and arsenic impacts to soils, vegetation communities and wildlife habitat in southwestern Montana uplands contaminated by smelter emissions: I. field evaluations. *Environ. Toxicol. Chem.* 14:1895–1903.

Gardner, R. H., R. V. O'Neill, J. B. Mankin, and K. D. Kumar. 1980. Comparative error analysis of six predator-prey models. *Ecology* 61:323–332.

Gardner, R. H., R. V. O'Neill, J. B. Mankin, and J. H. Carney. 1981. A comparison of sensitivity analysis and error analysis based on a stream ecosystem model. *Ecol. Modelling* 12:177–197.

Garg, P., R. D. Tripathi, U. N. Rai, S. Sinha, and P. Chandra. 1997. Cadmium accumulation and toxicity in submerged plant *Hydrilla verticillata* (L. F.) Royle. *Environ. Monitor. Assess.* 47:167–173.

Garten, C. T., Jr. 1980. Ingestion of soil by hispid cotton rats, white-footed mice, and eastern chipmunks. *J. Mammal.* 6:136–137.

Garten, C. T., Jr. 1999. Modeling the potential role of a forest ecosystem in phytostabilization and phytoextraction of ^{90}Sr at a contaminated watershed. *J. Environ. Radioactiv.* 43:305–323.

Garten, C. T., Jr. and J. R. Trabalka. 1983. Evaluation of models for predicting terrestrial food chain behavior of xenobiotics. *Environ. Sci. Technol.* 17:590–595.

Geraghty and Miller. 1994. Background Sample Investigation of Soils and Ground Water for the Portsmouth Uranium Enrichment Plant, Piketon, Ohio. U.S. Department of Energy, Piketon, OH.

Gersich, F. M., F. A. Blanchard, S. L. Applegath, and C. N. Park. 1986. The precision of daphnid (*Daphnia magna* Straus 1820) static acute toxicity tests. *Arch. Environ. Contam. Toxicol.* 15:741–749.

Gibbons, W. N. and K. R. Munkittrick. 1994. A sentinel monitoring framework for identifying fish population responses to industrial discharges. *J. Aquat. Eco. Health* 3:327–337.

Giesy, J. P., J. P. Ludwig, and D. E. Tillitt. 1994. Deformities in birds of the Great Lakes region: assigning causality. *Environ. Sci. Technol.* 28:128–135A.

Ginn, T. C. and R. A. Pastorok. 1992. Assessment and management of contaminated sediments in Puget Sound. In G. A. Burton, Jr. (Ed.), *Sediment Toxicity Assessment*. Lewis Publishers, Boca Raton, FL. 371–397.

Goddard, M. J. and D. Krewski. 1992. Interspecies extrapolation of toxicity data. *Risk Anal.* 12:315–317.

Goede, R. W. and B. A. Barton. 1990. Organismic indices and an autopsy-based assessment as indicators of health and condition of fish. *Am. Fish. Soc. Symp.* 8:93–108.

Goldstein, R. A. and J. W. Elwood. 1971. A two-compartment, three parameter model for the absorption and retention of ingested elements by animals. *Ecology* 52:935–939.

Good, I. J. 1983. *Good Thinking: The Foundations of Probability and Its Applications*. University of Minnesota Press, Minneapolis.

Goovaerts, P. 1997. *Geostats for Natural Resources Evaluation*. Oxford University Press. Oxford, U.K.

Greene, J. C., C. L. Bartels, W. J. Warren-Hicks, B. R. Parkhurst, G. L. Linder, S. A. Peterson, and W. E. Miller. 1988. Protocols for Short-Term Toxicity Screening of Hazardous Waste Sites. U.S. Environmental Protection Agency, Corvallis, OR.

Greenleaf-Jenkins, J. and R. J. Zasoski. 1986. Distribution, availability, and foliar accumulation of heavy metals from dewatered sludge applied to two acid forest soils. In S. D. West and R. J. Zasoski (Eds.), *Nutritional and Toxic Effects of Sewage Sludge in Forest Ecosystems*, College of Forest Resources, University of Washington, Seattle.

Greger, M., L. Kautsky, and T. Sandberg. 1995. A tentative model of Cd uptake in *Potamogeton pectinatus* in relation to salinity. *Environ. Exp. Bot.* 35:215–225.

Grothe, D. R., K. L. Dickson, and D. K. Reed-Judkins, Eds. 1996. *Whole Effluent Toxicity Testing: An Evaluation of Methods and Prediction of Receiving System Impacts.* SETAC Press, Pensacola, FL.

Grue, C. E., G. V. N. Powell, and M. J. McChesney. 1982. Care of nestlings by wild female starlings exposed to an organophosphate pesticide. *J. Appl. Ecol.* 19:327–335.

Grue, C. E., D. J. Hoffman, W. N. Beyer, and L. P. Franson. 1986. Lead concentrations and reproductive success in European starlings nesting within highway roadside verges. *Environ. Pollut.*, Ser. A. 42:157–182.

Grundel, R. and D. L. Dahlsten. 1991. The feeding ecology of mountain chickadees (*Parus gambelli*): Patterns of arthropod prey delivery to nestling birds, *Can. J. Zool.* 69:1793–1804.

Guillitte, O., J. Melin, and L. Wallberg. 1994. Biological pathways of radionuclides originating from the Chernobyl fallout in a boreal forest ecosystem. *Sci. Total Environ.* 157:207–215.

Gullet, P. A. 1987. Euthanasia. In M. Friend (Ed.), *Field Guide to Wildlife Diseases: General Field Procedures for Migratory Birds.* Res. Publ. 167. U.S. Fish and Wildlife Service, Washington, D.C. 59–63.

Gunderson, C. A., J. M. Kostuk, M. H. Gibbs, G. E. Napolitano, L. F. Wicker, J. E. Richmond, and A. J. Stewart. 1997. Multispecies toxicity assessment of compost produced in bioremediation of an explosives-contaminated sediment. *Environ. Toxicol. Chem.* 16:2529–2537.

Gunn, A. 1992. Use of mustard to estimate earthworm populations. *Pedobiologia* 36:65–67.

Gupta, M. and P. Chandra. 1998. Bioaccumulation and toxicity of mercury in rooted-submerged macrophyte *Vallisneria spiralis*. *Environ. Pollut.* 103:327–332.

Haim, A., I. Izhaki, and A. Golan. 1996. Rodent species diversity in pine forests recovering from fire. *Isr. J. Zool.* 42:353–359.

Halbrook, R. S., R. L. Brewer, Jr., and D. A. Buehler. 1999. Ecological risk assessment of a large river-reservoir. 7. Environmental contaminant accumulation and effects in great blue herons. *Environ. Toxicol. Chem.* 18(4):641–648.

Halbrook, R. S., L. Lewis, R. J. Aulerich, and S. J. Bursian. 1997. Mercury accumulation in mink fed fish collected from streams on the Oak Ridge Reservation. *Arch. Environ. Contam. Toxicol.* 33:312–316.

Halbrook, R. S., R. J. Aulerich, S. J. Bursian, and L. Lewis. 1999. Ecological Risk Assessment in a large river-reservoir: 8. Experimental study of the reproductive performance of mink. *Environ. Toxicol. Chem.* 18:641–648.

Hall, S. J. and M. J. C. Harding. 1997. Physical disturbance and marine benthic communities — The effects of mechanical harvesting of cockles on nontarget benthic fauna. *J. Appl. Ecol.* 34:497–517.

Hamby, D. M. 1996. Site remediation techniques supporting environmental restoration activities-a review. *Sci. of Total Environ.* 191:203–224.

Hamelink, J. L., P. F. Landrum, H. L. Bergman, and W. H. Bensen, Eds. 1994. *Bioavailability: Physical, Chemical and Biological Interactions*. SETAC Special Publication. Lewis Publishers, Boca Raton, FL.

Hamley, J. M. 1980. Sampling with gillnets. In T. Backiel and R.L. Welcomme (Eds.), *Guidelines for Sampling Fish in Inland Waters.* Food and Agricultural Organization of the United Nations, Rome, Italy. 37–53.

Hammel, W., L. Steubing, and R. Debus. 1998. Assessment of the ecotoxic potential of soil contaminants by using a soil-algae test. *Ecotoxicol. Environ. Saf.* 40:173–176.

Hammonds, J. S., F. O. Hoffman, and S. M. Bartell. 1994. An Introductory Guide to Uncertainty Analysis in Environmental and Health Risk Assessment. ES/ER/TM-35/R1. Oak Ridge National Laboratory, Oak Ridge, TN.

Hampton, N. L., R. C. Morris, and R. L. VanHorn. 1998. Methodology for conducting screening-level ecological risk assessments for hazardous waste sites. Part II: grouping ecological components. *Int. J. Environ. Pollut.* 9:47–61.

Hansch, C. and A. Leo. 1995. *Exploring QSAR: Fundamentals and Applications in Chemistry and Biology.* American Chemical Society, Washington, D.C.

Hansen, F. 1997. Policy for use of probabilistic analysis in risk assessment at the U.S. Environmental Protection Agency. Available at http://www.epa.gov/ncea/mcpolicy.htm. U.S. Environmental Protection Agency, Washington, D.C.

Hare, L., R. Carignan, and M. A. Huerta-Diaz. 1994. A field study of metal toxicity and accumulation by benthic invertebrates: Implications for the acid-volatile sulfide (AVS) model. *Limnol. Oceanog.* 39(7):1653–1668.

Harestad, A. S. and F. L. Bunnell. 1979. Home range and body weight — a reevaluation. *Ecology* 60:389–402.

Hartley, W. G. 1980. The use of electrical fishing for estimating stocks of freshwater fish. In T. Backiel and R.L. Welcomme (Eds.), *Guidelines for Sampling Fish in Inland Waters.* Food and Agricultural Organization of the United Nations, Rome, Italy. 91–95.

Hartwell, S. I., C. E. Dawson, D. H. Jordahl, and E. Q. Durell. 1995. Demonstrating a Method to Correlate Measures of Ambient Toxicity and Fish Community Diversity. CBRM-TX-95-1. Maryland Department of Natural Resources, Chesapeake Bay Research and Monitoring Division, Annapolis, MD.

Hattis, D. and D. E. Burmaster. 1994. Assessment of variability and uncertainty distributions for practical risk analyses. *Risk Anal.* 14:713–729.

Hatzinger, P. B. and M. Alexander. 1995. Effect of aging of chemicals in soil on their biodegradability and extractability. *Environ. Sci. Technol.* 29:537–545.

Haukka, J. 1991. Spatial distribution and formation of earthworm burrows. *Pedobiologia* 35:175–178.

Hays, R., L. C. Summers, and W. Seitz. 1981. Estimating Wildlife Habitat Variables. FWS/OBS-81/47. U.S. Fish and Wildlife Service, Washington, D.C.

He, Q. B. and B. R. Singh. 1994. Crop uptake of cadmium from phosphorus fertilizers: I. Yield and cadmium content. *Water Air Soil Pollut.* 74:251–265.

Heagy, P. A. and L. B. Best. 1983. Factors affecting feeding and brooding of brown thrasher nestlings. *Wilson Bull.* 95:297–303.

Health Council of the Netherlands. 1991. Quality Parameters for Terrestrial Ecosystems and Sediments. No. 91/17E. The Hague.

Heaton, S. N., S. J. Bursian, J. P. Giesy, D. E. Tillett, J. A. Render, P. D. Jones, D. A. Verbrugge, T. J. Kubiak, and R. J. Aulerich. 1995a. Dietary exposure of mink to carp from Saginaw Bay, Michigan. 1. Effects on reproduction and survival, and the potential risks to wild mink populations. *Arch. of Environ. Contam. and Toxicol.* 28:334–343.

Heaton, S. N., S. J. Bursian, J. P. Giesy, D. E. Tillett, J. A. Render, P. D. Jones, D. A. Verbrugge, T. J. Kubiak, and R. J. Aulerich. 1995b. Dietary exposure of mink to carp from Saginaw Bay, Michigan. 2. Hematology and liver pathology. *Arch. of Environ. Contam. Toxicol.* 29:411–417.

Hegdal, P. L. and B. A. Colvin. 1986. Radiotelemetry. In A. Y. Cooperrider, R. J. Boyd, and H. R. Stuart (Eds.), *Inventory and Monitoring of Wildlife Habitat.* U.S. Department of the Interior, Bureau of Land Management, Service Center, Denver, CO. 679–698.

Heiger-Bernays, W., C. Menzie, C. Montgomery, D. Edwards, and S. Panwels. 1997. A framework for biological and chemical testing of soil. In D. G. Linz and D. Nakles (Eds.), *Environmentally Acceptable Endpoints in Soil: Risk-Based Approach to Contaminated Site Management Based on Availability of Chemicals in Soil.* American Academy of Environmental Engineers, Annapolis, MD. 388–420.

Heimbach, U., P. Leonard, R. Miyakawa, and C. Able. 1994. Assessment of pesticide safety to the carabid beetle, *Poecilus cupreus*, using two different semifield enclosures. In M. H. Donker, H. Eijsackers, and F. Heimbackers (Eds.), *Ecotoxicology of Soil Organisms*. Lewis Publishers, Boca Raton, FL. 205–240.

Helfield, J. M. and M. L. Diamond. 1997. Use of constructed wetlands for urban stream restoration: a critical analysis. *Environ. Manage.* 21:329–341.

Hendriks, A. J., W.-C. Ma, J. J. Brouns, E. M. de Ruiter-Dijkman, and R. Gast. 1995. Modelling and monitoring organochlorine and heavy metal accumulation in soils, earthworms, and shrews in Rhine-Delta floodplains. *Arch. Environ. Contam. Toxicol.* 29:115–127.

Henning, M. H., N. M. Shear Weinberg, N. D. Wilson, and T. J. Iannuzzi. 1999. Distributions of key exposure factors controlling the uptake of xenobiotic chemicals by great blue herons (*Ardea herodius*) through ingestion of fish. *Hum. Ecol. Risk Assess.* 5(1):125–144.

Henriques, W. D. and K. R. Dixon. 1996. Estimating spatial distribution of exposure by integrating radiotelemetry, computer simulation, and geographic information systems (GIS) techniques. *Hum. Ecol. Risk Assess.* 2:527–538.

Hensler, G. L., and J. D. Nichols. 1981. The Mayfield method of estimating nesting success: a model, estimators, and simulation results. *Wilson Bull.* 93:42–53.

Herbes, S. E. and C. P. Allen. 1983. Lipid quantification of freshwater invertebrates: method modification for microquantification. *Can. J. Fish. Aquat. Sci.* 40:1315–1317.

Hesse, P. R. 1971. *A Textbook of Soil Chemical Analysis*. Chemical Publishing Co., New York,.

Heyer, W. R., M. A. Donnelly, R. W. McDiarmid, L. C. Hayek, and M. S. Foster. 1994. *Measuring and Monitoring of Biological Diversity: Standard Methods for Amphibians*. Smithsonian Institution Press, Washington, D.C.

Higgins, K. F., J. O. Oldemeyer, K. J. Jenkins, G. K. Clambey, and R. F. Harlow. 1994. Vegetation sampling and measurement. In T. A. Bookhout (Ed.), *Research and Management Techniques for Wildlife and Habitats*, 5th ed. The Wildlife Society, Bethesda, MD. 567–591.

Hill, A. B. 1965. The environment and disease: association or causation. *Proc. R. Soc. Med.* 58:295–300.

Hoff, D. J. and G. M. Henningsen. 1998. Extrapolating toxicity reference values in terrestrial and semi-aquatic wildlife species using uncertainty factors. Abstr., 37th Annual Meeting, Society of Toxicology, Reston, VA.

Holcombe, G. W., G. L. Phipps, and G. D. Veith. 1988. Use of aquatic lethality tests to estimate safe chronic concentrations of chemicals in initial ecological risk assessments. In G. W. Suter II and M. Lewis (Eds.), *Aquatic Toxicity and Hazard Assessment: Eleventh Symposium*. American Society for Testing and Materials, Philadelphia. 442–467.

Holmes, R. T. 1976. Body composition, lipid reserves and caloric densities of summer birds in a northern deciduous forest. *Am. Midl. Nat.* 96:281–290.

Hooda, P. S. and B. J. Alloway. 1993. Effects of time and temperature on the bioavailability of Cd and Pb from sludge-amended soils. *J. Soil. Sci.* 44:97–110.

Hope, B. K. 1995. A review of models for estimating terrestrial ecological receptor exposure to chemical contaminants. *Chemosphere* 30:2267–2287.

Hope, B. K. 1999. Assessment of risks to terrestrial receptors using uncertainty analysis — a case study. *Hum. Ecol. Risk Assess.* 5(1):145–170.

Hope, B., C. Loy, and P. Miller. 1996. Uptake and trophic transfer of barium in a terrestrial ecosystem. *Bull. Environ. Contam. Toxicol.* 56:683–689.

Host, G. E., R. R. Regal, and C. E. Stephan. 1991. Analysis of Acute and Chronic Data for Aquatic Life. PB93–154748. U.S. Environmental Protection Agency, Duluth, MN.

Houx, N. W. H. and W. J. M. Aben. 1993. Bioavailability of pollutants to soil organisms via the soil solution. *Sci. Total Environ. Suppl.* 1993:387–395.

Hubert, W. A. 1983. Passive capture methods. In L. A. Nielsen and D. L. Johnson (Eds.), *Fisheries Techniques*. American Fisheries Society, Bethesda, MD. 95–122.

Huckabee, J. W., F. O. Carton, and G. S. Kennington. 1972. Environmental Influence on Trace Elements in Hair of 15 Species of Mammals. ORNL/TM-3747. Oak Ridge National Laboratory, Oak Ridge, TN.

Huggett, R. J., R. A. Kinerle, P. M. Mehrle, and H. L. Bergman, Eds. 1992. *Biochemical, Physiological, and Histological Markers of Anthropogenic Stress*. Lewis Publishers, Boca Raton, FL.

Hughes, R. M. 1995. Defining acceptable biological status by comparing with reference conditions. In W. S. Davis and T. P. Simon (Eds.), *Biological Assessment and Criteria: Tools for Water Resource Planning and Decision Making*. Lewis Publishers, Boca Raton, FL. 31–47.

Hung, H. and D. Mackay. 1997. A novel and simple model of the uptake of organic chemicals by vegetation from air and soil. *Chemosphere* 35:959–977.

Hunter, B. A. and M. S. Johnson. 1982. Food chain relationships of copper and cadmium in contaminated grassland ecosystems. *Oikos* 38:108–117.

Hurlbert, S. H. 1984. Pseudoreplication and the design of ecological field experiments. *Ecol. Monogr.* 54:187–211.

Hutchinson, T. H., N. Scholz, and W. Guhl. 1998. Analysis of the ECETOC aquatic toxicity (EAT) database. IV. Comparative toxicity of chemical substances to freshwater versus saltwater organisms. *Chemosphere* 36:143–153.

IAEA (International Atomic Energy Agency). 1989. Evaluating the Reliability of Predictions Made Using Environmental Transfer Models. IAEA Safety Series 100. Vienna, Austria.

IAEA (International Atomic Energy Agency). 1994. Handbook of Parameter Values for the Prediction of Radionuclide Transfer in Temperate Environments. IAEA. Tech. Rep. Ser. No. 364. Vienna, Austria.

Ibrahaim, S. A., and F. W. Whicker. 1988. Comparative uptake of U and Th by native plants at a U production site. *Health Phys.* 54:413–419.

Iman, R. L. and J. C. Helton. 1988. An investigation of uncertainty and sensitivity analysis techniques for computer models. *Risk Anal.* 8:71–90.

Ingersoll, C. G., T. Dillon, and G. R. Biddinger, Eds. 1997. *Ecological Risk Assessment of Contaminated Sediments*. SETAC Press, Pensacola, FL.

Inoue, A. and K. Horikoshi. 1991. Estimation of solvent-tolerance of bacteria by the solvent parameter log *P*. *J. Ferment. Bioeng.* 71:194–196.

Ireland, M. P. 1983. Heavy metal uptake and tissue distribution in earthworms. In J. E. Satchell (Ed.), *Earthworm Ecology: From Darwin to Vermiculture*. Chapman & Hall, London. 247–265.

ISO (International Standards Organization). 1991. *Soil Determination of the Effect of Chemical Substances on the Reproduction of Earthworms*, as cited in CCME (1996b).

Isom, B. G. 1978. Benthic macroinvertebrates. In W. T. Mason, Jr. (Ed.), *Methods for the Assessment and Prediction of Mineral Mining Impacts on Aquatic Communities: A Review and Analysis Workshop Proceedings*. U.S. Fish and Wildlife Service, Harpers Ferry, WV. 67–74.

Jackson, D. R. and A. P. Watson. 1977. Disruption of nutrient pools and transport of heavy metals in a forested watershed near a lead smelter. *J. Environ. Qual.* 6:331–348.

Jackson, R. B., J. Canadell, J. R. Ehleringer, H. A. Mooney, O. E. Sala, and E. D. Schulze. 1996. A global analysis of root distributions for terrestrial biomes. *Oecologia* 108:489–511.

Jacobs, K. E. and W. D. Swink. 1982. Estimations of fish population size and sampling efficiency of electrofishing and rotenone in two Kentucky tailwaters. *North Am. J. Fish. Manage.* 2:239–248.

Jamil, K. and S. Hussain. 1992. Biotransfer of metals to the insect *Neochetina eichhornae* via aquatic plants. *Arch. Environ. Contam. Toxicol.* 22:459–463.

Janssen, M. P. M., A. Bruiuns, T. H. DeVries, and N. M. Van Straalen. 1991. Comparison of cadmium kinetics in four soil arthropod species. *Arch. Environ. Contam. Toxicol.* 20:305–312.

Janssen, R. P. T., L. Posthuma, R. Baerselman, H. A. Den Hollander, R. P. M. Van Veen, and W. J. G. M. Peijenburg. 1997. Equilibrium partitioning of heavy metals in Dutch field soils. II. Prediction of metal accumulation in earthworms. *Environ. Toxicol. Chem.* 16:2479–2488.

Jarvinen, A. W. and G. T. Ankley. 1999. *Linkage of Effects to Tissue Residues: Development of a Comprehensive Database for Aquatic Organisms*. SETAC Press, Pensacoła, FL.

Jenkins, D. W. 1979. Trace Elements in Mammalian Hair and Nails. EPA-600/4–79–049. U.S. Environmental Protection Agency, Las Vegas, NV.

Jenkins, K. D., C. R. Lee, and J. F. Hobson. 1995. A hazardous waste site at the Naval Weapons Station, Concord, CA. In G. Rand (Ed.), *Fundamentals of Aquatic Toxicology: Effects, Environmental Fate, and Risk Assessment*. Taylor & Francis, Washington, D.C. 883–901.

Jenni, K. E., M. W. Merkhofer, and C. Williams. 1995. The rise and fall of a risk-based priority system: lessons from DOE's environmental restoration priority system. *Risk Analysis* 15:397–410.

Jiang, Q. Q. and B. R. Singh. 1994. Effect of different forms and sources of arsenic on crop yield and arsenic concentration. *Water Air Soil Pollut.* 74:321–343.

Johnson, D. H. 1979. Estimating nest success: the Mayfield method and an alternative. *Auk* 96: 651–661.

Johnson, C. G. and L. R. Taylor. 1955. The development of large suction traps for airborne insects. *Ann. Appl. Biol.* 43: 51–62.

Johnson, G. D., D. J. Audet, J. W. Kern, L. J. LeCaptain, M. D. Strickland, D. J. Hoffman, and L. L. McDonald. 1999. Lead exposure in passerines inhabiting lead-contaminated floodplains in the Coeur d'Alene River Basin, Idaho, USA. *Environ. Toxicol. Chem.* 18:1190–1194.

Johnson, P. 1991. *The Birth of the Modern*. Harper Collins, New York.

Johnson, R. R., B. T. Brown, L. T. Haight, and J. M. Simpson. 1981. Playback recordings as a special avian census technique. *Stud. Avian Biol.* 6:68–75.

Jolly, G. M. 1981. Mark–recapture: What next? *Stud. in Avian Biol.* 6:137–138.

Jones, C., W. J. McShea, M. J. Conroy, and T. H. Kunz. 1996. Capturing mammals. In Wilson, D. E., F. R. Cole, J. D. Nichols, R. Rudran, and M. S. Foster (Eds.), *Measuring and monitoring biological diversity: Standard methods for mammals*. Smithsonian Institution Press. Washington, D.C. 115–155.

Jones, D. S., G. W. Suter II, and R. N. Hull. 1997. Toxicological Benchmarks for Screening Potential Contaminants of Concern for Effects on Sediment-Associated Biota: 1997 revision. ES/ER/TM-95/R3. Oak Ridge National Laboratory, Oak Ridge, TN.

Jones, D. S., L. W. Barnthouse, G. W. Suter II, R. A. Efroymson, J. M. Field, and J. J. Beauchamp. 1999. Ecological risk assessment in a large river-reservoir: 3. Benthic invertebrates. *Environ. Toxicol. Chem.* 18(4):599–609.

Jones, K. B. 1986. Amphibians and reptiles. In A. Y. Cooperrider, R. J. Boyd, and H. R. Stuart (eds.), Inventory and Monitoring of Wildlife Habitat. U.S. Department of the Interior, Bureau of Land Management, Service Center, Denver, CO. 267-290.

Jongbloed, H. R., T. P. Traas, and R. Luttik. 1996. A probabilistic model for deriving soil quality criteria based on secondary poisoning of top predators. II. Calculations for Dichlorodiphenyltrichloroethane (DDT) and cadmium. *Ecotoxicol. Environ. Saf.* 34: 279–306.

Jordan, M. J. and M. P. Lechevalier. 1975. Effects of zinc-smelter emissions on forest soil microflora. *Can. J. Microbiol.* 21:1855–1865.

Julliet, J. A. 1963. A comparison of four types of traps used for capturing flying insects. *Can. J Zool.* 41:219–223.

Kähkönen, M. A. and P. K. G. Manninen. 1998. The uptake of nickel and chromium from water by *Elodea canadensis* at different nickel and chromium exposure levels. *Chemosphere* 36:1381–1390.

Kähkönen, M. A., M. Pantsar-Kallio, and P. K. G. Manninen. 1997. Analysing heavy metal concentrations in the different parts of *Elodea canadensis* and surface sediment with PCA in two boreal lakes in southern Finland. *Chemosphere* 35:2645–2656.

Kammenga, J. E., C. A. M. Van Gestel, and J. Bakker. 1994. Patterns of sensitivity to cadmium and pentachlorophenol (among nematode species from different taxonomic and ecological groups). *Arch. Environ. Contam. Toxicol.* 27:88–94.

Kammenga, J. E., P. H. G. Van Koert, J. A. G. Riksen, G. W. Korthals, and J. Bakker. 1996. A toxicity test in artificial soil based on the life history strategy of the nematode *Plectus acuminatus*. *Environ. Toxicol. Chem.* 15:722–727.

Kaplan, E. L. and P. Meier. 1958. Nonparametric estimation from incomplete observations. *J. Amer. Stat. Assoc.* 53:457–481.

Kaplan, I., S.-T. Lu, R.-P. Lee, and G. Warrick. 1996. Polycyclic hydrocarbon biomarkers confirm selective incorporation of petroleum in soil and kangaroo rat liver samples near an oil well blowout site in the western San Joaquin Valley, California. *Environ. Toxicol. Chem.* 15:696–707.

Kapustka, L. A. 1997. Selection of phytotoxicity tests for use in ecological risk assessments. In W. Wang, J. W. Gorsuch, and J. S. Hughes (Eds.), *Plants for Environmental Studies*. Lewis Publishers, Boca Raton, FL. 515–548.

Kapustka, L. A., J. Lipton, H. Galbraith, D. Cacela, and K. LeJeune. 1995. Metal and arsenic impacts to soils, vegetation communities and wildlife habitat in southwest Montana uplands contaminated by smelter emissions: II. Laboratory phytotoxicity studies. *Environ. Toxicol. Chem.* 14:1905–1912.

Karickhoff, W. W. 1981. Semi-empirical estimation of sorption of hydrophobic pollutants on natural sediments and soils. *Chemosphere* 10:833–846.

Karlson, U. and W. T. Frankenberger, Jr. 1989. Accelerated rates of selenium volatilization from California soils. *Soil Sci. Soc. Am. J.* 53:749–753.

Karr, J. R. 1981. Surveying birds with mist nets. *Stud. Avian Biol.* 6:62–67.

Karr, J. R. and E. W. Chu. 1997. Biological monitoring: essential foundation for ecological risk assessment. *Hum. Ecol. Risk Assess.* 3:993–1004.

Karr, J. R., K. D. Fausch, P. L. Angermeier, P. R. Yant, and I. J. Schlosser. 1986. Assessing Biological Integrity in Running Waters; A Method and Its Rationale. Illinois Natural History Survey Special Pub. 5. Champaigne, IL.

Keddy, C. J., J. C. Greene, and M. A. Bonnell. 1995. Review of whole-organism bioassays: soil, freshwater sediment, and freshwater assessment in Canada. *Ecotoxicol. Environ. Chem.* 30:221–251.

Keith, L. H. 1994. Throwaway data. *Environ. Sci. Technol.* 28(8):389–390A.

Kelsey, J. W. and M. Alexander. 1997. Declining bioavailability and inappropriate estimation of risk of persistent compounds. *Environ. Toxicol. Chem.* 16:582–585.

Kelsey, J. W., B. D. Kottler, and M. Alexander. 1997. Selective chemical extractants to predict bioavailability of soil-aged organic chemicals. *Environ. Sci. Technol.* 31:214–217.

Kerans, B. L. and J. R. Karr. 1994. A benthic index of biotic integrity (B-IBI) for rivers of the Tennessee Valley. *Ecol. Appl.* 4:768–785.

Kerans, B. L., J. R. Karr, and S. A. Ahlstedt. 1992. Aquatic invertebrate assemblages – spatial and temporal differences among sampling protocols. *J. North Am. Benthol. Soc.* 11:377–390.

Kerr, D. R. and J. P. Meador. 1996. Modeling does response using generalized linear models. *Environ. Toxicol. Chem.* 15:395–401.

Kester, J. E., R. L. VanHorn, and N. L. Hampton. 1998. Methodology for conducting screening-level ecological risk assessments for hazardous waste sites. Part III: exposure and effects assessment. *Int. J. Environ. Pollut.* 9:62–89.

Keyes, B. E. and C. E. Grue. 1982. Capturing birds with mist nets: a review. *North Am. Bird Bander* 7:2–14.

Kier, L. B. and L. H. Hall. 1986. *Molecular Connectivity in Structure–Activity Analysis.* Research Studies Press, Letchworth, Hertfordshire, U.K., as cited in Dowdy and McKone (1997).

Kinerson, R. S., J. S. Mattice, and J. F. Stine. 1996. The Metals Translator: Guidance for Calculating a Total Recoverable Permit Limit from a Dissolved Criterion. EPA 823–B-96–007. Office of Water, U.S. Environmental Protection Agency, Washington, D.C.

Klapow, L. A. and R. H. Lewis. 1979. Analysis of toxicity data for California marine water quality standards. *J. Water Pollut. Control Fed.* 51:2054–2070.

Koch, A. L. 1966. The logarithm in biology 1. Mechanisms generating the log-normal distribution exactely. *J. Theoret. Biol.* 12:276–290.

Kocher, D. C. and F. O. Hoffman. 1991. Regulating environmental carcinogens: where do we draw the line? *Environ. Sci. Technol.* 25:1986–1989.

Kolar, C. S., P. L. Hudson, and J. F. Savino. 1997. Conditions for the return and simulation of the recovery of burrowing mayflies in western Lake Erie. *Ecol. Appl.* 7:665–676.

Kondolf, G. M. 1996. Salmon spawning habitat rehabilitation on the Merced River, California — An evaluation of project-planning and performance. *Trans. Am. Fish. Soc.* 125:899–912.

Kondolf, G. M. and E. R. Micheli. 1995. Evaluating stream restoration projects. *Environ. Manage.* 19:1–15.

Konemann, H. 1981. Fish toxicity tests with mixtures of more than two chemicals: a proposal for a quantitative approach and experimental results. *Toxicology* 19:229–238.

Kooijman, S. A. L. M. 1987. A safety factor for LC_{50} values allowing for differences in sensitivity among species. *Water Res.* 21:269–276.

Kowal, N. E. 1971. Models of elemental assimilation by invertebrates. *J. Theor. Biol.* 31:469–474.

Krebs, C. J. 1989. *Ecological Methodology*, Harper & Row, New York.

Kszos, L. A., A. J. Stewart, and P. A. Taylor. 1992. An evaluation of nickel toxicity to *Ceriodaphnia dubia* and *Daphnia magna* in a contaminated stream and in laboratory toxicity tests. *Environ. Toxicol. Chem.* 11:1001–1012.

Kufeld, R. C., J. H. Olterman, and D. C. Bowden. 1982. Mule deer. In D. E. Davis (Ed.), *Handbook of Census Methods for Terrestrial Vertebrates*. CRC Press, Boca Raton, FL. 259–261.

Kunz, T. H. 1988a. Methods of assessing the availability of prey to insectivorous bats. In T. H. Kunz (Ed.), *Ecological and Behavioral Methods for the Study of Bats*. Smithsonian Institution Press, Washington, D.C. 191–210.

Kunz, T. H. 1988b. *Ecological and Behavioral Methods for the Study of Bats*. Smithsonian Institution Press, Washington, D.C.

Kunz, T. H., R. Rudran, and G. Gurry-Glass. 1996. Human health concerns. In D. E. Wilson, F. R. Cole, J. D. Nichols, R. Rudran, and M. S. Foster (Eds.), *Measuring and Monitoring Biological Diversity: Standard Methods for Mammals*. Smithsonian Institution Press, Washington, D.C. 255–264.

Kunz, T. H., D. W. Thomas, G. C. Richards, C. R. Tidemann, E. D. Pierson, and P. A. Racey. 1996. Observational techniques for bats. In D. E. Wilson, F. R. Cole, J. D. Nichols, R. Rudran, and M. S. Foster (Eds.), *Measuring and Monitoring Biological Diversity: Standard Methods for Mammals*. Smithsonian Institution Press, Washington, D.C. 105–114.

Labieniec, P. A., D. A. Dzombak, and R. L. Siegrist. 1996. Risk variability due to uniform soil remediation goals. *J. Environ. Eng.* 122:612–621.

LaGoy, P. K. and C. O. Schulz. 1993. Background sampling: an example of the need for reasonableness in risk assessment. *Risk Anal.* 13:483–484.

Lakin, H. W. 1972. Selenium accumulation in soils and its absorption by plants and animals. *Geol. Soc. Am. Bull.* 83:181–90.

Lamberson, J. O., T. H. DeWitt, and R. C. Swartz. 1992. Assessment of sediment toxicity to marine benthos. In G. A. Burton, Jr. (Ed.), *Sediment Toxicity Assessment*. Lewis Publishers, Boca Raton, FL.

Lancia, R. A., J. D. Nichols, and K. H. Pollock. 1994. Estimating the number of animals in wildlife populations. In T. A. Bookhout (Ed.), *Research and Management Techniques for Wildlife and Habitats*, 5th ed. The Wildlife Society, Bethesda, MD. 215–253.

Larsson, P. 1984. Transport of PCBs from aquatic to terrestrial environments by emerging chironomids. *Environ. Pollut.* Ser. A 34:283–289.

Lasiewski, R. C. and W. A. Calder, Jr. 1971. A preliminary allometric analysis of respiratory variables in resting birds. *Resp. Physiol.* 11:152–166.

Laskowski, R., P. Kramarz, and P. Jepson. 1998. Selection of species for soil ecotoxicity testing. In H. Lokke and C. A. M. Van Gestel (Eds.), *Handbook of Soil Invertebrate Toxicity Tests*. John Wiley & Sons. Chichester, U.K. 21–32.

Lassiter, R. R. and T. G. Hallam. 1990. Survival of the fattest: implications for acute effects of lippophilic chemicals on aquatic populations. *Environ. Toxicol. Chem.* 9:585–595.

Lautenschlager, R. A. 1982. Deer (Track-Pellet). In D. E. Davis (Ed.), *Handbook of Census Methods for Terrestrial Vertebrates*. CRC Press, Boca Raton, FL. 249–250.

Layher, W. G. and O. E. Maughan. 1984. Comparison efficiencies of three sampling techniques for estimating fish populations in small streams. *Prog. Fish. Cult.* 46:180–184.

Lazim, M. N., M. A. Learner, and S. Cooper. 1989. The importance of worm identity and life-history in determining the vertical-distribution of tubificids (Oligochaeta) in a riverine mud. *Hydrobiologia* 178:81–92.

Lee, K. E. 1983. Earthworms of tropical regions — some aspects of their ecology and relationships with soils. In J. E. Satchell (Ed.), *Earthworm Ecology: From Darwin to Vermiculture*. Chapman & Hall, London. 179–193.

Lee, K. E. 1985. *Earthworms: Their Ecology and Relationships with Soils and Land Use*. Academic Press, Sydney.

LeJeune, K., H. Galbraith, J. Lipton, and L. A. Kapustka. 1996a. Effects of metals and arsenic on riparian communities in southwest Montana. *Ecotoxicol.* 5:297–312.

LeJeune, K., D. Beltman, D. Cacela, and D. Lane. 1996b. Site-specific soil metals toxicity and bioaccumulation evaluated using native taxa. In *Reprints in Environmental Toxicology and Chemistry*. SETAC '96. Annual Meeting of the Society of Environmental Toxicology and Chemistry, November 1996, Washington, D.C. Hagler Bailly Consulting. 23–32.

Li, X., Y. Feng, and N. Sawatsky. 1997. Importance of soil-water relations in assessing the endpoint of bioremediated soils. *Plant Soil* 192:219–226.

Licht, D. S, D. G. McCauley, J. R. Longcore, and F. Sepik. 1989. An improved method to monitor nest attentiveness using radiotelemetry. *J. Field Ornithol.* 60:251–258.

Lindberg, S. E., D. R. Jackson, J. W. Huckabee, S. A. Janzen, M. J. Levin, and J. R. Lund. 1979. Atmospheric emission and plant uptake of mercury from agricultural soils near the Almaden mercury mine. *J. Environ. Qual.* 8:572–578.

Linder, G., J. C. Greene, H. Ratsch, J. Nwosu, S. Smith, and D. Wilborn. 1990. Seed germination and root elongation toxicity tests in hazardous waste site evaluation: methods development and applications. In W. Wang, J. W. Gorsuch, and W. R. Lower (Eds.), *Plants for Toxicity Assessment*. ASTM STP 1091. American Society for Testing and Materials, Philadelphia. 177–187.

Linder, G., E. Ingham, C. J. Brandt, and G. Henderson. 1992. Evaluation of Terrestrial Indicators for Use in Ecological Assessments at Hazardous Waste Sites. EPA/600/R-92/183. U.S. Environmental Protection Agency, Office of Research and Development, Corvallis, OR.

Lindstrom, F. T., L. Boersma, and C. McFarlane. 1991. Mathematical model of plant uptake and translocation of organic chemicals: development of the model. *J. Environ. Qual.* 20:129–136.

Lishak, R. S. 1982. Thirteen-lined ground squirrel. In D. E. Davis (Ed.), *Handbook of Census Methods for Terrestrial Vertebrates*. CRC Press, Boca Raton, FL. 156–158.

Litaor, M. I., M. L. Thompson, G. R. Barth, and P. C. Molzer. 1994. Plutonium-239+240 and americium-241 in soils east of Rocky Flats, Colorado. *J. Environ. Qual.* 23:1231–1239.

LMES (Lockheed Martin Energy Systems). 1997. Preliminary Remediation Goals for Ecological Endpoints. ES/ER/TM-162/R2. Department of Energy. Oak Ridge, TN.

Lokke, H. 1994. Ecotoxicological extrapolation: Tool or toy. In M. Donker, H. Eijsackers, and F. Heimbach (Eds.), *Ecotoxicology of Soil Organisms*. Lewis Publishers, Boca Raton, FL. 411–425.

Lokke, H. and C. A. M. Van Gestel, Ed. 1998. *Handbook of Soil Invertebrate Toxicity Tests*. John Wiley & Sons. Chichester, U.K.

Long, E. R. and P. M. Chapman. 1985. A sediment quality triad: measures of sediment contamination, toxicity and infaunal community composition in Puget Sound. *Mar. Pollut. Bull.* 16:405–415.

Long, E. R. and D. D. MacDonald. 1998. Recommended uses of empirically-derived, sediment quality guidelines for marine and estuarine ecosystems. *Hum. Ecol. Risk Assess.* 4:1019–1039.

Long, E. R. and L. G. Morgan. 1991. The Potential for Biological Effects of Sediment-Sorbed Contaminants Tested in the National Status and Trends Program. NOSOMA 52. National Oceanic and Atmospheric Administration, Washington, D.C.

Long, E. R., D. D. MacDonald, S. L. Smith, and F. D. Calder. 1995. Incidence of adverse biological effects within ranges of chemical concentrations in marine and estuarine sediments. *Environ. Manage.* 19:81–97.

Longhurst, W. M. and G. E. Connolly. 1982. Deer (Pellet count). In D. E. Davis (Ed.), *Handbook of Census Methods for Terrestrial Vertebrates.* CRC Press, Boca Raton, FL. 247–248.

Lord, K. A., G. C. Brigs, M. C. Neale, and R. Manlove. 1980. Uptake of pesticides from water and soil by earthworms. *Pestic. Sci.* 11:401–408.

Ludwig, J. P., H. Kurita-Matsuba, H. J. Auman, M. E. Ludwig, C. L. Summer, J. P. Giesy, D. E. Tillit, and P. D. Jones. 1996. Deformities, PCBs, and TCDD-equivalents in double-crested cormorants (*Phalacrocorax auritus*) and Caspian terns (*Hydroprogne caspia*) of the upper Great Lakes 1986–1991: Testing a cause–effect hypothesis. *J. Great Lakes Res.* 22:172–197.

Luoma, S. N. and N. Fisher. 1997. Uncertainties in assessing contaminant exposure from sediments. In C. G. Ingersoll, T. Dillon, and G. R. Biddinger (Eds.), *Ecological Risk Assessment of Contaminated Sediments.* SETAC Press, Pensacola, FL. 211–238.

Lyman, W. J., W. F. Reehl, and D. H. Rosenblatt. 1982. *Handbook of Chemical Property Estimation Methods.* McGraw-Hill, New York.

Ma, W.-C. 1982. The influence of soil properties and worm-related factors on the concentration of heavy metals in earthworms. *Pedobiologia* 24:109–119.

Ma, W.-C. 1987. Heavy metal accumulation in the mole, *Talpa euorpa*, and earthworms as an indicator of metal bioavailability in terrestrial environments. *Bull. Environ. Contam. Toxicol.* 39:933–938.

Ma, W.-C. 1994. Methodological principles of using small mammals for ecological hazard assessment of chemical soil pollution, with examples of cadmium and lead. In M. H. Donker, H. Eijsackers, and F. Heimbach (Eds.), *Ecotoxicology of Soil Organisms.* Lewis Publishers, Boca Raton, FL. 357–371.

Ma, W.-C., J. Immerzeel, and J. Bodt. 1995. Earthworm and food interactions on bioaccumulation and disappearance in soil of polycyclic aromatic hydrocarbons: studies on phenanthrene and fluorene. *Ecotoxicol. Environ. Saf.* 32:226–232.

Ma, W.-C., T. Edelman, I. van Beersum, and T. Jans. 1983. Uptake of cadmium, zinc, lead, and copper by earthworms near a zinc-smelting complex: influence of soil pH and organic matter. *Bull. Environ. Contam. Toxicol.* 30:424–427.

Ma, W.-C., A. van Keunen, J. Immerzeel, and P. G. de Maagd. 1998. Bioaccumulation of polycyclic aromatic hydrocarbons by earthworms: assessment of equilibrium partitioning theory in in situ studies and water experiments. *Environ. Toxicol. Chem.* 17:1730–1737.

MacDonald, D. D. 1994. Approach to the Assessment of Sediment Quality in Florida Coastal Waters. Volume 1 — Development and Evaluation of Sediment Quality Assessment Guidelines. Florida Department of Environmental Protection, Tallahassee, FL.

MacDonald, D. D., B. L. Charlish, M. L. Haines, and K. Brydges. 1994. Approach to the Assessment of Sediment Quality in Florida Coastal Waters. Florida Department of Environmental Protection, Tallahassee, FL.

MacDonald, D. D., R. S. Carr, F. D. Calder, E. R. Long, and C. G. Ingersoll. 1996. Development and evaluation of sediment quality guidelines for Florida coastal waters. *Ecotoxicol.* 5:253–278.

MacIntosh, D. L., G. W. Suter II, and F. O. Hoffman. 1994. Uses of probabilistic exposure models in ecological risk assessments of contaminated sites. *Risk Anal.* 14:405–419.

Mackay, D. 1991. *Multimedia Environmental Models*. Lewis Publishers, Boca Raton, FL.

Mallory, M. L. and P. J. Weatherhead. 1992. A comparison of three techniques for monitoring avian nest attentiveness and weight change. *J. Field Ornithol.* 63:428–435.

Marcot, B. G. and R. Holthausen. 1987. Analyzing population viability of the spotted owl in the Pacific Northwest. *Trans. North Am. Wildl. Nat. Res. Conf.* 52:333–347.

Marinussen, M. P. J. C. and S. E. A. T. M. Van der Zee. 1997. Cu accumulation by *Lumbricus rubellus* as affected by total amount of Cu in soil, soil moisture and soil heterogeneity. *Soil. Biol. Biochem.* 29:641–647.

Marion, W. R., T. E. O'Meara, and D. S. Mohair. 1981. Use of playback recordings in sampling of elusive or secretive birds. *Stud. Avian Biol.* 6:81–85.

Markwell, R. D., D. W. Connell, and A. J. Gabrick. 1989. Bioaccumulation of lipophilic compounds in sediments by oligochaetes. *Water Res.* 23:1443–1450.

Martinucci, G. B., P. Crespi, P. Omodeo, G. Osella, and G. Traldi. 1983. Earthworms and TCDD (2,3,7,8–tetrachlorodibenzo-*p*-dioxin) in Seveso. In J. E. Satchell (Ed.), *Earthworm Ecology*. Chapman & Hall, New York. 275–283.

Maughan, J. T. 1993. *Ecological Assessment of Hazardous Waste Sites*. Van Nostrand Reinhold, New York.

Mayfield, H. F. 1975. Suggestions for calculating nest success. *Wilson Bull.* 87:456–466.

McCaffery, K. R. 1982. Deer trail survey. In D. E. Davis (Ed.), *Handbook of Census Methods for Terrestrial Vertebrates*. CRC Press, Boca Raton, FL. 257–258.

McCarty, L. S. and D. Mackay. 1993. Enhancing ecotoxicological modeling and assessment. *Environ. Sci. Technol.* 27:1719–1728.

McDaniels, T. L., L. J. Axelrod, N. S. Cavanagh, and P. Slovic. 1997. Perception of ecological risk to water environments. *Risk Anal.* 17:341–352.

McKay, D. and P. C. Singleton. 1974. Time Required to Reclaim Land Contaminated with Crude Oil — An Approximation. B-612. Agricultural Extension Service, University of Wyoming, Laramie.

McKim, J. M. 1985. Early life stage toxicity tests. In G. M. Rand and S. R. Petrocelli (Eds.), *Fundamentals of Aquatic Toxicology*. Hemisphere Publishing, Washington, D.C. 58–95.

McKone, T. E. 1993. CalTOX, a Multi-Media Total-Exposure Model for Hazardouse Waste Sites, Part II: The Dynamic Multi-Media Transport and Transformation Model. UCRL-CR111456PtII. Lawrence Livermore National Laboratory, Livermore, CA.

McKone, T. E. 1994. Uncertainty and variability in human exposure to soil contaminants through home-grown food: a Monte Carlo assessment. *Risk Anal.* 14:449–463.

McNab, B. K. 1963. Bioenergetics and the determination of home range size. *Am. Nat.* 97:133–140.

Means, J. C., S. G. Wood, J. J. Hassett, and W. L. Banwart. 1980. Sorption of polynuclear aromatic hydrocarbons by sediments and soils. *Environ. Sci. Technol.* 14:1524–1528.

Meier, J. R., L. W. Chang, S. Jacobs, J. Torsella, M. C. Meckes, and M. K. Smith. 1997. Use of plant and earthworm bioassays to evaluate remediation of soil from a site contaminated with polychlorinated biphenyls. *Environ. Toxicol. Chem.* 16:928–938.

Menzel, D. B. 1987. Physiological pharmacokinetic modeling. *Environ. Sci. Technol.* 21:944–950.

Menzie, C. A., D. E. Burmaster, D. S. Freshman, and C. Callahan. 1992. Assessment of methods for estimating ecological risk in the terrestrial component: a case study at the Baird and McGuire Superfund Site in Holbrook, Massachusetts. *Environ. Toxicol. Chem.* 11:245–260.

Menzie, C., M. H. Henning, J. Cura, K. Finkelstein, J. Gentile, J. Maughan, D. Mitchell, S. Petron, B. Potocki, S. Svirsky, and P. Tyler. 1996. A weight-of-evidence approach for evaluating ecological risks: report of the Massachusetts Weight-of-Evidence Work Group. *Hum. Ecol. Risk Assess.* 2:277–304.

Merritt, R. W. and K. W. Cummins. 1984. *An Introduction to the Aquatic Invertebrates of North America.* Kendall/Hunt, Dubuque, IA.

Metzger, J. N., R. A. Fjeld, J. A. Hammonds, and F. O. Hoffman. 1998. Evaluation of software for propagating uncertainty through risk assessment models. *Hum. Ecol. Risk Assess.* 4:263–290.

Meyer, F. P. and L. A. Barklay. 1990. Field Manual for the Investigation of Fish Kills. Resource Pub. 177. U.S. Fish and Wildlife Service, Washington, D.C.

Meylan, W. M., P. H. Howard, R. S. Boethling, D. Aronson, H. Pruntup, and S. Gouchie. 1999. Improved methods for estimating bioconcentration/bioaccumulation factor from octanol/water partition coefficient. *Environ. Toxicol. Chem.* 18:664–672.

Miles, L. J. and G. R. Parker. 1979. Heavy metal interaction for *Andropogon scoparius* and *Rudbeckia hirta* grown on soil from urban and rural sites with heavy metals additions. *J. Environ. Qual.* 8:443–449.

Miller, H. W., and D. H. Johnson. 1978. Interpreting results from nesting studies. *J. Wildl. Manage.* 42:471–476.

Miller, J. E., J. J. Hassett, and D. E. Koeppe. 1976. Uptake of cadmium by soybeans as influenced by soil cation exchange capacity, pH, and available phosphorus. *J. Environ. Qual.* 5:157–160.

Miller, J. E., J. J. Hassett, and D. E. Koeppe. 1977. Interactions of lead and cadmium on metal uptake and growth of corn plants. *J. Environ. Qual.* 6:18–20.

Mineau, P., B. T. Collins, and A. Baril. 1996. On the use of scaling factors to improve interspecies extrapolation of acute toxicity in birds. *Regul. Toxicol. Pharmacol.* 24:24–29.

Mitra, S. and R. M. Dickhut. 1999. Three-phase modeling of polycyclic aromatic hydrocarbon association with pore-water-dissolved organic carbon. *Environ. Toxicol. Chem.* 18:1144–1148.

Moore, D. R. J. and P. Y. Caux. 1997. Estimating low toxic effects. *Environ. Toxicol. Chem.* 16:794–801.

Morgan, M. G. 1978. Bad science and good policy analysis. *Science* 201:1225.

Morgan, M. G. and M. Henrion. 1990. *Uncertainty: A Guide to Dealing with Uncertainty in Quantitative Risk and Policy Analysis.* Cambridge University Press, Cambridge, U.K.

Morrison, M. L., C. J. Ralph, J. Verner, and J. R. Jehl, Jr. 1990. *Avian Foraging: Theory, Methodology, and Applications.* Stud. in Avian Biol. No. 13. Cooper Ornithological Society, Allen Press, Lawrence, KS.

Moskowitz, P. D., R. Pardi, V. M. Fthenakis, S. Holtzman, L. C. Sun, and B. Irla. 1996. An evaluation of three representative multimedia models used to support cleanup decision-making. *Risk Anal.* 16:279–288.

Mount, D. R., T. D. Dawson, and L. P. Burkhard. 1999. Implications of gut purging for tissue residues determined in bioaccumulation testing of sediment with *Lumbriculus variegatus. Environ. Toxicol. Chem.* 18:1244–1249.

Munn, R. E. 1975. *Environmental Impact Assessment, SCOPE 5.* Scientific Committee on Problems in the Environment, Paris.

Muramoto, S. 1989. Heavy metal tolerance of rice plants (*Oryza sativa* L.) to some metal oxides at the critical levels. *J. Environ. Sci. Health.* B24:559–568.

Murkin, H. R., D. A. Wrubleski, and F. A. Ried. 1994. Sampling invertebrates in aquatic and terrestrial habitats. In T. A. Bookhout (Ed.), *Research and Management Techniques for Wildlife and Habitats.* 5th ed. The Wildlife Society, Bethesda, MD. 349–369.

Myers, O. B. 1999. On aggregating species for risk assessment. *Hum. Ecol. Risk Assess.* 5:559–574.

Nabholz, J. V., R. G. Clements, and M. G. Zeeman. 1997. Information needs for risk assessment in EPA's Office of Pollution Prevention and Toxics. *Ecol. Appl.* 7:1094–1102.

Nagorsen, D. W. and R. L. Peterson. 1980. Mammal Collectors' Manual: A Guide for Collecting, Documenting, and Preparing Mammal Specimens for Scientific Research. Life Sciences Miscellaneous Publications. Royal Ontario Museum, Toronto.

Nagy, K. A. 1987. Field metabolic rate and food requirement scaling in mammals and birds. *Ecol. Monogr.* 57:11–128.

National Research Council. 1983. *Risk Assessment in the Federal Government: Managing the Process.* National Academy Press, Washington, D.C.

National Research Council. 1994. *Science and Judgement in Risk Assessment.* National Academy Press, Washington, D.C.

Naugle, D. E., J. A. Jenks, and B. J. Kernohan. 1996. Use of thermal infrared sensing to estimate density of white-tailed deer. *Wildl. Soc. Bull.* 24:37–43.

Neff, J. M., B. W. Cornaby, R. M. Vaga, T. C. Gulbransen, J. A. Scanlan, and D. J. Bean. 1988. An evaluation of the screening level concentration approach for validation of sediment quality criteria for freshwater and saltwater ecosystems. In W. J. Adams, G. A. Chapman, and W. G. Landis (Eds.), *Aquatic Toxicology and Hazard Assessment:* Vol. 10, ASTM STP 971, American Society for Testing and Materials, Philadelphia. 115–127.

Neuhauser, E. F., Z. V. Cukic, M. R. Malecki, R. C. Loehr, and P. R. Durkin. 1995. Bioconcentration and biokinetics of heavy metals in the earthworm. *Environ. Poll.* 89:293–301.

Newell, P. F. 1970. Mollusks: methods for estimating production and energy flow. In J. Phillipson (Ed.), *Methods of Study in Soil Ecology.* UNESCO, Paris. 285–291.

Newman, M. C., K. D. Greene, and P. M. Dixon. 1995. UNCENSOR v. 4.0. SREL — 44. Savannah River Ecology Laboratory, Aiken, SC.

Newman, M. C., P. M. Dixon, B. B. Looney, and J. E. Pinder. 1989. Estimating mean and variance of environmental samples with below detection limit observations. *Water Resour. Bull.* 35:905–916.

Newsted, J. L., J. P. Giesy, G. T. Ankley, D. E. Tillitt, R. A. Crawford, J. W. Gooch, P. D. Jones, and M. S. Denison. 1995. Development of toxicity equivalency factors for PCB congeners and the assessment of TCDD and PCB mixtures in rainbow trout. *Environ. Toxicol. Chem.* 14:861–871.

Nichols, J. D. and C. R. Dickman. 1996. Techniques for estimating abundance and species richness: capture-recapture methods. In D. E. Wilson, F. R. Cole, J. D. Nichols, R. Rudran, and M. S. Foster (Eds.), *Measuring and Monitoring Biological Diversity: Standard Methods for Mammals.* Smithsonian Institution Press, Washington, D.C. 217–225.

Nichols, J. D., B. R. Noon, S. L. Stokes, and J. E. Hines. 1981. Remarks on the use of mark-recapture methodology in estimating avian population size. *Stud. Avian Biol.* 6:121–136.

Nichols, J. W., C. P. Larsen, M. E. McDonald, G. J. Niemi, and G. T. Ankley. 1995. Bioenergetics-based model for accumulation of polychlorinated biphenyls by nestling tree swallows, *Tachycineta bicolor. Environ. Sci. Technol.* 29:604–612.

Nielsen, M. G. and G. Gissel-Nielsen. 1975. Selenium in soil animal relationships. *Pedobiologia* 15:65–67.

Niemi, G. J., P. DeVore, N. Detenbeck, D. Taylor, A. Lima, J. Pastor, J. D. Yount, and R. J. Naiman. 1990. Overview of case studies on recovery of aquatic ecosystems from disturbance. *Environ. Manage.* 14:571–587.

Nietfield, M. T., M. W. Barrett, and N. Silvey. 1994. Wildlife marking techniques. In T. A. Bookhout (Ed.), *Research and Management Techniques for Wildlife and Habitats*. 5th ed. The Wildlife Society, Bethesda, MD. 140–168.

NOAA (National Oceanic and Atmospheric Administration). 1995. The Utility of AVS/EqP in Hazardous Waste Site Evaluations. NOAA Technical Memorandum NOS ORCA 87. National Oceanic and Atmospheric Administration, Seattle, WA.

NOAA (National Oceanic and Atmospheric Administration). 1996a. OPA Regulations for Natural Resource Damage Assessment. 43 CFR Part 2. U.S. Government Printing Office, Washington, D.C.

NOAA (National Oceanic and Atmospheric Administration). 1996b. Damage Assessment and Restoration Program, Habitat Equivalency Analysis. Policy and Technical Paper Series No. 95-1. U.S. Government Printing Office, Washington, D.C.

NOAA Panel on Contingent Valuation. 1993. Report. *Fed. Regist.* 58:4601–4614.

Norberg, T. J. and D. I. Mount. 1985. A new fathead minnow (*Pimephales promelas*) subchronic toxicity test. *Environ. Toxicol. Chem.* 4:711–718.

Nuutinen, V., J. Pitkanen, E. Kuusela, T. Widbom, and H. Lohilahti. 1998. Spatial variation of an earthworm community related to soil properties and yield in a grass-clover field. *Appl. Soil Ecol.* 8:85–94.

Nwosu, J. U., H. Ratsch, and L. A. Kapustka. 1991. A method for on-site evaluation of phytotoxicity at hazardous waste sites. In J. W. Gorsuch, W. R. Lower, W. Wang, and M. A. Lewis (Eds.), *Plants for Toxicity Assessment:* Volume 2. ASTM STP 1115. American Society for Testing and Materials, Philadelphia. 333–334.

O'Connor, T. P. 1999. Sediment quality guidelines reply-to-reply. *SETAC News* 19:21–22.

Odell, D. K. 1982. California sea lion. In D. E. Davis (Ed.), *Handbook of Census Methods for Terrestrial Vertebrates*. CRC Press, Boca Raton, FL. 239–240.

ODEQ (Oregon Department of Environmental Quality). 1998. Guidance for Ecological Risk Assessment. Final April 1998. Updated November 1998. ODEQ, Salem.

Odum, E. P. 1993. Body masses and composition of migrant birds in the eastern United States. In J. B. Dunning, Jr. (Ed.), *CRC Handbook of Avian Body Masses*. CRC Press, Boca Raton, FL. 313–332.

OECD. 1992a. Report of the OECD Workshop on the Extrapolation of Laboratory Aquatic Toxicity Data to the Real Environment. OCDE/GD(92)169. Organization for Economic Cooperation and Development, Paris.

OECD. 1992b. Report of the OECD Workshop on Quantitative Structure Activity Relationships (QSARs) in Aquatic Effects Assessment. OCDE/GD (92)168. Organization for Economic Cooperation and Development, Paris.

OECD. 1998. *OECD Guidelines for Testing of Chemicals*. Organization for Economic Cooperation and Development, Paris.

Oelke, H. 1981. Limitations of the mapping method. *Stud. Avian Biol.* 6:114–118.

Office of Emergency and Remedial Response. 1988. Guidance for Conducting Remedial Investigations and Feasibility Studies under CERCLA. OSWER Directive No. 9355.3-01. U.S. Environmental Protection Agency, Washington, D.C.

Office of Emergency and Remedial Response. 1991. The Role of BTAGs in Ecological Risk Assessment. Pub. 9345.0-015. U.S. Environmental Protection Agency, Washington, D.C.

Office of Emergency and Remedial Response. 1992a. Briefing the BTAG: Initial Description of Setting, History, and Ecology of a Site. Publication 9345.0-051. U.S. Environmental Protection Agency, Washington, D.C.

Office of Emergency and Remedial Response. 1992b. Developing a Work Scope for Ecological Assessments. Publication 9345.0-051. U.S. Environmental Protection Agency, Washington, D.C.

Office of Emergency and Remedial Response. 1992c. Guidance for Data Useability in Risk Assessment. 9285.7-09A&B. U.S. Environmental Protection Agency, Washington, D.C.

Office of Emergency and Remedial Response. 1992d. The Role of Natural Resource Trustees in the Superfund Process. Publication 9345.0-051. U.S. Environmental Protection Agency, Washington, D.C.

Office of Emergency and Remedial Response. 1994a. Catalog of Standard Toxicity Tests for Ecological Risk Assessment. EPA 540-F-94-013. U.S. Environmental Protection Agency, Washington, D.C.

Office of Emergency and Remedial Response. 1994b. Field Studies for Ecological Risk Assessment. EPA 540-F-94-014. U.S. Environmental Protection Agency, Washington, D.C.

Office of Emergency and Remedial Response. 1994c. Selecting and Using Reference Information in Superfund Ecological Risk Assessments. EPA 540-F-94-015. U.S. Environmental Protection Agency, Washington, D.C.

Office of Emergency and Remedial Response. 1994d. Using Toxicity Tests in Ecological Risk Assessment. EPA 540-F-94-012. U.S. Environmental Protection Agency, Washington, D.C.

Office of Emergency and Remedial Response. 1996. Ecotox Thresholds. EPA 540/F-95/038. U.S. Environmental Protection Agency, Washington, D.C.

Office of Environmental Guidance. 1991. Natural Resource Trusteeship and Ecological Evaluation for Environmental Restoration at Department of Energy Facilities. DOE/EH-0192. U.S. Department of Energy, Washington, D.C.

Office of Environmental Guidance. 1993. Integrating Natural Resource Damage Assessment and Environmental Restoration Activities at DOE Facilities. U.S. Department of Energy, Washington, D.C.

Office of Environmental Policy and Assistance. 1996. Characterization of Uncertainties in Risk Assessment with Special Reference to Probabilistic Uncertainty Analysis. EH-413-68/0296. U.S. Department of Energy, Washington, D.C.

Office of Pesticide Programs. 1982. Pesticide Assessment Guidelines, Subdivision E, Hazard Evaluation: Wildlife and Aquatic Organisms. EPA-540/9-82-024. U.S. Environmental Protection Agency, Washington, D.C.

Office of Science and Technology. 1994. Interim Guidance on Determination and Use of Water Effect Ratios for Metals. EPA/823/B-94/001. U.S. Environmental Protection Agency, Washington, D.C.

Office of Water. 1998. National Sediment Bioaccumulation Conference: Proceedings. EPA 823-R-98-002. U.S. Environmental Protection Agency, Bethesda, MD.

Ohio EPA (Environmental Protection Agency). 1991. How Clean Is Clean Policy, Final. Division of Emergency and Remedial Response, Columbus, OH.

Ohlendorf, H. M. 1998. Evaluating bioaccumulation in wildlife food chains. In A. de Peyster and K. E. Day (Eds.), *Ecological Risk Assessment: A Meeting of Policy and Science*. SETAC Press, Pensacola, FL. 65–109.

Oliver, B. G. and A. J. Niimi. 1985. Bioconcentration factors for some halogenated organics for rainbow trout: limitations in their use for prediction of environmental residues. *Environ. Sci. Technol.* 19:842–849.

Oliver, B. G. and A. J. Niimi. 1988. Trophodynamic analysis of polychlorinated biphenyl congeners and other chlorinated hydrocarbons in the Lake Ontario ecosystem. *Environ. Sci. Technol.* 22:388–397.

O'Neill, R. V., R. H. Gardner, L. W. Barnthouse, G. W. Suter II, S. G. Hildebrand, and C. W. Gehrs. 1982. Ecosystem risk analysis: A new methodology. *Environ. Toxicol. Chem.* 1:167–177.

O'Neill, R. V., D. L. DeAngelis, J. B. Waide, and T. F. H. Allen. 1986. *A Hierarchical Concept of Ecosystems*. Princeton University Press, Princeton, NJ.

Opresko, D. M., B. E. Sample, and G. W. Suter II. 1993. Toxicological Benchmarks for Wildlife. ES/ER/TM-86. Oak Ridge National Laboratory, Oak Ridge, TN.

OPPT (Office of Pollution Prevention and Toxics). 1994. ECOSAR: Computer Program and User's Guide for Estimating the Ecotoxicity of Industrial Chemicals Based on Structure–Activity Relationships. EPA-748-R-93-002. U.S. Environmental Protection Agency, Washington, D.C.

Otte, M. L., J. Rozema, B. J. Beek, and R. A. Broekman. 1990. Uptake of arsenic by estuarine plants and interactions with phosphate, in the field (Rhine Estuary) and under outdoor experimental conditions. *Sci. Total Environ.* 97/98:839–854.

Owen, B. A. 1990. Literature-derived absorption coefficients for 39 chemicals via oral and inhalation routes. *Reg. Toxicol. Pharmacol.* 11:237–252.

Page, C. A., M. L. Diamond, M. A. Campbell, and S. McKenna. 1999. Life-cycle framework for assessment of site remediation options: case study. *Environ. Toxicol. Chem.* 18:801–810.

Paine, J. M., M. J. McKee, and M. E. Ryan. 1993. Toxicity and bioaccumulation of soil PCBs in crickets: comparison of laboratory and field studies. *Environ. Toxicol. Chem.* 12:2097–2103.

Parker, M. M. and D. H. van Lear. 1996. Soil heterogeneity and root distribution of mature loblolly pine stands in Piedmont soils. *Soil Sci. Soc. Am. J.* 60:1920–1925.

Parkhurst, B. R., W. Warren-Hicks, R. D. Cardwell, J. Volosin, T. Etchison, J. B. Butcher, and S. M. Covington. 1996. Aquatic Ecological Risk Assessment: A Multi-Tiered Approach. Project 91-AER-1. Water Environment Research Foundation, Arlington, VA.

Parkhurst, D. F. 1998. Arithmetic versus geometric means for environmental concentration data. *Environ. Sci. Technol.* 32(2):92–98A.

Parmelee, R. W., R. S. Wentsel, C. T. Phillips, M. Simini, and R. T. Checkai. 1993. Soil microcosm for testing the effects of chemical pollutants on soil fauna communities and trophic structure. *Environ. Toxicol. Chem.* 12:1477–1486.

Parmelee, R. W., C. T. Phillips, R. T. Checkai, and P. J. Bohlen. 1997. Determining the effects of pollutants on soil faunal communities and trophic structure using a refined microcosm system. *Environ. Toxicol. Chem.* 16:1212–1217.

Parr, J. F., P. B. Marsh, and M. Kla. 1983. *Land Treatment of Hazardous Wastes*. Noyes Data Corp., Park Ridge, NJ, as cited in Bysshe (1988).

Parsons, W. F. J. and D. Parkinson. 1986. Species composition, distribution, and abundance of Collembola colonizing reclaimed mine spoil in Alberta. *Pedobiologia* 29:33–45.

Pascoe, G. A. and J. A. DalSoglio. 1994. Planning and implementation of a comprehensive ecological risk assessment at the Milltown Reservoir-Clark Fork River superfund site, Montana. *Environ. Toxicol. Chem.* 13(12):1943–1956.

Pascoe, G. A., R. J. Blanchet, and G. Linder. 1996. Food chain analysis of exposures and risks to wildlife at a metals-contaminated wetland. *Arch. Environ. Contam. Toxicol.* 30:306–318.

Pascoe, G. A., R. L. Blanchet, G. Linder, D. Palawski, W. G. Brumbaugh, T. J. Canfield, N. E. Kemble, C. G. Ingersoll, A. Farag, and J. A. DalSoglio. 1994. Characterization of the ecological risks at the Milltown Reservoir-Clark Fork River Superfund site, Montana. *Environ. Toxicol. Chem.* 13:2043–2058.

Pascual, J. A. 1994. No effects of a forest spraying of malathion on breeding blue tips (*Parus caereuleus*). *Environ. Toxicol. Chem.* 13:1127–1131.

Pastorok, R. A., M. K. Butcher, and R. Dreas Nelson. 1996. Modeling wildlife exposure to toxic chemicals: trends and recent advances. *Hum. and Ecol. Risk Assess.* 2:444–480.

Paterson, S., D. Mackay, and C. McFarlane. 1994. A model of organic chemical uptake by plants from soil and the atmosphere. *Environ. Sci. Technol.* 28:2259–2266.

Paterson, S., D. Mackay, D. Tam, and W. Y. Shiu. 1990. Uptake of organic chemicals by plants: a review of processes, correlations and models. *Chemosphere* 21:297–331.

Pearson, W. H., E. Moksness, and J. R. Skalski. 1995. A field and laboratory assessment of oil spill effects on survival and reproduction of Pacific herring following the *Exxon Valdez* spill. In P. G. Wells, J. N. Butler, and J. S. Hughes (Eds.), Exxon Valdez *Oil Spill: Fate and Effects in Alaskan Waters*. American Society for Testing and Materials, Philadelphia. 626–661.

Perrins, C. M. 1965. Population fluctuations and clutch size in the great tit, *Parus major* L. *J. of Anim. Ecol.* 34:601–647.

Persaud, D. R., R. Jaagumagi, and A. Hayton. 1993. Guidelines for the protection and management of aquatic sediment quality in Ontario. Ontario Ministry of Environment and Energy, Toronto.

Peters, R. H. 1983. *The Ecological Implications of Body Size*. Cambridge University Press, Cambridge, U.K.

Peterson, M. J. and G. R. Southworth. 1994. Bioaccumulation studies: PCBs in caged clams. In J. G. Smith (Ed.), Second Report on the Oak Ridge K-25 Site Biological Monitoring and Abatement Program for Mitchell Branch. ESD Publication No. 3928. Oak Ridge National Laboratory, Oak Ridge, TN.

Phillips, D. J. H. 1978. Use of biological indicator organisms to quantitate organochlorine pollutants in aquatic environments: a review. *Environ. Pollut.* 13:281–317.

Phillips, R. L. 1982. Red fox (U.S.). In D. E. Davis (Ed.), *Handbook of Census Methods for Terrestrial Vertebrates*. CRC Press, Boca Raton, FL. 229–230.

Pietz, R. I., J. R. Peterson, J. E. Prater, and D. R. Zenz. 1984. Metal concentrations in earthworms from sewage sludge amended soils at a strip mine reclamation site. *J. Environ. Qual.* 13:651–654.

Pitkanen, J. and V. Nuutinen. 1997. Distribution and abundance of burrows formed by *Lumbricus terrestris* L and *Aporrectodea caliginosa* SAV in the soil profile. *Soil Biol. Biochem.* 29:463–467.

Pizl, V. and G. Josens. 1995. Earthworm communities along a gradient of urbanization. *Environ. Pollut.* 90:7–14.

Plafkin, J. L., M. T. Barbour, K. D. Porter, S. K. Gross, and R. M. Hughes. 1989. Rapid Bioassessment Protocols for Use in Streams and Rivers: Benthic Macroinvertebrates and Fish. EPA/444/4-89-001. U.S. Environmental Protection Agency, Washington, D.C.

Pokras, M. A., A. M. Karas, J. K. Kirkwood, and C. J. Sedgewick. 1993. An introduction to allometric scaling and its uses in raptor medecine. In P. T. Redig, J. E. Cooper, J. D. Remple, and D. B. Hunter (Eds.), *Raptor Biomedecine*. University of Minnesota Press, Minneapolis. 211–224.

Polisini, J. M., J. C. Carlisle, and L. M. Valoppi. 1998. Guidance for performing ecological risk assessments at hazardous waste sites and permitted facilities in California. In A. de Peyster and K. E. Day (Eds.), *Ecological Risk Assessment: A Meeting of Policy and Science*. SETAC Press, Pensacola, FL. 23–54.

Posthuma, L., C. A. M. van Gestel, C. E. Smit, D. J. Bakker, and J. W. Vonk. 1998. Validation of Toxicity Data and Risk Limits for Soils: Final Report. Report No. 607505. RIVM, Bilthoven, the Netherlands.

Pouyat, R. V., R. W. Parmelee, and M. M. Carreiro. 1994. Environmental effects of forest soil-invertebrate and fungal densities in oak stands along an urban-rural land use gradient. *Pedobiologia* 38:385–399.

Powell, R. L., R. A. Kimerle, G. T. Coyle, and G. R. Best. 1997. Ecological risk assessment of a wetland exposed to boron. *Environ. Toxicol. Chem.* 16:2409–2414.

Price, P. D., R. Pardi, V. M. Fthenakis, S. Holtzman, L. C. Sun, and B. Irla. 1996. Uncertainty and variation in indirect exposure assessments: an analysis of exposure to tetrachlorodibenzo-*p*-dioxin from a beef consumption pathway. *Risk Anal.* 16:263–277.

Prothro, M. G. 1993. Office of Water Technical Guidance on Interpretation and Implementation of Aquatic Life Metals Criteria. U.S. Environmental Protection Agency, Office of Water, Washington, D.C.

Quality Assurance Management Staff. 1994. Guidance for the Data Quality Objectives Process. EPA QA/G-4. U.S. Environmental Protection Agency, Washington, D.C.

Ralph, C. J. and J. M. Scott, Eds. 1981. *Estimating numbers of terrestrial birds. Stud. in Avian Biol.* 6. Cooper Ornithological Society. Allen Press, Lawrence, KS.

Ram, R. N. and J. W. Gillett. 1993. An aquatic/terrestrial foodweb model for polychlorinated biphenyls (PCBs). In *Environmental Toxicology and Risk Assessment*. ASTM STP 1179. American Society for Testing and Materials, Philadelphia. 192–212.

Rand, G. M. and J. R. Newman. 1998. The applicability of habitat evaluation methodologies in ecological risk assessment. *Hum. Ecol. Risk Assess.* 4:905–929.

Rankin, E. T. 1995. Habitat indices in water resource quality assessment. In S. Davis and T. P. Simon (Eds.), *Biological Assessment and Criteria: Tools for Water Resource Planning and Decision Making*. Lewis Publishers, Boca Raton, FL. 181–208.

Raw, F. 1959. Estimating earthworm populations using formalin. *Nature.* 148:1661–1662.

Redford, K. H. and J. G. Dorea. 1984. The nutritional value of invertebrates with emphasis on ants and termites as food for mammals. *J. Zool.* (London) 203:385–395.

Reilley, K. A., M. K. Banks, and A. P. Schwab. 1996. Dissipation of polycyclic aromatic-hydrocarbons in the rhizosphere. *J. Environ. Qual.* 25:212–219.

Reinecke, A. J. 1983. The ecology of earthworms in southern Africa. In J. E. Satchell (Ed.), *Earthworm Ecology: From Darwin to Vermiculture*. Chapman & Hall, London. 195–207.

Renner, R. 1998. Calculating the cost of natural resource damage. *Environ. Sci. Technol.* 38:86–90A.

Reynolds, J. B. 1983. Electrofishing. In L. A. Nielsen and D. L. Johnson (Eds.), *Fisheries Techniques*. American Fisheries Society, Bethesda, MD. 147–164.

Reynolds, R. T., J. M. Scott, and R. A. Nussbaum. 1980. A variable circular-plot method for estimating bird numbers. *Condor* 82:309–313.

Risk Assessment Forum. 1996. Summary Report for the Workshop on Monte Carlo Analysis. EPA/630/R-96/010. U.S. Environmental Protection Agency, Washington, D.C.

Risk Assessment Forum. 1997. Guiding Principles for Monte Carlo Analysis. EPA/630/R-97/001. U.S. Environmental Protection Agency, Washington, D.C.

Risk Assessment Forum. 1999. Report of the Workshop on Selecting Input Distributions for Probabilistic Assessments. EPA/630/R-98/004. U.S. Environmental Protection Agency, Washington, D.C.

Risk Assessment Program. 1997. Risk Assessment Program Quality Assurance Plan. ES/ER/TM-117/R1. Environmental Restoration Program, Oak Ridge, TN.

Robbins, C. T. 1993. *Wildlife Feeding and Nutrition*. Academic Press, San Diego, CA.

Robinson, S. C., R. J. Kendall, R. Robinson, C. J. Driver, and T. E. Lacher Jr. 1988. Effects of agricultural spraying of methyl parathion on cholinesterase activity and reproductive success in wild starlings (*Sturnus vulgaris*). *Environ. Toxicol. Chem.* 7:343–349.

Rolley, R. E. 1982. Moose (Alberta). In D. E. Davis (Ed.), *Handbook of Census Methods for Terrestrial Vertebrates*. CRC Press, Boca Raton, FL. 262–263.

Rose, K. A., E. P. Smith, R. A. Gardner, A. L. Brenkert, and S. G. Bartell. 1991a. Parameter sensitivities, Monte Carlo fitting, and model forecasting under uncertainty. *J. Forecasting* 10:117–133.

Rose, K. A., A. L. Brenkert, R. B. Cook, and R. H. Gardner. 1991b. Systematic comparison of ILWAS, MAGIC, and ETD watershed acidification models 1. Monte Carlo under regional variability. *Water Resour. Res.* 27:2577–2589.

Rosen, B. H. 1995. Use of periphyton in the development of biocriteria. In W. S. Davis and T. P. Simon (Eds.), *Biological Assessment and Criteria: Tools for Water Resource Planning and Decision Making*. Lewis Publishers, Boca Raton, FL. 209–215.

Rosenberg, K. V. and R. J. Cooper. 1990. Approaches to avian diet analysis. *Stud. Avian Biol.* 13:80–90.

Rubinstein, R. Y. 1981. *Simulation and Monte Carlo Method*. John Wiley & Sons, New York.

Ruckelshaus, W. D. 1984. Risk in a free society. *Risk Anal.* 4:157–162.

Rudman, R. and M. S. Foster. 1996. Conducting a survey to assess mammalian diversity. In D. E. Wilson, F. R. Cole, J. D. Nichols, R. Rudran, and M.S. Foster (Eds.), *Measuring and Monitoring Biological Diversity: Standard Methods for Mammals*. Smithsonian Institution Press. Washington. D.C. 71–80.

Rudman, R., T. H. Kunz, C. Southwell, P. Jarman, and A. P. Smith. 1996. Observational techniques for non-volant mammals. In D. E. Wilson, F. R. Cole, J. D. Nichols, R. Rudran, and M. S. Foster (Eds.), *Measuring and Monitoring Biological Diversity: Standard Methods for Mammals*. Smithsonian Institution Press. Washington. D.C. 81–104.

Ruhling, A., E. Baath, A. Nordgren, and B. Soderstrom. 1984. Fungi in metal-contaminated soil near the Gusum Brass Mill, Sweden. *Ambio* 13:34–36.

Rushton, S. P. 1986. Development of earthworm populations on pasture land reclaimed from open-cast coal mining. *Pedobiologia* 29:27–32.

Russell, R. W., F. A. P. C. Gobas, and G. D. Haffner. 1999. Role of chemical and ecological factors in trophic transfer of organic chemicals in aquatic food webs. *Environ. Toxicol. Chem.* 18:1250–1257.

Rutgers, M., I. M. van't Verlaat, B. Wind, L. Posthuma, and A. M. Breure. 1998. Rapid method for assessing pollution-induced community tolerance in contaminated soil. *Environ. Toxicol. Chem.* 17:2210–2213.

Ryder, R. A. 1986. Songbirds. In A. Y. Cooperrider, R. J. Boyd, and H. R. Stuart (Eds.), Inventory and Monitoring of Wildlife Habitat. U.S. Department of the Interior, Bureau of Land Management, Service Center, Denver, CO. 291–312.

Sadana, U. S. and B. Singh. 1987. Yield and uptake of cadmium, lead and zinc by wheat grown in a soil polluted with heavy metals. *J. Plant Sci. Res.* 3:11–17.

Sadiq, M. 1985. Uptake of cadmium, lead and nickel by corn grown in contaminated soils. *Water Air Soil Pollut.* 26:185–190.

Sadiq, M. 1986. Solubility relationships of arsenic in calcareous soils and its uptake by corn. *Plant and Soil* 91:241–248.

Safe, S. 1998. Hazard and risk assessment of chemical mixtures using the toxic equivalency factor (TEF) approach. *Environ. Health Perspect.* 106(54):1051–1058.

Salazar, M. and S. Salazar. 1998. Using caged bivalves as part of the exposure-dose-response triad to support an integrated assessment strategy. In A. de Peyster and K. E. Day (Eds.), *Ecological Risk Assessment: A Meeting of Policy and Science*. SETAC Press, Pensacola, FL. 167–192.

Salminen, J. E. and P. O. Sulkava. 1997. Decomposer communities in contaminated soil: is altered community regulation a proper tool in ecological risk assessment of toxicants? *Environ. Pollut.* 97:45–53.

Sample, B. E. 1996. Remediation of hazardous waste sites for protection of wildlife: development of remedial goal options. Abstr. of poster presented at the 17th Annual Meeting of the Society for Environmental Toxicology and Chemistry, November 17–21, 1996, Washington, D.C.

Sample, B. E. and C. A. Arenal. 1999. Allometric models for interspecies extrapolation for wildlife toxicity data. *Bull. Environ. Contam. Toxicol.* 62:653–663.

Sample, B. E. and G. W. Suter II. 1994. Estimating Exposure of Terrestrial Wildlife to Contaminants. ES/ER/TM-125. Oak Ridge National Laboratory, Oak Ridge, TN.

Sample, B. E. and G. W. Suter II. 1999. Ecological risk assessment of a large river-reservoir. 4. Piscivorous wildlife. *Environ. Toxicol. Chem.* 18:610–620.

Sample, B. E., D. M. Opresko, and G. W. Suter II. 1996. Toxicological Benchmarks for Wildlife. ES/ER/TM-86/R3. Oak Ridge National Laboratory, Oak Ridge, TN.

Sample, B. E., R. L. Hinzman, B. L. Jackson, and L. A. Baron. 1996. Preliminary Assessment of the Ecological Risks to Wide-Ranging Wildlife Species on the Oak Ridge Reservation: 1996 Update. DOE/OR/01-1407&D2. Oak Ridge National Laboratory, Oak Ridge, TN.

Sample, B. E., M. S. Aplin, R. A. Efroymson, G. W. Suter II, and C. J. E. Welsh. 1997a. Methods and Tools for Estimation of the Exposure of Terrestrial Wildlife to Contaminants. ORNL/TM-13391. Oak Ridge National Laboratory, Oak Ridge, TN.

Sample, B. E., G. W. Suter II, M. B. Schaeffer, D. S. Jones, and R. A. Efroymson. 1997b. Ecotoxicological Profiles for Selected Metals and Other Inorganic Chemicals. ES/ER/TM-210. Oak Ridge National Laboratory, Oak Ridge, TN.

Sample, B. E., J. Beauchamp, R. Efroymson, G. W. Suter II, and T. L. Ashwood. 1998. Development and Validation of Literature-Based Bioaccumulation Models for Small Mammals. ES/ER/TM-219. Oak Ridge National Laboratory, Oak Ridge, TN.

Sample, B. E., G. W. Suter II, J. J. Beauchamp, and R. A. Efroymson. 1999. Literature-derived bioaccumulation models for earthworms: development and validation. *Environ. Toxicol. Chem.* 18:2110–2120.

Sampling and Environmental Support Department. 1992. Requirements for Quality Control of Analytical Data for the Environmental Restoration Program. ES/ER/TM-16. Environmental Restoration Program, Oak Ridge, TN.

Samuel, M. D. and M. R. Fuller. 1994. Wildlife radiotelemetry. In T. A. Bookhout (Ed.), *Research and Management Techniques for Wildlife and Habitats,* 5th ed. The Wildlife Society, Bethesda, MD. 379–418.

Samuels, W. B. and A. Ladino. 1983. Calculation of seabird population recovery from potential oilspills in the mid-Atlantic region of the United States. *J. Ecol. Model.* 21:63–84.

Sanderson, J. T. and M. Van den Berg. 1999. Toxic equivalency factors (TEFs) and their use in ecological risk assessment: a successful method when used appropriately. *Hum. Ecol. Risk Assess.* 5:43–52.

SAS Institute. 1989. SAS/STAT User's Guide, Version 6, 4th ed., Vol. 2. SAS Institute, Cary, NC.

Satchell, J. E. 1970. Measuring population and energy flow in earthworms. In J. Phillipson (Ed.), *Methods of Study in Soil Ecology.* UNESCO, Paris. 261–267.

Sauvé, S., A. Dumestre, M. McBride, and W. Hendershot. 1998. Derivation of soil quality criteria using predicted chemical speciation of Pb^{2+} and Cu^{2+}. *Environ. Toxicol. Chem.* 17:1481–1489.

Savage-Rumbaugh, S. and R. Lewin. 1994. *Kanzi: The Ape on the Brink of the Human Mind.* John Wiley, New York.

Schemnitz, S. D. 1994. Capturing and handling wild animals. In T. A. Bookhout (Ed.), *Research and Management Techniques for Wildlife and Habitats*, 5th ed. The Wildlife Society, Bethesda, MD. 106–124.

Scheunert, I., E. Topp, A. Attar, and F. Korte. 1994. Uptake pathways of chlorobenzenes in plants and their correlation with N-octanol/water partition coefficients. *Ecotoxicol. Environ. Saf.* 27:90–104.

Schmoyer, R. L., J. J. Beauchamp, C. C. Brandt, and F. O. Hoffman. 1996. Difficulties with the lognormal model in mean estimation and testing. *Environ. Ecol. Stat.* 3:81–97.

Schoener, T. W. 1968. Sizes of feeding territories among birds. *Ecology* 49:123–141.

Schuler, F., P. Schmid, and C. Schlatter. 1997. The transfer of polychlorinated dibenzo-*p*-dioxins and dibenzofurans from soil into eggs of foraging chickens. *Chemosphere* 34:711–718.

Shacklette, H. T. and J. G. Boerngen. 1984. Elemental concentrations in soils and other surficial materials of the conterminous United States. Professional Paper 1270. U.S. Geological Survey, Washington, D.C.

Shariatpanahi, M. and A. C. Anderson. 1986. Accumulation of cadmium, mercury and lead by vegetables following long-term land application of wastewater. *Sci. Total Environ.* 52:41–47.

Sharma, R. P. and J. L. Shupe. 1977. Lead, cadmium, and arsenic residues in animal tissues in relation to those in their surrounding habitat. *Sci. Total Environ.* 7:53–62.

Shaw, B. P. and A. K. Panigrahi. 1986. Uptake and tissue distribution of mercury in some plant species collected from a contaminated area in India: its ecological implications. *Arch. Environ. Contam. Toxicol.* 15:439–446.

Shaw, D. G. and H. R. Bader. 1996. Environmental science in a legal context: the *Exxon Valdez* experience. *Ambio* 25:430–434.

Sheffield, S. R. 1995. Design and construction of an inexpensive, permanent terrestrial mesocosm for the study of environmental contaminants. Abstr., SETAC World Congress, Vancouver, British Columbia, Canada.

Sheppard, M. I. and D. H. Thibault. 1991. A four-year mobility study of selected trace elements and heavy metals. *J. Environ. Qual.* 20:101–114.

Sheppard, M. I., S. C. Sheppard, and B. D. Amiro. 1991. Mobility and plant uptake of inorganic ^{14}C and ^{14}C-labelled PCB in soils of high and low retention. *Health Phys.* 61:481–492.

Sheppard, M. I., D. H. Thibault, and S. C. Sheppard. 1985. Concentrations and concentration ratios of U, As, and Co in Scots pine grown in a waste-site soil and an experimentally contaminated soil. *Water Air Soil Pollut.* 26:85–94.

Sheppard, M. I., D. H. Thibault, and P. A. Smith. 1992. Effect of extraction techniques on soil pore-water chemistry. *Commun. Soil Sci. Plant Anal.* 23:1643–1662.

Sheppard, S. C. and W. G. Evenden. 1992. Bioavailability indices for uranium: effect of concentration in eleven soils. *Arch. Environ. Contam. Toxicol.* 23:117–124.

Sheppard, S. C. and W. G. Evenden. 1994. Simple whole-soil bioasay based on microarthropods. *Bull. Environ. Contam. Toxicol.* 52:95–101.

Sheppard, S. C. and M. I. Sheppard. 1989. Impact of correlations on stochastic estimates of soil contamination and plant uptake. *Health Phys.* 57:653–657.

Sheppard, S. C., W. G. Evenden, and B. D. Amiro. 1993. Investigation of the soil-to-plant pathway for I, Br, Cl, and F. *J. Environ. Radioact.* 21:9–32.

Sheppard, S. C., C. Gaudet, M. I Sheppard, P. M. Cureton, and M. P. Wong. 1992. The development of assessment and remediation guidelines for contaminated soils, a review of the science. *Can. J. Soil. Sci.* 72:359–394.

Shine, R. 1986. Diets and Abundances of Aquatic and Semi-Aquatic Reptiles in the Alligator Rivers Region. Technical Memorandum No. 16. Australian Government Publishing Service, Canberra, Australia.

Shore, R. F. 1995. Predicting cadmium, lead and fluoride levels in small mammals from soil residues and by species-species extrapolation. *Environ. Pollut.* 88:333–40.

Siegel, S. M. and B. Z. Siegel. 1988. Temperature determinants of plant-soil-air mercury relationships. *Water Air Soil Pollut.* 40:443–448.

Silva, M. and J. A. Downing. 1995. *CRC Handbook of Mammalian Body Masses.* CRC Press, Boca Raton, FL.

Simon, T. P. and J. Lyons. 1995. Application of the index of biotic integrity to evaluate water resource integrity in freshwater ecosystems. In W. Davis and T. P. Simon (Eds.), *Biological Assessment and Criteria: Tools for Water Resource Planning and Decision Making.* Lewis Publishers, Boca Raton, FL. 245–262.

Simonich, S. L. and R. A. Hites. 1994. Vegetation-atmosphere partitioning of polycyclic aromatic hydrocarbons. *Environ. Sci. Technol.* 28:939–943.

Sims, J. T. and J. S. Kline. 1991. Chemical fractionation and plant uptake of heavy metals in soils amended with co-composted sewage sludge. *J. Environ. Qual.* 20:387–395.

Skinner, D. J. 1986. Memorandum and order on defendant Beatrice Food Co.'s motion for immediate entry of final judgment under rule 54(b), September 17, 1986, Anderson v. Grace, Civil Action No. 82-1672-S. United States District Court, District of Massachusetts.

Slayton, D. and D. Montgomery. 1991. Michigan Background Soil Survey. Michigan Department of Natural Resources, Ann Arbor.

Sloof, W., J. A. M. van Oers, and D. de Zwart. 1986. Margins of uncertainty in ecotoxicological hazard assessment. *Environ. Toxicol. Chem.* 5:841–852.

Slovic, P. 1987. Perception of risk. *Science* 236:280–285.

Smith, A. T. 1982. Pika (*Ochatona*). In D. E. Davis (Ed.), *Handbook of Census Methods for Terrestrial Vertebrates.* CRC Press, Boca Raton, FL. 131–133.

Smith, E. P. and J. Cairns, Jr. 1993. Extrapolation methods for setting ecological standards for water quality: statistical and ecological concerns. *Ecotoxicology* 2:203–219.

Smith, E. P. and H. H. Shugart. 1994. Uncertainty in ecological risk assessment. Chap. 8 in Ecological Risk Assessment Issue Papers. EPA/630/R-94/009. U.S. Environmental Protection Agency, Washington, D.C.

Smith, W. P., D. J. Twedt, D. A. Wiedenfeld, P. B. Hamel, R. P. Ford, and R. J. Cooper. 1993. Point counts of birds in bottomland hardwood forests of the Mississippi alluvial valley: duration, minimum sample size, and points versus visits. Research Paper SO-274. U.S. Department ot Agriculture, Forest Service, SE Forest Experimental Station, New Orleans, LA.

Smock, L. A., J. E. Gladden, J. L. Riekenberg, L. C. Smith, and C. R. Black. 1992. Lotic macroinvertebrate production in three dimensions: channel surface, hyporheic, and floodplain environments. *Ecology* 73:876–886.

Solomon, K. R. 1996. Overview of recent developments in ecological risk assessment. *Risk Anal.* 16:627–633.

Solomon, K. R., D. B. Baker, R. P. Richards, K. R. Dixon, S. J. Klaine, T. W. La Point, R. J. Kendall, C. P. Weisskopf, J. M. Giddings, J. P. Giesy, L. W. Hall, and W. M. Williams. 1996. Ecological risk assessment for atrazine in North American surface waters. *Environ. Toxicol. Chem.* 15:31–76.

Sorenson, M. T. and J. A. Margolin. 1998. Ecological risk assessment guidance and procedural documents: An annotated compilation and evaluation of reference materials. *Hum. Ecol. Risk Assess.* 4:1085–1102.

Southwood, T. R. E. 1978. *Ecological Methods.* Chapman & Hall, New York.

Spowart, R. A. and F. B. Samson. 1986. Carnivores. In A. Y. Cooperrider, R. J. Boyd, and H. R. Stuart (Eds.), *Inventory and Monitoring of Wildlife Habitat.* U.S. Department of the Interior, Bureau of Land Management, Service Center, Denver, CO. 475–496.

Sprenger, M. D. and D. W. Charters. 1997. Ecological Risk Assessment Guidance for Superfund: Process for Designing and Conducting Ecological Risk Assessment, Interim Final. U.S. Environmental Protection Agency, Environmental Response Team, Edison, NJ.

Springett, J. A. 1981. A new method for extracting earthworms from soil cores, with a comparison of four commonly used methods for estimating populations. *Pedobiologia* 21:217–222.

Spurgeon, D. J. and S. P. Hopkin. 1995. Extrapolation of the laboratory-based OECD earthworm toxicity test to metal-contaminated field sites. *Ecotoxicology* 4:190–205.

Spurgeon, D. J. and S. P. Hopkin. 1996a. Risk assessment of the threat of secondary poisoning by metals to predators of earthworms in the vicinity of a primary smelting works. *Sci. Total. Environ.* 187:167–183.

Spurgeon, D. J. and S. P. Hopkin. 1996b. Effects of metal-contaminated soils on the growth, sexual development, and early cocoon production of the earthworm *Eisenia fetida*, with particular reference to zinc. *Ecotoxicol. Environ. Saf.* 35:86–95.

Stahl, W. R. 1967. Scaling of respiratory variables in mammals. *J. Appl. Physiol.* 22(3):453–460.

Starr, C. 1969. Social benefit versus technological risk. *Science* 165:1232–1238.

Stavric, B. and R. Klassen. 1994. Dietary effects on the uptake of benzo(*a*)pyrene. *Food Chem. Toxicol.* 8:727–734.

Stebbins, R. C. 1966. *A Field Guide to Western Reptiles and Amphibians.* Houghton Mifflin, Boston.

Stephan, C. E. 1993. Derivation of Proposed Human Health and Wildlife Bioaccumulation Factors for the Great Lakes Initiative. PB93-154672. National Technical Information Service, Springfield, VA.

Stephan, C. E., D. I. Mount, D. J. Hanson, J. H. Gentile, G. A. Chapman, and W. A. Brungs. 1985. Guidelines for Deriving Numeric National Water Quality Criteria for the Protection of Aquatic Organisms and Their Uses. PB85–227049. U.S. Environmental Protection Agency, Washington, D.C.

Stevens, D., G. Linder, and W. Warren-Hicks. 1989a. Field sampling design. In W. Warren-Hicks, B. R. Parkhurst, and J. S. Baker (Eds.), *Ecological Assessment of Harardous Waste Sites: A Field and Laboratory Reference Document.* EPA 600/3-89/013. Corvallis Environmental Research Laboratory, OR. 4-1–4-13.

Stevens, D., G. Linder, and W. Warren-Hicks. 1989b. Data interpretation. In W. Warren-Hicks, B. R. Parkhurst, and J. S. Baker (Eds.), *Ecological Assessment of Hazardous Waste Sites: A Field and Laboratory Reference Document.* EPA 600/3-9/013. Corvallis Environmental Research Laboratory, OR. 9-1–9-25.

Stockdill, S. M. J. and G. G. Cossens. 1966. The role of earthworms in pasture production and moisture conservation. *Proc. N. Z. Grassl. Assoc.* 1966:168–183, as cited in Syers and Springett (1983).

Streit, B. 1984. Effects of high copper concentrations on soil invertebrates (earthworms and oribatid mites). *Oecologia* 64:381–388.

Strojan, C. L. 1976. Forest litter decomposition in the vicinity of a zinc smelter. *Oecologia* (Berlin) 32:203–219.

Strojan, C. L. 1978. The impact of zinc smelter emissions on forest litter arthropods. *Oikos* 31:41–46.

Stubblefield, W. A., G. A. Hancock, W. H. Ford, and R. K. Ringer. 1995a. Acute and subchronic toxicity of naturally weathered *Exxon Valdez* crude-oil in mallards and ferrets. *Environ. Toxicol. Chem.* 14:1941–1950.

Stubblefield, W. A., G. A. Hancock, H. H. Prince, and R. K. Ringer. 1995b. Effects of naturally weathered *Exxon Valdez* crude-oil on mallard reproduction. *Environ. Toxicol. Chem.* 14:1951–1960.

Susser, M. 1986. Rules of inference in epidemiology. *Regul. Toxicol. Pharmacol.* 6:116–186.

Suter, G. W., II. 1989. Ecological endpoints. In W. Warren-Hicks, B. R. Parkhurst, and J. S. S. Baker (Eds.), *Ecological Assessment of Harardous Waste Sites: A Field and Laboratory Reference Document.* EPA 600/3–89/013. Corvallis Environmental Research Laboratory, Corvallis, OR. 2-1–2-28.

Suter, G. W., II. 1990. Use of biomarkers in ecological risk assessment. In J. F. McCarthy and L. L. Shugart (Eds.), *Biomarkers of Environmental Contamination.* Lewis Publishers, Ann Arbor, MI. 419–426.

Suter, G. W., II. 1993a. *Ecological Risk Assessment.* Lewis Publishers, Boca Raton, FL.

Suter, G. W., II. 1993b. A critique of ecosystem health concepts and indices. *Environ. Toxicol. Chem.* 12:1533–1539.

Suter, G. W., II. 1995a. Guide for Performing Screening Ecological Risk Assessments at DOE Dacilities. ES/ER/TM-153. Oak Ridge National Laboratory, Oak Ridge TN.

Suter, G. W., II. 1995b. Endpoints of interest at different levels of biological organization. In J. Cairns, Jr., and B. R. Niederlehner (Eds.), *Ecological Toxicity Testing: Scale, Complexity, and Relevance.* Lewis Publishers, Boca Raton, FL. 35–48.

Suter, G. W., II. 1996a. Abuse of hypothesis testing statistics in ecological risk assessment. *Hum. Ecol. Risk Assess.* 2:331–349.

Suter, G. W., II. 1996b. Toxicological benchmarks for screening contaminants of potential concern for effects on freshwater biota. *Environ. Toxicol. Chem.* 15:1232–1241.

Suter, G. W., II. 1996c. Risk Characterization for Ecological Risk Assessment of Contaminated Sites. ES/ER/TM-200. Oak Ridge National Laboratory, Oak Ridge, TN.

Suter, G. W., II. 1997. Guidance for Treatment of Variability and Uncertainty in Ecological Risk Assessment. ES/ER/TM-228. Oak Ridge National Laboratory, Oak Ridge, TN.

Suter, G. W., II. 1998a. Ecotoxicological effects extrapolation models. In M. C. Newman and C. L. Strojan (Eds.), *Risk Assessment: Logic and Measurement.* Ann Arbor Press, Ann Arbor, MI. 167–185.

Suter, G. W., II. 1998b. Retrospective assessment, ecoepidemiology, and ecological monitoring. In P. Calow (Ed.), *Handbook of Environmental Risk Assessment and Management.* Blackwell Scientific, Oxford, UK. 177–217.

Suter, G. W., II. 1998c. An overview perspective of uncertainty. In W. J. Warren-Hicks and D. R. J. Moore (Eds.), *Uncertainty Analysis in Ecological Risk Assessment*. SETAC Press, Pensacola, FL. 121–130.

Suter, G. W., II. 1999a. Developing conceptual models for complex ecological risk assessments. *Hum. Ecol. Risk Assess.* 5:375–396.

Suter, G. W., II. 1999b. A framework for assessment of ecological risks from multiple activities. *Hum. Ecol. Risk Assess.* 5:397–413.

Suter, G. W., II and A. E. Rosen. 1988. Comparative toxicology for risk assessment of marine fishes and crustaceans. *J. Environ. Sci. Technol.* 22:548–556.

Suter, G. W., II and C. L. Tsao. 1996. Toxicological Benchmarks for Screening Potential Contaminants of Concern for Effects on Aquatic Biota: 1996 Revision. ES/ER/TM-96/R2. Oak Ridge National Laboratory, Oak Ridge, TN

Suter, G. W., II, J. W. Gillett, and S. Norton. 1994. Characterization of exposure, Chap. 4 in Ecological Risk Assessment Issue Papers. EPA/630/R-94/009. U.S. Environmental Protection Agency, Washington, D.C.

Suter, G. W., II, R. J. Luxmoore, and E. D. Smith. 1993. Compacted soil barriers at abandoned landfill sites are likely to fail in the long term. *J. Environ. Qual.* 22:217–226.

Suter, G. W., II, D. S. Vaughan, and R. H. Gardner. 1983. Risk assessment by analysis of extrapolation error, a demonstration for effects of pollutants on fish. *Environ. Toxicol. Chem.* 2:369–378.

Suter, G. W., II, A. E. Rosen, E. Linder, and D. F. Parkhurst. 1987. Endpoints for responses of fish to chronic toxic exposures. *Environ. Toxicol. Chem.* 6:793–809.

Suter, G. W., II, A. E. Rosen, J. J. Beauchamp, and T. T. Kato. 1992. Results of Analysis of Fur Samples from the San Joaquin Kit Fox and Associated Water and Soil Samples from the Naval Petroleum Reserve No. 1, Tupman, California. ORNL/TM-12244. Oak Ridge National Laboratory, Oak Ridge, TN.

Suter, G. W., II, B. E. Sample, D. S. Jones, and T. L. Ashwood. 1994. Approach and Strategy for Performing Ecological Risk Assessments for the Department of Energy's Oak Ridge Reservation. ES/ER/TM-33/R1. Environmental Restoration Division, Oak Ridge, TN.

Suter, G. W., II, B. W. Cornaby, C. T. Hadden, R. N. Hull, M. Stack, and F. A. Zafran. 1995. An approach for balancing health and ecological risks at hazardous waste sites. *Risk Anal.* 15:221–231.

Suter, G. W., II, L. W. Barnthouse, R. E. Efroymson, and H. Jager. 1999. Ecological risk assessment of a large river-reservoir: 2. Fish community. *Environ. Toxicol. Chem.* 18:589–598.

Swartjes, F. A. 1997. Assessment of soil and groundwater quality in the Netherlands: criteria and remediation priority, paper presented at WasteTECH Symposium, March, Melbourne, Australia.

Swartz, M. R. C. 1999. Consensus sediment quality guidelines for polycyclic aromatic hydrocarbon mixtures. *Environ. Toxicol. Chem.* 18:780–787.

Swenson, J. E. 1982. Ospreys. In D. E. Davis (Ed.), *Handbook of Census Methods for Terrestrial Vertebrates*. CRC Press, Boca Raton, FL. 55–56.

Syers, J. K. and J. A. Springett. 1983. Earthworm ecology in grassland soils. In J. E. Satchell (Ed.), *Earthworm Ecology: From Darwin to Vermiculture*. Chapman & Hall, London. 67–83.

Talmage, S. S. and B. T. Walton. 1990. Comparative evaluation of several small species as monitors of heavy metals, radionuclides, and selected organic compounds. ORNL/TM-11605. Oak Ridge National Laboratory, Oak Ridge, TN.

Talmage, S. S. and B. T. Walton. 1993. Food chain transfer and potential renal toxicity of mercury to small mammals at a contaminated terrestrial field site. *Ecotoxicology* 2:243–256.

Tamis, W. L. M. and H. A. Udo de Haes. 1995. Recovery of earthworm communities (Lumbricidae) in some thermally and biologically cleaned soils. *Pedobiologia* 39:351–369.

Teer, J. G. 1982. White-tailed deer (Texas). In D. E. Davis (Ed.), *Handbook of Census Methods for Terrestrial Vertebrates*. CRC Press, Boca Raton, FL. 251–253.

Temple, P. J. and R. Wills. 1979. Sampling and analysis of plants and soil. In W. W. Heck, S. V. Krupa, and S. N. Linzon (Eds.), *Handbook of Methodology for the Assessment of Air Pollution Effects on Vegetation*. Air Pollution Control Association, Pittsburg. 13-1–13-23.

Teuschler, L. K. and R. C. Hertzberg. 1995. Current and future risk assessment guidelines, policy, and methods development for chemical mixtures. *Toxicology* 105:137–144.

Thomann, R. V. 1989. Bioaccumulation model of organic chemical distribution in aquatic food chains. *Environ. Sci. Technol.* 23:699–707.

Thomas, D. W. and R. K. LaVal. 1988. Survey and census methods. In T. H. Kunz (Ed.), *Ecological and Behavioral Study Methods for the Study of Bats*. Smithsonian Institution Press, Washington, D.C. 77-89.

Thompson, R. L. 1982. Red-cockaded woodpecker. In D. E. Davis (Ed.), *Handbook of Census Methods for Terrestrial Vertebrates*. CRC Press, Boca Raton, FL. 91–92.

Thornton, I. and P. Abrahams. 1983. Soil ingestion — a major pathway of heavy metals into livestock grazing contaminated land. *Sci. Total Environ.* 28:27–294.

Tillitt, D. E., J. P. Giesy, and G. T. Ankley. 1991. Characterization of the H4IIE rat hepatoma cell bioassay as a tool for assessing toxic potency of planar halogenated hydrocarbons in environmental samples. *Environ. Sci. Technol.* 25:87–92.

Tillitt, D. E., R. W. Gale, J. C. Meadows, J. L. Zajicek, P. H. Peterman, S. N. Heaton, P. D. Jones, S. J. Bursian, T. J. Kubiak, J. P. Giesy, and R. J. Aulerich. 1996. Dietary exposure of mink to carp from Saginaw Bay. 3. Characterization of dietary exposure to planar halogenated hydrocarbons, dioxin equivalents, and biomagnification. *Environ. Sci. Technol.* 30:283–291.

Tipping, E. 1994. WHAM — A chemical equilibrium model and computer code for waters, sediments, and soils incorporating a discrete site / electrostatic model of ion binding by humic substances. *Comput. Geosci.* 6:973–1023.

TNRCC. 1996. Draft Guidance for Conducting Ecological Risk Assessments under the Texas Risk Reduction Program. Office of Waste Management, Texas Natural Resource Conservation Commission, Austin.

Topp, E., I. Scheunert, A. Attar, and F. Korte. 1986. Factors affecting the uptake of ^{14}C-labeled organic chemicals by plants from soil. *Ecotoxicol. Environ. Saf.* 11:219–228.

Toxics Cleanup Program. 1994. Natural Background Soil Metals Concentrations in Washington State. Pub. #94-115. Washington State Department of Ecology, Olympia.

Trapp, S. 1995. Model for uptake of xenobiotics into plants. In S. Trapp and J. C. McFarlane (Eds.), *Plant Contamination: Modeling and Simulation of Organic Chemical Processes*. Lewis Publishers, Boca Raton, FL. 107–151.

Trapp, S. and M. Matthies. 1997. Modeling volatilization of PCDD/F from soil and uptake into vegetation. *Environ. Sci. Technol.* 31:71–74.

Travis, C. C. and A. D. Arms. 1988. Bioconcentration of organics in beef, milk, and vegetation. *Environ. Sci. Technol.* 22:271–274.

Travis, C. C. and J. M. Morris. 1992. On the use of 0.75 as an interspecies scaling factor. *Risk Anal.* 12:311–313.

Travis C. C. and R. K. White. 1988. Interspecific scaling of toxicity data. *Risk Anal.* 8:119–125.

Travis, C. C., S. A. Richter, E. A. C. Crouch, R. Wilson, and E. D. Klema. 1987. Cancer risk management. *Environ. Sci. Technol.* 21:415–420.

Troyer, M. E. and M. S. Brody. 1994. Managing Ecological Risks at EPA: Issues and Recommendations for Progress. EPA/600/R-94/183. U.S. Environmental Protection Agency, Washington, D.C.

Tufte, E. R. 1983. *The Visual Display of Quantitative Information*. Graphics Press, Cheshire, CT.

Tufte, E. R. 1990. *Envisioning Information*. Graphics Press, Cheshire, CT.

Tufte, E. R. 1997. *Visual Explanations*. Graphics Press, Cheshire, CT.

Tyler, G. 1984. The impact of heavy metal pollution on forests: a case study of Gusum, Sweden. *Ambio* 13:18–26.

Urban, D. J. and N. J. Cook. 1986. Hazard Evaluation, Standard Evaluation Procedure, Ecological Risk Assessment. EPA-540/9-85-001. U.S. Environmental Protection Agency, Washington, D.C.

U.S. Fish and Wildlife Service. 1987. Type B Technical Information Document: Guidance on Use of Habitat Evaluation Procedures and Suitability Index Models for CERCLA Applications. PB88-00151. U.S. Department of the Interior, Washington, D.C.

U.S. Fish and Wildlife Service. 1988. Proceedings of a Workshop on the Development and Evaluation of Habitat Suitability Criteria. Biological Report 88(11). Washington, D.C.

U.S. Fish and Wildlife Service and Canadian Wildlife Service. 1977. North American Bird Banding Manual. Vol. II. U.S. Department of the Interior, Washington, D.C.

Valoppi, L., M. Petreas, R. M. Donohoe, L. Sullivan, and C. A. Callahan. 2000. Use of PCB congener and homologue analysis in ecological risk assessment. In F. T. Price, K. V. Brix and N. K. Lane (Eds.), *Environmental Toxicology and Risk Assessment: Recent Achievements in Environmental Fate and Transport*. American Society for Testing and Materials, West Conshohocken, PA. (in press).

Van Brummelen, T. C., R. A. Verweij, S. A. Wedzinga, and C. A. M. Van Gestel. 1996. Polycyclic aromatic hydrocarbons in earthworms and isopods from contaminated forest soils. *Chemosphere* 32:315–341.

Van de Meent, D. and D. Toet. 1992. Dutch Priority Setting System for Existing Chemicals. Report No. 679120001. National Institute for Public Health and Environmental Protection, Bilthoven, the Netherlands.

Van den Berg, M. et. al. 1998. Toxic equivalency factors (TEFs) for PCBs, PCDDs, PCDFs for humans and wildlife. *Environ. Health Persp.* 106:775–792.

Van de Water, R. B. 1995. Modeling the Transport and Fate of Volatile and Semi-Volatile Organics in a Multimedia Environment, M.S. thesis, University of California, Los Angeles.

van Gestel, C. A. M. and W.-C. Ma. 1988. Toxicity and bioaccumulation of chlorophenols in earthworms, in relation to bioavailability in soil. *Ecotoxicol. Environ. Saf.* 15:289–297.

van Gestel, C. A. M. and W.-C. Ma. 1993. Development of QSAR's in soil ecotoxicology: earthworm toxicity and soil sorption of chlorophenols, chlorobenzenes and chloroanilines. *Water Air Soil Pollut.* 69:265–276.

van Gestel, C. A. M. and N. M. Van Straalen. 1994. Ecotoxicological test systems for terrestrial invertebrates. In M. H. Donker, H. Eijsackers and F. Heimbackers (Eds.), *Ecotoxicology of Soil Organisms*. Lewis Publishers, Boca Raton, FL. 205–240.

van Gestel, C. A. M., D. M. M. Adema, and E. M. Dirven-van Breeman. 1996. Phytotoxicity of some chloroanilines and chlorophenols, in relation to bioavailability in soil. *Water Air Soil Pollut.* 88:119–132.

van Gestel, C. A. M., W. Ma, and C. E. Smit. 1991. Development of QSARs in terrestrial ecotoxicology: earthworm toxicity and soil sorption of chlorophenols, chlorobenzenes, and chloroaniline. *Sci. Total. Environ.* 109/110:589–604.

Vangronsveld, J., J. V. Colpaert, and K. K. Van Tichelen. 1996. Reclamation of a bare industrial area contaminated by non-ferrous metals: physicochemical and biological evaluation of the durability of soil treatment and revegetation. *Environ. Pollut.* 94:131–140.

Vangronsveld, J., F. Van Assche, and H. Clijsters. 1995. Reclamation of a bare industrial area contaminated by non-ferrous metals: *in situ* metal immobilization and revegetation. *Environ. Pollut.* 87:51–59.

Van Hook, R. I. 1974. Cadmium, lead, and zinc distributions between earthworms and soils: potentials for biological accumulation. *Bull. Environ. Contam. Toxicol.* 12:509–512.

Van Hook, R. I. and A. J. Yates. 1975. Transient behavior of cadmium in a grassland arthropod food chain. *Environ. Res.* 9:76–83.

Van Leeuwen, K. 1990. Ecotoxicological effects assessment in the Netherlands. *Environ. Manage.* 14:779–792.

van Rhee, J. A. 1975. Copper contamination effects on earthworms by disposal of pig waste in pastures. In J. Vanek (Ed.), *Progress in Soil Zoology: Proceedings of the 5th International Colloquium*, W. Junk Publishers, The Hague, the Netherlands. 451–456.

van Straalen, N. M., and C. A. J. Denneman. 1989. Ecotoxicological evaluation of soil quality criteria. *Ecotoxicol. Environ. Saf.* 18:241–251.

van Straalen, N. M. and C. A. M. Van Gestel. 1993. Soil. Ch. 15 in P. Calow (Ed.), *Handbook of Ecotoxicology.* Blackwell Scientific, Oxford.

Varney, J. R. and D. J. Ellis. 1974. Telemetering egg for use in incubation and nesting studies. *J. Wildl. Manage.* 38:142–148.

Veith, G. D. and P. Kosian. 1983. Estimating bioconcentration potential from octanol/water partitioning coefficients. In D. R. Mackay, S. Patterson, S. Eisenreich, and M. Simmons (Eds.), *PCBs in the Great Lakes.* Ann Arbor Press, Ann Arbor, MI. 269–282.

Veith, G. D., D. J. Call, and L. T. Brook. 1983. Structure-toxicity relationships for fathead minnow, *Pimephales promelas*: narcotic industrial chemicals. *Can. J. Fish. Aquat. Sci.* 40:743–748.

Veith, G. D., D. L. DeFoe, and B. V. Bergstedt. 1979. Measuring and estimating the bioconcentration factor for chemicals in fish. *J. Fish. Res. Board Can.* 36:1040–1048.

Verner, J. 1985. Assessment of counting techniques. *Curr. Ornithol.* 2:247–302.

VROM. 1994. Environmental Quality Objectives in the Netherlands. Ministry of Housing, Spatial Planning, and the Environment, Amsterdam, the Netherlands.

Wagner, C. and H. Lokke. 1991. Estimation of ecotoxicological protection levels from NOEC toxicity data. *Water Res.* 25:1237–1242.

Wakely, J. S. 1987a. Avian territory mapping, sect. 6.3.4, U.S. Army Corps of Engineers Wildlife Resources Management Manual, Tech. Rept. EL-87-7. U.S. Army Engineer Waterways Experiment Station, Vicksburg, MS.

Wakely, J. S. 1987b. Avian line-transect methods, sect. 6.3.2 in U.S. Army Corps of Engineers Wildlife Resources Management Manual, Tech. Rept. EL-87-5. U.S. Army Engineer Waterways Experiment Station, Vicksburg, MS.

Wakely, J. S. 1987c. Avian plot methods, sect. 6.3.3 in U.S. Army Corps of Engineers Wildlife Resources Management Manual, Tech. Rept. EL-87-6. U.S. Army Engineer Waterways Experiment Station, Vicksburg, MS.

Warren-Hicks, W. J. and D. R. J. Moore, Eds. 1998. *Uncertainty Analysis in Ecological Risk Assessment.* SETAC Press, Pensacola, FL.

Warren-Hicks, W., B. R. Parkhurst, and S. S. Baker, Jr., Eds. 1989. *Ecological Assessments of Hazardous Waste Sites: A Field and Laboratory Reference Document.* EPA/600/3–89/013. U.S. Environmental Protection Agency, Corvallis, OR.

Washington-Allen, R. A., T. L. Ashwood, S. W. Christensen, H. Offerman, and P. Scarbrough-Luther. 1995. Terrestrial Mapping of the Oak Ridge Reservation: Phase 1. ES/ER/TM-152. Oak Ridge National Laboratory, Oak Ridge TN.

Watanabe, K., F. Y. Bois, and L. Zeise. 1992. Interspecies extrapolation: a reexamination of acute toxicity data. *Risk Anal.* 12:301–310.

Watkins, D. R., J. T. Ammons, J. L. Branson, and B. B. Burgoa. 1993. Final Report of the Background Soil Characterization Project at the Oak Ridge Reservation, Oak Ridge, Tennessee. DOE/OR/01-1175. Oak Ridge National Laboratory, Oak Ridge, TN.

Watson, A. P., R. I. Van Hook, D. R. Jackson, and D. E. Reichle. 1976. Impact of a Lead Mining-Smelting Complex on the Forest-Floor Litter Arthropod Fauna in the New Lead Belt Region of Southeast Missouri. ORNL/NSF/EATC-30. Oak Ridge National Laboratory, Oak Ridge, TN.

Weaver, R. W., J. R. Melton, D. Wang, and R. L. Duble. 1984. Uptake of arsenic and mercury from soil by bermudagrass *Cynodon dactylon. Environ. Pollut.* 33:133–142.

Webb, D. A. 1992. Background Metal Concentrations in Wisconsin Surface Waters. Wisconsin Department of Natural Resources, Madison.

Weber, C. I. 1991. Methods for Measuring the Acute Toxicity of Effluents to Freshwater and Marine Organisms, 4th ed., EPA-600/4–90/027. U.S. Environmental Protection Agency, Cincinnati, OH.

Weber, C. I., I. W. B. Horning, D. J. Klemm, T. W. Neiheisel, P. A. Lewis, E. L. Robinson, J. Menkedick, and F. Kessler. 1988. Short-Term Methods for Estimating the Chronic Toxicity of Effluents and Receiving Waters to Marine and Estuarine Organisms. EPA-600/4-87/028. U. S. Environmental Protection Agency, Cincinnati, OH.

Weber, C. I., W. H. Peltier, T. J. Norberg-King, I. W. B. Horning, F. Kessler, J. Menkedick, T. W. Neiheisel, P. A. Lewis, D. J. Klemm, W. H. Pickering, E. L. Robinson, J. Lazorchak, L. J. Wymer, and R. W. Freyberg. 1989. Short-Term Methods for Estimating the Chronic Toxicity of Effluents and Receiving Waters to Freshwater Organisms. EPA-600/4-89/001. U. S. Environmental Protection Agency, Cincinnati, OH.

Webster, J. A., and D. A. Crossley, Jr. 1978. Evaluation of two models for predicting elemental accumulation by arthropods. *Environ. Entomol.* 7:411–417.

Weeks, H. P., Jr. 1978. Characteristics of mineral licks and behavior of visiting white-tailed deer in southern Indiana. *Am. Midl. Nat.* 100:384–395.

Weeks, J. M. 1998. Effects of pollutants on soil invertebrates: links between levels. In G. Schuurmann and B. Markert (Eds.), *Ecotoxicology.* John Wiley & Sons, New York. 645–662.

Wemmer, C., T. H. Kunz, G. Lundie-Jenkins, and W. J. McShea. 1996. Mammalian sign. In D. E. Wilson, F. R. Cole, J. D. Nichols, R. Rudran, and M. S. Foster (Eds.), *Measuring and Monitoring Biological Diversity: Standard Methods for Mammals.* Smithsonian Institution Press, Washington, D.C. 157–176.

Whipple, C., Ed. 1987. *De Minimis Risk.* Plenum Press, New York.

Whitmore, R. C. 1982. Grasshopper sparrow. In Davis, D. E. (Ed.), *Handbook of Census Methods for Terrestrial Vertebrates.* CRC Press, Boca Raton, FL. 112.

Wiens, J. A. 1995. Recovery of seabirds following the *Exxon Valdez* oil spill. In P. G. Wells, J. N. Butler, and J. S. Hughes (Eds.), Exxon Valdez *Oil Spill: Fate and Effects in Alaskan Waters.* American Society for Testing and Materials, Philadelphia. 854–893.

Wiens, J. A. and K. R. Parker. 1995. Analyzing the effects of accidental environmental impacts: approaches and assumptions. *Ecol. Appl.* 5:1069–1083.

Wieser, W., G. Busch, and L. Buchel. 1976. Isopods as indicators of the copper content of soil and litter. *Oecologia* 23:107–114.

Wild, S. R., M. L. Berrow, S. P. McGrath, and K. C. Jones. 1992. Polynuclear aromatic hydrocarbons in crops from long-term field experiments amended with sewage sludge. *Environ. Pollut.* 76:25–32.

Wiley, M. L. and C-F. Tsai. 1983. The relative efficiencies of electrofishing vs. seines in Piedmont streams of Maryland. *North Am. J. of Fish. Manage.* 3:243–253.

Williams, D. D. and H. B. N. Hynes. 1973. The occurrence of benthos deep in the substratum of a stream. *Freshwater Biol. 4*:233–256.

Wilson, D. E., F. R. Cole, J. D. Nichols, R. Rudran, and M. S. Foster. 1996. *Measuring and Monitoring Biological Diversity: Standard Methods for Mammals.* Smithsonian Institution Press, Washington, D.C.

Windom, H. L., J. T. Byrd, R. G. Smith Jr., and F. Huan. 1991. Inadequacy of NASQAN data for assessing metal trends in the nation's rivers. *Environ. Sci. Technol.* 25:1137–1142.

Winterstein, S. R. 1992. Chi-square tests for intrabrood independence when using the Mayfield method. *J. Wildl. Manage.* 56:398–402.

Woodman, J. N. and E. B. Cowling. 1987. Airborne chemicals and forest health. *Environ. Sci. Technol.* 21:120–126.

Wolff, J. O. 1982. Snowshoe hare. In D. E. Davis (Ed.*), Handbook of Census Methods for Terrestrial Vertebrates.* CRC Press, Boca Raton, FL. 140–141.

Wright, J. F., P. D. Armitage, M. T. Furse, and D. Moss. 1989. Prediction of invertebrate communities using stream measurements. *Regul. Rivers: Res. Manage.* 4:147–155.

Xian, X. 1989. Effect of chemical forms of cadmium, zinc, and lead in polluted soils on their uptake by cabbage plants. *Plant Soil* 113:257–264.

Yang, R. S. H., Ed. 1994. *Toxicology of Chemical Mixtures: Case Studies, Mechanisms, and Novel Approaches.* Academic Press, New York.

Yeates, G. W., V. A. Orchard, T. W. Speir, J. L. Hunt, and M. C. C. Hermans. 1994. Impact of pasture contamination by copper, chromium, arsenic timber preservative on soil biological activity. *Biol. Fertil. Soils.* 18:200–208.

Yoder, C. O. and E. T. Rankin. 1995. Biological criteria program development and implementation. In W. S. Davis and T. P. Simon (Eds.), *Biological Assessment and Criteria: Tools for Water Resource Planning and Decision Making.* Lewis Publishers, Boca Raton, FL. 109–144.

Yoder, C. O. and E. T. Rankin. 1998. The role of biological indicators in a state water quality management process. *Environ. Monitor. Assess.* 51:61–88.

Yount, J. D. and G. J. Niemi. 1990. Recovery of lotic communities and ecosystems following disturbance: theory and application. *Environ. Manage.* 14:547–569.

Zabel, E. W., P. M. Cook, and R. E. Peterson. 1995. Toxic equivalency factors of polychlorinated dibenzo-p-dioxins, dibenzofurans, and biphenyl congeners based on early life stage mortality in rainbow trout (*Oncorhynchus mykiss*). *Aquat. Toxicol.* 31:315–328.

Zeeman, M. G. 1995. Ecotoxicity testing and estimation methods developed under section 5 of the Toxic Substances Control Act (TSCA). In G. Rand (Ed.), *Fundamentals of Aquatic Toxicology: Effects, Environmental Fate, and Risk Assessment.* Taylor and Francis, Washington, D.C. 703–715.

Zelles, L., I. Scheunert, and F. Korte. 1986. Comparison of methods to test chemicals for side effects on soil microorganisms. *Ecotoxicol. Environ. Saf.* 12:53–69.

Zhang, C. 1992. Rationale for extracting data from the USEPA ROD database printout and the development of the SUMROD database. Ontario Ministry of the Environment, Toronto, ON., as cited in Sheppard et al. (1992).

Zicsi, A. 1983. Earthworm ecology in deciduous forests in central and southeast Europe. In J. E. Satchell (Ed.), *Earthworm Ecology: From Darwin to Vermiculture*. Chapman & Hall, London. 171–177.

Index

A

B

C

F

G

H

I

J

L

M

Q

R

S

T

U

V

W